Vascular and Intravascular Imaging Trends, Analysis, and Challenges, Volume 2

Plaque characterization

Vascular and Intravascular Imaging Trends, Analysis, and Challenges, Volume 2

Plaque characterization

Petia Radeva
Universitat de Barcelona, Barcelona, Spain
and
Computer Vision Center, Bellaterra (Barcelona), Spain

Jasjit S Suri
ATHEROPOINT, California, USA

IOP Publishing, Bristol, UK

ISBN 978-0-7503-2002-3 (ebook)
ISBN 978-0-7503-1999-7 (print)
ISBN 978-0-7503-2001-6 (mobi)

DOI 10.1088/2053-2563/ab0820

Version: 20190801

IOP Expanding Physics
ISSN 2053-2563 (online)
ISSN 2054-7315 (print)

British Library Cataloguing-in-Publication Data: A catalogue record for this book is available from the British Library.

Published by IOP Publishing, wholly owned by The Institute of Physics, London

IOP Publishing, Temple Circus, Temple Way, Bristol, BS1 6HG, UK

US Office: IOP Publishing, Inc., 190 North Independence Mall West, Suite 601, Philadelphia, PA 19106, USA

To our families and friends for their infinite patience, love and support.

Contents

Preface xviii

Editor biographies xxi

List of contributors xxii

Section I Review on wall quantification, tissue characterization and coronary and carotid artery risk stratification

1 Coronary and carotid artery calcium detection, its quantification and grayscale morphology-based risk stratification in multimodality big data: a review 1-1

1.1 Introduction 1-2
1.2 Calcium detection in coronary and carotid arteries 1-6
1.2.1 Calcium detection in coronary arteries 1-6
1.2.2 Calcium detection in carotid arteries 1-9
1.3 Calcium area/volume quantification in coronary and carotid arteries 1-12
1.3.1 Calcium area/volume quantification in coronary arteries 1-12
1.3.2 Calcium area/volume quantification in carotid arteries 1-17
1.4 Metrics for performance evaluation for calcium detection algorithms and its validation 1-19
1.4.1 Statistical metrics for performance evaluation 1-19
1.4.2 Validation of calcium detection algorithms 1-21
1.5 Machine-learning-based risk stratification 1-22
1.5.1 Coronary risk assessment using ML-based approaches 1-22
1.5.2 Carotid risk assessment using ML-based approaches 1-24
1.6 Discussion 1-25
1.6.1 A note on the usage of calcium detection techniques in coronary and carotid arteries 1-25
1.6.2 A note on the usage of calcium quantification techniques in coronary and carotid arteries 1-25
1.6.3 A note on the use of statistical metrics for the evaluation of calcium detection algorithms 1-26
1.6.4 A note on feature selection in ML-based risk stratification for the coronary and carotid arteries 1-26
1.6.5 Recommended interventions for CVD patients 1-27

1.6.6 Atherosclerotic calcium in coronary and carotid imaging: ongoing challenges 1-28
1.7 Conclusions 1-28
References 1-29

2 Risk of coronary artery disease: genetics and external factors 2-1
2.1 Introduction 2-1
2.2 External factors 2-4
2.2.1 Ethnicity and CVD 2-4
2.2.2 Environmental factors and CVD 2-5
2.2.3 Air pollution and CVD risk 2-7
2.2.4 Nutrition and CVD risk 2-9
2.2.5 Family history and CVD risk 2-9
2.3 Genetics of coronary artery disease 2-10
2.3.1 Genetics of atherosclerosis 2-10
2.3.2 Genetics of diabetes 2-15
2.3.3 Genetics of rheumatoid arthritis 2-16
2.3.4 Anatomy of a 3D heart 2-17
2.4 Multimodal coronary imaging 2-19
2.4.1 Regular coronary artery 2-19
2.4.2 Coronary imaging using x-ray angiography 2-20
2.4.3 Coronary imaging using magnetic resonance angiography 2-21
2.4.4 Imaging coronary CT angiography 2-21
2.4.5 Coronary artery interpretation by OCT versus IVUS 2-22
2.5 Association of CVD with other prevalent diseases 2-23
2.5.1 Relationship between coronary artery and carotid disease 2-23
2.5.2 The relationship between diabetes and coronary artery disease 2-24
2.5.3 The link between rheumatic arthritis and cardiovascular disease 2-24
2.6 Treatments for cardiovascular disease 2-26
References 2-29

3 Wall quantification and tissue characterization of the coronary artery 3-1
3.1 Introduction 3-1
3.2 Physics of image acquisition 3-2
3.2.1 Image acquisition using optical coherence tomography 3-2
3.2.2 Image acquisition using intravascular ultrasound 3-3

3.2.3 Comparison of OCT and IVUS 3-4
3.3 Tissue characterization 3-5
3.3.1 OCT appearance of plaque tissues 3-6
3.3.2 Schools of thought on tissue characterization 3-7
3.3.3 Characterization using optical properties 3-8
3.3.4 Characterization using machine learning 3-12
3.3.5 Tissue characterization using deep learning 3-14
3.4 A link between carotid and coronary artery disease 3-14
3.4.1 Carotid intima–media thickness and CAD 3-14
3.4.2 Carotid plaque and CAD 3-15
3.4.3 Coronary IMT and carotid atheroma for CAD risk detection 3-16
3.4.4 Femoral and carotid IMT for CAD risk detection 3-16
3.4.5 Coronary calcium and carotid risk factors for risk detection 3-16
3.5 Wall quantification 3-17
3.5.1 Lumen measurement 3-17
3.5.2 Vessel wall measurement 3-18
3.5.3 Fibrous cap measurement 3-18
3.5.4 Measurement of calcium 3-19
3.5.5 Quantification of macrophages 3-19
3.5.6 The role of image registration 3-20
3.6 Risk assessment systems 3-22
3.7 Discussion 3-22
3.7.1 Benchmarking 3-23
3.7.2 A note on image acquisition hardware 3-23
3.7.3 A note on plaque component quantification 3-23
3.7.4 Validation of plaque characterization techniques 3-23
3.7.5 A note on the future of OCT 3-27
3.8 Conclusion 3-27
References 3-30

4 Rheumatoid arthritis: its link to atherosclerosis imaging and cardiovascular risk assessment using machine-learning-based tissue characterization 4-1

4.1 Introduction 4-2
4.2 Search strategy 4-3
4.3 Brief description of the pathogensis of rheumatoid arthritis 4-3
4.4 Atherosclerosis driven by rheumatoid arthritis 4-5

4.5 The role of platelets in atherothrombosis in RA 4-6
4.6 The role of amyloidosis in RA 4-6
4.7 Traditional CV risk factors in rheumatoid arthritis 4-7
4.7.1 Body mass index and physical inactivity 4-7
4.7.2 Lipids 4-7
4.7.3 Hypertension 4-8
4.7.4 Smoking 4-8
4.7.5 Insulin resistance and diabetes 4-8
4.7.6 Ankle–brachial index and arterial stiffness 4-9
4.8 RA-specific CV risk factors in rheumatoid arthritis 4-9
4.9 Conventional CV risk algorithms 4-9
4.10 Cardiovascular imaging in rheumatoid arthritis 4-11
4.10.1 Non-invasive imaging techniques 4-11
4.10.2 Invasive imaging techniques: IVUS and OCT 4-12
4.11 RA-driven atherosclerotic plaque wall tissue characterization: intelligence paradigm 4-12
4.11.1 Machine-learning-based tissue characterization 4-13
4.11.2 Deep-learning-based tissue characterization 4-15
4.12 Research agenda 4-16
4.13 Summary and conclusion 4-17
References 4-20

Section II Deep learning strategy for accurate lumen and carotid intima-media thickness measurement

5 A deep-learning fully convolutional network for lumen characterization in diabetic patients using carotid ultrasound: a tool for stroke risk 5-1

5.1 Introduction 5-1
5.2 Data demographics 5-5
5.3 Methodology 5-5
5.3.1 Pre-processing 5-6
5.3.2 The encoder 5-8
5.3.3 The decoder 5-8
5.4 Results 5-9
5.4.1 Experimental protocol 5-9
5.4.2 Experimental results 5-9

5.4.3 Performance evaluation 5-10
5.5 Discussion 5-17
5.5.1 Benchmarking 5-19
5.5.2 A short note on skip operation in FCN 5-21
5.5.3 A short note on manual tracings of LI borders 5-21
5.5.4 Strengths and weaknesses 5-21
5.6 Conclusion 5-22
References 5-29

6 Deep-learning strategy for accurate carotid intima–media thickness measurement: an ultrasound study on a Japanese diabetic cohort 6-1

6.1 Introduction 6-2
6.2 Data demographics and US acquisition 6-4
6.3 Methodology 6-5
6.3.1 Multiresolution as phase I 6-5
6.3.2 DL as phase II 6-6
6.3.3 Boundary extraction as phase III 6-7
6.3.4 Performance evaluation as phase IV 6-7
6.4 Experimental protocol and results 6-8
6.4.1 Experimental protocol 6-8
6.4.2 Results 6-8
6.5 Performance of the DL systems and variability analysis 6-11
6.5.1 Comparison of DL against expert manual tracing 6-13
6.5.2 Comparison of the DL against the sonographer's readings 6-13
6.5.3 Absolute and signed cIMT error analysis for DL1 and DL2 systems 6-14
6.5.4 DL versus previous methods 6-14
6.5.5 Interoperator variability of the DL systems: DL1 and DL2 6-17
6.5.6 Interobserver variability between the GT systems: GT1 and GT2 6-17
6.6 Statistical tests and risk analysis 6-18
6.6.1 Four statistical tests 6-18
6.6.2 Risk analysis by age 6-18
6.6.3 Risk stratification and ROC curves 6-18
6.7 Discussion 6-21
6.7.1 Benchmarking table 6-22

6.7.2 A short note on calibration 6-24
6.7.3 A special note on DL optimization 6-24
6.7.4 A special note on skip operation 6-25
6.7.5 Strengths, weaknesses and extensions 6-25
6.7.6 Hardware configuration 6-25
6.8 Conclusion 6-25
References 6-30

Section III Association of morphological and echolucency-based phenotypes with HbA1c

7 Echolucency-based phenotype in carotid atherosclerosis disease for risk stratification of diabetes patients 7-1

7.1 Introduction 7-2
7.2 Patient demographics and methodology 7-3
7.2.1 Patients demographics 7-3
7.2.2 Methodology 7-3
7.3 Results and statistical analysis 7-4
7.3.1 CC analysis of AAGSM and GSM_{conv} against HbA1c 7-6
7.3.2 CC between left and right CCA for AAGSM and GSM_{conv} 7-6
7.3.3 CC analysis of AAGSM–HbA1c and GSM_{conv}–HbA1c in males and females 7-6
7.3.4 Risk stratification based on AAGSM, and HbA1c and ROC analysis 7-6
7.3.5 Statistical tests 7-7
7.4 Discussion 7-8
7.4.1 A note on the HbA1c and AAGSM thresholds for risk stratification 7-9
7.4.2 Justification of the δth percentile value during GSM_δ measurement 7-9
7.4.3 A special note on age-adjustment pre-multiplier (M) selection 7-11
7.4.4 A note on the therapeutic implications of AAGSM 7-11
7.4.5 Benchmarking against the previous literature 7-12
7.4.6 Strengths, weaknesses and applications of AAGSM 7-12
7.5 Conclusion 7-13
References 7-13

8 Morphologic TPA (mTPA) and composite risk score for moderate carotid atherosclerotic plaque is strongly associated with HbA1c in a diabetes cohort 8-1

8.1 Introduction 8-2
8.2 Materials and methods 8-5
8.2.1 Patient demographics 8-5
8.2.2 Six phenotype measurements derived from carotid ultrasound scans 8-6
8.2.3 Statistical analysis 8-8
8.3 Results 8-8
8.3.1 Demographics and clinical characteristics of the patients 8-8
8.3.2 Visual display of six phenotypes using AtheroEdge™ 8-8
8.3.3 Correlation between operators and correlation between cIMT and mTPA for the left, right, and mean of the left and right carotid arteries 8-11
8.3.4 Logistic regression for the effect of the six phenotypes on HbA1c for the operator of AtheroEdge™ 8-12
8.4 Inter-operator variability and statistical tests 8-14
8.4.1 Inter-operator variability 8-14
8.4.2 Statistical tests 8-14
8.5 Discussion 8-17
8.5.1 A special note on mTPA and CRS 8-19
8.5.2 Benchmarking 8-19
8.5.3 A special note on the reproducibility of phenotypes 8-22
8.6 Conclusions 8-23
References 8-29

Section IV Deep learning strategy for accurate lumen and carotid intima-media thickness measurement

9 Plaque tissue morphology-based stroke risk stratification using carotid ultrasound: a polling-based PCA learning paradigm 9-1

9.1 Introduction 9-2
9.2 Demographics, data collection and preparation 9-4
9.2.1 Patient demographics 9-4
9.2.2 Data acquisition 9-4

9.2.3 Manual wall region extraction for the manual risk assessment system (mRAS) 9-5
9.2.4 Modeling the manual LD into two stratification classes: high risk and low risk 9-6
9.3 Risk assessment methodology 9-6
9.3.1 IMT far and near wall strip extraction 9-7
9.3.2 Assessment of stroke risk using a machine-learning system 9-8
9.3.3 Texture features 9-9
9.3.4 Support vector machine (SVM) and classification 9-10
9.3.5 Feature reduction technique using polling-based principal component analysis 9-10
9.3.6 Kernel optimization based on the machine-learning paradigm 9-11
9.4 Experimental protocol and results 9-12
9.4.1 Experiment 1: dominant feature selection and classification accuracy with changing PCA cutoff 9-12
9.4.2 Experiment 2: the role of data size in the performance of machine-learning 9-15
9.5 Performance evaluation 9-19
9.5.1 Precision-of-merit (PoM) analysis 9-20
9.5.2 Reliability analysis of the sRAS 9-21
9.5.3 Feature retaining power of the sRAS 9-23
9.5.4 Stability analysis of the sRAS 9-24
9.6 Discussion 9-25
9.6.1 About the risk assessment system 9-25
9.6.2 Justification for the three kinds of cross-validation protocols 9-26
9.6.3 Choice of biomarker (LD versus cIMT) 9-27
9.6.4 A note on wall segmentation validation 9-27
9.6.5 Benchmarking against the current literature 9-27
9.6.6 Summary of our contribution 9-28
9.6.7 Strengths, weaknesses and extensions 9-28
References 9-39

10 Multiresolution-based coronary calcium volume measurement techniques from intravascular ultrasound videos 10-1

10.1 Introduction 10-2
10.2 Patient demographics and data acquisition 10-4
10.2.1 Patient demographics 10-4

10.2.2 IVUS data acquisition 10-4
10.2.3 Coronary artery data size preparation 10-5
10.2.4 Region-of-interest estimation 10-5
10.3 Methodology 10-6
10.3.1 Overall system 10-6
10.3.2 Five multiresolution techniques 10-7
10.3.3 Four segmentation methods 10-13
10.4 Results 10-18
10.4.1 Calcium detection 10-18
10.4.2 Volume measurement 10-20
10.4.3 Percentage mean time improvement 10-20
10.5 Performance evaluation 10-24
10.5.1 Multiresolution error metrics against non-multiresolution technique 10-24
10.5.2 The mean Jaccard index (JI) and Dice similarity coefficient (DSC) 10-27
10.5.3 Manual scoring of detected calcium by a radiologist 10-28
10.5.4 Degradation ratio and quality assessment ratio 10-30
10.6 Discussion 10-33
10.6.1 Our system 10-33
10.6.2 Comparison of our down-sampling methods against other methods 10-34
10.6.3 A note on gating and registration 10-34
10.6.4 Bias correction 10-35
10.6.5 A note on time complexity and precision-of-merit 10-36
10.6.6 Benchmarking 10-37
10.6.7 Strengths, weaknesses and extensions 10-40
10.7 Conclusion 10-40
References 10-42

11 A cloud-based smart lumen diameter measurement tool for stroke risk assessment during multicenter clinical trials 11-1

11.1 Introduction 11-2
11.2 Materials and methods 11-5
11.2.1 Manual lumen diameter reading 11-5
11.2.2 Workflow architecture of the AtheroCloud ultrasound system 11-6
11.2.3 Engineering design of the AtheroCloud ultrasound system 11-7

11.2.4 Two application modes of AtheroCloud: routine mode and pharma trial mode 11-8
11.3 Results 11-9
11.3.1 Measurements and visualization 11-9
11.3.2 Performance evaluation of the AtheroCloud ultrasound system 11-9
11.3.3 PoM, FoM, CC and Bland–Altman plots 11-9
11.3.4 Cumulative frequency distribution of LD error and TLA error 11-11
11.3.5 Receiver operating characteristic 11-12
11.3.6 Statistical tests 11-12
11.4 Discussion 11-15
11.4.1 Our system 11-15
11.4.2 Benchmarking AtheroCloud against AtheroEdge™ 2.0 11-15
11.4.3 Strengths, weaknesses and extensions 11-15
11.5 Conclusion 11-16
References 11-19

Section V Micro-electro-mechanical-system (MEMS)

12 A MEMS-based manufacturing technique of vascular bed 12-1

12.1 Introduction 12-1
12.2 Microstructural anatomy of blood vessels 12-2
12.2.1 Arteries and veins 12-2
12.2.2 Capillaries 12-4
12.3 Modeling of blood vessels as a microsystem 12-5
12.3.1 Acoustic wave mechanosensors 12-5
12.3.2 Pressure mechanosensors 12-6
12.3.3 Microvalves and micropumps 12-8
12.4 Scaling laws of miniaturized blood vessels 12-10
12.4.1 Scaling in geometry 12-10
12.4.2 Scaling in fluid dynamics 12-11
12.5 Microfabrication of blood vessels 12-12
12.5.1 Soft lithography techniques 12-12
12.5.2 Self-assembly techniques 12-13
12.5.3 Sputtering techniques 12-13

12.6 Microvessel design 12-15
12.6.1 Design consideration 12-15
12.6.2 Mechanical design of a balloon angioplasty pressure sensor using finite element methods 12-15
12.7 Conclusion 12-19
References 12-20

Preface

Atherosclerosis is the leading cause of cardiovascular disease (CVD) and stroke. These diseases impose an immense financial burden and have the greatest impact in terms of morbidity. CVD is the cause of one in every three deaths in the USA and accounts for almost 45% of deaths in European countries. On average, per year, CVD causes 7.4 million deaths, while stroke causes 6.7 million. Between 2000 and 2030, it is estimated that about 35% of all CVD deaths in India will occur among 35- to 64-year-olds, which has been attributed to atherosclerosis. These facts raise different questions, such as: What are the most appropriate methods for calcium detection and its quantification for coronary and carotid arteries? What are the advantages and disadvantages of these methods and the risk stratification strategies? How can a combination of machine-learning and deep-learning techniques improve accuracy? How is rheumatoid arthritis (RA) associated with carotid atherosclerosis? How are plaque-based biomarker and carotid artery disease image-based phenotypes associated with HbA1c? How can the disease risk stratification accuracy and the speed of computation be to improved? Are there solutions to issues associated with multi-center clinical trials and routine vascular screening? How to establish a connection between the synthesis routes of micro-electro-mechanical systems (MEMSs) and their application to synthesize a multi-layered vascular bed with micro-scale level refinement?

In this book, we are pleased to witness several advanced clinical and medical imaging works that cover a wide spectrum of clinical disease issues, clinical intervention techniques, imaging modalities for plaque visualization and inspection, automatic analysis and clinical parameter extraction techniques, and advanced tools for the navigation of and intervention in both coronary and carotid lesions.

The book is organized into five sections: the first part is comprised of four review papers. The first paper presents a state-of-the-art review covering the methods for calcium detection and its quantification for coronary and carotid arteries, the advantages and disadvantages of these methods, and the risk stratification strategies. The review also presents different kinds of statistical models and gold standard solutions for the evaluation of software systems useful for calcium detection and quantification. The second and third review papers present comparisons between various methodologies used for tissue characterization, classification and measurement using OCT. The review also presents different ways to predict and stratify the risk associated with CVD based on plaque characterization and measurement. Based on comparative analysis between different schools of thought, a combination of machine-learning and deep-learning techniques has been verified to provide the best classification accuracy using OCT images. The review also discusses the physics of image acquisition using different imaging modalities followed by tissue characterization using three paradigms based on (i) optical feature measurement methodologies, (ii) machine-learning algorithms and (iii) deep-learning techniques. Quantification of vulnerable plaque components and risk stratification using the above mentioned paradigms are also discussed. The fourth review provides a brief understanding of the pathogenesis of RA and its association with carotid atherosclerosis imaged using B-mode ultrasound techniques. Lacunas in traditional risk scores and the role of machine-learning-based tissue characterization algorithms are discussed, which could facilitate cardiovascular risk assessment in RA patients.

As manual ultrasound (US)-based methods adapted for lumen diameter (LD) and carotid intima–media thickness (cIMT) measurement are tedious, error-prone and cause variability, an intelligence-based, novel, robust and clinically strong deep-learning (DL)-based strategy is the need of the hour. The second section of this book demonstrates the superior performance of DL systems over conventional methods and is comprised of two clinical papers. In the first chapter in this section, an automated DL-based system is presented, which consists of a combination of two systems: the encoder and decoder for lumen segmentation. The encoder employs a 13 layer convolution neural network (CNN) model for rich feature extraction and the decoder employs three up-sample layers of a fully convolutional network (FCN) for lumen segmentation. In the second chapter, a combination of DL and machine-learning (ML) paradigms are used for cIMT measurement. The first stage consists of a convolution-layer-based encoder for feature extraction and a FCN-based decoder for image segmentation. This stage generates the raw inner lumen borders and raw outer intra-adventitial borders. To smooth these borders, the DL system uses a cascaded second stage that consists of ML-based regression. The final outputs are the far wall lumen–intima (LI) and media–adventitia (MA) borders which were used for cIMT measurement. In both studies, the experimental results demonstrated the superior performance of the DL system over the conventional methods in the previously published literature.

The third section of this book investigates the association of plaque-based biomarker and carotid artery disease image-based phenotypes with HbA1c and comprises two clinical chapters. The first chapter investigates the association of carotid ultrasound echolucent plaque-based biomarkers with HbA1c, measured as an age-adjusted grayscale median (AAGSM) as a function of chronological age, total plaque area and conventional grayscale median (GSMconv). The study concluded that echolucent plaque in patients with diabetes can be more accurately characterized by risk stratification using AAGSM compared to GSMconv. In the second chapter, the association between six types of carotid artery disease image-based phenotypes and HbA1c in diabetes patients is explored. A set of six phenotypes (intima–media thickness measurements (cIMT (ave.), cIMT (max.), cIMT (min.)), bidirectional wall variability (cIMTV), morphology-based total plaque area (mTPA) and composite risk score (CRS)) were measured in an automated setting using AtheroEdge™. Among the six carotid phenotypes, all except for bidirectional wall variability showed a strong association with HbA1c. mTPA and CRS were equally strong phenotypes as cIMT. The CRS phenotype showed the strongest relationship to HbA1c.

The fourth section of this book presents studies performed to improve the risk stratification accuracy and the speed of computation. Further, a reliable, accurate, fast, completely automated, anytime-anywhere solution for multi-center clinical trials and routine vascular screening is discussed. This section comprises three clinical chapters. Several machine-learning systems have been previously developed for plaque wall risk assessment using morphology-based characterization. Even though these systems have the ability to perform risk stratification, they lack the ability to achieve higher performance due to their inability to select and retain dominant features. The first chapter introduces a polling-based principal component analysis (PCA) strategy which, when embedded with an ML-based framework, selects and retains dominant features and thus results in superior performance. As fast intravascular ultrasound (IVUS) video processing is required for calcium volume computation during the planning phase of percutaneous coronary interventional (PCI) procedures, the second chapter introduces the idea of embedding segmentation methods with nonlinear multiresolution techniques. To achieve this, four different segmentation methods for calcium volume measurement, namely threshold-based, fuzzy c-Means (FCM), K-means and hidden Markov random field (HMRF), are embedded with five different kinds of multiresolution techniques (bilinear, bicubic, wavelet, Lanczos and Gaussian pyramid). Among the 20 different combinations of multiresolution with calcium volume segmentation methods, the FCM embedded with wavelet-based multiresolution gave the best performance. Finally, the third chapter presents a completely automated, novel, smart, cloud-based, point-of-care system for (a) carotid LD, (b) stenosis severity index (SSI) and (c) total lumen area (TLA) measurement using B-mode ultrasound, which thus provides an anytime-anywhere solution for multi-center clinical trials and routine vascular screening.

The last section of this book is devoted to MEMSs, a kind of miniaturized system commonly being used in the domain of sensor technology and drug delivery devices in the healthcare industry. Although there is a significant amount of potential in the manufacturing routes of MEMS synthesis, their use tends to be limited to semiconductor device industries. In this section a very careful amalgamation has been carried out to form a connection between the synthesis routes of MEMSs and their application to synthesize multi-layered vascular bed with micro-scale level refinement. This technique can be used as a potential method for re-defining the construction of the multi-layered tissues of many organs.

In summary, this collection of chapters gives an overview of research on vascular and intravascular analysis, discussing in detail different scientific and clinical questions, and proposes advances in clinical treatment and medical imaging automatic analysis. We aim to give an overview of the active topics and problems in this field and encourage the community to continue in their search for scientific and clinical answers as to which are the most precise, objective, effective and efficient strategies for atherosclerotic diagnosis, treatment and follow-up, as it remains one of the most important health problems of humanity.

Petia Radeva
Jasjit S Suri

Editor biographies

Petia Radeva

Dr Petia Radeva (PhD 1993, Universitat Autònoma de Barcelona, Spain) is a senior researcher and full professor at the University of Barcelona. She received her PhD degree from the Universitat Autònoma de Barcelona in 1998. She is the head of the Computer Vision and Machine Learning Consolidated Research Group at the University of Barcelona and the head of MiLab of the Computer Vision Center (www.cvc.uab.es). Her current research interests include the development of learning-based approaches (in particular, deep learning methods) for computer vision and image analysis. Radeva has been an AIPR Fellow since 2015, and became an ICREA Academia researcher in 2014 for her outstanding research achievements. In 2015 she received the Aurora Pons Porrata award for her scientific merits as well as the Antonio Caparros award for the best technology transfer.

Jasjit S Suri

Jasjit S Suri, PhD, MBA, is an innovator, visionary, scientist and an internationally known world leader in the field of biomedical imaging and healthcare management. Dr Suri is a recipient of the Director General's Gold Medal (1980), was named a Fellow of the American Institute of Medical and Biological Engineering by the National Academy of Sciences, Washington, DC (2004), and received a Marquis Life Time Achievement Award (2018). Dr Suri is a board member in several organizations.

List of contributors

Dr Sumit K Banchhor
Chhattisgarh Swami Vivekanand Technical University, India

Dr Narendra D Londhe
National Institute of Technology Raipur, India

Dr Tadashi Araki
Toho University Ohashi Medical Center, Japan

Dr Luca Saba
University of Cagliari, Italy

Dr Narendra N Khanna
Indraprastha Apollo Hospital, India

Dr John R Laird
St Helena Hospital, CA, USA

Dr Ajay Gupta
Weill Cornell Medical College, NY, USA

Dr Andrew Nicolaides
Vascular Screening and Diagnostic Centre, UK

Dr Nobutaka Ikeda
National Center for Global Health and Medicine, Japan

Dr Ayman El-Baz
University of Louisville, KY, USA

Dr Shoaib Shafique
CorVasc Vascular Laboratory, IN, USA

Mr Mainak Biswas
National Institute of Technology Goa, India

Dr Venkatanareshbabu Kuppili
National Institute of Technology Goa, India

Dr Damodar Reddy Edla
National Institute of Technology Goa, India

Mr Harman S Suri
Brown University, RI, USA

Dr Aditya Sharma
University of Virginia, VA, USA

Dr Elisa Cuadrado Godia
IMIM—Hospital del Mar, Spain

Dr Tomaz Omerzu
University Medical Centre Maribor, Slovenia

Dr Narendra N Khanna
Apollo Hospital, India

Dr Arindam Bit
National Institute of Technology Raipur, India

Mr Ankush D Jamthikar
Visvesvaraya National Institute of Technology Nagpur, India

Dr Alberto Boi
University of Cagliari, Italy.

Dr Deep Gupta
Visvesvaraya National Institute of Technology Nagpur, India

Dr Matteo Piga
University Clinic and AOU of Cagliari, Italy

Dr Argiris A Giannopoulos
Imperial College London, UK

Dr Carlo Carcassi
University of Cagliari, Italy

Mr Harman Suri
Brown University, RI, USA

Dr Sophie Mavrogeni
Onassis Cardiac Surgery Center, Greece

Dr George D Kitas
Dudley Group NHS Foundation Trust, UK

Dr Vasileios Kotsis
Aristotle University of Thessaloniki, Greece

Dr Klaudija Viskovic
University Hospital for Infectious Diseases, Croatia

Mr Pankaj Jain
Indian Institute of Technology Varanasi (BHU), India

Dr Bikesh Singh
National Institute of Technology Raipur, India

Mr Md Maniruzzaman
Khulna University, Bangladesh

Mr Anudeep Puvvula
Annu's Hospitals for Skin and Diabetes, India

Dr Md Jahanur Rahman
University of Rajshahi, Bangladesh

Dr Sumit K Banchhor
Global Biomedical Technologies, FL, USA

Dr Jagjit S Teji
Ann and Robert H Lurie Children's Hospital of Chicago and Northwestern University, IL, USA

Section I

Review on wall quantification, tissue characterization and coronary and carotid artery risk stratification

Chapter 1

Coronary and carotid artery calcium detection, its quantification and grayscale morphology-based risk stratification in multimodality big data: a review

Sumit K Banchhor, Narendra D Londhe, Tadashi Araki, Luca Saba, Petia Radeva, Narendra N Khanna and Jasjit S Suri

Purpose of the review

Atherosclerosis is the leading cause of cardiovascular disease (CVD) and stroke. Typically, atherosclerotic calcium is found during the mature stage of atherosclerosis. It is therefore often a challenge to identify and quantify the calcium. This is due to the presence of multiple components of plaque build-up in the arterial walls. The American College of Cardiology/American Heart Association guidelines point to the importance of calcium in the coronary and carotid arteries and further recommend its quantification for the prevention of heart disease. It is therefore essential to stratify the CVD risk of the patient into low- and high-risk bins.

Recent findings

Calcium formation in the artery walls is multifocal in nature with sizes at the micrometre level. Thus, its detection requires high-resolution imaging. Clinical experience has shown that even though optical coherence tomography offers better resolution, intravascular ultrasound still remains an important imaging modality for coronary wall imaging. For a computer-based analysis system to be complete, it must be scientifically and clinically validated. This study presents a state-of-the-art review (condensation of 152 publications after examining 200 articles) covering the methods for calcium detection and its quantification for coronary and carotid arteries, the advantages and disadvantages of these methods, and the risk stratification strategies. The review also presents different kinds of statistical models and gold standard solutions for the evaluation of software systems useful for calcium detection and quantification. Finally, the review concludes with a possible vision for designing the next-generation system for better clinical outcomes.

1.1 Introduction

Atherosclerosis is the leading cause of CVD and stroke. These diseases impose an immense financial burden and have the greatest impact in terms of morbidity [1–4]. CVD is the cause of one in every three deaths in the USA and accounts for almost 45% of deaths in European countries [5]. On average, per year, CVD causes 7.4 million deaths, while stroke causes 6.7 million [6]. In India, due to a lack of healthcare facilities and awareness, CVD is more frequently observed in rural areas compared to urban areas [7]. Between 2000 and 2030, it is estimated that about 35% of all CVD deaths in India will occur among 35- to 64-year-olds [8], which has been attributed to atherosclerosis [9].

Atherogenesis is the process of plaque formation in the arteries [10]. During atherogenesis, plaques usually develop in the region where there is low endothelial shear stress. In this region, leucocytes such as monocytes and basophils attack the endothelium [11]. Monocytes migrate into the sub-endothelial region and become oxidised by low-density lipoprotein (LDL) cholesterol and become macrophages [12]. These macrophages become large foam cells containing oxidised LDL molecules [13, 14]. Foam cells, macrophages and intraplaque haemorrhages form a necrotic core; this lesion is called a fibroatheroma [15]. Microscopic calcium granules expand in this necrotic core and form a large lump of calcium deposits [16], as shown in figure 1.1. A fibrous cap separates the necrotic core from the vessel lumen [17]. If the plaque is small, the arteries will undergo positive remodelling and blood flow will be uninterrupted [18]. It has been observed that with an increase in the calcium content there is a decrease in the lipid core volume, leading to structural stabilisation of the plaque [19]. In contrast, the presence of juxtaluminal calcification elevates the local stress compared to when calcification is artificially covered with a 0.2 mm thick fibrous cap [20]. Progressive accumulation of lipids usually causes thinning of the fibrous cap [21], which may lead to plaque rupture. When the cap ruptures, platelets in the bloodstream attempt to heal the injury, which leads to the

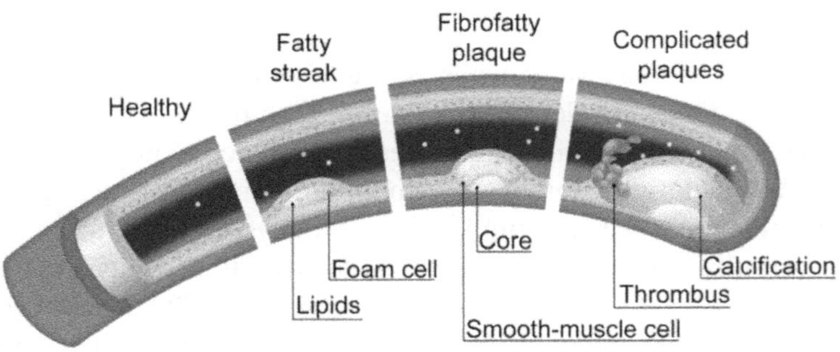

Figure 1.1. Calcified plaque formation in the arteries. (Courtesy of AtheroPoint™, Roseville, CA, USA.)

formation of a blood clot, or thrombus, which can block the artery [22]. If an artery is blocked, tissues are deprived of their blood supply, leading to cell death. If the coronary artery is blocked, the result is a myocardial infarction (MI). When a thrombus breaks off and travels through the bloodstream, it is called an embolus. If the embolus becomes lodged in a cranial artery, it leads to stroke [23].

In a prospective study of 40 patients, Joshi and his team [24] found a new way to detect plaque rupture non-invasively using F-sodium fluoride (F-NaF) PET radio-isotopes. Using coronary angiography and ultrasound, high F-NaF uptake was shown by both coronary and carotid arteries with microcalcifications and necrotic cores. The study demonstrates the need for more prospective trials to establish the relationship between high F-NaF uptake and plaque rupture [25], as the early detection of vulnerable plaque before rupture is very important.

Diabetic patients are at increased risk of atherosclerosis, particularly patients suffering from coronary artery disease (CAD) [26]. A large meta-analysis study carried out by Bulugahapitiya *et al* [27] involving 45 108 patients showed that patients with diabetes without prior MI had a 43% lower risk of CHD compared to patients without diabetes with prior MI. From a two-year retrospective analysis in Bangladesh consisting of 571 patients (333 in the diabetic and 238 in the non-diabetic group), Kabir *et al* [28] found that diameters of the left anterior descending (LAD), distal circumflex and right coronary arteries in diabetic patients were narrower than in non-diabetic subjects. As a result, the diabetic subjects needed longer stent lengths than non-diabetics. Another study carried out by Ertan *et al* [29] on 168 consecutive patients with CAD and 172 patients with normal coronary artery anatomy supported the previous work. The study showed that prediabetic patients have a smaller coronary size and diffuse coronary narrowing, and early detection of prediabetes may provide a more appropriate coronary lesion for percutaneous or surgical revascularization.

Atherosclerosis usually advances silently, and its clinical symptoms arise late in the CAD [9]. During atherosclerosis formation, the plaque usually consists of cholesterol, platelets and cellular waste products, while calcium builds up in the innermost layer of the artery [30]. Calcified plaques are only produced in the atheroma region, which lies in between the external elastic lamina (vessel region) and the internal elastic lamina (lumen region) [11], depicted in figure 1.2. Atherosclerotic

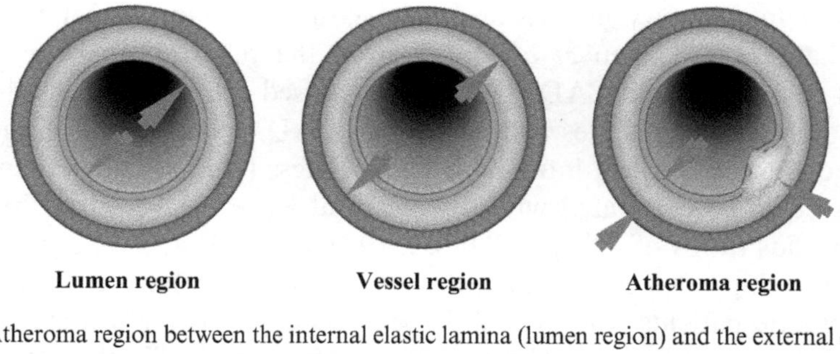

Figure 1.2. Atheroma region between the internal elastic lamina (lumen region) and the external elastic lamina (vessel region). (Courtesy of AtheroPoint™, Roseville, CA, USA.)

arteries limit the flow of oxygen-rich blood in the body and patients usually experience symptoms such as angina, shortness of breath, fatigue and lack of energy. However, in some cases, asymptomatic patients suffer from an MI or stroke without showing any preceding symptoms [31].

In the 20th century, researchers investigated the causes of atherosclerotic disease. Stryker [32] discussed five different cases of wall calcification in infants and suggested that calcification associated with fibroblast proliferation in the intima is the most frequent cause of coronary occlusion in infants. A similar relationship was seen in atherosclerotic diseases in childhood by Woolf [33] and Stary [34]. These studies indicate that the initial stages of atherogenesis can occur during childhood.

During their exploration of the origin of atherosclerotic disease, Hamby *et al* [35] found that patients with double- or triple-vessel disease are less susceptible to coronary artery calcification compared to patients with single-vessel disease. Furthermore, Kannel and Wolf [36] observed that atherosclerosis generally occurs not only as a result of genetic susceptibility, but also due to various other risk factors such as dyslipidemia, hypertension, adiposity, glucose intolerance, haemostatic factors, cigarette smoking, inflammatory markers and a sedentary lifestyle [37]. Even in the absence of these risk factors, patients with genetic hyperlipidaemia have shown an increased incidence of CAD. In the presence of other cardiovascular risk factors, lower levels of lipids can also cause atherosclerosis [38, 39]. Hirsch *et al* [40] found a spatial association between unesterified cholesterol and hydroxyapatite, which shows that there may be more than one mechanism of calcium deposition in atherosclerosis. One year later, Doherty and Detrano [41] showed that Gla-containing proteins and other proteins normally associated with bone metabolism play a significant role in the process of atherosclerotic calcification. Guyton and Klemp [42] suggested that the early core is associated with the accumulation of vesicular lipids rich in free cholesterol. However, later in core development, lipid deposits become more diverse. In such scenarios, early detection and risk stratification of calcium in the arteries is important, as there are few benefits of diagnosis at the advanced stages of atherosclerosis.

During atherosclerosis formation, different arterial beds usually share the same risk of stenosis [43]. For this reason, stenosis in one artery also boosts the chances of stenosis in other arteries [44]. Previous studies [45–47] have also shown that plaque accumulation in coronary and carotid arteries has the same genetic makeup, as shown in figure 1.3. Cohen *et al* [48] showed the relationship between carotid ultrasound parameters and CAD. The study analysed 150 patients, in which 71.3% of patients had carotid plaques and 57.1% had CAD. Independent of age and sex, carotid plaques with a mean intima–media thickness (IMT) greater than 0.75 mm were observed to be correlated with disease in at least one vessel in the coronary artery with odds ratios of 2.8 ($p = 0.03$), 2.19 ($p = 0.073$) and 2.22 ($p = 0.058$). A similar relationship between carotid atherosclerosis and coronary artery calcification in asymptomatic patients with type 2 diabetes mellitus (T2DM) was examined by Jeevarethinam *et al* [49]. In a cohort of 262 asymptomatic T2DM patients, cIMT and coronary artery calcium were examined. Using binary logistical regression,

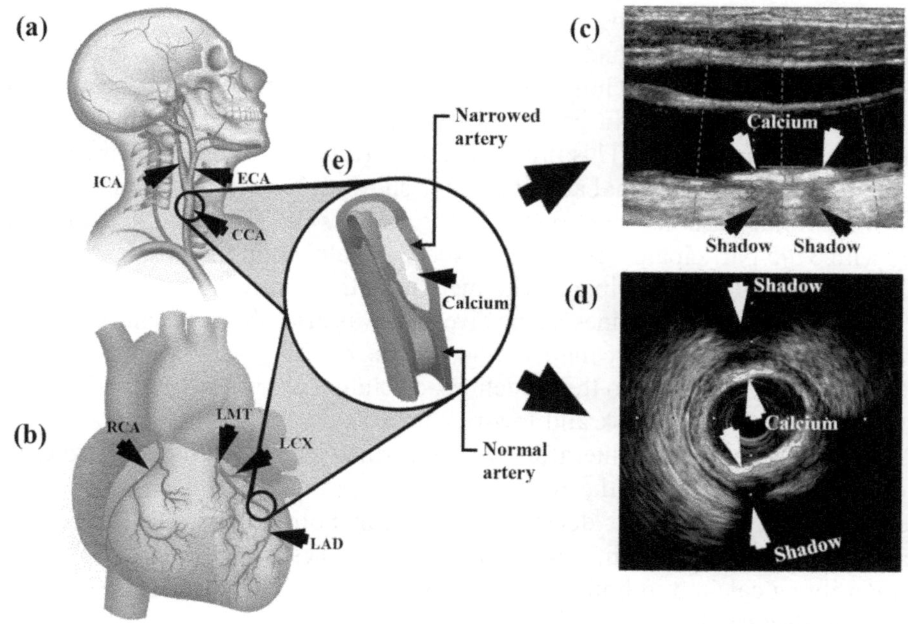

Figure 1.3. (a) and (b) Illustrations of the coronary and carotid artery, respectively. (c) and (d) Ultrasound images of the coronary and carotid artery, respectively, with calcium indicated by the arrows. (e) The calcified plaque narrows the cross section and causes abnormal blood flow in the arteries. (Courtesy of AtheroPoint™, Roseville, CA, USA.)

carotid plaques significantly predicted the severe coronary artery calcium (CAC) burden with an odds ratio of 3.26 (2.05–5.19). Recently, a total of 49 asymptomatic male marathon runners who underwent carotid ultrasound and CT angiography were assessed by Burgstahler *et al* [50]. The goal of the study was to evaluate the diagnostic accuracy of carotid ultrasound to predict coronary atherosclerosis. Between carotid ultrasound and coronary atherosclerosis, the study observed a sensitivity of 54.55% (95% CI 32.2–75.6), a specificity of 80.8% (CI 60.6–93.4), a positive predictive value of 70.6% (CI 44.1–89.9) and a negative predictive value of 67.7% (CI 48.6–83.3), with a positive likelihood ratio of 2.84 (CI 1.18–6.82) and a negative likelihood ratio of 0.56 (CI 0.34–0.92). Therefore, when a patient is diagnosed with calcium accumulation in the carotid artery, they should immediately undergo coronary atherosclerosis tests.

In support of these studies, the American College of Cardiology/American Heart Association (ACC/AHA) [51] and the European Society of Cardiology/European Society of Anaesthesiology (ESC/ESA) [52] guidelines also point out the importance of calcium in the arteries and further recommend its measurement for the prevention of heart disease and stroke [53, 54]. Recent studies have evaluated the recommendations made by these two sets of guidelines. Nasir *et al* [53], in 2015, applied the ACC/AHA guidelines in a Multi-Ethnic Study of Atherosclerosis (MESA) study with 4758 participants. According to the guidelines, 50% of participants were recommended for statin therapy of which 41% of participants had no coronary artery calcification. It was observed that patients with no calcium had a 10 year risk of 4.7% even when they had a 10 year atherosclerosis cardiovascular disease (ASCVD) risk of <20%. Thus, patients with a 10 year ASCVD risk between 5%–20% with no calcium can be stratified in the low-risk bin and can be deferred from taking statin therapy. This proves the importance of calcium measurement in the arteries. Recently, Mahabadi *et al* [54] showed the differences in the statin therapy recommendations based on the two guidelines (ACC and ECS) in an MESA study with 3745 participants. It was observed that calcium in the arteries was absent in 53% and 43% of the participants who met the statin therapy criteria recommended by the ACC and ESC guidelines, respectively. It was also observed that participants with a higher calcium score had a higher rate of CHD and CVD. The study concluded that, in addition to the guidelines, calcium score improves stratification of the participants into high-risk and low-risk bins.

A detailed analysis of the literature was performed using the PubMed and Google Scholar search engines. In the next section, we present a detailed survey of the different modalities used for detecting calcium in both the coronary and carotid arteries. In the second section, we discuss various studies carried out in the literature for quantifying calcium in both the coronary and carotid arteries. This section also includes studies carried out for fast calcium quantification using a multiresolution paradigm, and discusses the role of connected component analysis (CCA) for removing the false calcium signal in order to avoid overestimation during calcium quantification. The third section presents a detailed survey of the techniques used for performance evaluation and the methods used for validating the results. Finally, we present an exhaustive survey of various risk stratification studies using the machine-learning (ML) paradigm.

1.2 Calcium detection in coronary and carotid arteries

Imaging modalities are broadly divided into two categories: (a) non-invasive (such as computed tomography (CT), echocardiography and magnetic resonance imaging (MRI)) and (b) invasive (such as angiography, intravascular ultrasound (IVUS) and optical coherence tomography (OCT)). These imaging modalities play a vital role in the diagnosis, treatment and monitoring of patients suffering from CAD.

1.2.1 Calcium detection in coronary arteries

1.2.1.1 Using computer tomography

Studies have shown that CT, with recent advances in temporal and spatial resolution, plays an important role in assessing the coronary artery [55]. Sandercock *et al* [56] showed the usefulness of CT diagnosis and observed that, under certain defined circumstances, non-invasive methods including electron beam CT (EBCT) and multidetector CT (MDCT) can be employed for the detection of coronary artery calcium. EBCT is typically performed using echocardiography (ECG) triggering and MDCT using an x-ray tube in the presence of multiple detector rings. Modern MDCT scanners can achieve high temporal (75–150 ms) and spatial (0.5 mm) resolutions, even with a low patient heart rate [57]. Recently, in the

MESA study carried out by Bittencourt *et al* [58], 6781 patients underwent non-contrast cardiac CT to evaluate their calcium score. The results showed improvement in the prediction of CVD events, indicating that non-contrast cardiac CT can be considered a biomarker for the detection of myocardial infarction. It was observed that CT can provide a calcium score [59] in the artery but at the expense of a higher radiation dose, which could compromise patient safety [60].

1.2.1.2 Using echocardiography

Echocardiography has also been used for the detection of calcium in the aortic valves and aortic walls. Nucifora *et al* [61] used an echocardiography-derived calcium score (ECS) to predict the presence of severe CAC and obtained a high sensitivity and specificity of 87% for both. In a similar study by Pressman *et al* [62], global cardiac calcification (scored by echocardiography) showed a moderate correlation with CAC. The results showed that an echo score $\geqslant 5$ had a 60% positive predictive value for CAC > 400. Acharya further evaluated echocardiography images using a Gaussian mixture model (GMM) classifier to stratify CAD in patients [63]. The efficiency was close to 100%. These studies indicate the importance of echocardiographic evaluations for the detection of calcium in arteries.

1.2.1.3 Using angiography

Angiographic calcium can detect moderate calcification, but only during the cardiac cycle before contrast injection, whereas severe calcification, which affects both sides of the arterial lumen, can be detected without cardiac motion. In a comparative study of 183 patients, angiography identified less than half (45%) of the patients with any detected coronary calcification [64]. CAC can be easily detected using angiography, but this method has potential implications for percutaneous coronary intervention (PCI) outcomes [65, 66].

1.2.1.4 Using magnetic resonance imaging

In the late 20th century, MRI emerged as a radiation-free, safe technique for the diagnosis of CAD. A study carried out by Kaufman *et al* [67] investigated the impact of nuclear MRI on CVD. Three years later, Awad *et al* [68] used MRI to assess subcortical lesions in the elderly population. The study included 240 MRI scans among patients over 50 years of age. It was concluded that subcortical lesions can be used as an index of chronic cerebrovascular disease in elderly patients. Mohiaddin *et al* [69] also used MRI to measure both regional aortic compliance and total arterial compliance in 70 healthy volunteers, 13 athletes and 17 patients with CAD. Regional aortic compliance was higher than normal in athletes, whereas it was lower than normal in patients with CAD. Despite its benefits, the long acquisition imaging time in MRI can cause anxiety in some patients during image acquisition [70].

1.2.1.5 Using intravascular ultrasound

With the innovation of high-frequency sound waves (20–30 MHz), IVUS has emerged as a safer modality for the identification and location of calcium in stenotic arteries [71, 72]. The grayscale IVUS-based acquisition system consists of three parts: (i) a catheter, (ii) a pullback device and (iii) a scanning console [73]. The IVUS catheter carries an ultrasound transducer at its tip, which can both transmit and receive ultrasound signals. Before starting acquisition, the catheter is first manually advanced to the distal end of the coronary artery. Typically, the catheter is first inserted along with a guide wire from the femoral artery up to the site of occlusion in the coronary artery. Then, using the pullback device, the catheter is automatically pulled back at a speed of 0.5 mm s^{-1}. The pullback device is connected to a computer via a cable. The reflected ultrasound amplitude is used to create cross-sectional images which are stored for post-processing [73]. The echogenicity of different plaques is different. Plaques can be characterised as hypoechoic, isoechoic or hyperechoic [74]. Lipids and thrombi are usually hypoechoic, whereas the fibrous cap and calcium are hyperechoic [75, 76].

In the literature, several theories have been proposed for using ultrasound scanned images for accurate quantification of lipid and calcified plaques [77]. Kovalski *et al* [78] proposed an algorithm that uses active contour principles to identify the lumen–intima (LI) border and the media–adventitia (MA) border. Later, the features were used to reconstruct the coronary artery in 3D. The 3D structure further helped in better understanding of coronary artery geometry and plaque deposition. Depending on the calcium location within the plaque, calcium can be further quantified as deep or superficial [79]. The potential of IVUS to estimate CAC was compared to histology by Friedrich *et al* [80]. This study showed high sensitivity (90%) and specificity (100%) for the detection of dense calcium. Mintz *et al* [81] and Tuzcu *et al* [64] further compared IVUS to CA and found that IVUS had a higher sensitivity in detecting calcification compared to CA. These studies showed a higher accuracy in detecting CAC compared to histology.

With the advancement in IVUS technology, integrated backscattered IVUS (IB-IVUS) and IVUS-Virtual Histology (IVUS-VH^{TM}) further enhanced CAC detection and quantification. To improve the quantitative assessment obtained by ultrasound signals, IB-IVUS uses the time domain information from radiofrequency (RF) signals [82]. Furthermore, IVUS-VH^{TM} adopted a spectral analysis of ultrasound signals for plaque characterisation to stratify different plaque components by using different coloured maps [83]. This showed a higher predictive accuracy (96.7%–100%) compared to histology [84].

1.2.1.6 Using optical coherence imaging (OCT)

In comparison to IVUS, OCT has a much better resolution (10–20 μm) as it measures the amplitude of the backscattered light and is one step ahead of IVUS in assessing coronary vessels [85]. A physical overview of an OCT system is shown in figure 1.4. Unlike IVUS, OCT provides fast data acquisition (2.5 s), yielding detailed images of the vessel lumen, neointimal tissue and strut distribution [86, 87]. Recently, Wang *et al* [88] evaluated OCT and IVUS against coronary angiography for the assessment of target lesion calcification. Of the 440 calcium lesions, coronary angiography detected 40.2%, IVUS detected 82.7% and OCT detected 76.8%, respectively.

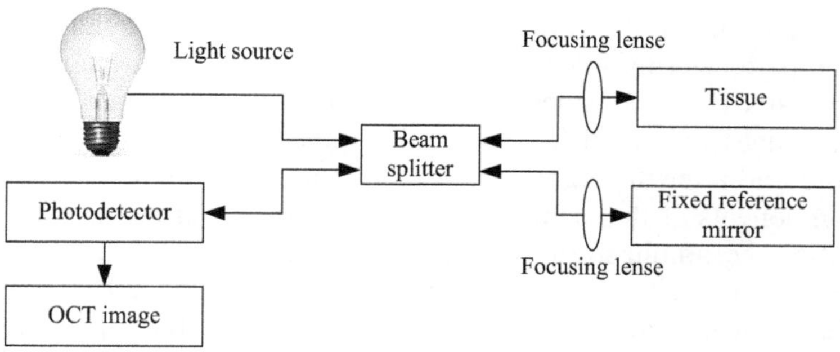

Figure 1.4. Physical overview of an OCT system.

1.2.2 Calcium detection in carotid arteries

1.2.2.1 Using computer tomography

With advancements in CT technologies, its application in carotid artery disease management has increased tremendously [89]. CT is usually employed for assessing calcium in the carotid arteries and has been shown to be a useful tool for plaque tissue characterisation [90] and in the prediction of stroke risk [91–93]. In *in vitro* [94] and *in vivo* studies [95], de Weert *et al* retrospectively evaluated the performance of 16 slice MDCT for the assessment of carotid plaque components (calcifications, fibrous tissue and lipid). These studies tried to quantify the atherosclerotic carotid plaque components and compared the results against histology. In an *in vitro* study [94], the calcified and lipid areas on MDCT and histology correlated well ($R^2 = 0.83$ and $R^2 = 0.68$, respectively). Similarly, in an *in vivo* study [95], the results showed a good correlation ($R^2 > 0.73$) between MDCT and histology, except for lipid core areas, which only had a good correlation ($R^2 > 0.77$) in mild calcified (0%–10%) plaques.

In another retrospective study consisting of 122 carotid arteries, Saba *et al* [96] observed no correlations between MDCT angiography-assessed carotid artery plaque volumes in the presence of ulceration. The same group [97] further evaluated the application of semi-automated techniques for the detection and measurement of carotid artery wall plaque. By carrying out a study using MDCTA in 22 patients, the authors demonstrated that the proposed semi-automatic method based on the level set model (LSM) can automatically measure the thickness of the plaque. By analysing 70 patients, the same group [98] tried to study the correlation between plaque in the carotid arteries (using a 16 detector row CT scanner) and cerebral microbleeds (CMB) in the brain (with a 1.5T MR imaging system). The results suggested an association between the presence of carotid artery fatty plaque, cerebrovascular symptoms and CMB, and concluded that the presence of CMB may represent an indication of the severity of cerebrovascular symptoms. Three years later, Saba *et al* [99] proposed an automatic mean shift-based algorithm for labelling calcified plaques in ICA using CT images taken from 75 patients. Independent of the number and size of calcium regions, the proposed approach provided reasonably accurate labelling of calcified plaques.

In the CT assessment of carotid plaques, the analysis of the attenuation value is a fundamental parameter in order to classify the type of the plaque components. In a retrospective study of 68 patients (192 slides), Saba *et al* [100] examined the attenuation values measured in Hounsfield units (HU) of the region-of-interest (ROI) before and after the administration of contrast medium. The study showed that the components of the plaque in ROI sampling, performed in the CT dataset acquired after the administration of contrast medium, had a greater degree of heterogeneity compared to the baseline measurement. This effect was observed because, during acquisition, different amounts of contrast were observed for different carotid artery plaque components.

1.2.2.2 Using magnetic resonance imaging

MRI is generally used to assess the soft tissue characteristics of carotid atherosclerotic plaques based on morphological features [101]. Most previous plaque characterisation work was focused on wall thickness measurements [102].

Merickel *et al* [103] used the functional and structural information of plaque and computed two different measurements. First, the authors computed the ratio of the plaque component volume with respect to the total wall volume. Later, they measured the difference in the cross-sectional area between the diseased lumen and the normal lumen. The first measurement provided an estimation of stenosis progression, while the second measurement provided an extent of blockage in the lumen. The study successfully demonstrated significant segregation between atherosclerotic tissues and calcified plaque.

In an *in vivo* study, Toussaint *et al* [104] showed that T2-weighted MRI can also discriminate lipid cores, fibrous caps and calcifications in human atheromatous plaques. The authors carried out an *in vivo* study on seven lesions from six patients, prior to surgery. Further, the authors repeated the same protocol *in vitro*. For each plaque component, the study observed a high correlation between *in vitro* and *in vivo* measurements by adapting the linear regression.

Coombs *et al* [105] showed the capability of a 3D MRI in identifying fatty plaque, fibrous plaque and calcified plaque. Twenty-one carotid endarterectomy tissue sections were analysed by both MRI and histology. The study observed different signal characteristics for different plaque components, leading to the conclusion that 3D gradient-echo MRI can distinguish and identify atherosclerotic plaque components.

Recently, Lee *et al* [106] showed the importance of high-risk carotid plaques in choosing the treatment strategy for carotid stenosis patients. From 2014 to 2016, the study collected data from 15 patients who underwent angiography for stenosis measurements. The authors further analysed intraplaque haemorrhage (IPH) using MRI. The results showed a significant relationship between IPH and ischaemic symptoms. Despite its benefits, challenges remain for patients with marked arrhythmia and metal implants [107].

1.2.2.3 Using B-mode ultrasound

A carotid B-mode ultrasound acquisition system is shown in figure 1.5. Using the rationale that variable echogenicity is produced by different tissues, Lal *et al* [77]

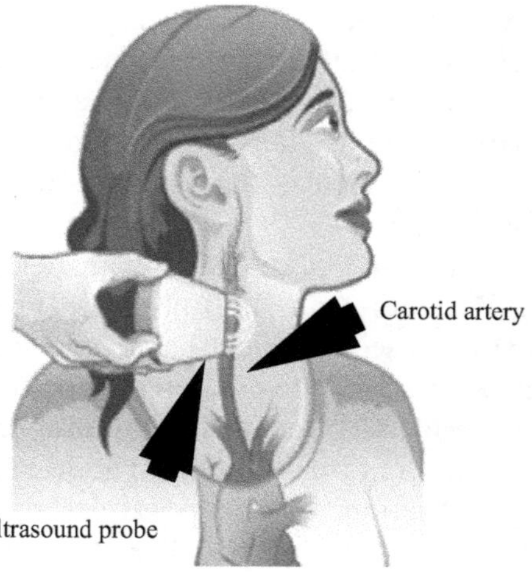

Figure 1.5. Carotid B-mode ultrasound acquisition system. (Courtesy of Atheropoint^{TM}, Roseville, CA, USA.)

used computer-assisted duplex ultrasound (DU) scanned images to quantify the echogenicity of blood, lipid, fibromuscular tissue and calcium. Pixel-based distribution analysis (PDA) was used to quantify these components in 10 healthy tissues and 20 carotid artery plaques from 19 patients. The results showed a strong correlation with the histologic readings and this proved that PDA can accurately quantify calcium components in control subjects.

Molinari *et al* [108] proposed an ultrasound-based technique for automatic characterisation of different plaque components (such as thrombi, lipids, fibrous tissue and calcium) in the carotid artery. For plaque boundary segmentation, an automated technique CULEX [109] was used. Twenty plaque specimen results were compared against histology. The results demonstrated that the proposed methodology can effectively identify plaque components. The studies showed that echogenicity produced by different plaque components in B-mode ultrasound images can be used as a biomarker for the accurate quantification of calcium components in the carotid arteries. Hitchner *et al* [110] further tried to estimate the role of IVUS in the characterisation of carotid plaque components. The study explored the relationship between microemboli and plaque tissues. In a group of 38 high-risk patients, microemboli were analysed by comparing the pre- and postoperative diffusion-weighted MRI images. Using univariate and multivariate logistic regression, the area of fibrous tissue and calcification was observed to be related to the microemboli.

Many studies have detected arterial calcification by ultrasound, but the diagnostic accuracy is still not well-validated. To validate the accuracy of ultrasound examination, Jashari *et al* [111] performed a comparative study of atherosclerotic calcification detection using two imaging modalities: carotid ultrasound and cone beam CT (CBCT). A pool of 88 patients (94 carotid arteries) who underwent pre-endarterectomy ultrasound examination were chosen for this study. Initially, atherosclerotic calcification was determined using carotid B-mode ultrasound. Calcium was identified from its high echogenicity and posterior shadow. After endarterectomy, the calcium volume was computed using CBCT. To determine the accuracy, the calcium volumes acquired from both imaging modalities were compared. The results showed that carotid ultrasound could accurately (sensitivity of 96%) detect the presence of calcified atherosclerotic lesions having a volume $\geqslant 8$ mm^3.

In comparison to pixel intensity-based tissue characterisation (usually based on an intensity threshold), Pazinato *et al* [112] proposed image descriptors (such as statistical moments, texture-based, gradient-based and local binary patterns) for carotid ultrasound images to classify five different types of tissues, such as blood, lipids, muscle, fibrous material and calcium. The proposed classification consisted of the following pipeline: (i) image normalisation, (ii) multiscale feature extraction and (iii) machine-learning classification. The proposed descriptor was computed while using the pixel neighbourhood information. The study outperformed a standard threshold-based method by showing a 19% increase in accuracy. Thus, MRI is generally used for soft tissue component characterisation [113].

1.3 Calcium area/volume quantification in coronary and carotid arteries

1.3.1 Calcium area/volume quantification in coronary arteries

For an optimal interventional procedure, a cardiologist must know the exact location, position and volume of the calcified plaque in the coronary arteries [114]. Several prospective cohort studies have tried to quantify the area/volume in the coronary artery using IVUS (see table 1.1). Weissman *et al* [115] measured plaque volume in 19 patients before and after atherectomy. The volume of the calcified plaque was calculated using the modified Simpson's rule [116]. To access calcium along the length of the vessel, Scott *et al* [117] presented a two-layered technique to quantify calcium in the coronary arteries. In this study, the total and calcified plaque luminal circumferential length was first measured, and then the plaque area was computed using the standard Simpson's rule [118]. The study accurately reflected coronary calcium as determined using histology. Previous studies lacked the automation of the calcium detection process.

In 2008, Santos Filho *et al* [119] proposed an automated calcium quantification technique by finding the optimised threshold using the iterative Otsu's method [120]. In their study, the calcified region was distinguished from other bright regions by identifying the presence of the acoustic shadow. Zhang *et al* [121] outperformed the Santos method by proposing an automated detection algorithm to detect calcification using snakes and contourlet transform. The study utilised a 2D Renyi's entropy algorithm to produce the ROI from which the contours of calcification were obtained. The study outperformed the Santo method by 2.76% and 14.53% in terms of sensitivity and specificity, respectively. Gao *et al* [122] showed that there were two reasons as to why the performance of previous methods was inferior for detecting and computing the calcium volume: (i) the detection of the ROI did not consider the concept of acoustic shadowing and (ii) refinement of the calcified plaque relied on

Table 1.1. Prospective studies on calcium computation in the coronary arteries.

Year	Authors	Modality	Techniques	Metric	N	Fr/ Video	Benchmark	Statistical analysis with CI	Entire video usage	Validation
1995	Weissman *et al* [115]	Coronary	Modified Simpson's rule	Volume	19	✗	None	*t*-test (CI: 95%)	✗	Manual
2000	Scott *et al* [117]	Coronary	Standard Simpson's rule	Area	6	≈ 195	None	*t*-test (CI: 95%)	✗	Histology
2008	Santos *et al* [119]	Coronary	Adaptive thresholding	Area	14	✗	Otsu *et al* [120]	✗	✗	Manual
2010	Zhang *et al* [121]	Coronary	Snakes and contourlet transform	Area	11	≈ 8	Santos *et al* [119]	✗	✗	Simulation
2014	Gao *et al* [122]	Coronary	RMM, MRF and GSA	Area	8	≈ 125	None	✗	✗	Manual
2015	Araki *et al* [123]	Coronary	Shape-based	Volume	100	≈ 2549	None	✗	✓	Scoring
2016	Araki *et al* [114]	Coronary	FCM, K-means and HMRF	Volume	15	≈ 2040	Santos *et al* [119] and Araki *et al* [123]	*Z*-test and Wilcoxon test (CI: 95%)	✓	Scoring

N: number of patients; Fr/Video: number of IVUS frames per video; RMM: Rayleigh mixture model; MRF: Markov random field; GSA: graph search algorithm; FCM: fuzzy c-means; HMRF: hidden Markov random field; Manual: manual tracings of calcium; Histology: calcium determined by histology; Simulation: simulation of an IVUS image with calcium; Scoring: calcium scoring by trained expert; CI: confidence interval.

grey intensities. The authors provided an automated framework on 996 *in vivo* IVUS images acquired from eight patients. The technique was composed of the Rayleigh mixture model for performing pixel classification, the Markov random field for the detection of angular location, and the graph search algorithm to detect the borders of the calcified plaque. The study achieved a high sensitivity and specificity of 94.68% and 95.82%, respectively.

Besides Weissman *et al* [115] and Araki [123], the above mentioned studies had achieved accurate calcium area quantification but lacked volume computation. Weissman *et al* [115] study did not mention the number of frames utilised for the calcium volume computation. Further, the study did not perform calcium quantification. By utilising the entire IVUS video, Araki *et al* [123] used a shape-based approach for detection of the largest calcium region in each frame of the video. The study was performed on 100 patients resulting in an accuracy of 81%. The major drawback of this approach was selecting the largest connected calcium while ignoring the loosely unconnected small lesions. Since calcium is multifocal in nature, true calcium estimation is possible only if all the calcified components in the arteries are considered. By using the above concept, the same group [114] had utilised three segmentation techniques (fuzzy c-means (FCM), K-means and hidden Markov random field (HMRF)) for the automated detection of multifocal calcium regions in each frame throughout the IVUS video. K-means showed the best performance with an accuracy of 92.80%. As the number of IVUS frames per videos is usually large (~2040 frames/video), the proposed studies suffered from prolonged computational time.

High computation speed is a basic requirement of any automated calcium detection technique. By adapting multiresolution techniques (the so-called down sampling mode), it is possible to speed up the computation. Several prospective studies have tried multiresolution techniques for the detection and measurement of calcium in a coronary artery (see table 1.2). Recently, Banchhor *et al* [124] applied a set of five different multiresolution-based techniques (bilinear, bicubic, wavelet, Lanczos and Gaussian pyramid), on a set of four kinds of segmentation methods (threshold, FCM, K-means and HMRF). By carrying out a study on 38 760 IVUS frames acquired from 19 patients, the study observed an improvement in the mean computational time. It was observed that the FCM detection technique when embedded with wavelet-based multiresolution paradigm produced the best performance. Even though the study resolves the computational time issue, it did not take into account the noise in IVUS images. This leads to a bias of overestimation in the final detected calcium volume. To overcome this limitation, the same group [125] proposed an automated connected component analysis (CCA)-based approach to remove the noise, as shown in figure 1.6. The study was based on the assumption that isolated calcium size cannot be smaller than 100 pixel^2. Using the CCA-based approach, the study observed an improvement of 38.54% in the mean overall performance. The threshold-based classifier embedded with Lanczos multiresolution was found to be an optimal combination. Among the different automated techniques proposed, the quantification of coronary calcium volume using complete IVUS videos can assist the cardiologist during the planning of PCI procedures.

Table 1.2. Survey on calcium detection techniques using multiresolution paradigm in coronary arteries.

Year	Authors	*N*	F/Video	Objective	Multiresolution technique	Performance evaluation	Accuracy	Benchmark	Statistical analysis with CI	MR comparisons	Bias correction
2006	Liang *et al* [126]	Not mentioned	Not mentioned	Arterial wall strain	Gaussian pyramid	Correlation coefficient	N/A	None	×	None	None
2009	Srinivas *et al* [127]	Not mentioned	Not mentioned	Pre-rupture plaque	Wavelets	None	N/A	None	×	None	None
2010	Zhang *et al* [121]	11	≈ 8	Calcium area	Contourlet transform	MD, RMD, MSD, RDD, ARD, TD, LD, RARD, RTD and RLD	N/A	Santo *et al* [119]	×	None	None
2012	Katouzian *et al* [128]	5	1158 frames	LI border	Wavelets	TP, FP, RMSE	N/A	None	×	None	None
2012	Lazrag and Naceur [129]	Not mentioned	20 simulated, 30 real frames	Vascular boundary	Contourlet transform	None	N/A	Zhang *et al* [121]	×	None	None
2012	Lazrag and Naceur [130]	Not mentioned	Not mentioned	Speckle reduction	Wavelets	PSNR, NMSE	N/A	None	×	None	None
2015	Puertas *et al* [131]	Not mentioned	Not mentioned	Stacked sequential learning	Gaussian pyramid	Accuracy, overlapping, recall and precision	N/A	None	Friedman and Nemenyi test (CI: 95%)	None	None

(*Continued*)

2016	Banchhor *et al* [124]	19	≈ 2040	Calcium volume	Bilinear, bicubic, wavelets, Lanczos and Gaussian pyramid	PTI, PoM, MVS, DR and QAR	$PTI^*_{\forall S}$, $PTI^*_{\forall M}$, PoM^*_{VL}, PoM^*_{FL}, PoM^*_{DR} and PoM^*_{QAR}	Zhang *et al* [121], Lazrag and Naceur [130]	×	Yes	Yes

N: number of patients; F/Video: number of IVUS frames per video; MR: multiresolution, $PTI^*_{\forall S}$: percentage mean time improvement for multiresolution technique; $PTI^*_{\forall M}$: percentage mean time improvement for segmentation method; PoM: precision-of-merit; VL: volume level; FL: frame level; MVS: mean volume similarity; MVSIBC: mean volume similarity increase with bias correction; DR: degradation ratio; QAR: quality assessment ratio; CI: confidence interval; IMT: intima–media thickness; PSNR: peak signal-to-noise ratio; NMSE: normalised mean square error; MD: mean distance error; RMD: relative mean distance error; MSD: mean signed distance error; RDD: relative difference degree; ARD: arc difference; TD: thickness difference; LD: length difference; RARD: relative arc difference; RTD: relative thickness difference; RLD: relative length difference; TP: true positive; FP: false positive; RMSE: root mean square error; N/A: not available.

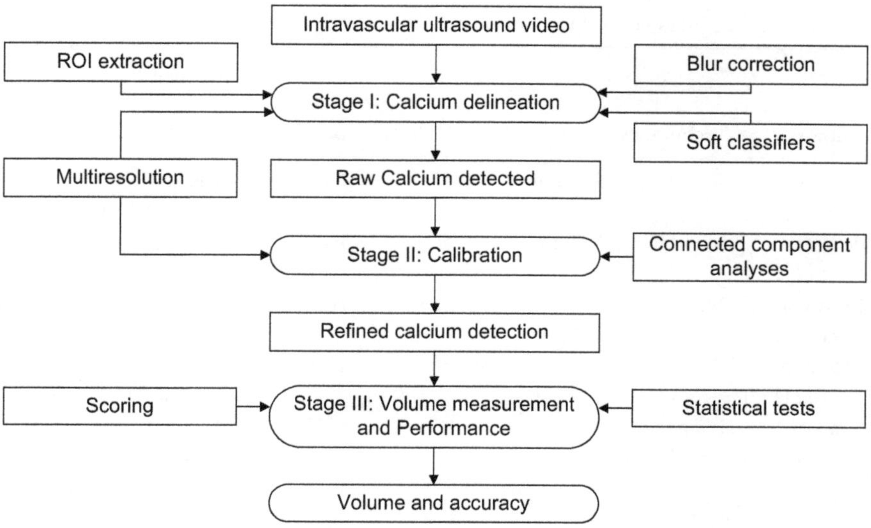

Figure 1.6. A well-balanced system for calcium detection using the CCA-based approach.

Multiresolution techniques and the CCA-based approach can further improve the speed and accuracy of different calcium detection techniques.

1.3.2 Calcium area/volume quantification in carotid arteries

Several prospective cohort studies have tried to quantify area/volume in the carotid artery (see table 1.3). We have divided our calcium area/volume measurement strategy based on the modality chosen. Key authors in the CT-based paradigm are first discussed followed by key authors in the ultrasound-based paradigm.

1.3.2.1 CT-based measurements

Denzel *et al* [132] examined 92 CT-based internal carotid artery (ICA) endarterectomy specimens with stenosis greater than 70%. The results showed that the calcium scores computed, using the method discussed in Agatston *et al* [59], enabled precise *in vitro* measurements from ICA plaques (consisting of calcified plaques, lipid and combined plaques). The authors observed a high mean correlation ($R = 0.628$, $p < 0.001$) between the calcium score and radiological classification for slight, moderate and marked calcifications, but there was no *in vivo* validation of the plaque components. De Weert *et al* [133] also tried to estimate the volume of plaque and its components using MDCTA images. The authors analysed 56 carotid arteries using three observers. The observers manually drew the vessel contour based on the HU threshold. Since MDCTA cannot differentiate atherosclerotic plaque and tunica media, the technique can potentially lead to overestimation of plaque volume. Marquering *et al* [134] tried to explore the relationship between carotid calcium volume and degree of stenosis from CT angiography images using the Pearson correlation coefficient. The study observed a weak correlation between calcium volume and stenosis (a sensitivity of 47% and a specificity of 52%). The authors

Table 1.3. Prospective studies on calcium computation in the carotid arteries.

Year	Authors	Techniques	Metric	N	Fr/Video	Benchmark	Statistical analysis with CI	Validation
2004	Denzel *et al* [132]	Agatston	Scoring	89	92	None	CK test (CI: 95%)	None
2008	de Weert *et al* [133]	Manual	Volume	56	56	None	*t*-test (CI: 95%)	None
2011	Marquering *et al* [134]	Manual	Volume	90	159	None	*t*-test (CI: 95%)	None
2012	Molinari *et al* [137]	Bicubic	Area	NM	365 frames	None	*t*-test (CI: 95%)	None
2016	Anzidei *et al* [135]	ImageJ software	Volume	62	✖	None	KS, MW, Wilcoxon, *t* and CK tests (CI: 95%)	None
2017	Gepner *et al* [136]	Agatston	Scoring	4955	✖	None	✖	None

N: number of patients; Fr/Video: number of IVUS frames per video; Manual: manual tracings of calcium; CK: Cohen's kappa; KS: Kolmogorov–Smirnov; MW: Mann–Whitney; CI: confidence interval.

concluded that the calcium volume measurement cannot estimate the degree of stenosis in the carotid arteries. Anzidei *et al* [135] examined the relationship between head and neck radiation therapy treatment (HNXRT) and the CT volume in carotid artery plaque, fatty plaques and mixed plaque components. A pool of 100 patients was analysed at baseline, and two years later, 62 patients (who underwent HNXRT) were reanalysed. In these two years, the volumes of carotid artery plaques (533 mm^3), fatty plaques (103 mm^3) and mixed plaque components (328 mm^3) were observed as 746 mm^3, 202 mm^3 and 419 mm^3, respectively. The study observed an increase in carotid artery plaque volume (particularly the fatty plaque component) with patients who underwent HNXRT.

Gepner *et al* [136] compared the CT carotid plaque score with the coronary artery calcium score. In a multi-ethnic cohort of 6814 patients, the results were analysed at baseline and after a follow-up of three years. For predicting CVD, the hazard ratios (HRs) for CAC scores and carotid plaque scores were HR = 1.78 (95% CI, 1.16–1.98; $p < 0.001$) and HR = 1.27 (95% CI, 1.16–1.40; $p < 0.001$), respectively. Similarly, for predicting CHD, the HRs for the CAC scores and carotid plaque scores were HR = 2.09 (95% CI, 1.84–2.38; $p < 0.001$) and HR = 1.35 (95% CI, 1.21–1.51; $p < 0.001$), respectively. The results showed that CAC scores proved to be a stronger predictor of CVD compared to the carotid plaque scores.

1.3.2.2 Ultrasound-based measurements

Not much has been proposed for the quantification of calcium area and volume using the ultrasound-based paradigm. The focus has been more on wall thickness measurement, instead of component quantification. Keeping this in mind, Tsiaparas *et al* [138] proposed a multiresolution approach for carotid atherosclerotic tissue classification. Out of the four decomposition schemes (discrete wavelet transforms, stationary wavelet transforms, wavelet packets and Gabor transform), wavelet packets followed by Haar function produced the best performance (82.5% and 77.5%).

The scale-space strategy introduced by Suri's group dominates several fundamental carotid artery wall measurement paradigms. The basic idea was to apply a higher order Gaussian derivative filter to extract the edges of the wall (see Molinari *et al* [137]). This was accomplished in the multiresolution framework to ensure that the scales were not too high. By carrying out a study on 365 B-mode longitudinal carotid images, CAMES observed an 8.4% increase in the accuracy compared to their previous integrated approach using feature-based extraction and classifier (CALEX) [139]. The result proved CAMES as a clinical tool for accurate cIMT measurements in large multicenter clinical trials.

1.4 Metrics for performance evaluation for calcium detection algorithms and its validation

Metrics play an important role in evaluating the performance of calcium detection algorithms. Furthermore, it is important to discuss the types of strategies one can adopt to establish the gold standard when benchmarking calcium detection methods. These two factors are presented in this section.

1.4.1 Statistical metrics for performance evaluation

Any proposed calcium detection algorithm should have performance indices such as accuracy, reliability and robustness. To choose or compare, we must evaluate the performance by choosing the correct metrics. The basic evaluation measure is the supervised evaluation in which the output of the algorithm is compared against the ground truth (GT). The ground truth may be a set of binary images or calcium area/volume reflecting true calcium detection either derived by a human expert or from any prior established study.

Depending on the type of variable, the number of classes in the study and the distribution of the data, statistical tests commonly used in the literature are the *t*-test, *z*-test, Mann–Whitney test, Kolmogorov–Smirnov (KS) test, ANOVA test, Chi-squared test, Friedman test and Wilcoxon test [94, 98]. Bland–Altman plots are also used to display the average bias or the average of the differences between the two readings [140].

Given a set of GT images, the preferred performance evaluation metrics are the Jaccard index (JI), Dice similarity coefficient (DSC), signal-to-noise ratio (SNR) and contrast-to-noise ratio (CNR). JI and DSC are the simplest ways to quantify the proximity between two binary images containing calcium regions. The JI and DSC always lie between 0 and 1, where 0 and 1 correspond to the lowest and highest similarity between the GT and the segmented binary image, respectively. If A is the ground truth binary image and B is the segmented binary image, then the JI formula can be given as [125]

$$J_{AB} = \frac{|A \cap B|}{|A \cup B|} \text{ and } D_{AB} = \frac{2|A \cap B|}{|A| + |B|}, \tag{1.1}$$

where $|A \cap B|$ and $|A \cup B|$ indicate the sum of all the related and unrelated pixels between the GT and the segmented binary image, respectively. Similarly, $|A|$ and $|B|$ indicate the sum of all the pixels of the GT and the segmented binary image, respectively. Here, $0 \leqslant J_{AB} \leqslant 1$ and $0 \leqslant D_{AB} \leqslant 1$.

Araki *et al* [114] used these two metrics to quantify the degree of similarity between their proposed three automated detection metrics and the threshold-based method. Banchhor *et al* [124] used the same measure to compare segmented binary results obtained with and without the use of multiresolution techniques. SNR is a measure to compare the signal strength over the background noise and CNR is a measure to determine the image quality [141]. SNR can be mathematically given as [141]

$$\text{SNR} = \frac{S_L(i, j) - S_B(k, l)}{\sqrt{2} \,.\, \sigma_B(k, l)}, \tag{1.2}$$

where $S_L(i, j)$ is the mean signal strength in the ROI with a lesion at the location (i, j). $S_B(k, l)$ and $\sigma_B(k, l)$ are the mean signal strength and standard deviation of the background ROI without a lesion at the location (k, l), respectively. Similarly, CNR can be mathematically given as [142]

$$\text{CNR} = \sqrt{\frac{(\mu_L(i, j) - \mu_B(k, l))^2}{\sigma_L(i, j) + \sigma_B(k, l)}}, \tag{1.3}$$

where $\mu_L(i, j)$ and $\mu_B(k, l)$ are the mean signal strength in the ROI with a lesion at the location (i, j) and the background ROI without a lesion at the location (k, l), respectively. Similarly, $\sigma_L(i, j)$ and $\sigma_B(k, l)$ are the standard deviation of the signal strength in the ROI with a lesion at the location (i, j) and the background ROI without a lesion at location (k, l), respectively.

Unlike regional image-based performance, one can use performance metrics given the area or volume measurements. Typically adapted metrics are the correlation coefficient (CC), precision-of-merit (PoM), figure-of-merit (FoM) and receiver operating characteristic (ROC), The CC is a measure to predict the degree to which changes in one variable predict the change of another. PoM and FoM are measures to compare individual and mean readings, respectively [94]. Mathematically, PoM can be depicted as [143]

$$\text{PoM} = \frac{1}{N}\sum_{j=1}^{N} 100 - \left[\left(\frac{|\text{Auto}(j) - \text{Manual}(j)|}{\text{Manual}(j)}\right) * 100\right], \tag{1.4}$$

where Auto(j) is automatically computed by any system and Manual(j) values are obtained from manual measurements on the jth image of the database of N images.

The central tendency of the error can also be computed using FoM [143–145], which is given as

$$\mathrm{FoM} = 100 - \left[\left(\frac{|\overline{\mathrm{Auto}} - \overline{\mathrm{Manual}}|}{\overline{\mathrm{Manual}}}\right) * 100\right], \tag{1.5}$$

where

$$\overline{\mathrm{Auto}} = \frac{1}{N}\sum_{i=1}^{N}\mathrm{Auto}(i) \tag{1.6}$$

$$\overline{\mathrm{Manual}} = \frac{1}{N}\sum_{i=1}^{N}\mathrm{Manual}(i). \tag{1.7}$$

Here N represents the total number of images in the database.

ROC is a graphical representation of sensitivity and specificity where a higher AUC confirms superior performance [125]. ROC needs the computation of the true positive rate (TPR), false positive rate (FPR), positive predictive value (PPV) and negative predictive value (NPV). True positives (TPs) and false negatives (FNs) are defined as the number of times true calcium is correctly and incorrectly identified with respect to the manually computed calcium for the cut-off risk threshold. Similarly, true negatives (TNs) and false positives (FPs) are defined as the number of times true calcium is correctly and incorrectly identified for the cut-off risk threshold. TPR, FPR, PPV and NPV can be mathematically formulated as [125]

$$\mathrm{TPR} = \frac{\mathrm{TP}}{(\mathrm{TP} + \mathrm{FN})} \tag{1.8}$$

$$\mathrm{FPR} = \frac{\mathrm{TN}}{(\mathrm{TN} + \mathrm{FP})} \tag{1.9}$$

$$\mathrm{PPV} = \frac{\mathrm{TP}}{(\mathrm{TP} + \mathrm{FP})} \tag{1.10}$$

$$\mathrm{NPV} = \frac{\mathrm{TN}}{(\mathrm{TN} + \mathrm{FN})}. \tag{1.11}$$

1.4.2 Validation of calcium detection algorithms

Validation is an important component of a system as it explains how reliable the calcium detection methodology is. The two most common methods used for validation are: (i) giving a score to the evaluated results by an expert (scoring strategy) and (ii) comparing against histology. In scoring, the radiologist scores each calcium detected frame on a scale from 0 to 5 (where 5 indicates true calcium detection and 0 indicates false calcium detection) [124]. To avoid inter-observer variability, the scoring should be performed by more than one expert. Also, each radiologist should repeat their tracings to avoid intra-observer variability. To avoid any bias, the current score should be performed blinded against the score done in the past. In the end, all the scores should be averaged to find the mean score for the computed automated values and GT, and compared between them, which forecasts the performance of the calcium detection algorithm. During scoring, factors such as the effect of lighting conditions, radiologist fatigue, experience and type of image resolution (DICOM versus JPEG) must also be taken into consideration [94].

The second most common method of validation is comparing the results of detection algorithms against histology. Scott *et al* [117], computed the coronary calcium from IVUS as a percentage of the coronary luminal surface. For the histology analysis, the arteries were decalcified and cut at 3 mm intervals, which is the smallest distance practical for processing tissue for light microscopy. Calcified areas were measured by computerised planimetry and the calcium volume was computed using Simpson's rule. The study found a high degree of correlation ($r = 0.84$, $p < 0.0001$) between the computed reading against the calcium area estimated using histological analysis.

1.5 Machine-learning-based risk stratification

Plaque risk assessment in diseased arteries is beneficial for cardiologists prior to any interventional procedures such as PCI. Further, procedures such as stent deployment or bypass shunting can be better planned [73] if plaque severity is known *a priori.* IVUS is more popular than other medical imaging modalities as it can provide an accurate tomography of the vulnerable plaque, which can be used for measurement of morphological features [53, 71]. Several studies have been conducted to stratify the vascular risk using a machine-learning (ML) paradigm (table 1.4).

1.5.1 Coronary risk assessment using ML-based approaches

Araki *et al* [158] demonstrated the use of the fusion of shape-based features with geometric-based coronary calcium volume for risk stratification on 92 patients with stable angina pectoris. Normalised calcium volume was computed using a geometric-based segmentation strategy. This strategy was used to isolate the calcium lesion by suppressing the non-calcium region. In comparison to shape-based features (AUC = 0.58), the fusion of the calcium volume features with shape-based features showed an improvement of 6.2%. Araki *et al* [154] further showed a CADx system for coronary risk stratification that utilised a support vector machine (SVM), which when trained using 56 plaque texture features yielded a classification accuracy of 94.95%. The major limitation of this study was a lack of feature selection adaptation; thus, it was never optimised. Later, the same group [155] modified their CADx system by using a dominant principal component analysis (PCA)-based polling technique for feature selection leading to an improvement of 3.48% (~98.43%) over their previous method. Recently, Banchhor *et al* [145] designed a CADx risk stratification tool by fusing plaque texture-based features with the wall-based measurement features (see figure 1.7). By using a PCA-based polling strategy, dominant features were selected from the

Table 1.4. Survey of risk stratification techniques in the literature.

Year	Authors	Arterial type	Data size	Feature type(s)	Total features	Feature selection technique(s)	Classifier(s)	Cross-validation accuracy
2003	Christodoulou *et al* [146]	Carotid	230	Tex	61	Mean, SD, Distance	MNN	73.1%
2005	Kyriacou *et al* [147]	Carotid	274	Tex, Wall	10	N/A	NN	71.2%
2007	Mongiakakou *et al* [148]	Carotid	54	SF, Law's	21	ANOVA	HNN	99.1%
2009	Kyriacou *et al* [149]	Carotid	274	Tex	10	N/A	SVM	73.7%
2012	Acharya *et al* [150]	Carotid	346	Tex	4	*t*-test	SVM, AdaBoost	SVM—82.4%, AdaBoost—81.7%
2012	Acharya *et al* [151]	Carotid	346	Tex, Wall	3	*t*-test	SVM	83%
2013	Acharya *et al* [152]	Carotid	492	DWT, HOS, Tex	7	*t*-test	SVM	91.7%
2014	Pedro *et al* [153]	Carotid	146	Rayleigh Mixture	16	N/A	EAI	77%
2016	Araki *et al* [154]	Coronary	2865	Tex	56	N/A	SVM	94.95%
2016	Araki *et al* [155]	Coronary	2865	Tex	56	PCA	SVM	98.43%
2017	Araki *et al* [156]	Carotid	407	Tex	16	N/A	SVM	FW—98.00%, NW—98.00%
2017	Saba *et al* [157]	Carotid	407	Tex	16	PCA	SVM	FW—98.55%, NW—98.83%
2017	Banchhor *et al* [145]	Coronary	4930	Only Tex, Tex fused with Wall	65	PCA	SVM	Only Tex—89.98%, Tex fused with Wall—95.41%

Tex: plaque texture-based; Wall: wall-based; NN: neural network; SF: statistical feature; SD: standard deviation; MNN: modular neural network: HNN: hybrid neural network; SVM: support vector machine; EAI: enhanced activity index; PCA: principal component analysis; DWT: discrete wavelet transform; HOS: higher order spectra; NW: near-wall; FW: far-wall; N/A: not applicable.

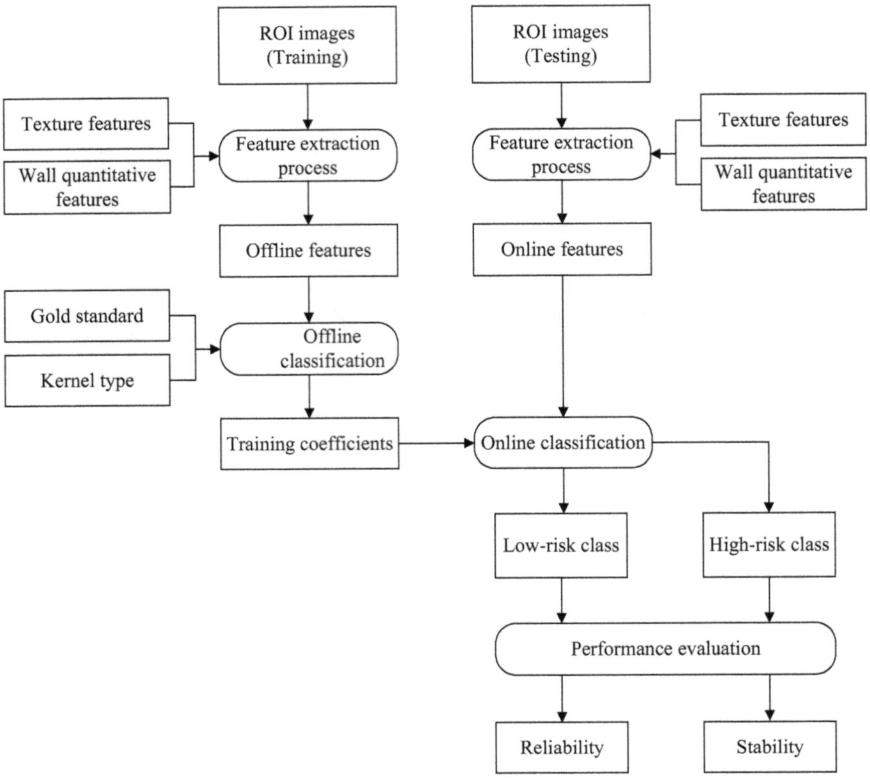

Figure 1.7. Improved coronary risk assessment system using a machine-learning paradigm.

pool of 65 different features. In an offline classification (using a training classifier such as SVM), the system used a combination of (i) training ground truth (or gold standard) risk labels (acquired from carotid plaque burden such as intima–media thickness) and (ii) dominant image-based training features extracted from training data sets to produce offline training coefficients [154, 155]. These training coefficients were then utilised to transform the online test features from the test images using the same SVM test classifier for predicting the risk of CAD patients and stratifying them into high- and low-risk bins. As compared to stand-alone plaque texture-based features, the proposed CADx system exhibited an improvement of ~6% in accuracy for coronary risk stratification into high- and low-risk bins.

1.5.2 Carotid risk assessment using ML-based approaches

Not much has been demonstrated in carotid tissue characterisation compared to IVUS-based risk assessment strategies. Araki *et al* [156] proposed a CADx system using ultrasonic echolucent carotid wall plaque morphology by independently evaluating the near and far walls. Adapting the classification paradigm of SVM, the CADx system showed an accuracy of 98.05% and 97.53% for the far-wall and near-wall, respectively. Using the spirit of coronary risk stratification for feature selection in [155], Saba *et al* [157] modified their CADx system by selecting dominant features using the PCA-based strategy. This brought an increase in the risk stratification accuracies of 98.28% and 93.92% to the far-wall and near-wall. For both coronary and carotid risk stratification, a PCA-based embedded system using greyscale morphology established a powerful paradigm for risk assessment and thus can be adapted to the clinical setting. In the future, a multimodal approach may be developed for validating detected calcium, which would be a useful diagnostic component for better CVD management.

1.6 Discussion

1.6.1 A note on the usage of calcium detection techniques in coronary and carotid arteries

Beyond traditional risk factors, the computation of calcium in both coronary and carotid arteries provides valuable prognostic information about the extent of cardiovascular disease. In section 1.2, we observed that previous studies had utilised two different techniques for calcium detection in coronary and carotid arteries, including non-invasive (CT, echocardiography and MRI) and invasive (angiography, IVUS and OCT) methods. MDCT offers increased spatial and temporal resolution and thus exhibits the potential for calcium detection and quantification. MRI is a radiation-free, safe technique and is generally used to assess soft tissue component characterisation [113]. Compared to other modalities, IVUS is safe, economic and easy to use with real-time diagnosis. Due to its optical properties, OCT has emerged as a valuable modality and has shown success in clinical demonstrations.

Even though there are advantages, both invasive and non-invasive modalities have some limitations. Previous studies have shown the utility of CT in the detection of coronary and carotid plaque components. Although CT provides calcium scores in the artery, higher radiation could compromise patient safety. From the literature, we have observed that many carotid studies have used MRI for accurate detection of soft plaque and its components. However, challenges remain for patients with marked arrhythmia and metal implants [107, 108]. It was also observed that grayscale IVUS-based studies required a shorter acquisition time and that near-real-time diagnosis is instrumental in detecting calcium in the coronary arteries. Clinical experience has shown that even though OCT offers better resolution, it has limitations in the estimation of the true vessel size for the assessment of plaque burden. OCT also lacks clinical standardisation; as a result, IVUS still remains important in the cardiologist's armamentarium [87]. Furthermore, PCI treatments involve risk to a patient's life, therefore, cardiologists usually diagnose arterial stenosis by observing all the imaging modalities adopted for the coronary and carotid artery.

1.6.2 A note on the usage of calcium quantification techniques in coronary and carotid arteries

Several studies in the literature have tried to quantify calcium in both the coronary and carotid arteries, as discussed in sections 1.3.1 and 1.3.2. In coronary arteries, initially the techniques [115, 117] were not automated, but later automated studies slowly achieved high accuracy. The main limitation of the automated studies [119, 121, 122] is the lack of calcium volume quantification methods. Weissman *et al* [115] and Araki *et al* [123] did compute the calcium volume, but they either lacked quantification [115] or ignored quantification of multifocal small lesions [123]. The above two limitations were taken care of by Araki *et al* [114], whose study proposed K-means as the best approach in terms of accuracy (92.80%). The only limitation of Araki's study [114] was its computational time. This limitation was taken care of by Banchhor *et al* [124] by adapting multiresolution-based techniques. The results proved that best performance was achieved when the FCM detection technique was embedded with the wavelet-based multiresolution paradigm. Further, to improve the accuracy, the same group adapted a CCA-based approach. The study proved that the threshold-based classifier, when embedded with Lanczos multiresolution, produced the highest accuracy (94.06%) in quantifying the calcium volume in a complete coronary artery video.

On the other hand, in the carotid arteries, researchers have computed either plaque or calcium scores and tried to determine which score could be used for predicting stenosis in CVD patients. Denzel *et al* [132] and Gepner *et al* [136] proved that in comparison to the plaque score, the calcium score was a stronger predictor of stenosis in CVD patients.

1.6.3 A note on the use of statistical metrics for the evaluation of calcium detection algorithms

The statistical tests mostly used in the literature are: the *t*-test [115, 117, 133–135, 137], *z*-test [114], Wilcoxon test [114, 135], Friedman test [131], Nemenyi test [131], Cohen's Kappa-test [132, 135], Kolmogorov–Smirnov-test [135] and Mann–Whitney-test [135] with a 95% confidence interval. All these statistical tests are two-tailed tests and p-values < 0.05 are considered significant. According to the available data size and nature of the sample data, researchers must strategically choose their statistical tests in order to improve the accuracy, reliability and robustness of the results, and statistical significance [114, 124, 125, 140, 143]. Furthermore, logistic regression [144] must be performed to study the odds ratios of image phenotypes and the ranking order of the image phenotypes for the risk of associations.

1.6.4 A note on feature selection in ML-based risk stratification for the coronary and carotid arteries

Several studies in the literature have tried to perform risk stratification of CVD patients into high-risk and low-risk bins, as discussed in sections 1.5.1 and 1.5.2. In coronary risk assessment, earlier studies usually employed either texture features [154] or shape-based features [158] in their ML-based approaches. To obtain these features, a large dataset is important. In 2014, Araki *et al* [158] showed that the accuracy of risk stratification can be improved by fusing the shape-based features with the geometric-based coronary calcium volume features. Fusing more features can affect the risk stratification accuracy. Usually, among the pool of all features, only a set of dominant features helps the system in stratifying the patient's risk. To overcome this limitation, Araki *et al* [155] used a PCA-based polling strategy and observed an improvement in accuracy with respect to their previous CADx system [154]. A study performed by Araki *et al* [155] proved the importance of dominant feature selection techniques. Recently, by adapting the classification paradigm of SVM, Banchhor *et al* [145] performed risk stratification on a large database (4930 US scans). The authors did prove that the fusion of texture-based and wall-based features along with the dominant feature selection using PCA-based polling strategy is highly accurate for risk stratification. This risk stratification approach using PCA-based polling strategy also showed promising results for carotid risk assessment using ML-based approaches. Saba *et al*, in their two studies [155, 157], proved that using a PCA-based polling strategy brought about an increase in the risk stratification accuracies in CVD patients.

1.6.5 Recommended interventions for CVD patients

Patients with coronary artery calcium (CAC) scores are mostly recommended to undergo a lifestyle change. The CAC score is measured using the Agatston method [59]. Recently, studies found that even in the absence of luminal narrowing [159] and in patients with low lifetime risk [160], CAC was the strongest predictor of incident CHD. Bittencourt and Riella [161] in 2016 discussed the limitation of CAC density which can be computed by dividing the Agatston CAC score by the CAC volume. The CAC density was found to be inversely associated with the incidence of cardiovascular events.

If the 10 year risk of atherosclerotic CVD is less than 20%, the patients are deferred to go for statin therapy [46]. Waheed *et al* [162] in 2016, also found that statin therapy is favourable in patients with high CAC. The study found that patients with low CAC might not benefit from statin therapy within 5 years. From a MESA study, Miedema *et al* [163] in 2014 observed that the use of aspirin had a favourable risk/benefit estimation if the patient's CAC score is greater than 100. The study concludes that patients with very low CAC score can receive harm from the use of aspirin. For heavy calcified plaque, apart from statin and aspirin therapy, rotational atherectomy also had emerged as a measure to open the narrowed arteries for increasing the blood flow. Li *et al* [164] in 2016 suggested that for a calcified plaque with severe calcified coronary lesions, rotational atherectomy (RA) using a cutting balloon is a more safe and effective measure compared to RA with a conventional plain balloon. In extreme cases, patients had to undergo interventional procedures such as shunting or stenting. Compared to coronary artery bypass grafting, coronary stenting is less expensive but in multiple diseased arteries, CABG is mostly preferred [165, 166].

Calcium measurement is not beneficial in every case. Recently, Messenger *et al* [167] showed that calcium scanning is associated with radiation equivalent to the dose of a mammogram. The study carried out by Nasir *et al* [53] and Mahabadi *et al* [54] also revealed that patients who are suffering from cardiovascular disease or who are already taking some medical therapy and have a high calcium score receive no benefit if the calcium measurement is performed.

1.6.6 Atherosclerotic calcium in coronary and carotid imaging: ongoing challenges

The following are some of the key challenges observed during the quantification of calcium in coronary and carotid arteries. Obtaining a well-annotated dataset from various medical institutes is a challenging task, as it must pass through multiple guidelines and protocols from the institutional review board, which consumes an excessive amount of time. Calcified plaques are only produced in the atheroma region, which lies between the internal elastic lamina (IEL) interface and the external elastic lamina (EEL) interface (see figure 1.2). To prepare the ground truth, manual tracings of both the IEL and EEL borders are required. There are many crucial factors that can affect the performance of manual tracings. These factors include the operator's background and experience, image resolution, the type of hardware system specification used, the time of day, lighting conditions, operator fatigue, internet speed and the extent of changes needed during tracings [143]. Further, the video produced by an IVUS scanner consists of a very large number of frames (average of 2040 frames per video) [114], so manual tracing of all the frames is tedious and prone to error. Usually, studies validate their obtained results with histology (as a ground truth). Since histological studies are performed on human cadavers, this is a time-consuming, expensive [168] and extremely tedious process.

1.7 Conclusions

This state-of-the-art review provides an engineering perspective on calcium detection, its quantification, and morphology-based risk stratification methods in the coronary and carotid arteries. Different imaging modalities were covered, however, ultrasound was the primary focus. Thirty-four automated methods were covered and compared in the form of benchmarking tables. Speed issues were presented in the form of multiresolution paradigms. Verification and validation strategies were also presented. Finally, machine-learning-based risk stratification studies were discussed, indicating a need for a more robust multimodal approach for CAD systems. Finally, the review covered the ongoing challenges for improving collaborative efforts to undertake more meaningful basic research, leading to clinical delivery. Clearly, there is a need for multidisciplinary roles, and better and closer collaboration is needed between several departments.

Conflict of interest

None declared.

Acknowledgements

Reprinted from Banchhor S K, Londhe N D, Araki T, Saba L, Radeva P, Khanna N N and Suri J S 2018 Calcium detection, its quantification, and grayscale morphology-based risk stratification using machine learning in multimodality big data coronary and carotid scans: a review *Comput. Biol. Med.* **101** 184–98, with permission from Elsevier.

Funding

The author(s) received no financial support for the research, authorship and/or publication of this chapter.

References

[1] Ajay V S and Prabhakaran D 2010 Coronary heart disease in Indians: implications of the INTERHEART study *Indian J. Med. Res.* **132** 561–6

[2] Nichols M *et al* 2014 Cardiovascular disease in Europe 2014: epidemiological update *Eur. Heart J.* **35** 2950–9

[3] Fernandez R *et al* 2015 Risk factors for coronary heart disease among Asian Indians living in Australia *J. Transcult. Nurs.* **26** 57–63

[4] Mozaffarian D *et al* 2016 Executive summary: heart disease and stroke statistics—2016 update: a report from the American Heart Association *Circulation* **133** 447–54

[5] Wilkins E *et al* 2017 *European Cardiovascular Disease Statistics* (Brussels: European Heart Network)

[6] WHO CVD website http://www.who.int/mediacentre/factsheets/fs317/en/ (Accessed: 2 April 2018)

[7] Prabhakaran D, Jeemon P and Roy A 2016 Cardiovascular diseases in India: current epidemiology and future directions *Circulation* **133** 1605–20

[8] Leeder S *et al* 2004 *A Race Against Time: The Challenge of Cardiovascular Disease in Developing Economies* (New York: Columbia University)

[9] Rose G 1991 ABC of vascular diseases. Epidemiology of atherosclerosis *BMJ* **303** 1537–9

[10] Li H *et al* 2017 Inflammatory biomarkers of coronary heart disease *Front. Biosci.* **22** 504–15

[11] Libby P *et al* 2009 Inflammation in atherosclerosis: from pathophysiology to practice *J. Am. Coll. Cardiol.* **54** 2129–38

[12] Mallika V, Goswami B and Rajappa M 2007 Atherosclerosis pathophysiology and the role of novel risk factors: a clinicobiochemical perspective *Angiology* **58** 513–22

[13] Butoi E, Gan A M and Manduteanu I 2014 Molecular and functional interactions among monocytes/macrophages and smooth muscle cells and their relevance for atherosclerosis *Crit. Rev. Eukaryot. Gene Expr.* **24** 341–55

[14] Drechsler M, Duchene J and Soehnlein O 2015 Chemokines control mobilization, recruitment, and fate of monocytes in atherosclerosis significance *Arterioscler. Thromb. Vasc. Biol.* **35** 1050–5

[15] Virmani R *et al* 2003 Pathology of the thin-cap fibroatheroma *J. Interv. Cardiol.* **16** 267–72

[16] Otsuka F *et al* 2014 Has our understanding of calcification in human coronary atherosclerosis progressed? *Arterioscler. Thromb. Vasc. Biol.* **34** 724–36

[17] Kockx M M *et al* 1998 Apoptosis and related proteins in different stages of human atherosclerotic plaques *Circulation* **97** 2307–15

[18] Burke A P *et al* 2002 Morphological predictors of arterial remodeling in coronary atherosclerosis *Circulation* **105** 297–303

[19] Wong K K *et al* 2012 Effect of calcification on the mechanical stability of plaque based on a three-dimensional carotid bifurcation model *BMC Cardiovasc. Disord.* **12** 7

[20] Teng Z *et al* 2014 How does juxtaluminal calcium affect critical mechanical conditions in carotid atherosclerotic plaque? An exploratory study *IEEE Trans. Biomed. Eng.* **61** 35–40
[21] Falk E, Shah P K and Fuster V 1995 Coronary plaque disruption *Circulation* **92** 657–71
[22] Friedman M 1971 The coronary thrombus: its origin and fate *Hum. Pathol.* **2** 81–128
[23] Carr S *et al* 1996 Atherosclerotic plaque rupture in symptomatic carotid artery stenosis *J. Vasc. Surg.* **23** 755–66
[24] Joshi N V *et al* 2014 18F-fluoride positron emission tomography for identification of ruptured and high-risk coronary atherosclerotic plaques: a prospective clinical trial *Lancet* **383** 705–13
[25] Thomas G S and Haraszti R A 2014 A new frontier in atherosclerotic coronary imaging *Lancet* **383** 674–5
[26] Raggi P *et al* 2004 Prognostic value of coronary artery calcium screening in subjects with and without diabetes *J. Am. Coll. Cardiol.* **43** 1663–9
[27] Bulugahapitiya U *et al* 2009 Is diabetes a coronary risk equivalent? Systematic review and meta-analysis *Diabet. Med.* **26** 142–8
[28] Kabir C S, Haq M M and Cader F A 2017 Disparity in coronary artery diameter in diabetic and nondiabetic subjects undergoing percutaneous coronary intervention in Bangladesh: a 2-year retrospective analysis *Bangladesh Heart J.* **32** 23–8
[29] Ertan C *et al* 2014 Association of prediabetes with diffuse coronary narrowing and small-vessel disease *J. Cardiol.* **63** 29–34
[30] Libby P, Ridker P M and Hansson G K 2011 Progress and challenges in translating the biology of atherosclerosis *Nature* **473** 317–25
[31] Degrell P *et al* 2015 Screening for coronary artery disease in asymptomatic individuals: why and how? *Arch. Cardiovas. Dis.* **108** 675–82
[32] Stryker W A 1946 Arterial calcification in infancy with special reference to the coronary arteries *Am. J. Pathol.* **22** 1007–31
[33] Woolf N E 1978 The origins of atherosclerosis *Postgrad. Med. J.* **54** 156–62
[34] Stary H C 1990 The sequence of cell and matrix changes in atherosclerotic lesions of coronary arteries in the first forty years of life *Eur Heart J.* **11** 3–19
[35] Hamby R I *et al* 1974 Coronary artery calcification: clinical implications and angiographic correlates *Am. Heart. J.* **87** 565–70
[36] Kannel W B and Wolf P A 2006 Peripheral and cerebral atherothrombosis and cardiovascular events in different vascular territories: insights from the Framingham Study *Curr. Atheroscler. Rep.* **8** 317–23
[37] Patel A K *et al* 2016 A review on atherosclerotic biology, wall stiffness, physics of elasticity, and its ultrasound-based measurement *Curr. Atheroscler. Rep.* **18** 83
[38] Munnur R K, Nerlekar N and Wong D T 2016 Imaging of coronary atherosclerosis in various susceptible groups *Cardiovasc. Diagn. Ther.* **6** 382–95
[39] Blankstein R *et al* 2017 The implication of coronary artery calcium testing for cardiovascular disease prevention and diabetes *Endocrinol. Metab.* **32** 47–57
[40] Hirsch D *et al* 1993 Colocalization of cholesterol and hydroxyapatite in human atherosclerotic lesions *Calcif. Tissue Int.* **52** 94–8
[41] Doherty T M and Detrano R C 1994 Coronary arterial calcification as an active process: a new perspective on an old problem *Calcif. Tissue Int.* **54** 224–30
[42] Guyton J R and Klemp K F 1996 Development of the lipid-rich core in human atherosclerosis *Arterioscler. Thromb. Vasc. Biol.* **16** 4–11

[43] Allam A H *et al* 2017 Is coronary calcium scoring too late? Total body arterial calcium burden in patients without known CAD and normal MPI *J. Nucl. Cardiol.* **25** 1990–8
[44] Razzouk L *et al* 2015 Co-existence of vascular disease in different arterial beds: peripheral artery disease and carotid artery stenosis—data from Life Line Screening® *Atherosclerosis* **241** 687–91
[45] Craven T E *et al* 1990 Evaluation of the associations between carotid artery atherosclerosis and coronary artery stenosis. a case-control study *Circulation* **82** 1230–42
[46] Geroulakos G A *et al* 1994 The carotid intima–media thickness as a marker of the presence of severe symptomatic coronary artery disease *Eur. Heart J.* **15** 781–5
[47] Davis P H *et al* 1999 Increased carotid intimal–medial thickness and coronary calcification are related in young and middle-aged adults: the Muscatine study *Circulation* **100** 838–42
[48] Cohen G I *et al* 2013 Relationship between carotid disease on ultrasound and coronary disease on CT angiography *JACC Cardiovasc. Imaging* **6** 1160–7
[49] Jeevarethinam A *et al* 2017 Relationship between carotid atherosclerosis and coronary artery calcification in asymptomatic diabetic patients: a prospective multicenter study *Clin. Cardiol.* **40** 752–8
[50] Burgstahler C *et al* 2017 Coronary and carotid atherosclerosis in asymptomatic male marathon runners *Scand. J. Med. Sci. Sports* **28** 1397–403
[51] Benjamin E J *et al* 2017 Heart disease and stroke statistics—2017 update: a report from the American Heart Association *Circulation* **135** e146–603
[52] Kristensen S D *et al* 2014 ESC/ESA Guidelines on non-cardiac surgery: cardiovascular assessment and management: the Joint Task Force on non-cardiac surgery: cardiovascular assessment and management of the European Society of Cardiology (ESC) and the European Society of Anaesthesiology (ESA) *Eur. Heart J.* **35** 2383–431
[53] Nasir K *et al* 2015 Implications of coronary artery calcium testing among statin candidates according to American College of Cardiology/American Heart Association cholesterol management guidelines: MESA (multi-ethnic study of atherosclerosis) *J. Am. Coll. Cardiol.* **66** 1657–68
[54] Mahabadi A A *et al* 2017 CAC score improves coronary and CV risk assessment above statin indication by ESC and AHA/ACC primary prevention guidelines *JACC: Cardiovasc. Imaging* **10** 143–53
[55] Rumberger J 2004 Tomographic plaque imaging with CT *Plaque Imaging: Pixel to Molecular Level Studies in Health Technology and Informatics* ed J S Suri *et al* (Amsterdam: IOS Press) pp 182–207
[56] Sandercock P, Molyneux A and Warlow C 1985 Value of computed tomography in patients with stroke: Oxfordshire Community Stroke Project *Br. Med. J.* **290** 193–7
[57] Achenbach S and Raggi P 2010 Imaging of coronary atherosclerosis by computed tomography *Eur. Heart. J.* **31** 1442–8
[58] Bittencourt M S *et al* 2016 Left ventricular area on non-contrast cardiac computed tomography as a predictor of incident heart failure—the Multi-Ethnic Study of Atherosclerosis *J. Cardiovasc. Comput. Tomogr.* **10** 500–6
[59] Agatston A S *et al* 1990 Quantification of coronary artery calcium using ultrafast computed tomography *J. Am. Coll. Cardiol.* **15** 827–32
[60] Shinbane J S and Budoff M J 2004 Computed tomographic cardiovascular imaging. In studies in health technology and informatics *Plaque Imaging: Pixel to Molecular Level Studies in Health Technology and Informatics* ed J S Suri *et al* (Amsterdam: IOS Press) pp 148–81

[61] Nucifora G *et al* 2009 Usefulness of echocardiographic assessment of cardiac and ascending aorta calcific deposits to predict coronary artery calcium and presence and severity of obstructive coronary artery disease *Am. J. Cardiol.* **103** 1045–50

[62] Pressman G S *et al* 2011 Can total cardiac calcium predict the coronary calcium score? *Int. J. Cardiol.* **146** 202–6

[63] Acharya U R *et al* 2013 Automated classification of patients with coronary artery disease using grayscale features from left ventricle echocardiographic images *Comput. Methods Programs Biomed.* **112** 624–32

[64] Tuzcu E M *et al* 1996 The dilemma of diagnosing coronary calcification: angiography versus intravascular ultrasound *J. Am. Coll. Cardiol.* **27** 832–8

[65] Virmani R, Farb A and Burke A P 1994 Coronary angioplasty from the perspective of atherosclerotic plaque: morphologic predictors of immediate success and restenosis *Am. Heart J.* **127** 163–79

[66] Vavuranakis M *et al* 2001 Stent deployment in calcified lesions: can we overcome calcific restraint with high-pressure balloon inflations? *Catheter. Cardiovasc. Interv.* **52** 164–72

[67] Kaufman L *et al* 1983 The potential impact of nuclear magnetic resonance imaging on cardiovascular diagnosis *Circulation* **67** 251–7

[68] Awad I A *et al* 1986 Incidental subcortical lesions identified on magnetic resonance imaging in the elderly. I. Correlation with age and cerebrovascular risk factors *Stroke* **17** 1084–9

[69] Mohiaddin R H *et al* 1989 Regional aortic compliance studied by magnetic resonance imaging: the effects of age, training, and coronary artery disease *Heart* **62** 90–6

[70] van der Wall E E *et al* 1995 Magnetic resonance imaging in coronary artery disease *Circulation* **92** 2723–39

[71] Schoenhagen P and Nissen S 2002 Understanding coronary artery disease: tomographic imaging with intravascular ultrasound *Heart* **88** 91–6

[72] Giannoglou G D *et al* 2007 A novel active contour model for fully automated segmentation of intravascular ultrasound images: *in vivo* validation in human coronary arteries *Comput. Biol. Med.* **37** 1292–302

[73] Katouzian A *et al* 2012 A state-of-the-art review on segmentation algorithms in intravascular ultrasound (IVUS) images *IEEE Trans. Inf. Technol. Biomed.* **16** 823–34

[74] Gupta A *et al* 2015 Plaque echolucency and stroke risk in asymptomatic carotid stenosis *Stroke* **46** 91–7

[75] Polak J F *et al* 1998 Hypoechoic plaque at US of the carotid artery: an independent risk factor for incident stroke in adults aged 65 years or older. Cardiovascular Health Study *Radiology* **208** 649–54

[76] AbuRahma A F, Wulu J T and Crotty B 2002 Carotid plaque ultrasonic heterogeneity and severity of stenosis *Stroke* **33** 1772–5

[77] Lal B K *et al* 2002 Pixel distribution analysis of B-mode ultrasound scan images predicts histologic features of atherosclerotic carotid plaques *J. Vasc. Surg.* **35** 1210–7

[78] Kovalski G *et al* 2000 Three-dimensional automatic quantitative analysis of intravascular ultrasound images *Ultrasound Med. Biol.* **26** 527–37

[79] Mintz G S *et al* 2001 American College of Cardiology clinical expert consensus document on standards for acquisition, measurement and reporting of intravascular ultrasound studies (IVUS) *J. Am. Coll. Cardiol.* **37** 1478–92

[80] Friedrich G J *et al* 1994 Detection of intralesional calcium by intracoronary ultrasound depends on the histologic pattern *Am. Heart J.* **128** 435–41

[81] Mintz G S *et al* 1995 Patterns of calcification in coronary artery disease. a statistical analysis of intravascular ultrasound and coronary angiography in 1155 lesions *Circulation* **91** 1959–65
[82] Kawasaki M *et al* 2002 *In vivo* quantitative tissue characterization of human coronary arterial plaques by use of integrated backscatter intravascular ultrasound and comparison with angioscopic findings *Circulation* **105** 2487–92
[83] Nair A *et al* 2007 Automated coronary plaque characterization with intravascular ultrasound backscatter: *ex vivo* validation *EuroIntervention* **3** 113–20
[84] Funada R *et al* 2009 The potential of RF backscattered IVUS data and multidetector-row computed tomography images for tissue characterization of human coronary atherosclerotic plaques *Int. J. Cardiovasc. Imaging* **25** 471–8
[85] Akhtar M and Liu W 2016 Use of intravascular ultrasound vs optical coherence tomography for mechanism and patterns of in-stent restenosis among bare metal stents and drug eluting stents *J. Thorac. Dis.* **8** E104–8
[86] Gan Y *et al* 2016 Automated classification of optical coherence tomography images of human atrial tissue *J. Biomed. Opt.* **21** 101407
[87] Dash D 2017 Optical coherence tomography is a kid on the block: I would choose intravascular ultrasound *Indian Heart J.* **69** 407–10
[88] Wang X *et al* 2017 *In vivo* calcium detection by comparing optical coherence tomography, intravascular ultrasound, and angiography *JACC Cardiovasc. Imaging* **10** 869–79
[89] Saba L and Suri J S 2013 *Multi-detector CT Imaging: Abdomen, Pelvis, and CAD Applications* vol 2 (Boca Raton, FL: CRC Press)
[90] Reynoso E *et al* 2017 Spectral signal density of carotid plaque using dual-energy computed tomography *J. Neuroimaging* **27** 511–6
[91] Nandalur K R *et al* 2005 Calcified carotid atherosclerotic plaque is associated less with ischemic symptoms than is noncalcified plaque on MDCT *Am. J. Roentgenol.* **184** 295–8
[92] Nandalur K R *et al* 2006 Carotid artery calcification on CT may independently predict stroke risk *Neuro Radiol.* **186** 547–52
[93] Nandalur K R *et al* 2007 Composition of the stable carotid plaque: insights from a multidetector computed tomography study of plaque volume *Stroke* **38** 935–40
[94] de Weert T T *et al* 2005 *In vitro* characterization of atherosclerotic carotid plaque with multidetector computed tomography and histopathological correlation *Eur. Radiol.* **15** 1906–14
[95] de Weert T T *et al* 2006 *In vivo* characterization and quantification of atherosclerotic carotid plaque components with multidetector computed tomography and histopathological correlation *Arterioscler. Thromb. Vasc. Biol.* **26** 2366–72
[96] Saba L *et al* 2012 Association between carotid artery plaque volume, composition, and ulceration: a retrospective assessment with MDCT *Am. J. Roentgenol.* **199** 151–6
[97] Saba L *et al* 2012 Analysis of carotid artery plaque and wall boundaries on CT images by using a semi-automatic method based on level set model *Neuroradiology* **54** 1207–14
[98] Saba L *et al* 2012 Association between carotid artery plaque type and cerebral microbleeds *Am. J. Neuroradiol.* **33** 2144–50
[99] Saba L *et al* 2015 Automated calcium burden measurement in internal carotid artery plaque with CT: a hierarchical adaptive approach *Int. Angiol.* **34** 290–305
[100] Saba L *et al* 2016 Carotid artery intra-plaque attenuation variability using computed tomography *Neurovasc. Imaging* **2** 12

[101] Adame I M *et al* 2004 Automatic segmentation and plaque characterization in atherosclerotic carotid artery MR images *MAGMA* **16** 227–34
[102] Suri J S, Yuan C and Wilson D L (ed) 2005 *Plaque Imaging: Pixel To Molecular Level* (Amsterdam: IOS Press)
[103] Merickel M B *et al* 1988 Identification and 3-D quantification of atherosclerosis using magnetic resonance imaging *Comput. Biol. Med.* **18** 89–102
[104] Toussaint J F *et al* 1995 T2-weighted contrast for NMR characterization of human atherosclerosis *Arterioscler. Thromb. Vasc. Biol.* **15** 1533–42
[105] Coombs B D *et al* 2001 Structure of plaque at carotid bifurcation: high-resolution MRI with histological correlation *Stroke* **32** 2516–21
[106] Lee J, Kil J, Kim D W and Kang S D 2017 Usefulness of plaque magnetic resonance imaging in identifying high-risk carotid plaques irrespective of the degree of stenosis *J. Cerebrovasc. Eendovasc. Neurosurg.* **19** 291–300
[107] Stradiotti P *et al* 2009 Metal-related artifacts in instrumented spine. Techniques for reducing artifacts in CT and MRI: state of the art *Eur. Spine J.* **18** 102–8
[108] Molinari F *et al* 2010 Carotid plaque characterization with contrast-enhanced ultrasound imaging and its histological validation *J. Vasc. Ultrasound* **34** 175–84
[109] Molinari F *et al* 2007 Accurate and automatic carotid plaque characterization in contrast enhanced 2-D ultrasound images *Conf. Proc. IEEE Eng. Med. Biol. Soc.* **2007** 335–8
[110] Hitchner E *et al* 2014 Intravascular ultrasound as a clinical adjunct for carotid plaque characterization *J. Vasc. Surg.* **59** 774–80
[111] Jashari F *et al* 2015 Atherosclerotic calcification detection: a comparative study of carotid ultrasound and cone beam CT *Int. J. Mol. Sci.* **16** 19978–88
[112] Pazinato D V *et al* 2016 Pixel-level tissue classification for ultrasound images *IEEE J. Biomed. Health Inform.* **20** 256–67
[113] Dweck M R *et al* 2016 MR imaging of coronary arteries and plaques *JACC: Cardiovasc. Imaging* **9** 306–16
[114] Araki T *et al* 2016 Reliable and accurate calcium volume measurement in coronary artery using intravascular ultrasound videos *J. Med. Syst.* **40** 51
[115] Weissman N J *et al* 1995 Three-dimensional intravascular ultrasound assessment of plaque volume after successful atherectomy *Am. Heart J.* **130** 413–9
[116] Schiller N B *et al* 1989 Recommendations for quantitation of the left ventricle by two-dimensional echocardiography *J. Am. Soc. Echocardiogr.* **2** 358–67
[117] Scott D S *et al* 2000 Pathologic validation of a new method to quantify coronary calcific deposits *in vivo* using intravascular ultrasound *Am. J. Cardiol.* **85** 37–40
[118] von Birgelen C *et al* 1997 Simpson's rule for the volumetric ultrasound assessment of atherosclerotic coronary arteries: a study with ECG-gated three-dimensional intravascular ultrasound *Coron. Artery Dis.* **8** 363–9
[119] Santos Filho E *et al* 2008 Detection and quantification of calcifications in intravascular ultrasound images by automatic thresholding *Ultrasound Med. Biol.* **34** 160–5
[120] Otsu N 1979 A threshold selection method from gray-level histograms *IEEE Trans. Syst. Man Cybern.* **9** 62–6
[121] Zhang Q *et al* 2010 Automatic segmentation of calcifications in intravascular ultrasound images using snakes and the contourlet transform *Ultrasound Med Biol.* **36** 111–29
[122] Gao Z *et al* 2014 Automated detection framework of the calcified plaque with acoustic shadowing in IVUS images *PLoS One* **9** e109997

[123] Araki T *et al* 2015 Shape-based approach for coronary calcium lesion volume measurement on intravascular ultrasound imaging and its association with carotid intima-media thickness *J. Ultrasound Med.* **34** 469–82
[124] Banchhor S K *et al* 2016 Five multiresolution-based calcium volume measurement techniques from coronary IVUS videos: A comparative approach *Comput. Methods Programs Biomed.* **134** 237–58
[125] Banchhor S K *et al* 2017 Well-balanced system for coronary calcium detection and volume measurement in a low resolution intravascular ultrasound videos *Comput. Biol. Med.* **84** 168–81
[126] Liang Y *et al* 2006 Estimation of arterial wall strain based on IVUS image registration *28th Annual Int. Conf. of the IEEE in Engineering in Medicine and Biology Society. EMBS'06* pp 3218–21
[127] Srinivas S *et al* 2009 Multiresolution analysis of intravascular ultrasound harmonic signals to image pre-rupture plaques *2009 IEEE Int. Ultrasonics Symp. (IUS)* pp 2363–6
[128] Katouzian A *et al* 2012 Applications of multiscale overcomplete wavelet-based representations in Intravascular Ultrasound (IVUS) images *Ultrasound Imaging* (Berlin: Springer) pp 313–36
[129] Lazrag H and Naceur M S 2012 Combination of the level-set methods with the contourlet transform for the segmentation of the IVUS images *J. Biomed. Imaging* **2012** 439597
[130] Lazrag H and Naceur M S 2012 Wavelet filters analysis for speckle reduction in intravascular ultrasound images *2012 6th Int. Conf. in Sciences of Electronics, Technologies of Information and Telecommunications (SETIT)* pp 375–9
[131] Puertas E, Escalera S and Pujol O 2015 Generalized multi-scale stacked sequential learning for multi-class classification *Pattern Anal. Appl.* **18** 247–61
[132] Denzel C *et al* 2004 Carotid artery calcium: accuracy of a calcium score by computed tomography—an *in vitro* study with comparison to sonography and histology *Eur. J. Vasc. Endovasc. Surg.* **28** 214–20
[133] de Weert T T *et al* 2008 Assessment of atherosclerotic carotid plaque volume with multidetector computed tomography angiography *Int. J. Cardiovasc. Imaging* **24** 751–9
[134] Marquering H A *et al* 2011 The relation of carotid calcium volume with carotid artery stenosis in symptomatic patients *Am. J. Neuroradiol.* **32** 1182–7
[135] Anzidei M *et al* 2016 Longitudinal assessment of carotid atherosclerosis after radiation therapy using computed tomography: a case control study *Eur. Radiol.* **26** 72–8
[136] Gepner A D *et al* 2017 Comparison of carotid plaque score and coronary artery calcium score for predicting cardiovascular disease events: The Multi-Ethnic Study of Atherosclerosis *J. Am. Heart Assoc.* **6** e005179
[137] Tsiaparas N N *et al* 2011 Comparison of multiresolution features for texture classification of carotid atherosclerosis from B-mode ultrasound *IEEE Trans. Inf. Technol. Biomed.* **15** 130–7
[138] Molinari F *et al* 2012 Completely automated multiresolution edge snapper—a new technique for an accurate carotid ultrasound IMT measurement: clinical validation and benchmarking on a multi-institutional database *IEEE Trans. Image Process.* **21** 1211–22
[139] Molinari F, Zeng G and Suri J S 2010 Intima–media thickness: setting a standard for a completely automated method of ultrasound measurement *IEEE Trans. Ultrason. Ferroelectr. Freq. Control* **57** 1112–24

[140] Banchhor S K *et al* 2017 Relationship between automated coronary calcium volumes and a set of manual coronary lumen volume, vessel volume and atheroma volume in Japanese diabetic cohort *J. Clin. Diagn. Res.* **11** TC09

[141] Suri J *et al* 2005 Image quality assessment via segmentation of breast lesion in x-ray and ultrasound phantom images from Fischer's full field digital mammography and ultrasound (FFDMUS) system *Technol. Cancer Res. Treat.* **4** 83–92

[142] Ma T *et al* 2015 Multi-frequency intravascular ultrasound (IVUS) imaging *IEEE Trans. Ultrason. Ferroelectr. Freq. Control* **62** 97–107

[143] Saba L *et al* 2018 Intra- and inter-operator reproducibility of automated cloud-based carotid lumen diameter ultrasound measurement *Indian Heart J.* **70** 649–64

[144] Ivanova N, Postadzhiyan A and Apostolova M D 2012 An application of logistic regression and multifactor dimensionality reduction analyses for detecting genotype-phenotype interactions associated with developing of atherosclerosis in Bulgarian cohort *Biotechnol. Biotechnol. Equip.* **26** 191–9

[145] Banchhor S K *et al* 2017 Wall-based measurement features provides an improved IVUS coronary artery risk assessment when fused with plaque texture-based features during machine learning paradigm *Comput. Biol. Med.* **91** 198–212

[146] Christodoulou C I *et al* 2003 Texture-based classification of atherosclerotic carotid plaques *IEEE Trans. Med. Imaging* **22** 902–12

[147] Kyriacou E *et al* 2005 Ultrasound imaging in the analysis of carotid plaque morphology for the assessment of stroke *Stud. Health Technol. Inform.* **113** 241–75

[148] Mougiakakou S G *et al* 2007 Computer-aided diagnosis of carotid atherosclerosis based on ultrasound image statistics, laws' texture and neural networks *Ultrasound Med. Biol.* **33** 26–36

[149] Kyriacou E *et al* 2009 Classification of atherosclerotic carotid plaques using morphological analysis on ultrasound images *Appl. Intell.* **30** 3–23

[150] Acharya R U *et al* 2012 Symptomatic vs. asymptomatic plaque classification in carotid ultrasound *J. Med. Syst.* **36** 1861–71

[151] Acharya U R *et al* 2012 Atherosclerotic risk stratification strategy for carotid arteries using texture-based features *Ultrasound Med. Biol.* **38** 899–915

[152] Acharya U R *et al* 2013 Understanding symptomatology of atherosclerotic plaque by image-based tissue characterization *Comput. Methods Programs Biomed.* **110** 66–75

[153] Pedro L M *et al* 2014 Asymptomatic carotid disease—a new tool for assessing neurological risk *Echocardiography* **31** 353–61

[154] Araki T *et al* 2016 A new method for IVUS-based coronary artery disease risk stratification: a link between coronary and carotid ultrasound plaque burdens *Comput. Methods Programs Biomed.* **124** 161–79

[155] Araki T *et al* 2016 PCA-based polling strategy in machine learning framework for coronary artery disease risk assessment in intravascular ultrasound: a link between carotid and coronary grayscale plaque morphology *Comput. Methods Programs Biomed.* **128** 137–58

[156] Araki T *et al* 2017 Stroke risk stratification and its validation using ultrasonic Echolucent Carotid Wall plaque morphology: a machine learning paradigm *Comput. Biol. Med.* **80** 77–96

[157] Saba L *et al* 2017 Plaque tissue morphology-based stroke risk stratification using carotid ultrasound: a polling-based PCA learning paradigm *J. Med. Syst.* **41** 98

[158] Araki T *et al* 2014 Effect of geometric-based coronary calcium volume as a feature along with its shape-based attributes for cardiological risk prediction from low contrast intravascular ultrasound *J. Med. Imaging Health Inform.* **4** 255–61

[159] Cho I *et al* 2017 Prognostic implications of coronary artery calcium in the absence of coronary artery luminal narrowing *Atherosclerosis* **262** 185–90

[160] Joshi P H *et al* 2016 Coronary artery calcium predicts cardiovascular events in participants with a low lifetime risk of cardiovascular disease: the Multi-Ethnic Study of Atherosclerosis (MESA) *Atherosclerosis* **246** 367–73

[161] Bittencourt M S and Riella L V 2016 Coronary artery calcium density for the prediction of mortality in CKD patients: one size does not fit all *Atherosclerosis* **250** 180–2

[162] Waheed S *et al* 2016 Collective impact of conventional cardiovascular risk factors and coronary calcium score on clinical outcomes with or without statin therapy: The St Francis Heart Study *Atherosclerosis* **255** 193–9

[163] Miedema M D *et al* 2014 Use of coronary artery calcium testing to guide aspirin utilization for primary prevention: estimates from the multi-ethnic study of atherosclerosis *Circ.: Cardiovasc. Qual. Outcomes* **7** 453–60

[164] Li Q *et al* 2016 Intensive plaque modification with rotational atherectomy and cutting balloon before drug-eluting stent implantation for patients with severely calcified coronary lesions: a pilot clinical study *BMC Cardiovasc. Disord.* **16** 112

[165] Serruys P W *et al* 2001 Comparison of coronary-artery bypass surgery and stenting for the treatment of multivessel disease *N. Engl. J. Med.* **344** 1117–24

[166] Lin Y K *et al* 2011 Cost-effectiveness of clinical pathway in coronary artery bypass surgery *J. Med. Syst.* **35** 203–13

[167] Messenger B *et al* 2016 Coronary calcium scans and radiation exposure in the multi-ethnic study of atherosclerosis *Int. J. Cardiovas. Imaging* **32** 525–9

[168] Tot T 2010 Cost-benefit analysis of using large-format histology sections in routine diagnostic breast care *Breast* **19** 284–8

Chapter 2

Risk of coronary artery disease: genetics and external factors

Ankush D Jamthikar, Alberto Boi, Luca Saba, Deep Gupta, John R Laird, N N Khanna and Jasjit S Suri

Cardiovascular disease (CVD), caused by atherosclerosis, is a major cause of death around the world. Globally, each year 17.7 million people die due to CVD. Arterial stenosis is the result of plaque deposition throughout the periphery of the coronary arterial wall, the progression of which leads to catastrophic events. Vulnerable plaque components such as thin-cap fibroatheroma (TCFA), macrophage infiltration, a large necrotic core and thrombus are the microstructural plaque components and can only be detected with high-resolution imaging modalities. Optical coherence tomography (OCT) is the light-based alternative to the intravascular ultrasound (IVUS) technique and has high axial and spatial resolution. Characterization of plaque components is a vital and at the same time challenging task. Various approaches have been suggested in the literature to characterize and classify the above-mentioned plaque components. This review presents a comparison between various methodologies used for tissue characterization, classification and measurement using OCT. Moreover, this chapter also presents different ways to predict and stratify the risk associated with CVD based on plaque characterization and measurement. Based on comparative analysis among different schools of thought, a combination of machine-learning and deep-learning techniques has been determined to provide the best classification accuracy using OCT images.

2.1 Introduction

CVD has been shown to be a prominent killer in the majority of populations around the world. Globally, the possibility of having at least one type of CVD in the general population is high. The severity of heart disease results in a hike in the death rate. As per the statistics of the World Health Organization (WHO), each year 17.7 million

doi:10.1088/2053-2563/ab0820ch2

people die due to CVD. Coronary artery disease (CAD) makes a large contribution to the mortality rate, as it results in heart attack and stroke [1]. The rise in morbidity and mortality rates has increased the financial burden regionally and globally. Despite the efforts of the medical community to prevent heart disease and reduce the overall expenditure using economical diagnostic solutions, the medical cost has become uncontrollable. The global financial burden due to CVD reported in 2010 was around $863 million. In the USA, the total financial burden in 2016 was $555 billion, which is expected to rise to more than a trillion dollars by 2035 [2, 3].

The primary cause of atherosclerosis is plaque formation in the vessel walls of coronary arteries, as shown in figure 2.1. The deposition of plaque within the coronary vessel layer results in hardening of the arteries, which is known as atherosclerosis. The progression of atherosclerosis leads to plaque rupture, which may block a coronary artery and reduce the oxygenated blood supply to the heart, resulting in heart attack or stroke [4, 5]. It is obvious to say that the coronary arteries which carry the maximum amount of blood are more susceptible to plaque deposition. Atherosclerotic plaque mostly develops in the proximal left anterior descending coronary artery (LAD), proximal and distal right coronary artery (RCA), and left circumflex coronary artery (LCx). Moreover, coronary calcification in the LDA is frequently located in the neighborhood of the septal branch. Hence, it is essential to understand the geometric location of the coronary arteries when dealing with the problems associated with atherosclerosis.

It has been observed that the risk of atherosclerosis is higher in females than in males. This is mainly because of the lower lumen size in females compared to males. The lumen size also varies with age and ethnicity [6–8]. Diabetes also has a high degree of correlation with coronary artery size, resulting in heart attack [9, 10]. In addition, there are other risk factors which contribute to CVD, such as smoking, diabetes, low-density lipoprotein (LDL) cholesterol, consumption of alcohol and sedentary lifestyles [3].

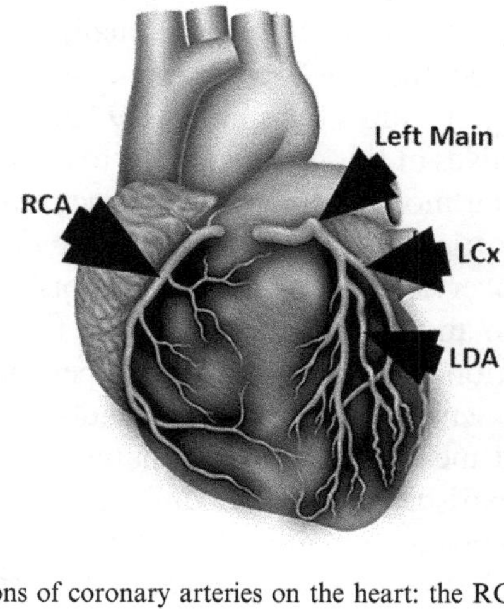

Figure 2.1. Geometrical locations of coronary arteries on the heart: the RCA, LDA and LCx. (Courtesy of AtheroPoint, Roseville, CA, USA.)

Conventional pathological blood tests provide the level of LDL cholesterol and hemoglobin to understand the progress of CVD. However, these methods are limited in accuracy. Moreover, they do not provide any information about the geometry of the plaque formation within the arteries.

Hence, to obtain a complete intracoronary view and to quantify the plaque within the vessel wall, robust and accurate imaging techniques are required. In the past two decades, various medical imaging techniques have been used for analyzing atherosclerosis, such as magnetic resonance imaging (MRI), computed tomography (CT) and intravascular ultrasound (IVUS) [11]. CT and MRA are non-invasive techniques and are used for understanding the coronary artery score and to identify classified and nonclassified lessons. In CT, patients are exposed to radiation, which may affect the health of patients. Also, the quality of images generated by CT is poor and affected by artifacts. All these techniques do not differentiate between plaque components in the vessel wall, mainly due to their low spatial resolution [12–15]. This limits their ability to stratify the risk at earlier stages. IVUS and OCT are intravascular techniques and provide a cross-sectional view of the coronary artery. Although IVUS is a preferable choice compared to CT and MRI, it does not achieve the criteria to identify different microstructural vulnerable plaque components. Thus, a high-resolution imaging modality is required which is low-cost compared to the above-mentioned imaging techniques and can provide detailed information about the plaque.

Unlike IVUS, optical coherence tomography (OCT) uses light instead of sound and has been shown to be the best choice for identifying various tissue layers within the vessel walls. OCT was first introduced by Huang *et al* [16] to analyze the retina and atherosclerotic plaque composition. The intravascular application of OCT using a catheter-based system was initially suggested by Brezinski *et al* [17]. The significant advantage of OCT is its high resolution, of the order of a few microns (10–20 μm). This helps in the categorization of the various tissue layers, such as the intima, media and adventitia of the coronary artery, along with the plaque components, such as a thin fibrous cap, necrotic layer, calcified nodules, macrophages and adipose tissue [18]. OCT has being compared to IVUS for the visualization atherosclerotic plaque and it has been determined that OCT is a low-cost and safe technique which provides more deterministic features than IVUS [19, 20].

The fundamental objectives of this chapter are (i) to analyze the potential of OCT compared to other imaging modalities for microstructural tissue characterization, (ii) to determine the application of OCT in coronary arterial tissue characterization, (iii) to provide a detailed overview of the different schools of thought in the literature on plaque characterization and measurement, and (iv) to perform risk stratification based on plaque composition to deal with CAD. As per the best of our knowledge, this is the first review to benchmark various methodologies for plaque characterization, classification and measurement. This chapter starts with an overview of OCT physics and a comparison to other modalities. A brief overview of CAD is provided, followed by discussion of plaque characterization and measurement in the latter part of the chapter. A summary is presented in the discussion section, followed by the conclusion of this review.

2.2 External factors

2.2.1 Ethnicity and CVD

2.2.1.1 Risk of CVD in black and white populations

CVD is the leading cause of death in the world, irrespective of socio-economic status, gender, age and ethnicity. This section discusses the prevalence of CVD in people belonging to black and white ethnicities. Jolly *et al* [21] presented a study which examined the prevalence of CVD in younger black and white people. This study indicated a high prevalence ratio (1.9) at lower ages between black and white people that reduced with the progression of age. The study also indicated that among total CVD deaths more than a quarter of deaths occurred in black (28%) compared to white people (13%) for ages < 65 years. Irrespective of age group, black people were more likely to manifest CVD, heart failure and stroke. The death rate from myocardial infarction (MI) is high among black people compared to white people, but at older ages rates in the white population are dominant. The percentage of death was higher in black people compared to white people irrespective of varying age group. Reinier *et al* [22] also validated the claims made in the earlier literature by indicating a two-fold increase in the death rate in black men and women than in their white counterparts (175 and 90 in 100 000 compared to 84 and 40 in 100 000). Furthermore, the mortality rate was reported to be higher irrespective of gender and age group. He also reported on the severity of CVD risk factors in black people compared to white people, such as diabetes (52% versus 33%), hypertension (77% versus 65%) and chronic renal insufficiency (34% versus 19%). In addition to the prevalent congestive heart failure, left ventricular hypertrophy and QT interval were also reported to be high in black people compared to white people.

A very recent study by Alenghat *et al* [23] demonstrated an association between connective tissue disease (CTD) and atherosclerosis CVD (ASCVD), and reported a higher prevalence of ASCVD in patients suffering from CTD than in those without CTD (29.7% versus 14.7%) for African American people compared to white people. With progression in age, ASCVD increases in patients suffering from CTD more than in the general population. When considering rheumatoid arthritis (RA) independently, African American people show high ASCVD rates compared to white people. Furthermore, irrespective of ethnicity, RA increases the risk of ASCVD in all populations with a high prevalence ratio (4.1 versus 3.6). Recently, Hamer *et al* [24] reported a 3 year follow-up study to identify the prevalence of CVD in black African people, and reported that they have a higher prevalence of CVD than white people (49.1% versus 32%). Ferdinand [25] reported that black men and women are more likely to die due to heart disease than white men and women. However, in this study the authors indicted an increase in life expectancy rate over the last decade in both black and white people. Sundquist *et al* [26] presented a study which compared the CVD risk factors (type II diabetes, physical inactivity, abdominal obesity, hypertension, smoking and high non-HDL cholesterol) with gender and ethnicity. This study presented some key points such as (i) besides smoking, all the risk factors are more prevalent in black women compared to black men, (ii) besides smoking and LDL cholesterol, all other risk factors were high in black women compared to white women, and (iii) besides obesity and LDL cholesterol, all other risk factors were high in black men compared to white men.

2.2.1.2 Geography and CVD

CVD, as a leading cause of death, also varies based on geography. It is prevalent in both developed and developing nations. In 2001, around 80% of CVD deaths occurred in low- and middle-income countries [27]. Despite the availability of a high standard of medical facilities, heart attack or MI is commonly observed in developed nations. In the USA, one in every three deaths occurs due to CVD. A similar situation has been reported in European nations, with a mortality rate of 45% due to CVD [28, 29]. Further, CVD also has a high impact on low- and middle-income countries. The overall contribution to the cardiovascular (CV) death rate by India, China and other developing nations has been reported to be 75% of the total deaths [30, 31].

South Asian countries such as India have a high mortality rate due to coronary heart disease (CHD). As per the statistics [31], Indian people are more likely to be affected by CVD than their European counterparts. In comparison to Western countries, 52% of people below the age of 70 in India are suffering from CHD. Future projections indicate that, by 2030, the total deaths due to CHD may reach up to 3070 per 100 000 people in India, which is higher than in China and Brazil [32, 33]. Although Indian society is diverse, CHD is prevalent in almost every Indian state and region, irrespective of socio-economic status. The sedentary lifestyle of Indian people living in both urban and suburban areas increases their possibility of being affected by CVD. The prevalence of CVD is high in people living in rural areas because of a lack of medical facilities and awareness [34]. The Medical Certification of Cause of Death Report 2015 by the Office of Registrar General of India provided mortality rate statistics and showed that 32.2% of the total deaths in India are due to CVD. Thirty-three Indian states were categorized in this report and Telangana state was identified as having high CVD rates (53%) compared to Assam (6.9%) [35].

2.2.2 Environmental factors and CVD

Environmental factors can modify and alter the risk associated with CVD. Various environmental factors have an impact on cardiovascular diseases and can modify or alter the risk associated with CVD [36]. Even in the absence of family history and genetics, environmental factors can increase the risk of heart disease. Three primary environmental factors influence CVD risk [36]: (1) the natural environment, (2) the personal environment and (3) the social environment. In the natural environment, changes in circadian rhythm have been reported to influence cardiovascular function, with the heart rate and blood pressure lower at night than during the sleep–wakeup period. The day/night cycle variation regulates cardiovascular genes and proteins as well as cardiovascular functioning and influences cardiac events [37, 38]. Furthermore, myocardial infarction was reported to occur prevalently between 6 pm to 12 pm, and between 3 am to 6 am, with a higher risk in the early morning than during the night [39, 40]. Seasonal variation, changes in sunlight exposure and regular physical activity affect cardiovascular responses and influence cardiac disease. These variations lead to mortality due to CVD, which was reported to be higher in winter than in summer [41, 42]. Atherosclerosis causes myocardial infarction due to plaque rupture and the formation of thrombosis. Cold temperatures in winter can lead to plaque rupture and hence contributes to MI [36, 43]. The mortality rates are observed to be higher in both England (20 000 yr^{-1}) and the USA (53% of cases of MI) during winter irrespective of the age of the person [44, 45].

Another important factor that influences CVD risk is sunlight. Exposure to sunlight reduces the probability of being affected by CVD. The main reason for the high risk in winter is the relatively lower exposure to sunlight. In a study of around 200 districts in the United Kingdom, a negative association was reported between CVD and sunlight. Sunlight exposure synthesizes vitamin D in cardiac tissues. In the winter days are shorter and due to increased cloud cover less ultraviolet radiation reaches the surface, which results in less synthesis of vitamin D causing a deficiency of vitamin D. Variation in vitamin D level is associated with variation in blood pressure level. In addition to this, vitamin D manages other cardiovascular functions. A deficiency of vitamin D results in an increased risk of cardiovascular events. From the findings of meta-analysis, it was found that vitamin D intake reduces all causes of CVD mortality.

In addition to sunlight, high altitude has also been reported to influence CVD risk. People living at higher altitudes (>1500 m) are less susceptible to CVD events, as these people are well adapted to cold temperature and high altitudes. It has been reported that people living at higher altitudes have low LDL cholesterol levels and high HDL cholesterol levels, and hence the risk of CVD in these people is less. Furthermore, coronary heart disease and myocardial infarction have an inverse relationship with altitude and hence the death rate decreases with an increase in altitude. Recently, a high altitude study in Switzerland reported that, with every 1000 m increase in altitude, the risk of heart disease reduces by 22%. Greenery is another important factor that reduces the CVD risk and mortality. The greener the place, the lower the chances of heart disease. People living in high-vegetation areas have a low risk of diabetes, stroke and CVD. In a prospective study of England, it has been reported that the mortality rate in less green areas is twice than that in highly vegetated areas. Also, in Perth, Australia, the rate of hospitalization due to CVD was found to be 37% less for occupants of green areas. Another study in Ontario, Canada, reported a high survival rate for people living in highly green areas. A similar claim was also validated in a study conducted in the USA. Collectively, all these natural environmental factors are responsible for modifying the risk of CVD and hence are essential factors to consider. This may either promote or prevent various diseases including CVD.

2.2.3 Air pollution and CVD risk

2.2.3.1 Mortality

Air pollution is a major risk factor and has deleterious effects on human health. The initial epidemiological research on the effects of air pollution was started after two severe catastrophic events, the London fog episode in 1952 and the Meuse Valley, Belgium, incident in December 1930 [46, 47]. After this, constant efforts have been made to reduce air pollution. In 2010, as per the Global Burden of Disease study, the total deaths attributed to air pollution were 3.1 million of 52.8 million all-cause and all-age mortality [48]. Furthermore, air pollution is an important contributor to cardiovascular disease. Several studies have reported an association between air pollution and CVD [49–51]. In 2016, of the total CVD deaths (> 17 million), more than 3 million deaths were caused by air pollution [52]. The mortality rate due air pollution is higher in developing nations such as China. It has been reported that short-term exposure to $PM_{2.5}$ of 10 $\mu g\ m^{-3}$ increases the mortality rate by 1% [53].

2.2.3.2 Air pollutants

Air pollution is an amalgamation of compound gases (nitrogen oxide, carbon monoxide, sulfur dioxide and ozone), lipids and particulate matters. Particulate matter (PM) itself is a mixture of solid and liquid airborne particles of varying size and area [54]. Based on the aerodynamic diameter of the PM, the US Environmental Protection Agency (USEPA) has categorized airborne PM into three types (figure 2.2): PM_{10} (diameter < 10 μm), $PM_{2.5}$ (diameter < 2.5 μm) and ultrafine particles (UFPs, with diameter < 100 nm) [55]. PM_{10} and $PM_{2.5}$ particles are also called coarse particles and are able to penetrate into the respiratory tracheobronchial tract. The $PM_{2.5}$ particles (or fine particles) were found to be able to penetrate deep into the human lungs and the UFPs into human alveoli [56]. These air pollutants are generated from (i) natural sources such as volcanos, wildfires, lightning or land dust, and (ii) anthropogenic sources such as combustion of fossil fuels from traffic, industry and power generation, and household cooking and heating [57].

2.2.3.3 Role of air pollutants and CVD risk

Particulate matter ($PM_{2.5}$ and PM_{10}) is an important air pollutant which is strongly associated with respiratory disease and CVD mortality, as shown in figure 2.3 [54]. Studies conducted in 204 counties of the US with a 4 year follow-up period and in Europe in five major cities with a 10 year follow-up period indicated an increase in the hospitalization rate due to ischemic heart disease (IHD) [59, 60]. Short-term as well as long-term exposure to these pollutants increases the risk of coronary events. It has been reported that a 10 $\mu g\ m^{-3}$ increase in mean $PM_{2.5}$ increases the daily CVD mortality risk by 0.4%–1% [61]. Recently, the European Study of Cohorts for Air Pollution Effects (ESCAPE) group presented a meta-analysis of 11 cohorts for long-term exposure to PMs and reported a 13% increase in the risk of coronary events with an 5 $\mu g\ m^{-3}$ annual increase of $PM_{2.5}$ [62]. Also, for a 10 $\mu g\ m^{-3}$ increase in PM_{10}, a 12% increase in the risk of coronary events is reported. Heart failure (HF) is a major type of CVD and long-term exposure to PM and nitrous oxide (NO), increases the risk of HF.

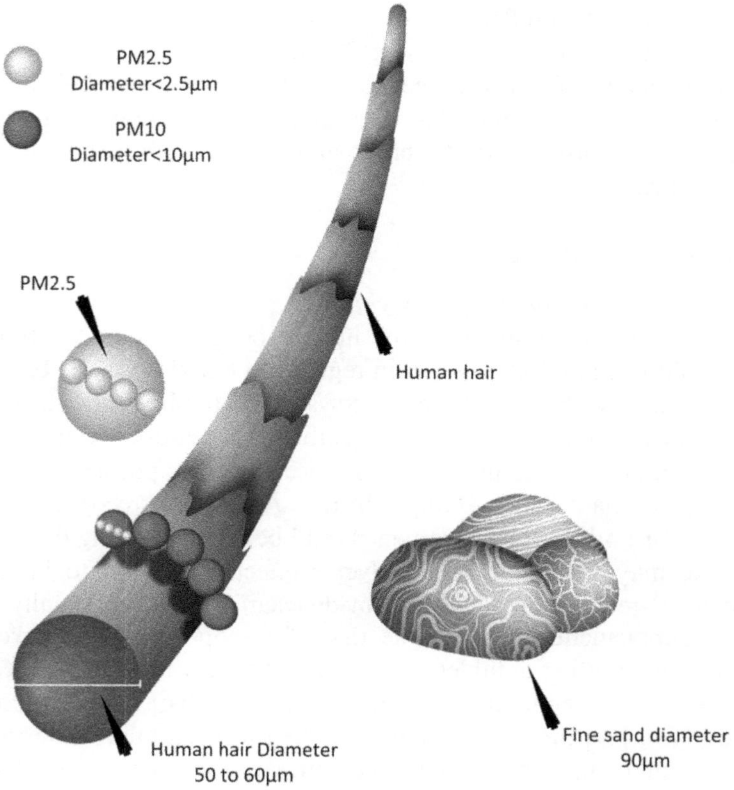

Figure 2.2. Particulate matter air pollution size distribution. (Courtesy of AtheroPoint, Roseville, CA, USA.)

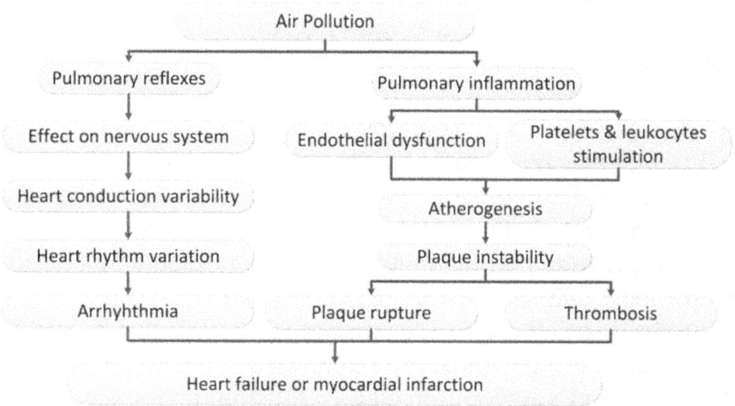

Figure 2.3. Possible mechanistic effects of air pollution on cardiovascular morbidity and mortality.

2.2.3.4 Air pollutants and inflammation

There exists a positive correlation between long-term exposure to pollutants such as PM and atherosclerosis [64]. When PM or UFPs enter the respiratory system, in the lungs these pollutants generate the oxidative radicals and lead to oxidative stress.

Air pollutants can lead to inflammatory oxidative stress in two ways: (i) when the amount of oxidizing species increases in the lungs cells, or (ii) by decreasing the number of antioxidants. The PM influences this inflammatory process and provokes pro-inflammatory oxidative stress. The increase in inflammation generates cytokines that further lead to vascular dysfunction and atherosclerotic plaque formation, progression and rupture.

2.2.4 Nutrition and CVD risk

2.2.4.1 Nutrition and coronary artery calcium

Calcium is an important mineral in the human body. It helps to strengthen the bones, joints and teeth and also assists in regulating blood pressure, blood clotting, muscular function and the overall immune system of the body. Sources of calcium are milk products, legumes, leafy green vegetables and seafoods. Calcium deficiency leads to a reduction of bone mass and may lead to osteoporosis, which weakens bone, reducing its mass and making it brittle. As per the recommendations, the calcium intake for both men and women should be 1000–1300 mg day^{-1} [65]. High calcium intake may also have some adverse effects that lead to kidney failure, nausea and heart-related disease. Calcium deficient people are usually advised to take calcium supplements to avoid the risk of osteoporosis. Excessive intake of calcium supplements either in tablet or in liquid from may lead to unexpected health issues such as hypercalcemia and milk alkali syndrome [66]. A recent cohort study by Anderson *et al* [67] presented a high prevalence of coronary artery calcium in people with calcium supplementary intake compared to those with calcium intake through their daily food. CAC risk increased by 22% in people with high calcium supplement intake. Furthermore, the authors reported high calcium intake in females through calcium supplements, which may increase the risk of CAC. The common daily practice of having tea or coffee was also studied in relation to the prevalence of CAC level.

2.2.4.2 Nutrition and Mediterranean diet

A low mortality rate was reported in Mediterranean countries, mainly due to their healthy dietary habits. Several short-term and long-term clinical studies including cohort studies reported significant reduction in CVD risk factors such LDL cholesterol, blood pressure and inflammatory markers such as C-reactive proteins. Mediterranean foods such as olive oil and nuts, when compared to a low-fat diet, were reported to be associated with a low risk of cardiovascular disease [68]. The use of olive oil compared to seed oil and vegetable fats has also been determined to reduce CAD in a recent study presented by Oikonomou *et al* [69], with a hazard ratio of 0.70 ($p = 0.047$).

2.2.5 Family history and CVD risk

Heart disease is prevalent if a person has a family history of CVD. If the immediate relatives of the patient have a CVD history then there is a high chance of the person being affected by CVD [70]. A family history of premature cardiovascular events is one of the independent risk factors associated with CVD mortality [71]. Several researchers have found a strong link between family history and heart disease [72–75]. The risk associated with CVD mortality varies with family members. It has been reported that lifetime CVD risk doubles if any single-degree relative has had CVD events in the past [76]. Furthermore, CVD increases by a factor of four if an immediate relative had a premature CVD event before the age of 55 [77]. Although family history is an important risk marker, the Framingham study also incorporated age, cholesterol level, gender, smoking, HDL cholesterol level and blood pressure to assess the 10 year risk of CVD. Hence, it is essential to include the family genetics related to CVD in risk prediction methods by correlating with CVD risk factors. Nasir *et al* [78] reported a correlation between family history and atherosclerotic risk markers such CAC, carotid intima–media thickness (cIMT) and inflammatory markers such as C-reactive protein. The CAC is prevalent if any family member in the past has experienced CHD events [78]. Parental and sibling family history were also compared by the authors and they reported a high prevalence of CAC in the latter case compared to parental CHD family history (78% versus 64%). For first-degree relatives and second-degree family members such as grandparents, aunts and uncles, the prevalence of CAC was associated with CHD related family history for these two groups (26.6% versus 26.5%) [79]. In addition to CAC, a family history of CHD also has a strong association with cIMT. Juonala *et al* [80] presented a study showing the association between cIMT and family history and reported a significant increase in cIMT in patients with a family history of CHD. Furthermore, the Framingham heart study reported an increase in cIMT in people with a parental family history of premature coronary heart disease at < 60 years of age [81].

2.3 Genetics of coronary artery disease

2.3.1 Genetics of atherosclerosis

Atherosclerosis is an inflammatory disease, a major coronary artery disorder, which promotes lipid metabolism and development of lesions by depositing a fatty substance called plaque within the arterial peripheral wall [82]. Although the starting point of atherosclerosis is the subject of ongoing research, dysfunction of endothelial cells is considered an initial stage of plaque build-up. Endothelial dysfunction possibly occurs because of various risk factors such as high LDL cholesterol, smoking, consuming alcohol and tobacco [83]. The improper functioning of endothelium allows the penetration of LDL cholesterol within the intima layer where it becomes oxidized. Oxidized LDL cholesterol activates the endothelial adhesion molecules, such as vascular cell adhesion molecule-1 (VCAM-1), and increases the permeability of leukocytes, in particular monocyte and T-lymphocyte recruitment. Further, chemokines support the penetration of these white blood cells in the intima layer [84]. Once in the intima, monocytes are transformed into macrophages which take up the oxidized cholesterol and become lipid-laden foam cell. Foam cells, also known as fatty streaks, are an important marker for atheroma regions in atherosclerosis [85]. Foam cells generate cytokines which are responsible for smooth muscle cell proliferation from the media layer to the intima layer and

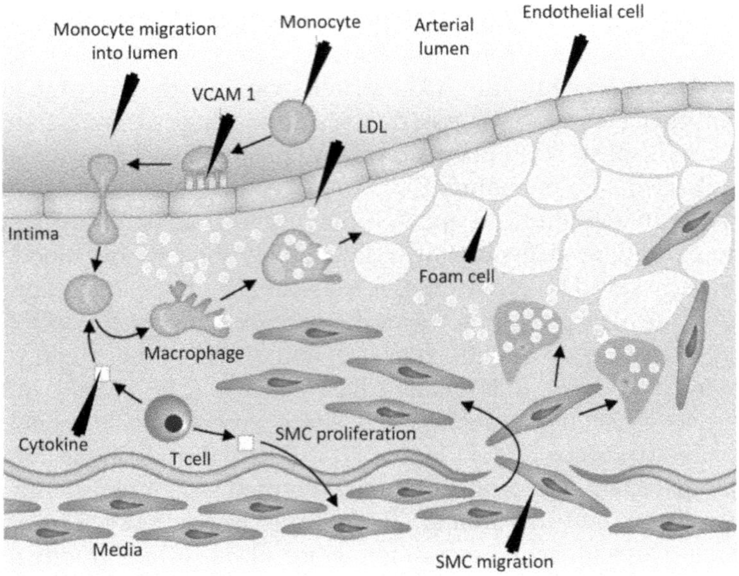

Figure 2.4. Plaque components within the vessel wall. (Reproduced with permission from [87]. Courtesy of AtheroPoint, Roseville, CA, USA.)

also increase leukocyte recruitment. Continuation of this inflammatory process leads to the death of foam cells forming a necrotic core. The necrotic core contains lipid-rich macrophages, foam cell debris and smooth muscle cells [86]. The entire atherosclerosis process is depicted in figure 2.4.

2.3.1.1 Role of cholesterol

Cholesterol is the main risk factor in the progression of atherosclerosis [83]. Cholesterol is a fatty substance produced by the liver or generated from our daily diet, which is further packaged into small particles called lipoproteins. The human body needs cholesterol to generate hormones, vitamin D and a substance which helps to digest food, called bile. There are two types of lipoproteins, low-density lipoprotein (LDL cholesterol) and high-density lipoprotein (HDL cholesterol) (figure 2.5). LDL travels through the blood and deposits cholesterol in the blood where it is needed. Excessive LDL cholesterol starts building up in the walls of the arteries, forming a fatty streak called plaque. High levels of LDL cholesterol increase the risk of CVD [88]. The progression of plaque deposition narrows the coronary artery reducing oxygenated blood supply to the heart. LDL is the primary contributor to plaque build-up. The liver also produces HDL cholesterol, which removes the deposition of LDL cholesterol, preventing the plaque deposition and hence HDL is called good cholesterol. LDL cholesterol has a direct link to coronary artery disease and efforts are usually made to lower the percentage of LDL in the blood using statin therapy.

Recent guidelines by the European Society of Cardiology state that an LDL cholesterol level < 70 mg dl^{-1} reduces the possibility of atherosclerosis and CAD [89].

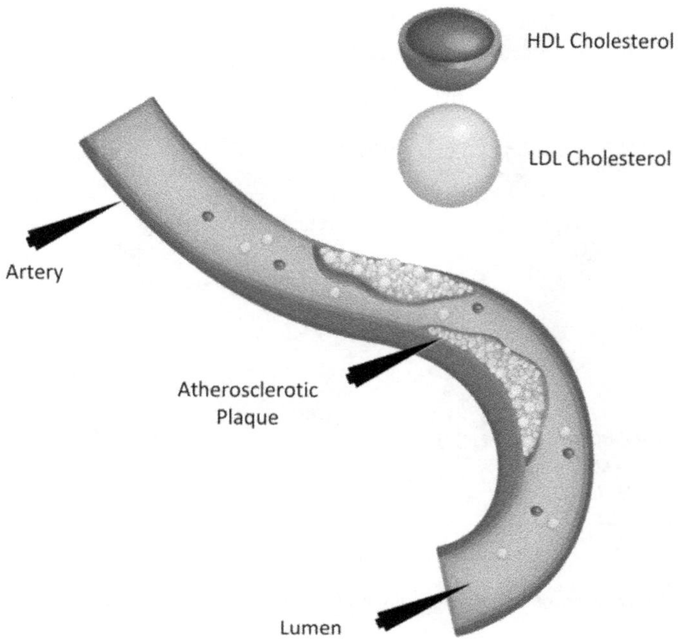

Figure 2.5. LDL versus HDL cholesterol. (Courtesy of AtheroPoint, Roseville, CA, USA.)

In the coronary artery vessel wall, LDL cholesterol is mostly deposited in the tunica intima and becomes oxidized and activates the endothelial cells. A recent study [90] explored the use of an FD-OCT system to visualize the morphology of the unstable plaque components in patients with low LDL cholesterol levels. It has been reported that of 280 patients, a fibrous plaque was prominent in subjects with low LDL cholesterol levels, which provides highly stabilized plaque morphology.

2.3.1.2 The role of monocytes and macrophages

Monocytes are the elemental composition of the white blood cells or leukocytes and hence a part of the human immune system. A recent cohort study of 951 patients summarized the high correlation between monocytes and CVD [91]. Dysfunction of endothelial cells allows the LDL cholesterol to penetrate into the intima layer where it is trapped and oxidized. Oxidized cholesterol accumulated within the tunica intima activates the endothelial cells by developing adhesion molecules over the endothelial surface. Blood monocytes then are attracted towards the endothelial surface and chemokines, such as monocyte chemoattractant protein (MCP)-1, support the monocytes to enter the intima layer [92]. In the intima layer, the monocytes are transformed into macrophages and, using scavenger receptors, engulf the oxidized LDL cholesterol to form foam cells. Macrophages are also responsible for migration of SMCs from the media to intima, which leads to SMC proliferation [93]. Plaque vulnerability is attributed to a thin fibrous cap, a large lipid core containing macrophages and collagen fibers, and occlusive thrombus. Monocytes play an important role in the destabilization of these plaque components [94].

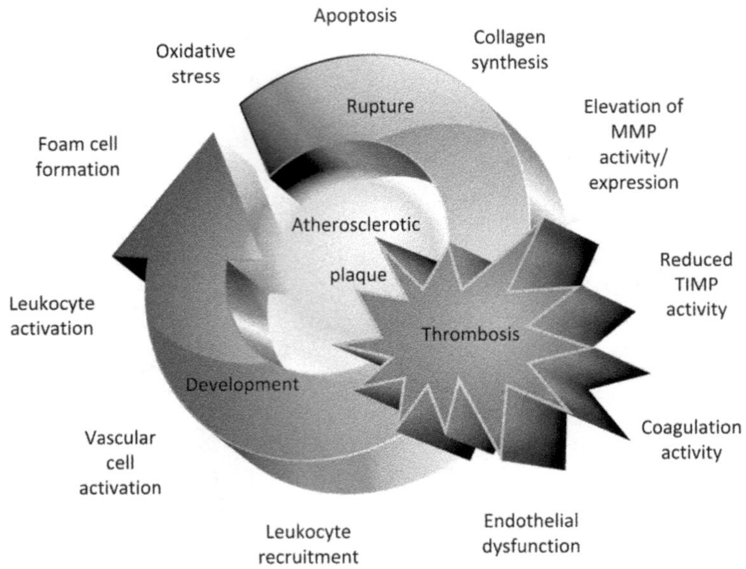

Figure 2.6. Atherosclerosis based on cytokines. (Courtesy of AtheroPoint, Roseville, CA, USA.)

2.3.1.3 The role of cytokines

Cytokines are signaling cells that facilitate communication between different cells and allow the movement of immune cells towards inflammatory sites (figure 2.6). Cytokines regulate both the immune and inflammatory functions. In other words, they act as intercellular mediators. Once released they are able to bind to specific receptors (i.e. cytokine receptors) on the surfaces of other cells [95]. Most cytokine signals are self-activated (autocrine action) and some of them act on neighboring cells (paracrine action). Pro-inflammatory cytokines boost the progression of macrophages within the intimal layer and develop the disease, whereas anti-inflammatory cytokines exert antiatherogenic activities. Cells which are involved in cytokine production are lymphocytes, macrophages, dendritic cells, endothelial cells, epithelial cells and connective fibrous tissues. Each of these follows different cytokine signaling to communicate with other cells. Atherosclerosis is the inflammatory process and the cytokines involved in acute inflammation are tumor necrosis factor (TNF), interleukins and chemokine. All cells are able to generate and respond to a particular type of cytokine. Interleukins are a type of cytokine and, to date, 37 have been found, denoted by IL followed the number of interleukin (IL-1 and IL-2). Macrophages are a major factor in atherosclerosis, which generate TNF and IL-1 cytokine once activated. Cytokines generated by monocytes are known monokines [96].

2.3.1.4 Penetration of LDL/monocytes into endothelial cells and the intimal wall

Damage to the endothelial cells allows the penetration of LDL cholesterol into the intima layer where they become oxidized and activate the endothelial receptors. The activation of endothelial receptors secretes chemokines, which will be received by the chemokine receptors on the monocytes [98]. This allows the migration of monocytes from blood to the intima layer where they become macrophages.

2.3.1.5 The role of macrophages and foam cells

The macrophages, once in the intima layers, are trapped and become activated, generating cytokine signals to recruit a number of macrophages into the intima layer. Macrophages are a type of phagocyte, a major component of white blood cells. A phagocyte usually engulfs the pathogen in the vessel and differentiates between pathogens using enzymes within them. Here oxidized LDL cholesterol is nothing other than a pathogen in the intima layer and hence the macrophages engulf it and are transformed into foam cells. The development of foam cells results in two important points, first it coincides with T-helper cells appearing in the intima and generating the pro-inflammatory cytokines, which recruit a number of macrophages, forming more foam cells [5]. The formation of a number of foam cells increases the intima thickness, which is a potential risk factor for atherosclerosis. The second important role performed by foam cells is allowing SMCs to migrate from the tunica media layer to the tunica intima, forming a fibrous cap covering the lipid core. Moreover, it promotes the SMC proliferation [99]. Accumulation of foam cells over a period of time results in plaque formation.

2.3.1.6 The role of SMCs in the formation of lipid

The tunica media is the primary source of SMCs. SMCs play an important role in the initiation and advancement of coronary artery disease. The formation of foam cells enables the migration of SMCs from the media layer to the intima layer. A previous study has also shown the initial presence of SMCs in the intima layer giving rise to intimal thickening areas. The proliferation of SMCs in the intima results in an enlargement of the intimal area, reducing the effective size of the lumen. Cytokines produced by SMCs result in inflammatory responses to secret and activate monocytes and lymphocytes, further increasing the proliferation of SMCs into the intima. This at first results in the growth of the intima layer towards the adventitia and after a certain threshold it starts thickening in the opposite direction, towards the lumen [99, 100]. Platelet-derived growth factor (PDGF), transforming growth factor (TGF)-β, interleukin (IL)-1 and interferon gamma (IFN-γ) are the cytokines produced by SMCs. SMCs are considered to the high producers of extracellular matrix, which generates fibrous tissues and adds collagen fibers to form a thin fibrous cap to protect the lipid region from being exposed to blood. Increased synthesis of collagen fibers due to SMC proliferation leads to hardening of the plaque.

2.3.1.7 The role of rupture of walls

Apoptosis of foam cells releases all their lipid contents, resulting in the formation of a plaque. The progression of the plaque increases the pressure within the tunica intima which may lead to rupture. The thickness of the thin fibrous cap (FC) covering this entire lipid core affects the likelihood of rupture. An FC less than

65 μm thick increases the risk of plaque rupture. Once the plaque is ruptured it leads to occlusive thrombosis, which involves clotting agents, impeding blood flow.

2.3.2 Genetics of diabetes

2.3.2.1 Normal scenario

After food intake that contains carbohydrates, chemicals in our small intestine break them down into single sugar molecule. Then, the intestinal cells absorb the glucose and pass it into the bloodstream. As soon as the blood reaches the pancreas, the β-cells within the pancreas detect the rising glucose levels. In order to control the glucose levels and to allow target cells to uptake the glucose, the pancreas generates a specific hormone known as insulin. Insulin facilitates glucose uptake by different target cells. When insulin binds to receptors on the cell surface, glucose starts to enter the target body cell, where it is stored in the form of a chain known as glycogen. Now the cells can use the glucose to produce energy to function properly. In the absence of glucose for energy production, there is an increase in generation of another hormone, known as glucagon, which enters cells such as those of the liver, breaks the glycogen chains and allows the glucose to circulate through the blood to the respective target cell. This entire process is also known as homeostasis.

2.3.2.2 Diabetes mellitus

In diabetes, the body fails to maintain homeostasis. Diabetes is characterized by high-level blood glucose, also known as hyperglycemia. Diabetes mellitus (DM) increases the risk of macro-vascular and micro-vascular diseases, including atherosclerosis and CVD (figure 2.7). Furthermore, hyperglycemia in DM may result in serious life-threatening problems [102]. There are two types of chronic diabetes conditions: (i) type 1 diabetes and (ii) type-2 diabetes. Pregnant women can also experience a transient form of diabetes called 'gestational diabetes', which is resolved after the baby is born. Pre-diabetes is diagnosed when blood sugar is at the boundary line, higher than normal and below diabetic.

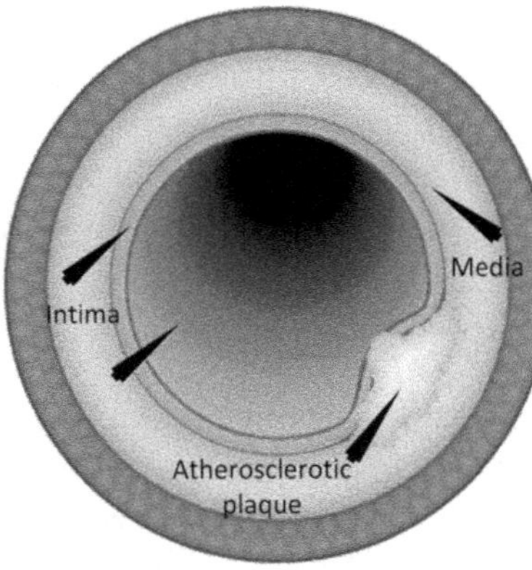

Figure 2.7. Diabetes driven atherosclerosis. (courtesy of AtheroPoint, Roseville, CA, USA.)

2.3.2.3 Types of diabetes

In type 1 diabetes, the β-cells of the pancreas are destroyed by the immune system by mistake. The reason for this happening is unclear, however, genetic factors are believed to play a major role. Because of this insulin production is reduced, less insulin binds to the receptors, there is less glucose intake into target cells and hence the glucose level in the blood increases. Type 1 diabetes is normally managed by insulin injection. Type 1 diabetes is, therefore, insulin dependent. In type-2 diabetes, something goes wrong either with the receptors or with signaling inside the target cells. The cells are not responsive to insulin and hence cannot import glucose and this glucose stays in the blood. Type-2 diabetes is insulin resistant. In type-2 diabetes, body cells resist the insulin and its effects. In insulin resistance, insulin cannot unlock the cells to let glucose in. This raises the blood glucose level in the bloodstream which is also called hyperglycemia. Hyperglycemia leads to the following symptoms: excessive hunger, thirst, increased urine volume and unexplained weight loss.

2.3.2.4 Vascular changes due to diabetes

In diabetes, there is high blood glucose and endothelial cells take up this glucose. Endothelial cells do not need insulin to take up glucose. With so much glucose inside the endothelium, the cells produce reactive oxygen species (ROS). The increase in ROS leads to the formation of the advanced glycolated product (AGP) and protein kinase C (PKC) which leads to endothelial dysfunctioning. PKC enables the receptors for the monocytes on the endothelial cells to increase the vascular permeability of white blood cells, which further leads to plaque build-up and atherosclerosis. Impaired glucose tolerance (IGT) is a pre-diabetic state of hyperglycemia that is associated with insulin resistance and CVD. IGT may precede type-2 diabetes for many years. According to the WHO and ADA, IGT is defined as a two hour glucose level of 140–199 mg dl^{-1} on 75 g oral glucose.

2.3.3 Genetics of rheumatoid arthritis

Rheumatoid arthritis (RA) is a disorder of the joints, which connect two or more bones together to provide support, movement and flexibility to the human body. RA is an autoimmune disease in which the immune system attacks itself by mistake, primarily in the synovial joints [103]. The synovial joints mainly consist of bone, the joint capsule, the synovial membrane also known as synovium, and articular cartilage, which allows the smooth movement of the bones. Synovium produces synovial fluid which primarily helps in lubricating bones for free motion. Inflammation of synovium is the main characteristic of RA, also known as synovitis, and causes pain and swelling. Furthermore, bone and cartilage destruction is also attributed to RA (figure 2.8). The synovial membrane is made of up of a thin cellular layer, also known as synovial fibroblasts or fibroblast-like synoviocytes (FLS), which play a vital role in the inflammation and pathogenesis of RA [104]. Although the exact cause of RA is unknown, the progression of RA has been analyzed based on various cells present in the synovial membrane. $CD4^{+}$ T-cells play an influential role and have been reported as a driving force in the pathogenesis of RA [105].

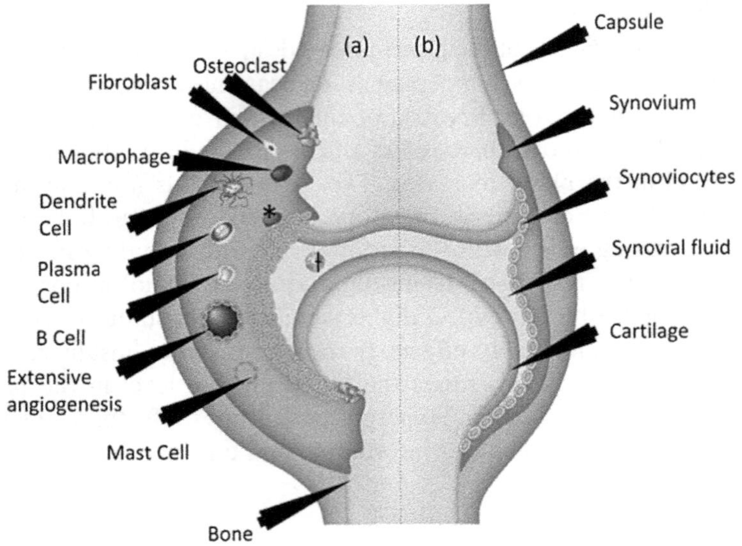

Figure 2.8. Schematic view of a normal joint affected by RA (a) and a normal joint (b). (Courtesy of AtheroPoint, Roseville, CA, USA.)

$CD4^+$ T-cells promote inflammation and generate interleukin 17 (IL-17), which allows the recruitment of macrophages. Furthermore, T-cells also produce FLS cells and also help in the expression of the receptor activator of NF-κB ligand (RANKL) cells which stimulates osteoclast activity for bone erosion [106, 107]. In addition to T-cells, macrophage infiltration is also an important process in the progression of RA and has a main role in secreting inflammatory cytokines such as IL-1, IL-6 and TNF-α [108, 109].

Macrophages are responsible for stimulating FLS cells which lead to their activation and proliferation, which further assist in RANKL expression. Cytokines (i.e. IL-1, IL-6 and TNF-α) together with RANKL cells lead to osteoclast activity for bone erosion [107]. The activation and proliferation of FLS cells also has an important role in the secretion of proteases, which influences cartilage degradation [108]. Synovial fibroblasts have another important feature, in that they lead to symmetrical RA by migrating from one joint to another. The synovial membrane also contains plasma cells in small amounts, which assists the inflammation process through cytokines and antibodies [108]. In the synovial fluid neutrophils produce proteases and ROS which cause bone and cartilage degradation and contributes to inflammation.

2.3.4 Anatomy of a 3D heart

The human heart is the most important muscular organ of the body, and receives deoxygenated blood from veins and pumps oxygen-rich blood back to different parts of the body via arteries. The oxygenated nutrition-rich blood is delivered to the body tissues via the aorta, arteries, arterioles and finally through capillaries. Capillaries are very thin hair-like blood vessels which deliver oxygenated blood to the body tissues and collect carbon-dioxide-rich waste. They also provides the interconnections between arterioles and venules. The deoxygenated blood then travels towards the heart via venules, veins and the vena cava. For purification, the blood is then transferred towards the lungs via the pulmonary arteries, via the right atrium and ventricle. In the lungs, the blood is purified with abundant oxygen supply, releasing carbon dioxide. This entire circulatory system is maintained by the heart.

The human heart has four chambers, the two upper chambers are called atria and the two lower chambers are the ventricles. The right and left atria are separated by an interatrial septum. Similarly, the two ventricles are separated by an interventricular septum. The blood flows from the atria to the ventricles using muscular valves. A tricuspid valve exists between the right atrium and right ventricle, which has three openings which open towards the right ventricle and prevent the backward flow of the blood supply. The left atrium and ventricle are separated by a bicuspid valve, also known as the mitral valve. At the opening of the aorta and pulmonary artery, two more valves are present. These are known as the aortic valve and pulmonary valve. Due to their shape, these valves are also known as semilunar valves. The opening and closing of such a valve creates a sound which is normally used to count the heart beats in a minute. The thickness of the wall of each of the four chambers depends on the function it performs or distance the blood is pushed by them. The thickness of both atria is small as they are only pushing the blood to the ventricles. The wall thickness of the left ventricle is greater than the right ventricle. This is because the left ventricle pumps blood to the whole body whereas the right ventricle pumps the blood a smaller distance to the lungs.

2.3.4.1 Types of arteries feeding oxygenated blood to the heart

In order to function, the heart tissues require an oxygen-rich blood supply which is provided via the coronary arteries. The left main coronary and right coronary arteries are the major blood vessels of the heart. The left main coronary artery, which originates from the left coronary sinus of Valsalva, is further divided into the left circumflex (LCx) and left anterior descending artery (LAD). Both of these arteries originate from the base of the aorta, called the coronary ostia [110]. As these arteries lie on the surface of the heart, they are also referred as pericardial arteries. The LAD artery has three main sections, proximal, mid and distal, which are further divided into diagonal arteries. The LCx coronary artery passes through the atrial–ventricular sulcus and branches into obtuse marginal branches.

2.3.4.2 Size of coronary arteries

The size of the coronary arteries varies based on gender, age and diseases such as diabetes. It has been reported that that the dimensions of normal coronary arteries such as LAD and LM are smaller in female compared to male patients when measured using ultrasound [111, 112]. The assessment of coronary artery dimensions, such as lumen area and diameter, is essential to estimate the risk at the earlier stages of disease. A study of 257 patients indicated that the mean left main coronary artery and the luminal area are smaller in women compared to men (table 2.1).

Table 2.1. Coronary artery dimensions based on gender [111].

Metric	Men	Women
Luminal area (mm^2)	16.65 ± 4.04	14.0 ± 3.24
External elastic lamina area (mm^2)	20.58 ± 4.91	17.21 ± 3.97
Averaged luminal diameter (mm^2)	4.26 ± 0.55	3.92 ± 0.45

Table 2.2. Coronary artery dimensions in diabetic patients [9].

Coronary arteries	Diabetic	Non-diabetic
Proximal LAD (mm)	2.99 ± 0.44	3.14 ± 0.50
Mid LAD (mm)	2.90 ± 0.38	3.10 ± 0.42
Distal LAD (mm)	2.25 ± 0.39	2.42 ± 0.45
Proximal LCX (mm)	2.98 ± 0.21	3.01 ± 0.25
Distal LCX (mm)	2.35 ± 0.40	2.49 ± 0.43
Proximal RCA (mm)	3.0 ± 0.28	3.28 ± 0.25
Mid RCA (mm)	2.97 ± 0.26	3.91 ± 0.25
Distal RCA (mm)	2.43 ± 0.51	2.87 ± 0.32

Body surface area also has an important effect on coronary dimensions, perhaps more than gender. The dimensions of the major coronary arteries also vary with diabetes. According to a study presented using a cohort of 571 patients [9], it has been observed that coronary narrowing is greater for diabetic patients (table 2.2).

2.3.4.3 Normal heart motion

Normal heart function and its motion can be tracked using an electrocardiogram (ECG) at the preliminary stage. ECG is a composite electrical signal which provides information about the cardiovascular cycle. The cardiovascular cycle starts with the P-wave on the ECG, indicating atrial depolarization due to the firing of the sinoatrial node. The QRS complex on the ECG represents ventricular depolarization. During this phase the ventricles are contracted and transfer blood to the lungs and towards the rest of the body. The relaxation of the ventricles on the ECG is indicated by the T-wave. In order to compute the normal heart rate, the R to R interval is usually considered an important parameter.

2.4 Multimodal coronary imaging

2.4.1 Regular coronary artery

From histology, it has been found that the normal coronary artery wall has a well-delineated layered structure containing four important sections, a lumen which carries the actual blood flow, and the intima, media and adventitia layers. Each of these layers has distinct features (figure 2.9).

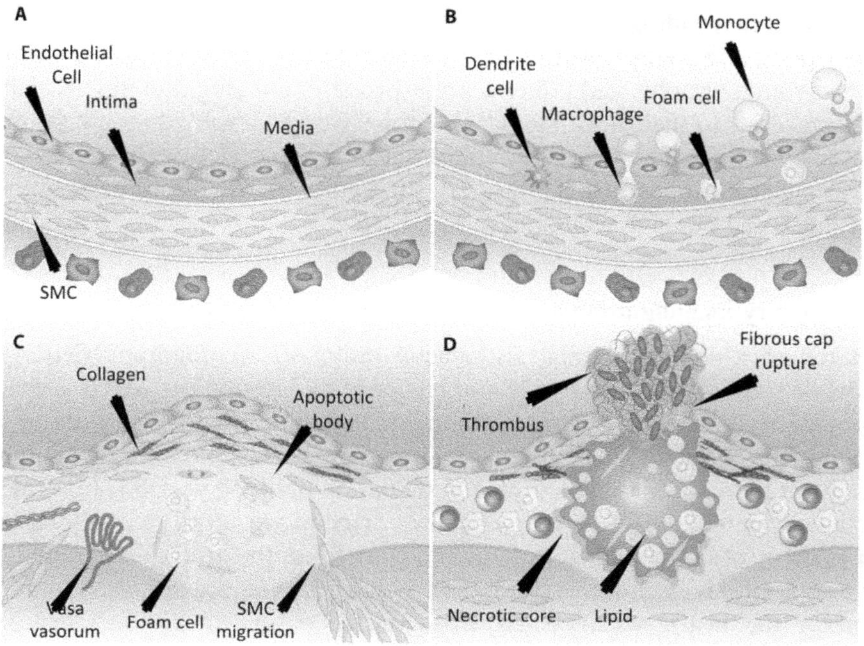

Figure 2.9. Overview of plaque formation and rupture. (Reprinted with permission from [113]. Courtesy of AtheroPoint, Roseville, CA, USA.)

The tunica intima is the innermost layer of the coronary vessel wall, followed by the tunica media and adventitia. The tunica intima of the vessel wall can be identified as a highly backscattering signal-rich layer. The normal thickness of the intima layer ranges between 60 to 65 μm [114]. The next important layer is the tunica media. As the name suggests, it is the middle layer of the vessel wall structure and has abundant smooth muscle cells which play a vital role in the pathogenesis of atherosclerosis [115]. Intima–media thickness is a crucial risk predictor of atherosclerosis. An increase in the thickness of these combined layers reduces the lumen diameter and hence obstructs the effective blood flow area. The adventitia layer is the outer most muscular layer of the vessel wall.

2.4.2 Coronary imaging using x-ray angiography

Various imaging modalities have been used to assess the risk associated with CAD. Coronary angiography was the first and is routinely used by cardiologists to analyze blockages within coronary vessel walls. In general, it uses a guidewire to align a catheter at the opening of the coronary artery. To obtain a clear picture of blood vessels, a dye/contrast agent is usually injected through the catheter into the arteries and then x-rays are used to capture images. The narrowing of the coronary arteries can be visualized easily and with accuracy using angiography, and hence it is considered as the primary choice compared to non-invasive techniques. In addition to all these benefits, cardiac catheterization using coronary angiography has the following shortfalls: (i) it does not provide any information about the coronary wall thickness and diameter of the artery, (ii) the absence of arterial narrowing may also result in plaque formation within the vessel periphery, which cannot be captured using x-ray angiography, (iii) patients are exposed to ionizing radiation which may prove hazardous to their health and (iv) coronary angiography does not permit the measurement of plaque vulnerability and the risks associated with plaque rupture [11]. Moreover, the expensive nature of this modality has encouraged researchers to pursue advanced imaging techniques.

2.4.3 Coronary imaging using magnetic resonance angiography

Compared to x-ray angiography, magnetic resonance angiography (MRA) is a worthwhile solution with lower cost and higher spatial resolution. MRA is generally performed to understand the anatomical structure of the heart without using radiation. Moreover, it is a non-invasive imaging technique with the capability of imaging soft tissues. This enables the differentiation of blood vessels [117]. There are two challenges with MRA. First, coronary arteries are mostly circuitous with a larger number of twists and turns, and hence to detect the arterial structure, a high spatial resolution is essential. Second, the coronary arteries are subjected to the large motion of the heart, which is higher than the vessel dimensions, hence it is challenging to acquire images with high contrast and large resolution. In addition, MRA has the advantage of visualization of mediastinal structures of the coronary arteries [118] but with lower spatial resolution (0.2–0.3 mm).

2.4.4 Imaging coronary CT angiography

Coronary CT angiography (CCTA) is a non-invasive technique and has shown remarkable progress when compared to the conventional x-ray angiography discussed in the section above. CCTA performs an examination of heart which indicates any plaque build-up or blockage in the coronary arteries. For obtaining the images, iodine-rich contrast agents are, in general, injected through the arm and images are captured when this liquid flows through the coronary arteries. These images can be used to produce three-dimensional images of the heart. CCTA has the potential to distinguish various coronary arteries and hence has proved to be advantageous over conventional 'luminography' [119]. The advancements of CT scanners allow them to also detect arterial stenosis due to calcium and predict the future risk associated with CAD. CCTA images are used to determine the plaque morphology and are able to classify them into calcified or non-calcified lesions. Compared to calcified plaque, the non-calcified plaque burden is essential to predict the future possibility of atherosclerosis, which may lead to myocardial infarction [120, 121]. To quantify coronary calcium deposition, the Agatston score is usually followed [122]. Morphological changes such formation of necrotic core or thin-cap fibroatheroma (TCFA) and macrophages cannot be observed using conventional x-ray angiography, as the internal vessel wall is not visible and hence CCTA can be employed in such scenarios.

2.4.5 Coronary artery interpretation by OCT versus IVUS

OCT imaging of the coronary vessel wall layers and lumen requires knowledge of the optical and morphological appearance of different tissues within the artery. A regular coronary artery is easily demarcated into three layers, the intima, media and adventitia, by their reflective properties in OCT (figure 2.10). The tunica intima and tunica adventitia are highly reflective when compared to the tunica media and can be easily identified by OCT. The mean thickness of the media layer is reported as 200 μm, which lies in the capturing range of OCT [123]. The media layer is delineated by the internal elastic lamina (EEL) and external elastic lamina (EEL), and is usually observed as a dark band. The IEL separates the lumen of the artery from the intima layer, whereas the EEL forms a separation between the media and adventitia layers. The region between these two borders provides the space for initial atherosclerotic plaque formation [125]. At present, the IEL and EEL can only be detected by OCT rather than IVUS with utmost accuracy. Intimal thickening provides the absolute risk prediction for CADs and hence requires accurate characterization. The ultrasound technique, due to its lower resolution, has been proved to be incapable of defining the sharp boundary between the intima and media layers. OCT provides better visualization of both of these layers and, because of the high correlation between intima and intima–media thickness, it delineates the IMT more accurately than IVUS (figure 2.11). Previous studies have demonstrated the good association between OCT findings and histological analysis for the detection of IMT and the other laminas discussed above [124].

The high resolution of OCT allows the identification of the IEL and EEL, and the intima, media and adventitia layers, which is difficult with other imaging modalities [126]. The media is seen as a dark band delimited by the IEL and EEL [127]. The media is identified by a homogeneous layer that has low backscattering, and the adventitia by a highly backscattering layer. A dark band denotes the low signal muscular media layer between the IEL and EEL. Estimation of lumen diameter is the first step in characterization and is important in estimating coronary artery disease severity. The correlation of intima–media thickness is higher when analyzed using OCT and compared using histology.

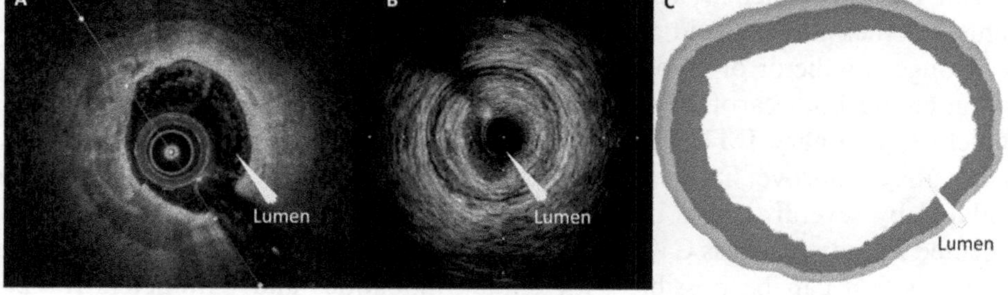

Figure 2.10. The coronary artery. (A) OCT image courtesy of Luca Saba University of Cagliari, Italy. (B) IVUS image courtesy of Toho University, Japan. (C) Histology image designed by AtheroPoint (TM), Roseville, CA. The combination is a concept and not necessarily for mapping the three modalities: OCT, IVUS and histology.

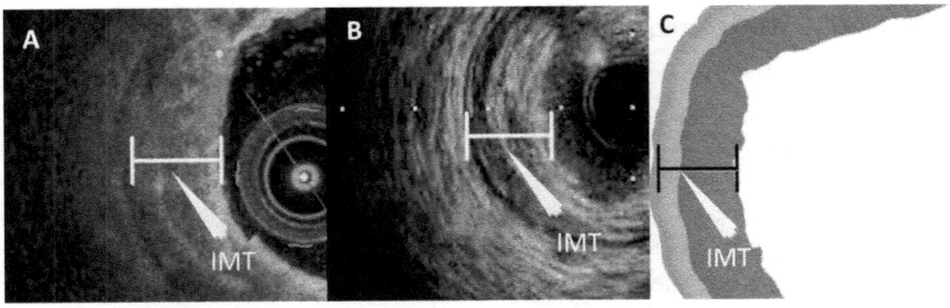

Figure 2.11. Intima–media thickness. (A) OCT image courtesy of Luca Saba University of Cagliari, Italy. (B) IVUS image courtesy of Toho University, Japan. (C) Histology image designed by AtheroPoint (TM), Roseville, CA. The combination is a concept and not necessarily for mapping the three modalities: OCT, IVUS and histology.

2.5 Association of CVD with other prevalent diseases

2.5.1 Relationship between coronary artery and carotid disease

Coronary and carotid artery disease are closely related to each other and atherosclerosis has been reported as a common disorder of these arteries. In the last two decades, several cohort studies, meta-analyses and clinical trials have indicated the association of carotid and coronary artery disease, and projected cIMT and the plaque burden as the important biomarkers for CVD risk assessment [128–130]. O'Leary *et al* [131] have shown the significant association of IMT in both the internal carotid artery (ICA) and common carotid artery (CCA) with myocardial infarction and stroke. Polak *et al* [132] investigated the link between carotid arteries and reported the differences in IMT values from CCA and ICA. However, both of them were independently associated with CVD risk. Ogata *et al* [133] have shown a link between carotid and coronary atherosclerosis and demonstrated the significant correlation between cIMT and left coronary atherosclerosis. In addition to cIMT, carotid plaque is also considered to be an important, perhaps more powerful biomarker than cIMT to predict CV risk [134, 135]. Carotid plaque is defined as a structure that protrudes into the lumen by at least 0.5 mm or 50% of the surrounding IMT value or an IMT thickness > 1.5 mm. Recently, Sun *et al* [136] presented a study which indicated carotid plaque as a predictor of systematic cardiac events. Furthermore, carotid plaque area, when compared to cIMT, was reported as the stronger predictor of myocardial infarction in women compared to men [137].

Combining both carotid biomarkers (cIMT and plaque burden) and CVD risk predictors (coronary IMT, coronary artery calcium, SYNTAX score), CVD risk factors may improve CVD risk prediction at earlier stages, and this has been validated by several studies in the last few decades [138, 139]. Amato *et al* [140] presented a study to assess CV risk using multiple modalities and concluded that CV risk prediction can be possible with unique modalities and parameters for both carotid and coronary arteries. It also indicated a strong correlation between carotid and coronary IMT. Recently, Polak *et al* [141] presented a study in which the coronary artery calcium and cIMT were highly correlated with each other and when added to Framingham risk scores improved the risk prediction for CHD. Furthermore, a similar group also presented a cohort study considering both black and white populations to assess the relation between cardiovascular risk factors and IMT measured from different segments of the carotid artery. This study demonstrated a high CVD risk factor-based variability in common carotid IMT measurements compared to internal and carotid bulb segments [142]. A very recent study presented by Chung *et al* [143] demonstrated a correlation between carotid plaque score and SYNTAX score to find the relation between carotid and coronary lesions.

2.5.2 The relationship between diabetes and coronary artery disease

The pathogenesis of diabetes-related atherosclerosis involves several general mechanisms. The first relates to metabolic factors including dyslipidemia, hypertension, increase free fatty acids and hyperglycemia from insulin resistance with insulin deficiency—all of these contribute to atherosclerosis. Among other things, hyperglycemia increases the oxidative stress and glycation. This increase the free radicals, and increases the lipid and peroxidation contributing to foam cell formation into arterial walls. Insulin resistance plays a role by contributing to endothelial dysfunction through loss of nitric oxide, an important precursor to atherosclerosis. Diabetes promotes platelet aggregation, which is the result of increased inflammatory response that augments the generation of growth factors and stimulates the proliferation and migration of SMCs, both of which are associated with thrombosis. Diabetes is considered as a prothrombotic state, which can lead to an imbalance in the atherosclerotic lesion and plaque instability. Diabetes-related atherosclerosis increases the risk of CVD, including myocardial infarction.

2.5.3 The link between rheumatic arthritis and cardiovascular disease

RA is a chronic inflammatory disease which is related to the joints (figure 2.12). As inflammation is the major factor responsible for heart disease, RA results in heart-related diseases. Several studies have reported an elevation in the risk of atherosclerosis in patients suffering from RA [144, 145]. Coronary artery disease happens both in patients with and without RA, but the inflammation rate is higher in the former case. Cardiovascular risk increases 1.5–2 fold in RA patients with a 50% higher mortality rate compared to the general population [146, 147]. The growth of mortality and morbidity indicates the need for special attention towards the risk factors of RA, similar to diabetes mellitus, that lead to CVD. In order to assess the risk of CVD it is essential to analyze the risk factors for both RA and CVD. Traditional risk factors are important to understand the progression and pathogenesis of RA, but they do not stratify the CVD risk in RA patients [148]. This poses the requirement to search for newly advanced risk calculators which stratify the risk accurately and predict it well before the occurrence of any severe CV events. In general, to assess the risk associated with CVD, risk calculators such as systematic coronary risk evaluation (SCORE) are being followed, which considers the age and gender of the person [149]. The SCORE provides a considerable risk assessment in the general population, but underestimates the findings in RA patients. To overcome

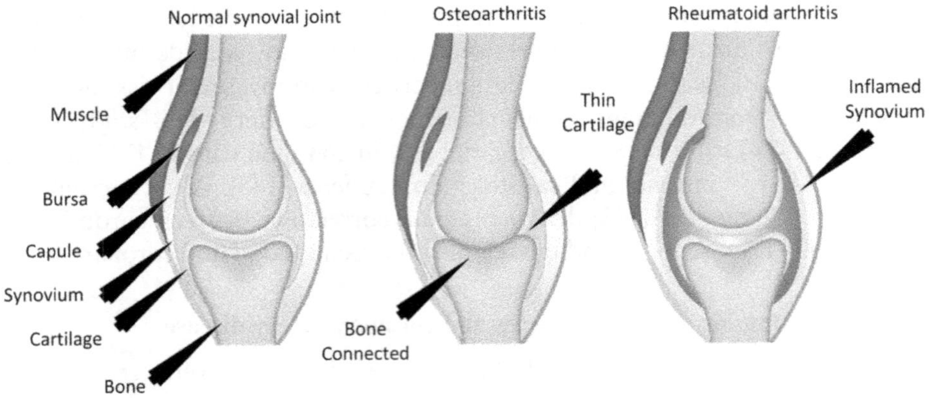

Figure 2.12. Normal joint, oesteoarthitis and rheumatoid arthritis. (Courtesy of AtheroPoint, Roseville, CA, USA.)

this flaw, EULAR guidelines suggest applying a multiplication factor of 1.5 to the SCORE result [150]. This modification was also reported to produce a less accurate risk assessment than it does for the general population. An angiographic approach was also followed to assess the risk associated with CVD in RA patients by Alizade *et al* [151]. The authors compared the SYNTAX score to the serum rheumatoid factor level and based on this comparison stratified the risk into either the low (SYNTAX score < 22) or moderate/high group (SYNTAX score > 22).

The prevalent link between coronary atherosclerosis and RA indicates a major association between coronary calcification and severity of RA (figure 2.13). Giles *et al* [152] reported a higher coronary calcification in RA patients compared to control patients. Furthermore, the study also reported a high prevalence of calcification in men compared to women (a prevalence ratio of 1.19) suffering from rheumatoid arthritis. Coronary calcification was assessed using CT images and the Agatston score. Chung *et al* [153] also predicted the CAC in RA patients and reported a higher value of CAC in RA patients, but this was attributed to high age and triglyceride levels. However, both the cases and controls showed a similar incidence and progression rate for CAC. Recently, Wahlin *et al* [154] also presented a similar study which shows the relation between RA inflammation and CAC.

Carotid IMT and plaque measured using ultrasound are the important predictors of CVD risk in the general population. In addition to CAC, carotid artery parameters have also been used for CV risk prediction in RA patients. Corrales *et al* [156] presented the CVD risk stratification using both CAC and carotid biomarkers such carotid IMT and plaque. Based on the EULAR guidelines, the risk was evaluated using a modified SCORE calculator and the risk was stratified into three groups: low, moderate and high. The authors concluded that carotid ultrasound measurements were more sensitive to atherosclerosis than CAC values. González-Gay *et al* [155] presented a study and reported an algorithm which uses carotid ultrasound to stratify the risk when the mSORE results does not provide accurate risk assessment. Another study presented by Patel *et al* [157] compared the cIMT in RA patients and non-RA patients and reported a high value in the prior

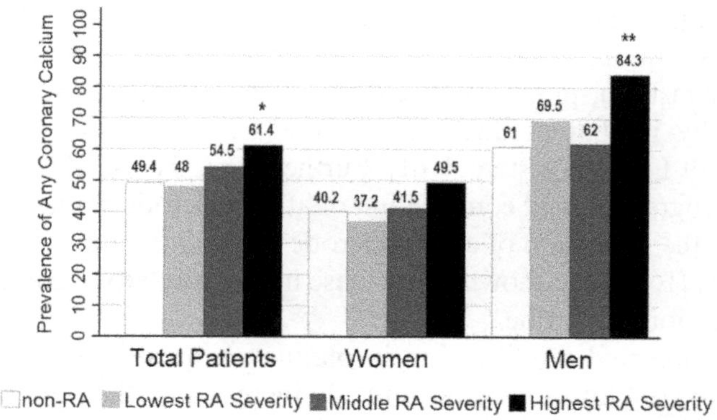

Figure 2.13. Coronary artery calcification in RA patients. (Reprinted with permission from [152]. Courtesy of Biomed Central Ltd.)

case (0.86 versus 0.53 mm). Rheumatoid arthritis is a systematic inflammatory disease, in which erythrocyte sedimentation rate (ESR) and C-reactive protein (CRP) are considered as measures of systematic inflammation. Rincón *et al* [158] presented a study which identified these inflammatory measures such as ESR and CRP in RA patients using carotid IMT and plaque to predict the CV risk. This associated carotid atherosclerosis with RA and showed a linear relation between the increase in cIMT and ESR values.

2.6 Treatments for cardiovascular disease

The increased risk of atherosclerosis driven CVD that leads to mortality and morbidity is a major concern of the world's research community. Cholesterol is a fatty substance which, if deposited within the coronary artery, leads to myocardial infarction and may result in mortality. Cholesterol produces steroid hormones, bile acids and vitamin D, which are essential for the human body. Furthermore, most of the cells have cholesterol as a vital element in their respective cell membrane. The human body takes cholesterol from the food consumed. Also, the liver generates cholesterol within the human body and it is transported to other parts via the bloodstream. Elevated LDL cholesterol level is the primary risk factor that causes atherosclerosis and leads to heart attack. In contrast, HDL cholesterol prevents the deposition of these LDL cholesterols within the coronary arteries. Reducing the LDL cholesterol level also minimizes the risk of CVD. There are various ways to reduce the LDL cholesterol level and prevent CVD, such as maintaining a nutritious low-fat diet, performing daily exercise and avoiding a sedentary lifestyle. In addition to this, proper medications can also minimize the risk of CVD. The main objective of treatment techniques for CVD is to prevent the occurrence of heart events and to improve the quality and lifespan of a person.

Statin therapy is one of the most studied and tested treatments, and is normally used to reduce the cholesterol level in blood and prevent atherosclerosis driven CVD. At present, more than 200 million people, including 30 million people daily, take statins as a preventive measure for CVD [159]. The LDL cholesterol level in blood determines the amount of statins to be used for CVD prevention [160]. Statins are a group of drugs, among which atorvastatin is the most common, followed by simvastatin, pravastatin, pitavastatin and rosuvastatin. Statins prevent the synthesis of cholesterol by inhibiting HMG-CoA reductase in the liver. This lowers the generation rate of LDL cholesterol [161]. Furthermore, the use of statins has multiple effects on the progression and composition of atherosclerotic plaque. Many researchers have shown the regression of atherosclerotic plaque after applying statin therapy.

Ibrahimi *et al* [162] have shown an increase in atherosclerotic plaque echogenicity with the application of atatine.

Zheng *et al* reported modification in plaque composition after statin therapy with a reduction in fibrous tissue and an increase in the calcium level [163]. However, this effect was attributed to the intensity of statin application and the long-term follow-up period. Okumura *et al* [164] presented a study to understand the effect of pitavastatin on lipid present in the blood, and reported a significant reduction in carotid IMT with an overall increase in HDL cholesterol level. Furthermore, the authors also demonstrated that the effect of high-intensity lipid-lowering pitavastatin therapy was strongly associated with patients who are free from diabetes and metabolic syndrome. Taylor *et al* [165] analyzed the effect of two statins, atorvastatin and pravastatin, and reported that the reduction in cIMT is higher for atorvastatin compared to pravastatin over a follow-up period of one year. Earlier use of statins may result in improvement in the cardiac health of patients and lowers the risk of CVD. Recently, Nishiguchi *et al* [166] reported a significant increase in the fibrous-cap thickness when statins, in particular pitavastatin, were used earlier, almost 3 weeks from the baseline, compared to 36 weeks from the baseline (figure 2.14).

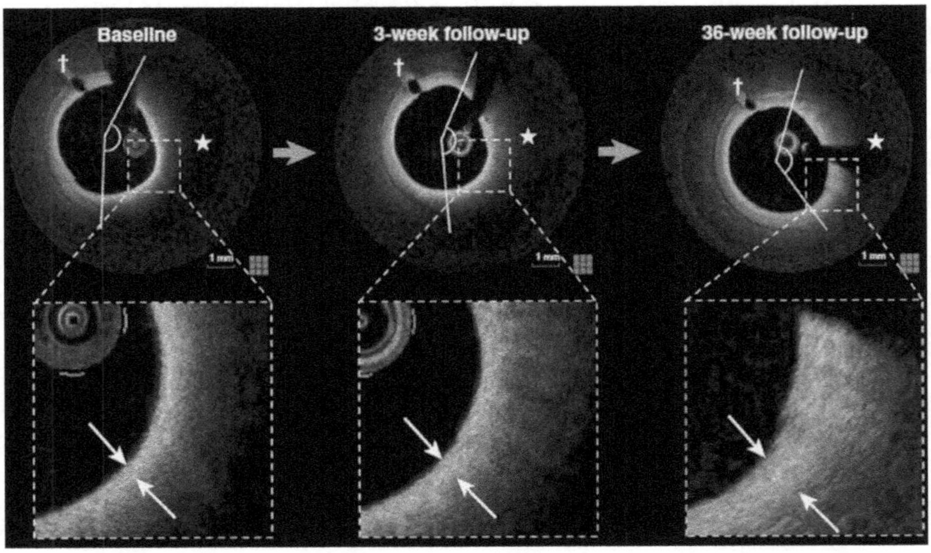

Figure 2.14. Fibrous-cap thickness (white arrows) increased between baseline (130 mm) and 3 week follow-up (160 mm) and between 3 week follow-up and 36 week follow-up (370 mm). (Reprinted with permission from [166]. Copyright Elsevier).

In order to reduce the risk of CVD, the LDL cholesterol level should be less than 70 mg dl^{-1} [167]. Statins alone are not sufficient to achieve this targeted level and hence there is the need to use an additional lipid-lowering agent with statins. Ezetimibe is used to treat hypercholesterolemia, a high cholesterol level in the blood, when diet and exercise are not enough (figure 2.15). In the human body cholesterol can be found in two forms: one which is synthesized endogenously in the liver and another which is absorbed in the intestine [168]. Ezetimibe belongs to the compounds which inhibit the absorption of cholesterol in the small intestine and decrease cholesterol transmission to the liver [169].

This leads to the transfer of blood LDL cholesterol to the liver reducing the cholesterol level in the bloodstream. Adding ezetimibe to statins reduces the blood cholesterol level by 23%–24%, achieving the targeted level of LDL-C < 70 mg dl^{-1} [171, 172]. The use of two or more lipid lowering therapies has been reported to show better reduction in LDL cholesterol levels as well other inflammatory parameters. Furthermore, a significant impact on the plaque composition when compared to single statin therapy has also been reported by many researchers. A similar kind of dual lipid

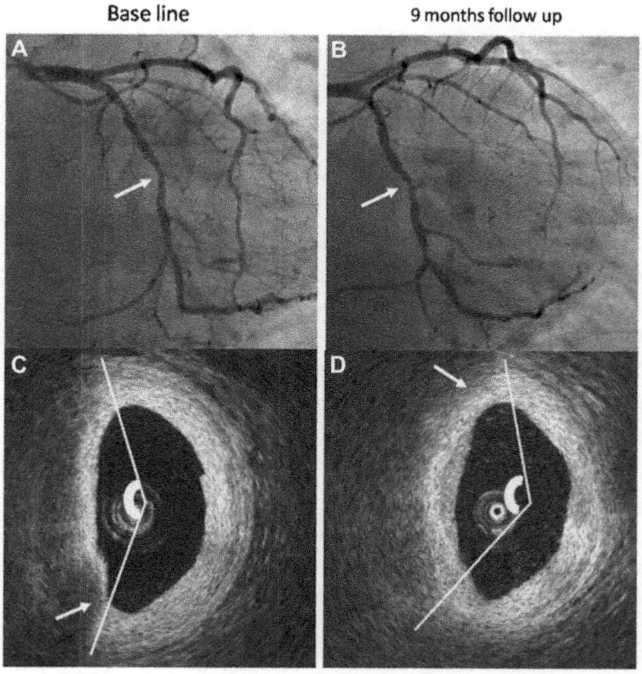

Figure 2.15. Representative case of the ezetimibe *þ* statin group. (A) Angiography showing mild stenosis at the mid portion of the left circumflex artery (white arrow). (B) Angiogram at 9 month follow-up. The lesion was not significantly changed (white arrow). (C) OCT image of the left circumflex artery lesion at baseline. The minimum lumen area site and minimum fibrous-cap thickness site were the same portion. The lumen area, minimum fibrous-cap thickness (white arrow) and angle of lipid (yellow line) were 1.85 cm^2, 0.07 mm and 150.6°, respectively. (D) OCT image of the lesion at 9 month follow-up. The minimum lumen area, minimum fibrous-cap thickness (white arrow) and angle of lipid (yellow line) were 1.40 cm^2, 0.17 mm and 145.1°, respectively. (Reprinted with permission from [170]. Copyright Elsevier.)

therapy using OCT was recently reported by Habara *et al* [170] to evaluate the role of statin therapy alone against the combination fluvastatin and ezetimibe for analysis of fibrous-cap thickness. High-resolution optical coherence tomography (OCT) has the potential to image the various vulnerable plaque components such as thin-cap fibroatheroma, FCT and macrophage infiltration. FCT is considered as a vulnerable plaque that if ruptured may lead to myocardial infarction [5]. The study reported three important findings related to the fluvastatin–ezetimibe combination compared to single fluvastatin: (i) the reduction in LDL cholesterol level, (2) the increase in the FCT and (3) the decrease in lipid angle were significantly larger in the former case compared to the latter. Lee *et al* [173] compared ezetimibe and simvastatin against pravastatin to evaluate the effect on plaque composition. This study reported a significant reduction in LDL-C level and fibro fatty plaque using the dual therapy compared to single pravastatin. In addition, proprotein convertase subtilisin/kexin type 9 (PCSK9) is another enzyme which, when added to statins, reduces the LDL-C level significantly [174].

References

[1] World Health Organization 2017 *Cardiovascular Disease Fact Sheet* www.who.int/mediacentre/factsheets/fs317/en/ (Accessed: 1 July 2018)

[2] American Heart Association and American Stroke Association 2017 *Cardiovascular Disease: A Costly Burden for America. Projections through 2035* https://healthmetrics.heart.org/cardiovascular-disease-a-costly-burden/

[3] Writing G M *et al* 2016 Heart disease and stroke statistics—2016 update: a report from the American Heart Association *Circulation* **133** e38

[4] Mehta V and Tzima E 2016 A turbulent path to plaque formation *Nature* **540** 531

[5] Bentzon J F, Otsuka F, Virmani R and Falk E 2014 Mechanisms of plaque formation and rupture *Circ. Res.* **114** 1852–66

[6] Dodge J T, Brown B G, Bolson E L and Dodge H T 1992 Lumen diameter of normal human coronary arteries. Influence of age, sex, anatomic variation, and left ventricular hypertrophy or dilation. *Circulation* **86** 232–46

[7] Raut B K, Patil V N and Cherian G 2017 Coronary artery dimensions in normal Indians *Indian Heart J.* **69** 512–4

[8] Zhou F-F *et al* 2017 Coronary artery diameter is inversely associated with the severity of coronary lesions in patients undergoing coronary angiography *Cell. Physiol. Biochem.* **43** 1247–57

[9] Kabir C S, Haq M M and Cader F A 2017 Disparity in coronary artery diameter in diabetic and nondiabetic subjects undergoing percutaneous coronary intervention in Bangladesh: a 2-year retrospective analysis *Bangladesh Heart J.* **32** 23–8

[10] Ertan C *et al* 2014 Association of prediabetes with diffuse coronary narrowing and small-vessel disease *J. Cardiol.* **63** 29–34

[11] Tarkin J M *et al* 2016 Imaging atherosclerosis *Circ. Res.* **118** 750–69

[12] Kramer C M and Anderson J D 2007 MRI of atherosclerosis: diagnosis and monitoring therapy *Expert Rev. Cardiovasc. Ther.* **5** 69–80

[13] Sandfort V, Lima J A and Bluemke D A 2015 Noninvasive imaging of atherosclerotic plaque progression *Circ.: Cardiovasc. Imaging* **8** e003316

[14] Saremi F and Achenbach S 2015 Coronary plaque characterization using CT *Am. J. Roentgenol.* **204** W249–60
[15] Choe Y H 2005 Noninvasive imaging of atherosclerotic plaques using MRI and CT. *Korean Circ. J.* **35** 1–14
[16] Huang D *et al* 1991 Optical coherence tomography *Science* **254** 1178
[17] Tearney G J *et al* 1997 *In vivo* endoscopic optical biopsy with optical coherence tomography *Science* **276** 2037–9
[18] Bezerra H G, Costa M A, Guagliumi G, Rollins A M and Simon D I 2009 Intracoronary optical coherence tomography: a comprehensive review: clinical and research applications *JACC: Cardiovasc. Interv.* **2** 1035–46
[19] Li J *et al* 2014 Integrated IVUS-OCT for real-time imaging of coronary atherosclerosis *JACC: Cardiovasc. Imaging* **7** 101–3
[20] van der Sijde J N *et al* 2017 Safety of optical coherence tomography in daily practice: a comparison with intravascular ultrasound *Eur. Heart J. Cardiovasc. Imaging* **18** 467–74
[21] Jolly S, Vittinghoff E, Chattopadhyay A and Bibbins-Domingo K 2010 Higher cardiovascular disease prevalence and mortality among younger blacks compared to whites *Am. J. Med.* **123** 811–8
[22] Reinier K *et al* 2015 Distinctive clinical profile of blacks versus whites presenting with sudden cardiac arrest *Circulation* **132** 380–7
[23] Alenghat F J 2016 The prevalence of atherosclerosis in those with inflammatory connective tissue disease by race, age, and traditional risk factors *Sci. Rep.* **6** 20303
[24] Hamer M *et al* 2015 Progression of cardiovascular risk factors in black Africans: 3 year follow up of the SABPA cohort study *Atherosclerosis* **238** 52–4
[25] Ferdinand K C 2016 Cardiovascular risk reduction in African Americans: current concepts and controversies *Glob. Cardiol. Sci. Pract.* **2016** e201602
[26] Sundquist J, Winkleby M A and Pudaric S 2001 Cardiovascular disease risk factors among older black, Mexican American, and white women and men: an analysis of NHANES III, 1988–1994 *J. Am. Geriatr. Soc.* **49** 109–16
[27] Gaziano T A, Bitton A, Anand S, Abrahams-Gessel S and Murphy A 2010 Growing epidemic of coronary heart disease in low-and middle-income countries *Curr. Probl. Cardiol.* **35** 72–115
[28] Benjamin E J *et al* 2017 Heart disease and stroke statistics—2017 update: a report from the American Heart Association *Circulation* **135** e146–603
[29] Wilkins E *et al* 2017 *European Cardiovascular Disease Statistics 2017* (Brussels: European Heart Network)
[30] World Health Organization *WHO Cardiovascular disease* www.who.int/mediacentre/factsheets/fs317/en/ (Accessed: October 1, 2017)
[31] Gupta R, Mohan I and Narula J 2016 Trends in coronary heart disease epidemiology in India *Ann. Glob. Health* **82** 307–15
[32] Chauhan S and Aeri B T 2015 The rising incidence of cardiovascular diseases in India: assessing its economic impact *J. Prev. Cardiol.* **4** 735–40
[33] Prabhakaran D, Jeemon P and Roy A 2016 Cardiovascular diseases in India *Circulation* **133** 1605–20
[34] Gupta R, Guptha S, Sharma K K, Gupta A and Deedwania P 2012 Regional variations in cardiovascular risk factors in India: India Heart Watch *World J. Cardiol.* **4** 112

[35] Office of the Registrar General & Census Commissioner, India: Medical Certification of Cause of Death 2015 (New Delhi: Ministry of Home Affairs)
[36] Bhatnagar A 2017 Environmental determinants of cardiovascular disease *Circ. Res.* **121** 162–80
[37] Martino T A and Sole M J 2009 Molecular time: an often overlooked dimension to cardiovascular disease *Circ. Res.* **105** 1047–61
[38] Martino T *et al* 2004 Day/night rhythms in gene expression of the normal murine heart *J. Mol. Med.* **82** 256–64
[39] Muller J E *et al* 1985 Circadian variation in the frequency of onset of acute myocardial infarction *N. Engl. J. Med.* **313** 1315–22
[40] Cohen M C, Rohtla K M, Lavery C E, Muller J E and Mittleman M A 1997 Meta-analysis of the morning excess of acute myocardial infarction and sudden cardiac death *Am. J. Cardiol.* **79** 1512–6
[41] Tanaka H *et al* 2000 Seasonal variation in mortality from ischemic heart disease and cerebrovascular disease in Okinawa and Osaka: the possible role of air temperature *J. Epidemiol.* **10** 392–8
[42] Barnett A G, De Looper M and Fraser J F 2008 The seasonality in heart failure deaths and total cardiovascular deaths *Aust. N. Z. J. Public Health* **32** 408–13
[43] Dai M-X *et al* 2014 The impact of intermittent and repetitive cold stress exposure on endoplasmic reticulum stress and instability of atherosclerotic plaques *Cell. Physiol. Biochem.* **34** 393–404
[44] Spencer F A, Goldberg R J, Becker R C and Gore J M 1998 Seasonal distribution of acute myocardial infarction in the second National Registry of Myocardial Infarction *J. Am. Coll. Cardiol.* **31** 1226–33
[45] Pell J and Cobbe S 1999 Seasonal variations in coronary heart disease *QJM* **92** 689–96
[46] Logan W 1953 Mortality in the London fog incident, 1952 *Lancet* **261** 336–8
[47] Nemery B, Hoet P H and Nemmar A 2001 The Meuse Valley fog of 1930: an air pollution disaster *Lancet* **357** 704–8
[48] Lim S S *et al* 2012 A comparative risk assessment of burden of disease and injury attributable to 67 risk factors and risk factor clusters in 21 regions, 1990–2010: a systematic analysis for the Global Burden of Disease Study 2010 *Lancet* **380** 2224–60
[49] Pope C A *et al* 1995 Particulate air pollution as a predictor of mortality in a prospective study of US adults *Am. J. Respir. Crit. Care Med.* **151** 669–74
[50] Dockery D W *et al* 1993 An association between air pollution and mortality in six US cities *N. Engl. J. Med.* **329** 1753–9
[51] Pope C A *et al* 2004 Cardiovascular mortality and long-term exposure to particulate air pollution: epidemiological evidence of general pathophysiological pathways of disease *Circulation* **109** 71–7
[52] Gakidou E *et al* 2016 Global, regional, and national comparative risk assessment of 84 behavioural, environmental and occupational, and metabolic risks or clusters of risks, 1990–2016: a systematic analysis for the Global Burden of Disease Study *Lancet* **390** 1345–422
[53] Atkinson R, Kang S, Anderson H, Mills I and Walton H 2014 Epidemiological time series studies of PM2. 5 and daily mortality and hospital admissions: a systematic review and meta-analysis *Thorax* **69** 660–5

[54] Brook R D *et al* 2010 Particulate matter air pollution and cardiovascular disease: an update to the scientific statement from the American Heart Association *Circulation* **121** 2331–78
[55] Epa U 2004 *Air Quality Criteria for Particulate Matter* (Research Triangle Park, NC: US Environmental Protection Agency)
[56] Daigle C C *et al* 2003 Ultrafine particle deposition in humans during rest and exercise *Inhal. Toxicol.* **15** 539–52
[57] Lee B-J, Kim B and Lee K 2014 Air pollution exposure and cardiovascular disease *Toxicol. Res.* **30** 71
[58] Brook R D *et al* 2004 Air pollution and cardiovascular disease *Circulation* **109** 2655
[59] Dominici F *et al* 2006 Fine particulate air pollution and hospital admission for cardiovascular and respiratory diseases *J. Am. Med. Assoc.* **295** 1127–34
[60] von Klot S *et al* 2005 Ambient air pollution is associated with increased risk of hospital cardiac readmissions of myocardial infarction survivors in five European cities *Circulation* **112** 3073–9
[61] Pope C A III and Dockery D W 2006 Health effects of fine particulate air pollution: lines that connect *J. Air Waste Manag. Assoc.* **56** 709–42
[62] Cesaroni G *et al* 2014 Long term exposure to ambient air pollution and incidence of acute coronary events: prospective cohort study and meta-analysis in 11 European cohorts from the ESCAPE Project *Brit. Med. J.* **348** f7412
[63] Goovaerts P, Auchincloss A and Diez-Roux A 2006 Performance comparison of spatial and space–time interpolation techniques for prediction of air pollutant concentrations in the Los Angeles area *Society for Mathematical Geology XIth Int. Congress: 2006* pp S13–11
[64] Newby D E *et al* 2014 Expert position paper on air pollution and cardiovascular disease *Eur. Heart J.* **36** 83–93
[65] Daly R M and Ebeling P R 2010 Is excess calcium harmful to health? *Nutrients* **2** 505–22
[66] Felsenfeld A J and Levine B S 2006 Milk alkali syndrome and the dynamics of calcium homeostasis *Clin. J. Am. Soc. Nephrol.* **1** 641–54
[67] Anderson J J *et al* 2016 Calcium intake from diet and supplements and the risk of coronary artery calcification and its progression among older adults: 10 year follow up of the Multi Ethnic Study of Atherosclerosis (MESA) *J. Am. Heart Assoc.* **5** e003815
[68] Estruch R *et al* 2006 Effects of a Mediterranean-style diet on cardiovascular risk factors: a randomized trial *Ann. Intern. Med.* **145** 1–11
[69] Oikonomou E *et al* 2017 Dietary consumption of olive oil and cardiovascular outcome in patients with coronary artery disease *J. Am. Coll. Cardiol.* **69** 146
[70] Kolber M R and Scrimshaw C 2014 Family history of cardiovascular disease *Can. Fam. Phys.* **60** 1016
[71] Friedlander Y *et al* 1998 Family history as a risk factor for primary cardiac arrest *Circulation* **97** 155–60
[72] Barrett-Connor E and Khaw K 1984 Family history of heart attack as an independent predictor of death due to cardiovascular disease *Circulation* **69** 1065–9
[73] Schildkraut J M, Myers R H, Cupples L A, Kiely D K and Kannel W B 1989 Coronary risk associated with age and sex of parental heart disease in the Framingham Study *Am. J. Cardiol.* **64** 555–9
[74] Nilsson P M, Nilsson J A and Berglund G 2004 Family burden of cardiovascular mortality: risk implications for offspring in a national register linkage study based upon the Malmö Preventive Project *J. Intern. Med.* **255** 229–35

[75] McCusker M E, Yoon P W, Gwinn M, Malarcher A M, Neff L and Khoury M J 2004 Family history of heart disease and cardiovascular disease risk-reducing behaviors *Genet. Med.* **6** 153
[76] Ciruzzi M *et al* 1997 Frequency of family history of acute myocardial infarction in patients with acute myocardial infarction. Argentine FRICAS (Factores de Riesgo Coronario en America del Sur) Investigators *Am. J. Cardiol.* **80** 122–7
[77] Silberberg J S, Wlodarczyk J, Fryer J, Robertson R and Hensley M J 1998 Risk associated with various definitions of family history of coronary heart disease. The Newcastle Family History Study II *Am. J. Epidemiol.* **147** 1133–9
[78] Nasir K *et al* 2004 Coronary artery calcification and family history of premature coronary heart disease: sibling history is more strongly associated than parental history *Circulation* **110** 2150–6
[79] Taylor A J, Bindeman J, Bhattarai S, Feuerstein I M and O'Malley P G 2004 Subclinical calcified atherosclerosis in men and its association with a family history of premature coronary heart disease in first- and second-degree relatives *Prev. Cardiol.* **7** 163–7
[80] Juonala M, Viikari J S A, Räsänen L, Helenius H, Pietikäinen M and Raitakari O T 2006 Young adults with family history of coronary heart disease have increased arterial vulnerability to metabolic risk factors *Arterioscler. Thromb. Vasc. Biol.* **26** 1376
[81] Wang T J *et al* 2003 Carotid intima–media thickness is associated with premature parental coronary heart disease: the Framingham Heart Study *Circulation* **108** 572–6
[82] Libby P 2003 Vascular biology of atherosclerosis: overview and state of the art *Am. J. Cardiol.* **91** 3–6
[83] Ross R 1999 Atherosclerosis—an inflammatory disease *N. Engl. J. Med.* **340** 115–26
[84] Libby P 2012 History of discovery: inflammation in atherosclerosis *Arterioscler. Thromb. Vasc. Biol.* **32** 2045–51
[85] Riccioni G *et al* 2003 Atherosclerotic plaque formation and risk factors *Int. J. Immunopathol. Pharmacol.* **16** 25–31
[86] Fok P-W 2011 Growth of necrotic cores in atherosclerotic plaque *Math. Med. Biol.* **29** 301–27
[87] Pathogenesis of Atherosclerosis [http://sphweb.bumc.bu.edu/otlt/mph-modules/ph/ph709_heart/ph709_heart3.html]
[88] Libby P, Ridker P M and Hansson G K 2011 Progress and challenges in translating the biology of atherosclerosis *Nature* **473** 317–25
[89] Catapano A, Reiner Z and De Backer G European Atherosclerosis Society (EAS) 2011 ESC/EAS Guidelines for the management of dyslipidaemias: the Task Force for the management of dyslipidaemias of the European Society of Cardiology (ESC) and the European Atherosclerosis Society (EAS) *Atherosclerosis* **217** 3–46
[90] Kataoka Y *et al* 2015 Plaque microstructures in patients with coronary artery disease who achieved very low low-density lipoprotein cholesterol levels *Atherosclerosis* **242** 490–5
[91] Rogacev K S *et al* 2012 CD14++ CD16+ monocytes independently predict cardiovascular events: a cohort study of 951 patients referred for elective coronary angiography *J. Am. Coll. Cardiol.* **60** 1512–20
[92] Libby P and Clinton S K 1993 The role of macrophages in atherogenesis *Curr. Opin. Lipidol.* **4** 355–63
[93] Moore K J and Tabas I 2011 Macrophages in the pathogenesis of atherosclerosis *Cell* **145** 341–55

[94] Ghattas A, Griffiths H R, Devitt A, Lip G Y and Shantsila E 2013 Monocytes in coronary artery disease and atherosclerosis: where are we now? *J. Am. Coll. Cardiol.* **62** 1541–51
[95] Dinarello C A 2007 Historical insights into cytokines *Eur. J. Immunol.* **37**
[96] Zhang J-M and An J 2007 Cytokines, inflammation and pain *Int. Anesthesiol. Clin.* **45** 27
[97] Ait-Oufella H, Taleb S, Mallat Z and Tedgui A 2011 Recent advances on the role of cytokines in atherosclerosis *Arterioscler. Thromb. Vasc. Biol.* **31** 969–79
[98] Geissmann F, Manz M G, Jung S, Sieweke M H, Merad M and Ley K 2010 Development of monocytes, macrophages, and dendritic cells *Science* **327** 656–61
[99] Lusis A J 2000 Atherosclerosis *Nature* **407** 233–41
[100] Doran A C, Meller N and McNamara C A 2008 Role of smooth muscle cells in the initiation and early progression of atherosclerosis *Arterioscler. Thromb. Vasc. Biol.* **28** 812–9
[101] Pedicino D, Giglio A F, Galiffa V A, Trotta F and Liuzzo G 2012 Type 2 diabetes, immunity and cardiovascular risk: a complex relationship *Pathophysiology and Complications of Diabetes Mellitus* (London: InTech)
[102] Zaccardi F, Webb D R, Yates T and Davies M J 2015 Pathophysiology of type 1 and type 2 diabetes mellitus: a 90-year perspective *Postgrad. Med. J.* **92** 63–9
[103] Weyand C M and Goronzy J J 1997 Pathogenesis of rheumatoid arthritis *Med. Clin.* **81** 29–55
[104] Cooles F A and Isaacs J D 2011 Pathophysiology of rheumatoid arthritis *Curr. Opin. Rheumatol.* **23** 233–40
[105] Firestein G S 2003 Evolving concepts of rheumatoid arthritis *Nature* **423** 356
[106] Choy E 2012 Understanding the dynamics: pathways involved in the pathogenesis of rheumatoid arthritis *Rheumatology* **51** v3–11
[107] McInnes I B and Schett G 2011 The pathogenesis of rheumatoid arthritis *N. Engl. J. Med.* **365** 2205–19
[108] Smolen J S, Aletaha D, Koeller M, Weisman M H and Emery P 2007 New therapies for treatment of rheumatoid arthritis *Lancet* **370** 1861–74
[109] Smolen J S and Steiner G 2003 Therapeutic strategies for rheumatoid arthritis *Nature Rev. Drug Discov.* **2** 473
[110] Coronary Anatomy and Blood Flow [http://cvphysiology.com/Blood%20Flow/BF001]
[111] Kim S G *et al* 2004 The importance of gender on coronary artery size: *in vivo* assessment by intravascular ultrasound *Clin. Cardiol.* **27** 291–4
[112] Sheifer S E *et al* 2000 Sex differences in coronary artery size assessed by intravascular ultrasound *Am. Heart J.* **139** 649–52
[113] Patel A K *et al* 2016 A review on atherosclerotic biology, wall stiffness, physics of elasticity, and its ultrasound-based measurement *Curr. Atheroscler. Rep.* **18** 83
[114] Dionne A *et al* 2015 Coronary wall structural changes in patients with Kawasaki disease: new insights from optical coherence tomography (OCT) *J. Am. Heart Assoc.* **4** e001939
[115] Falk E 2006 Pathogenesis of atherosclerosis *J. Am. Coll. Cardiol.* **47** C7–12
[116] Unal G, Bucher S, Carlier S, Slabaugh G, Fang T and Tanaka K 2008 Shape-driven segmentation of the arterial wall in intravascular ultrasound images *IEEE Trans. Inf. Technol. Biomed.* **12** 335–47
[117] Stuber M and Weiss R G 2007 Coronary magnetic resonance angiography *J. Magn. Reson. Imaging* **26** 219–34

[118] Gerretsen S C *et al* 2007 Magnetic resonance imaging of the coronary arteries *Cardiovasc. J. Africa* **18** 248–59
[119] Eckert J, Schmidt M, Magedanz A, Voigtländer T and Schmermund A 2015 Coronary CT angiography in managing atherosclerosis *Int. J. Mol. Sci.* **16** 3740–56
[120] Bauer R W, Thilo C, Chiaramida S A, Vogl T J, Costello P and Schoepf U J 2009 Noncalcified atherosclerotic plaque burden at coronary CT angiography: a better predictor of ischemia at stress myocardial perfusion imaging than calcium score and stenosis severity *Am. J. Roentgenol.* **193** 410–8
[121] Oberoi S *et al* 2014 Reproducibility of noncalcified coronary artery plaque burden quantification from coronary CT angiography across different image analysis platforms *Am. J. Roentgenol.* **202** W43–9
[122] Agatston A S, Janowitz W R, Hildner F J, Zusmer N R, Viamonte M and Detrano R 1990 Quantification of coronary artery calcium using ultrafast computed tomography *J. Am. Coll. Cardiol.* **15** 827–32
[123] Regar E *et al* 2016 Identifying stable coronary plaques with OCT technology *Contin. Cardiol. Educ.* **2** 77–88
[124] Kume T *et al* 2005 Assessment of coronary intima–media thickness by optical coherence tomography *Circ. J.* **69** 903–7
[125] Araki T *et al* 2016 A new method for IVUS-based coronary artery disease risk stratification: a link between coronary and carotid ultrasound plaque burdens *Comput. Methods Programs Biomed.* **124** 161–79
[126] Terashima M, Kaneda H and Suzuki T 2012 The role of optical coherence tomography in coronary intervention *Korean J. Intern. Med.* **27** 1
[127] Prati F *et al* 2009 Expert review document on methodology, terminology, and clinical applications of optical coherence tomography: physical principles, methodology of image acquisition, and clinical application for assessment of coronary arteries and atherosclerosis *Eur. Heart J.* **31** 401–15
[128] Craven T E *et al* 1990 Evaluation of the associations between carotid artery atherosclerosis and coronary artery stenosis. A case-control study *Circulation* **82** 1230–42
[129] Salonen J T and Salonen R 1991 Ultrasonographically assessed carotid morphology and the risk of coronary heart disease *Arterioscler. Thromb. Vasc. Biol.* **11** 1245–49
[130] Chambless L E *et al* 1997 Association of coronary heart disease incidence with carotid arterial wall thickness and major risk factors: the Atherosclerosis Risk in Communities (ARIC) Study, 1987–1993 *Am. J. Epidemiol.* **146** 483–94
[131] O'Leary D H, Polak J F, Kronmal R A, Manolio T A, Burke G L and Wolfson S K Jr 1999 Carotid-artery intima and media thickness as a risk factor for myocardial infarction and stroke in older adults *N. Engl. J. Med.* **340** 14–22
[132] Polak J F *et al* 2010 Associations of carotid artery intima–media thickness (IMT) with risk factors and prevalent cardiovascular disease *J. Ultrasound Med.* **29** 1759–68
[133] Ogata T, Yasaka M, Yamagishi M, Seguchi O, Nagatsuka K and Minematsu K 2005 Atherosclerosis found on carotid ultrasonography is associated with atherosclerosis on coronary intravascular ultrasonography *J. Ultrasound Med.* **24** 469–74
[134] Park T H 2016 Evaluation of carotid plaque using ultrasound imaging *J. Cardiovasc. Ultrasound* **24** 91–5
[135] Naqvi T Z and Lee M-S 2014 Carotid intima–media thickness and plaque in cardiovascular risk assessment *JACC: Cardiovasc. Imaging* **7** 1025–38

[136] Sun J *et al* 2017 Carotid plaque lipid content and fibrous cap status predict systemic CV outcomes: the MRI substudy in AIM-HIGH *JACC: Cardiovasc. Imaging* **10** 241–9

[137] Johnsen S H *et al* 2007 Carotid atherosclerosis is a stronger predictor of myocardial infarction in women than in men: a 6-year follow-up study of 6226 persons: the Tromsø Study *Stroke* **38** 2873–80

[138] Polak J F, Pencina M J, Herrington D and O'leary D H 2011 Associations of edge-detected and manual-traced common carotid intima–media thickness measurements with Framingham risk factors *Stroke* **42** 1912–6

[139] Gepner A D *et al* 2017 Comparison of carotid plaque score and coronary artery calcium score for predicting cardiovascular disease events: the Multi Ethnic Study of Atherosclerosis *J. Am. Heart Assoc.* **6** e005179

[140] Amato M *et al* 2007 Carotid intima–media thickness by B-mode ultrasound as surrogate of coronary atherosclerosis: correlation with quantitative coronary angiography and coronary intravascular ultrasound findings *Eur. Heart J.* **28** 2094–101

[141] Polak J F, Szklo M and O'Leary D H 2017 Carotid intima–media thickness score, positive coronary artery calcium score, and incident coronary heart disease: The Multi Ethnic Study of Atherosclerosis *J. Am. Heart Assoc.* **6** e004612

[142] Polak J F *et al* 2010 Segment-specific associations of carotid intima–media thickness with cardiovascular risk factors: the Coronary Artery Risk Development in Young Adults (CARDIA) study *Stroke* **41** 9–15

[143] Chung J-W 2017 Association between carotid artery plaque score and SYNTAX score in coronary artery disease patients *Gen. Med.* **5** 301

[144] Gonzalez-Juanatey C, Llorca J, Testa A, Revuelta J, Garcia-Porrua C and Gonzalez-Gay M A 2003 Increased prevalence of severe subclinical atherosclerotic findings in long-term treated rheumatoid arthritis patients without clinically evident atherosclerotic disease *Medicine* **82** 407–13

[145] Klingenberg R and Lüscher T F 2015 Rheumatoid arthritis and coronary atherosclerosis: two cousins engaging in a dangerous liaison *Eur. Heart J.* **36** 3423–5

[146] Liao K P 2017 Cardiovascular disease in patients with rheumatoid arthritis *Trends Cardiovasc. Med.* **27** 136–40

[147] John H and Kitas G 2012 Inflammatory arthritis as a novel risk factor for cardiovascular disease *Eur. J. Intern. Med.* **23** 575–9

[148] Del Rincón I, Williams K, Stern M P, Freeman G L and Escalante A 2001 High incidence of cardiovascular events in a rheumatoid arthritis cohort not explained by traditional cardiac risk factors *Arthritis Rheumatol.* **44** 2737–45

[149] Conroy R *et al* 2003 Estimation of ten-year risk of fatal cardiovascular disease in Europe: the SCORE project *Eur. Heart J.* **24** 987–1003

[150] Agca R *et al* 2017 EULAR recommendations for cardiovascular disease risk management in patients with rheumatoid arthritis and other forms of inflammatory joint disorders: 2015/2016 update *Ann. Rheum. Dis.* **76** 17–28

[151] Alizade E *et al* 2015 The relationship between rheumatoid factor levels and coronary artery lesion complexity and severity in patients with stable coronary artery disease *Postępy Kardiol. Interwencyjnej (Adv. Interv. Cardiol.)* **11** 26

[152] Giles J T *et al* 2009 Coronary arterial calcification in rheumatoid arthritis: comparison with the Multi-Ethnic Study of Atherosclerosis *Arthritis Res. Ther.* **11** R36

[153] Chung C P *et al* 2013 Progression of coronary artery atherosclerosis in rheumatoid arthritis: comparison with participants from the Multi-Ethnic Study of Atherosclerosis *Arthritis Res. Ther.* **15** R134

[154] Wahlin B, Meedt T, Jonsson F, Henein M Y and Wållberg-Jonsson S 2016 Coronary artery calcification is related to inflammation in rheumatoid arthritis: a long-term follow-up study *BioMed Res. Int.* **2016** 1261582

[155] González-Gay M A, González-Juanatey C and Llorca J 2012 Carotid ultrasound in the cardiovascular risk stratification of patients with rheumatoid arthritis: when and for whom? *Ann. Rheum. Dis.* **71** 796–8

[156] Corrales A *et al* 2013 Cardiovascular risk stratification in rheumatic diseases: carotid ultrasound is more sensitive than Coronary Artery Calcification Score to detect subclinical atherosclerosis in patients with rheumatoid arthritis *Ann. Rheum. Dis.* **72** 1764–70

[157] Patel S *et al* 2017 A study of carotid intimomedial thickness as a primary marker of atherosclerosis in patients with rheumatoid arthritis *Int. Cardiovasc. Forum J.* **9** 31–5

[158] del Rincón I, Williams K, Stern M P, Freeman G L, O'leary D H and Escalante A 2003 Association between carotid atherosclerosis and markers of inflammation in rheumatoid arthritis patients and healthy subjects *Arthritis Rheumatol.* **48** 1833–40

[159] Blaha M J and Martin S S 2013 How do statins work?: changing paradigms with implications for statin allocation *J. Am. Coll. Cardiol.* **62** 2382–91

[160] Expert Panel on Detection, Evaluation, and Treatment of High Blood Cholesterol in Adults 2001 Executive summary of the Third Report of the National Cholesterol Education Program (NCEP) expert panel on detection, evaluation, and treatment of high blood cholesterol in adults (Adult Treatment Panel III) *J. Am. Med. Assoc.* **285** 2486

[161] Hill J S and Qiu G 2008 Understanding how statins work: the path to better treatments for heart disease and more *Future Cardiol.* **4** 5–8

[162] Ibrahimi P, Jashari F, Bajraktari G, Wester P and Henein M Y 2015 Ultrasound assessment of carotid plaque echogenicity response to statin therapy: a systematic review and meta-analysis *Int. J. Mol. Sci.* **16** 10734–47

[163] Zheng G, Li Y, Huang H, Wang J, Hirayama A and Lin J 2015 The effect of statin therapy on coronary plaque composition using virtual histology intravascular ultrasound: a meta-analysis *PLoS One* **10** e0133433

[164] Okumura K *et al* 2015 High HDL cholesterol level after treatment with pitavastatin is an important factor for regression in carotid intima–media thickness *Heart Vessels* **30** 154–61

[165] Taylor A J, Kent S M, Flaherty P J, Coyle L C, Markwood T T and Vernalis M N 2002 ARBITER: arterial biology for the investigation of the treatment effects of reducing cholesterol: a randomized trial comparing the effects of atorvastatin and pravastatin on carotid intima medial thickness *Circulation* **106** 2055–60

[166] Nishiguchi T *et al* 2017 Effect of early pitavastatin therapy on coronary fibrous-cap thickness assessed by optical coherence tomography in patients with acute coronary syndrome: The ESCORT Study *JACC: Cardiovasc. Imaging* **11** 829–38

[167] Grundy S M *et al* 2004 Implications of recent clinical trials for the national cholesterol education program adult treatment panel III guidelines *Circulation* **110** 227–39

[168] Silbernagel G, Baumgartner I and März W 2015 Ezetimibe plus a statin after acute coronary syndromes *N. Engl. J. Med.* **373** 1473

[169] Patel J, Sheehan V and Gurk-Turner C 2003 Ezetimibe (Zetia): a new type of lipid-lowering agent *Baylor University Medical Center Proc.: 2003* (London: Taylor and Francis) pp 354–8

[170] Habara M *et al* 2014 Impact on optical coherence tomographic coronary findings of fluvastatin alone versus fluvastatin+ ezetimibe *Am. J. Cardiol.* **113** 580–7
[171] Ballantyne C M, Blazing M A, King T R, Brady W E and Palmisano J 2004 Efficacy and safety of ezetimibe co-administered with simvastatin compared with atorvastatin in adults with hypercholesterolemia *Am. J. Cardiol.* **93** 1487–94
[172] Morrone D *et al* 2012 Lipid-altering efficacy of ezetimibe plus statin and statin monotherapy and identification of factors associated with treatment response: a pooled analysis of over 21,000 subjects from 27 clinical trials *Atherosclerosis* **223** 251–61
[173] Lee J-H *et al* 2016 Early effects of intensive lipid-lowering treatment on plaque characteristics assessed by virtual histology intravascular ultrasound *Yonsei Med. J.* **57** 1087–94
[174] Robinson J G, Heistad D D and Fox K A 2015 Atherosclerosis stabilization with PCSK-9 inhibition: An evolving concept for cardiovascular prevention *Atherosclerosis* **243** 593–7

IOP Publishing

Vascular and Intravascular Imaging Trends, Analysis, and Challenges, Volume 2
Plaque characterization
Petia Radeva and Jasjit S Suri

Chapter 3

Wall quantification and tissue characterization of the coronary artery

Ankush D Jamthikar, Alberto Boi, Luca Saba, Deep Gupta, John R Laird, N N Khanna and Jasjit S Suri

3.1 Introduction

The previous chapter discusses the biology and genetics of atherosclerosis, which is the main type of cardiovascular disease (CVD) and the prevalent cause of mortality around the world. The formation of lipid-rich, fibrous and calcified plaque in the vessel wall causes atherosclerosis and hence in order to predict the risk at earlier stages, it is essential to characterize and quantify these plaque components. Among the different plaque types, thin-cap fibroatheroma, fibrous cap (FC) thickness, a necrotic core and macrophage infiltration in the FC are considered as 'vulnerable plaque' components that determine the severity of the targeted lesion in the coronary vessel wall [1, 2]. Each of these plaque components has unique acoustic, optical and texture features that distinguish it from the others. Furthermore, they provide different vital information when visualized using medical imaging modalities such as computer (CT), magnetic resonance imaging (MRI), ultrasound and optical coherence tomography (OCT).

This chapter discusses the physics of image acquisition using different imaging modalities, followed by tissue characterization using three paradigms based on (i) optical feature measurement methodologies, (ii) machine-learning algorithms and (iii) deep-learning techniques. Quantification of vulnerable plaque components and risk stratification using the above mentioned paradigms is also discussed in the latter part of this chapter.

doi:10.1088/2053-2563/ab0820ch3

3.2 Physics of image acquisition

3.2.1 Image acquisition using optical coherence tomography

OCT is the light-based counterpart of intravascular ultrasound (IVUS) which provides cross-sectional visualization of the targeted tissue [3]. OCT employs a near infrared light source in the 1250–1350 nm spectral range to capture back-reflections from various vessel wall layers [4]. Unlike ultrasound, it is challenging to detect the light waves because of their high-speed ($3 \times 10^8 \text{m s}^{-1}$) and hence OCT uses the interferometry principle to visualize and quantify the optical echoes coming from tissue layers at varying depths.

Low-coherence interferometry is a dominant technique used by OCT to capture the magnitude and echo time lag of the backscattered signal. Figure 3.1 depicts an overview of an interferometer-principle-based frequency domain (FD)-OCT system. A light wave from the broadband source passes through the beam splitter, where it is divided into two lights beams, one for the reference arm and the other for the measurement arm of the catheter [5, 6]. The catheter focuses the light wave to the target tissue and scans for the backscatter signals from different depths. Another arm of the interferometer contains the reference mirror at a fixed distance. The low-coherence interferometry principle correlates the light beam backscattered from the reference mirror and that from the targeted tissue. This results in the generation of interference patterns with intensity variation at the optical beam splitter, which is further detected by the photodetector [7].

Primarily, OCT has two methods for imaging, namely time-domain (TD-)OCT and FD-OCT [8]. TD-OCT employs a single frequency light source with a movable reference mirror arrangement for acquisition of backscattered signals at various depths [9]. The TD-OCT method has two major limitations: (i) the image acquisition rate is low due to the mechanical arrangement of the reference mirror and (2) the occlusion balloon catheter technique is employed to flush the blood, which leads to risks such as to balloon injury [10, 11].

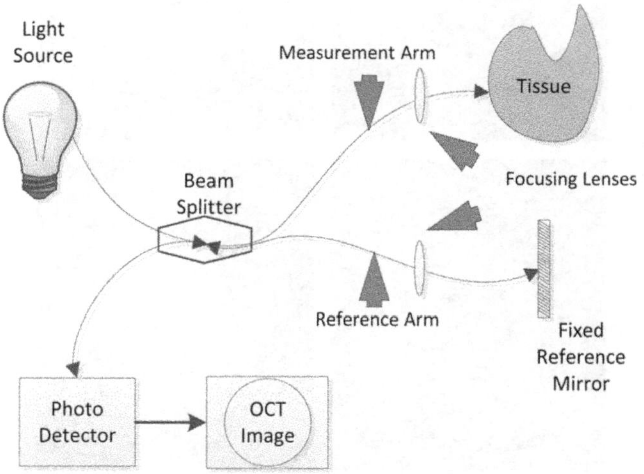

Figure 3.1. Physical overview of an FD-OCT system with a fixed reference mirror.

OCT is capable of generating the cross-sectional 2D images of the target lesion. The amplitude scan (A-scan) is the elementary method which captures the axial back-scatter intensity variation from the targeted tissue. The intensity variation depends on the refraction index of the tissue layer. In order to obtain a complete cross-sectional view (figure 3.2) of the artery, successive axial A-scans are required. The primary advantage of using OCT over other imaging modalities such as CT and IVUS is its ability to provide high-resolution images. The axial and lateral resolutions of OCT are 15–20 μm and 20–40 μm, respectively, which is far better than the IVUS resolutions (150–300 μm for axial and 200–300 μm for lateral) [4, 10, 12].

3.2.2 Image acquisition using intravascular ultrasound

IVUS has been widely used in coronary imaging to characterize plaque components, identify calcified lesions and understand plaque morphology, which can help in risk stratification related to coronary artery disease (CAD). As the name indicates, IVUS uses a sound frequency above the normal human hearing range (above 20 kHz). Commercially available IVUS systems use a sound source with a frequency range between 10–40 MHz to capture the images within the cardiac or any other vascular arteries [13]. The IVUS uses a small probe with a piezoelectric crystal placed at the front end to generate high-frequency sound waves. The probe captures the echoes of the transmitted sound waves in successive A-scan modes and produces a cross-sectional image of the target lesion. To obtain a complete 360° visualization of the vessel, two arrangements are proposed in the literature, one uses a mechanical set-up and the other makes use of an electronic system.

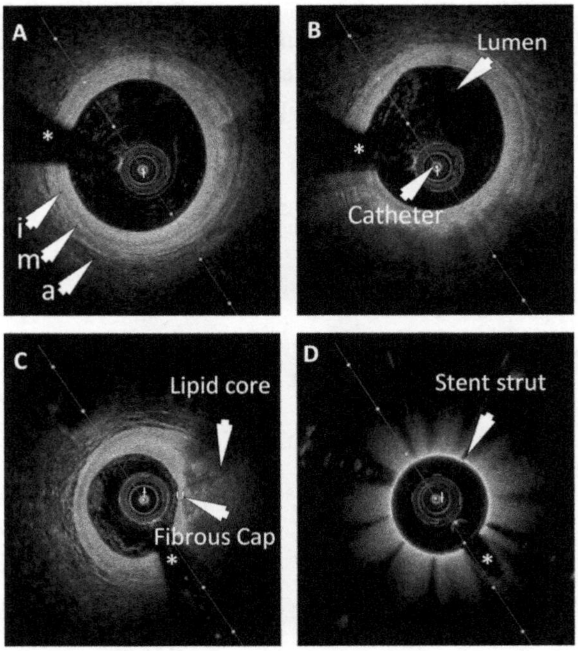

Figure 3.2. OCT images for types of atherosclerotic plaque. (Courtesy of AtheroPoint, Roseville, CA, USA.)

The mechanical set-up uses a fixed catheter with a rotating transducer at the tip which circulates at 360° and captures the image. The electronic arrangement uses a fixed catheter with around 64 stationary transduces, each mounted on the circumference of the catheter tip. The responses received from all these transducers are added to obtain the final cross-sectional image. The IVUS image is generated by combining the echoes back-reflected from multiple targets in the tissue layers. In the coronary artery, the vessel wall is made of three types of structural layers, the intima, media and adventitia. Sound waves behave differently when they pass through each of these layers, mainly because of their unique acoustic properties.

Ultrasound is reflected by the intima layer with a larger change in acoustic impedance. The media layer contains homogeneous smooth cells and hence sound waves can penetrate easily without many reflections. The adventitia layer is the storage of a large number collagen fibers, which provides a large back-reflection of ultrasound. An IVUS image of the normal coronary artery can be easily interpreted as having three layers: the intima layer with high reflections and bright echoes, the media layer with fewer reflections and a dark zone, and finally the adventitia with high reflections and bright echoes. The resolution of IVUS is around 120–150 μm which is useful for the detection of stent apposition in post-PCI [14]. The larger penetration depth of IVUS facilitates visualization throughout the circumference of the vessel at larger depths.

3.2.3 Comparison of OCT and IVUS

Both of the imaging modalities are counterparts of each other. They can be compared based on the application in which they are employed. Both OCT and IVUS show similar behavior in the tissue layers (table 3.1). OCT analyzes the optical properties (attenuation and backscatter coefficients) of the tissue layers, whereas IVUS relies on acoustic interpretations. As discussed in the previous section, the intima and adventitia have high back-reflections and less penetration resulting in bright regions in IVUS images.

The same is true for OCT images. The vital difference between IVUS and OCT can be observed in their resolutions (table 3.2). OCT has a nearly ten times higher resolution than IVUS [15]. This means that with OCT more detailed structures of

Table 3.1. Acoustic behavior of tissue layers and plaque components.

Tissue component	Reflection properties	Interpretation of IVUS image
Intima	Larger reflection, less penetration	Bright echoes
Media	Low reflection	Dark zone
Adventitia	High reflection	Bright region
Calcium	Extreme back-reflection, less penetration	Bright region
Fibrotic plaque	High reflection	Bright region
Fat	Less reflection, more penetration	Dark zone

Table 3.2. Comparison between IVUS and FD-OCT [7, 8, 11, 19].

Characteristics	OCT	IVUS
Energy source	Near infrared	Ultrasound (10–40 MHz)
Wavelength, μm	1.3	35–80
Resolution, μm	15–20 (axial); 20–40 (lateral)	100–200 (axial); 200–300 (lateral)
Frame rate, frames/s	100	30
Pull-back, mm s^{-1}	20	0.5–1
Max. scan diameter	7	15
Tissue penetration	2–3.5	10
Balloon occlusion	Optional	Not necessary

the vessel walls can be observed, which is not possible with IVUS, and this is a deciding factor in most interventional applications. More information about the plaque is essential to accurately diagnose the CAD and predict the risk associated with it. Some vital parameters of the plaque, such as fibrous cap thickness, can only be detected using OCT because of its high resolution (10–15 μm) [16]. Macrophages and necrotic cores are visualized more clearly by OCT than IVUS. It has been stated previously that IVUS is not the ideal approach for lipid-rich plaque characterization [17]. Moreover, it has been shown that OCT provides intra-coronary imaging with high speed compared to IVUS. In addition to its high resolution (ten times higher than IVUS), OCT uses a fiber optic cable for its set-up and hence is relatively inexpensive, nonetheless providing high-resolution images with great accuracy [12].

The shortfall of OCT is its inability to penetrate deep within the vessel walls, which limits the measurement of the whole volume of large plaques. The second important limitation is the requirement of a blood free region in OCT imaging [3]. The penetration depth of OCT is in the range of 1–2 mm, whereas for IVUS it is 10 mm. With its low penetration depth OCT cannot image the lipid pool or calcium deposition, but it has been found that most of the admissible features are deposited within 500 μm regions [3, 18]. Both of these limitations are addressed in a recently developed FD-OCT system, the C7XR Imaging System (LightLab Imaging). This system injects Dextran, a blood clearing solution, through the guiding catheter. With this, an improvement in penetration depth of 3.5 mm has been observed [17].

3.3 Tissue characterization

Coronary plaque tissue characterization is the process of identifying various arterial tissues to analyze disease or normal artery layers. This is important to understand which particular layers and plaque components play a vital role in plaque rupture. This section mainly deals with the methodologies commonly used to identify the plaque components.

3.3.1 OCT appearance of plaque tissues

As discussed above, atherosclerosis is the result of plaque deposition. Before characterizing the different plaque components, we must understand and identify the various constituents of plaque. Each layer of the vessel wall and the plaque components within them have distinct optical and acoustic properties which are usually captured using OCT and IVUS, respectively. Lumen segmentation and boundary estimation is the first step in almost all methodologies, followed by delineation of the intima–media, media and adventitia. Once the layers are identified, the next step is to determine the plaque region and its components. The plaques which are prone to rupture are known as 'vulnerable plaques'. Such plaques are characterized by three types, which are a fibrous plaques, lipid-laden plaques and calcified plaques. Each of these can be uniquely identified in OCT images by their optical characteristics. Validating OCT findings with histopathology, Yabushita *et al* [20] presented three major criteria to characterize these atherosclerotic plaque components. This study also reports high sensitivity and specificity for lipid-rich plaque when compared to MRI, spectroscopy and IVUS.

The high resolution of OCT allows the identification of microstructural changes within the vessel wall, which starts with intimal thickening. The atheroma is the region between the internal elastic lamina (IEL) and external elastic lamina (EEL). The progression of plaque deposition in the intima layer leads to the formation of an atheroma, which contains lipid-rich contents, fibrous tissues, calcium nodules and macrophages. The inflammation of the atheroma region results in the intrusion of plaque within the lumen which creates a thrombus. Thus, the detection of atheromas using OCT is important. OCT identifies various plaque components with back-scattered light waves reflected from various penetration depths. The penetration depths vary with the plaque tissues, being higher in fibrous plaque and the lowest in the thrombus. Calcium and lipid plaques have intermediate values of penetration depths [23].

3.3.1.1 Calcification

Coronary calcification is a marker of cardiac events. Calcification of coronary arteries reduces the elasticity and thus the plaque becomes prone to rupture. Coronary calcification can be identified using OCT as sharply delineated borders with a signal poor region. There are basically two types of calcified nodules, one protrudes into the lumen and the other does not protrude into the lumen. A superficial calcification protruding into the lumen is considered to be a vulnerable plaque component. Matsumoto *et al* [24] classified superficial calcifications into two types based on the thickness of the measured and actual lumen surface (figure 3.3). The analysis and measurement of calcium become challenging when the calcium thickness is greater than the maximum penetration depth of OCT. In addition to this, calcifications are often mischaracterized as lipid plaques, because of the lower signal attenuation from various structures, which also characterizes lipid components. In such a situation, the sharp borders of the calcium are considered as the determining factor [25].

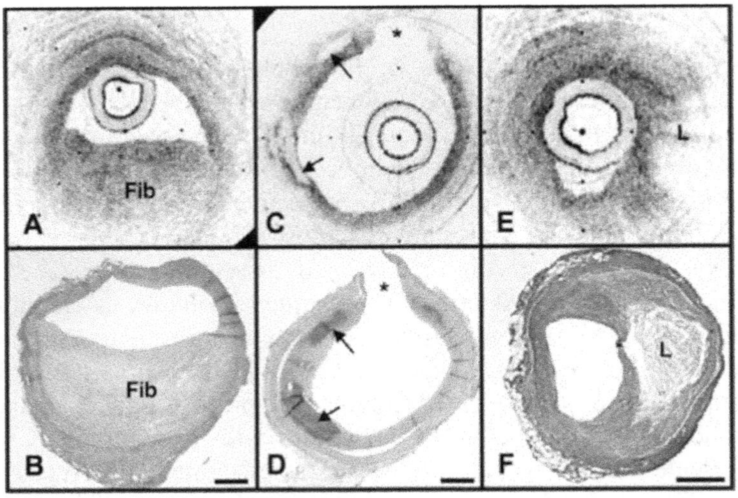

Figure 3.3. OCT versus histology findings of plaque types. (Reprinted with permission from [21, 22]. Copyright SPIE.)

3.3.1.2 Fibrous plaque

Fibrous plaque can be characterized as a homogeneous region with high back-scattering signals. They are composed of smooth muscle cells, solid fibrous or fibrocellular tissues, and collagen fibers [26].

3.3.1.3 The necrotic core and lipid-rich plaque

The necrotic core can be characterized by OCT as a signal poor region with fewer delineated borders than a calcified lesion, and has very little or no backscattering signal. Necrotic cores contain lipid-rich tissues and are usually covered by a fibrous cap [27]. Necrotic cores are present in tissues at larger depths than the penetration depth of OCT and present no backscattering signal, and hence cannot be diagnosed. Hence OCT is suitable for identifying the lipid-cores near the luminal surface. It has been reported that OCT cannot quantify the area or volume of necrotic cores at large depths. It has been reported that the attenuation coefficients are usually high for unstable vulnerable plaque components, in particular the lipid-rich necrotic core [28, 29]. A thickness of fibrous cap $< 65\ \mu$m distinguishes a thin-cap fibroatheroma from a thick-cap fibroatheroma. The index of plaque attenuation (IPA) and lipid-core burden index (LCBI) are the two metrics used for assessing the types of plaque, specifically the lipid core (table 3.3).

3.3.2 Schools of thought on tissue characterization

Three major techniques are predominantly used for plaque characterization which are based on (i) optical properties, (ii) machine-learning approaches and (iii) deep-learning techniques. Each of these three paradigms provides information about the components of the plaque and coronary vessel wall with distinct advantages and disadvantages. The primary objective of intravascular imaging techniques is to

Table 3.3. OCT appearance of plaque components [20, 30].

Plaque type	OCT appearance
Fibrous plaque	Homogeneous and signal rich
Fibro-calcified plaque	Signal poor, sharply delineated borders
Lipid-rich plaque	Signal poor, diffused borders

accurately predict and prevent the risk associated with coronary artery disease at an early stage. There are two possible approaches by which we can predict the risk associated with atherosclerosis and plaque formation: (i) by identifying the atheroma region, its exact location and spread within the vessel wall and (ii) by identifying the different plaque components responsible for atheroma progression. An atheroma is a combination of lipid-rich and fibrous plaque, composed of oxidized LDL cholesterol, macrophages and foam cells [31]. The location of atheromas has been identified as being between the IEL and EEL. The deposition of calcium in the coronary vessel wall is considered as a major cause of atheroma progression which reduces the lumen diameter [32]. Various approaches have been presented in the literature to identify and delineate the atheroma region. Because of its high resolution (10–15 μm), OCT has proven to be the primary choice of researchers, in comparison to IVUS, for lumen identification and boundary extraction. The work presented in [18] can accurately delineate the IEL and EEL and extract the atheroma location from OCT images. The study also verified the characteristics of lipid-rich and fibrotic plaque identified by OCT.

Vulnerable plaques are prone to rupture and may intrude into the lumen of the coronary artery. Thin-cap fibroatheroma (TCFA) is widely accepted as a vulnerable plaque component. The main function of the fibrous cap is to prevent an intrusion of calcified plaque into the blood. As the atherosclerotic plaque has clotting agents, interaction with blood results in occlusive thrombosis. Thrombosis leads to a shortage of blood supply to the heart and causes myocardial infarction. In order to identify various plaque components using intravascular OCT, various approaches have been proposed in the last few decades. Kume *et al* [16, 33–35] analyzed the potential of OCT for identifying TCFA, intima–media thickness, arterial thrombus and visualization of neointima formation. The results obtained from the reported study were validated using histological analysis and compared against the IVUS results. The high resolution of OCT provides a clear visualization of atherosclerotic plaque when compared to IVUS.

3.3.3 Characterization using optical properties

Each atherosclerotic plaque tissue component exhibits unique optical properties, such as the attenuation and backscatter coefficients. When the OCT signal is projected towards a target lesion, part of the signal is attenuated by the tissue and the remaining signal is scattered in multiple directions. An informative OCT signal model as a function of depth is required to understand this process and to determine the scattering coefficients. A curve fitting approach has been followed in the literature to fit the A-scan signal with either an exponential or linear model for estimating the attenuation coefficients [36]. This study identified the scattering coefficient and anisotropy factor using multiple scattering models from OCT images. If a single scattering effect for the projected light signal is assumed, then the OCT signal can be represented by the Lambert–Beer law

$$i(x) = i_0 e^{-u_t x}, \tag{3.1}$$

where $i(x)$ indicates the intensity of the A-line OCT signal at distance x, i_0 is the product of the intial signal intensity and backscatter coefficient, and u_t is the attenuation coefficient [37]. Furthermore, the confocal properties of the optical lens, such as the position of the lens in the targeted lesion and the depth of focus, also have an effect on the OCT signal, that is specified by an axial point spread function, which is given by [38, 39]

$$A(x) = \frac{1}{\sqrt{\left(\frac{x - x_0}{z_0}\right) + 1}}, \tag{3.2}$$

where $(x - x_0)$ is the distance between the focal point at x_0 and the tissue depth at x. z_0 is termed as the Rayleigh length, which is half of the depth of focus of the optical lens of the OCT system. Thus, the OCT signal can be modeled by combining equations (3.1) and (3.2) as follows:

$$i(x) \propto \frac{e^{-u_t x}}{\sqrt{\left(\frac{x - x_0}{z_0}\right) + 1}}. \tag{3.3}$$

After choosing the model, either with single scattering or multiple scattering [40], it is necessary to fit this model with an average of at least 50–100 A-line scans that provide smooth signal for curve fitting with u_t as a fitting parameter [39, 41].

Until 2008, atherosclerotic plaque characterization was performed using a single attenuation coefficient, and it was proved that a similar approach can be useful to distinguish between normal and atherosclerotic tissue layers [36, 41]. Xu *et al* [42], for the first time, presented an approach by combining both attenuation and backscatter coefficients to characterize three atherosclerotic plaque types, i.e. calcified, fiber and lipid-rich, using a single scattering model. The authors used a least-square curve fitting approach to match both the A-line signal and an exponential fitting model:

$$p(x) \approx \frac{1}{2} p_i \mu_b L_c A(x) e^{\frac{-2\mu_t x}{n}}, \tag{3.4}$$

where p_i is the initial power incident from the optical source, μ_b is the backscattering coefficient, L_c is the coherence length of the optical source, $A(x)$ is the beam divergent function, μ_t is the attenuation coefficient and n is the refractive index of the tissue. Since the study was performed on a phantom with a known refractive index and with a very small attenuation coefficient, the power from the backscattered signal from the phantom can be given as

$$p_0(x) \approx \frac{1}{2} P_i \mu_b L_c A(x), \tag{3.5}$$

with μ_{b0} the backscatter coefficient of the phantom used. The OCT signal obtained from the A-line scan was fitted using a curve fitting approach with the following model:

$$\log\left(\frac{p(x)}{p_0(x)}\right) = \log\left(\frac{\mu_b}{\mu_{b0}}\right) - \frac{2\mu_t x}{n}. \tag{3.6}$$

Atherosclerotic plaque characterization using manual tracings and assessment of OCT images leads to intra-observer variability and is also a time-consuming process (figure 3.4). To overcome this problem an automated plaque characterization was presented, which characterized the calcium, necrotic core, macrophage infiltration and fibrous tissues by determining the attenuation coefficient from OCT images [37]. The study also used a single scattering model based on the Lambert–Beer equation and performed curve fitting with an A-line OCT signal to determine the extinction coefficient. The curve fitting is followed for different tissue types using a window with variable size. This study classified the atherosclerotic plaque tissues based on the attenuation coefficient with a high-value attenuation coefficient ($\mu_t \geqslant 10\ \text{mm}^{-1}$) for the necrotic core and macrophages compared to fibrous and calcified tissues ($\mu_t \approx 2$–$5\ \text{mm}^{-1}$). However, the study was limited by its ability to find the backscatter coefficient.

A similar single scattering homogeneous model was also adopted in [43] to characterize atherosclerotic plaque lesions. However, instead of fitting the model to the complete A-line OCT signal, the A-line was partitioned into various small homogeneous tissue layers and then fitted with the following model:

$$(x) \propto T(x) \sum_{m=0}^{p} I_{0,m} \cdot e^{-\mu_{t,m} \cdot d} \cdot \prod (x - x_m), \tag{3.7}$$

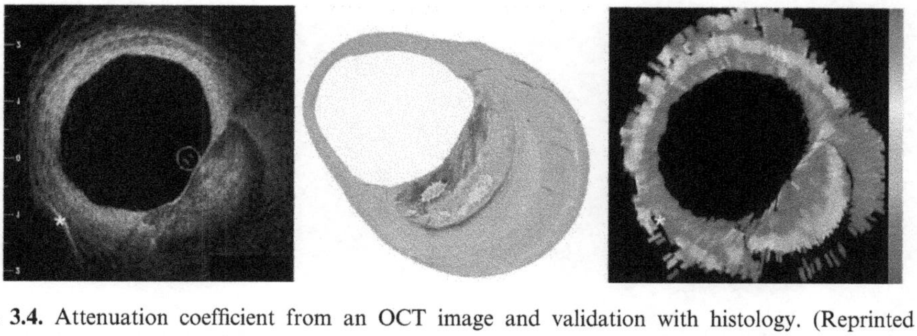

Figure 3.4. Attenuation coefficient from an OCT image and validation with histology. (Reprinted with permission from [37]. Copyright SPIE.)

where x_m is the mth layer starting point and $\prod(x - x_m)$ is the variable length window whose value remains one within the partitioned homogeneous layer, otherwise it is zero. The optimal value of of p is decided by fitting the above model with the OCT signal for multiple values of p and checking for the minimum sum of square error (figure 3.5).

The exponential curve fitting approach provided by Lambert–Beer's law, discussed above, provides a single uniform value of attenuation coefficient over a certain homogeneous depth in a tissue layer. Furthermore, the averaging of several A-lines that is required prior to curve fitting smoothens the signal and hence provides the global value of attenuation with low spatial resolution. Vermeer *et al* [44] proposed a model in which the attenuation coefficient was estimated as a function of depth in the tissue layer:

$$l(x) = l_0 e^{-\int_0^x \mu(z)dz}, \tag{3.8}$$

where $l(x)$ is the irradiance of light at depth x and l_0 is the initial incident light beam irradiance. A constant value of attenuation coefficient will reduce the above expression to Lamber–Beer's law in equation (3.1). The authors computed the attenuation coefficient for each pixel value in an image using

$$\mu[i] \approx \frac{I[i]}{2\Delta \sum_{i+1}^{\infty} I[i]}, \tag{3.9}$$

where $\mu[i]$ and $I[i]$ are the attenuation coefficient and intensity value at level i and Δ is the pixel size.

The depth-resolved approach presented by Vermeer *et al* was the first automated approach without the use of curve fitting implementation. Furthermore, it used invertible mapping to the attenuation coefficient from the OCT signal model. As it

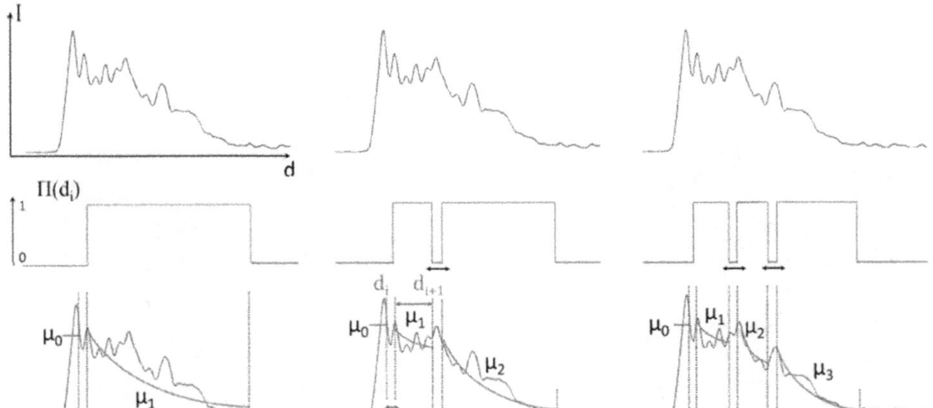

Figure 3.5. OCT signal fitted with a curve fitting model with a varying value of p. (Reproduced with permission from [43]. Copyright Optical Society of America.)

was depth-dependant, it eliminated the low-resolution limitation of the conventional curve fitting approach. However, the authors did not take the confocal function and system fall-off rate into account and hence it was applicable only when the focal plane was above the sample [45]. Smith *et al* [45] presented an extended version with improvement in the accuracy of the attenuation coefficient estimation by adding both the confocal function and system fall-off rate. Recently, attenuation coefficient imaging was implemented for tissue characterization, such as to characterize the lipid plaque and macrophage infiltration as well as for the classification of human atrial tissues [46–48]. Both of these approaches used the single scattering OCT model with a point spread function [37].

Recently, the work presented by Liu *et al* [49] proposed an approach which characterizes six types of tissue using the attenuation coefficient, backscatter coefficient and pixel-wise intensity in an OCT image (table 3.4). The initial lumen segmentation is performed using a deformable model and further two prominent layers, such as the intima and media, and their boundaries are identified. The aim of the paper was to create a ground truth based on peak information and image quantization. Eight statistical features, the attenuation and backscatter coefficient, along with image intensities were used for characterization.

3.3.4 Characterization using machine learning

With the advent of machine-learning tools, it has become possible to obtain accurate characterization of the plaque in the vessel wall [50–53]. Recently, Ughi *et al* [43] proposed a method to automatically characterize atherosclerotic plaques using a machine-learning framework. Three different plaque types, i.e. fibrotic, lipid-rich and calcified, are characterized and classified using the supervised classification algorithm of random forest (RF). A set of texture-based and geometrical features along with attenuation coefficients were used for classifying the input OCT image into three types. The overall classification accuracy was reported as 81.5% with a per class accuracy of 89.5%, 72.1% and 79.5% for fibrotic, lipid-rich and calcified plaques, respectively. There are three potential limitations associated with this paper: (i) the ground truth depends on the manual analysis, which may lead to

Table 3.4. Optical coefficient values for various tissues types.

Tissue type	Attenuation coefficient (mm^{-1})	Backscatter coefficient (mm^{-1})
Healthy vessel wall	2–5	–
Intimal thickening	2–5	–
Calcification	5.7 ± 1.4	4.9 ± 1.5
Fibrous plaque	6.4 ± 1.2	18.4 ± 6.4
Lipid pool	13.7 ± 4.5	28.1 ± 8.9
Necrotic core	⩾ 10	–
Macrophage infiltration	> 12 mm	–

inter-observer variability and thus weakens the ground truth, (ii) the atherosclerotic plaque is considered as homogeneous, which may not be true, as plaque itself is the combination of different lipid components, and (iii) coronary artery walls are considered as homogeneous, although multiple scattering can be possible from the artery wall layers.

Another method, proposed in [54], uses an A-line modeling approach for plaque characterization. The proposed algorithm first identifies the intimal thickening and further classifies the tissue layers into fibrous and lipid-containing plaque using linear discriminant analysis (LDA). The feature set contains 11 morphological features extracted from the IVOCT image. This study was evaluated using histology as the ground truth. The proposed method has a limitation that the effect of blood is not considered and hence, in practice, the results may be affected by blood artifacts. In practice, one pull-back contains multiple B-scan images, hence its difficult to analyze the atherosclerotic plaque manually. In order to tackle this problem, researchers have used multiple features from OCT images to automate this task. These features are based on texture and geometric patterns or optical information obtained from the OCT imaging. The optical properties of biological tissue play an important role in atherosclerotic plaque component identification [55, 56]. The attenuation and backscatter coefficients derived from the OCT image have been verified as the important features of atherosclerotic plaque characterization.

In order to obtain more accuracy in characterization and classification, machine-learning approaches have been followed by researchers across the world. Once trained, the machine-learning approach is used to automate the characterization process from OCT images. The previously discussed study by Ughi *et al* in [43] used the RF for plaque classification. A feature vector obtained using texture and geometrical properties with estimated values of the attenuation coefficient were used as an input to the RF classifier. The RF is an extension of the decision tree mechanism and combines multiple weak tree classifiers to provide the final classification. RF classifies the incoming OCT images into any of the four output classes with an overall classification accuracy of 81.5%. The classification accuracy is lower and leads to false results due to the limited dataset for training. The accuracy can further be improved using a large feature set instead of only seven.

The work presented by Athanasiou *et al* [57] verified RF as the best feature classifier to classify the coronary artery tissue layers and plaque components. Their paper provides a characterization of three tissue types: calcified tissue, lipid tissue and fibrous tissue. The algorithm proposed by the paper utilized a simple image processing algorithm with a machine-learning approach. The initial lumen segmentation employs the most commonly used Otsu's segmentation method. The segmented image is then partitioned into three different clusters using the K-means segmentation method. The dimensionality of the texture-based feature vector of 32 is reduced using a wrapper feature selection method. Only dominant features are applied to the classifier. The RF was compared against a neural network and support vector machine and was determined to be the best among these techniques. The paper has a few serious limitations: (i) it failed to characterize the media layer of the coronary artery wall, (ii) vulnerable plaques such as TCFA and macrophage infiltration cannot be determined using the proposed method, (3) manual analysis as the gold standard is a function of variability and (4) the proposed method does not provide information on plaque components.

3.3.5 Tissue characterization using deep learning

The evolution of deep learning has become a primary choice for researchers to obtain more information from medical images. In a short period of time, deep learning has been used in a variety of applications for the classification of biological images. The convolutional neural network (CNN) is a building block for the deep-learning technique. A recent study proposed by Abdolmanafi *et al* [58] used the deep-learning approach for coronary artery layer characterization. The proposed technique identifies the intima and media layers of the coronary artery wall from the OCT image. The performance of three different classifiers (SVM, RF and CNN) was compared and RF was found to be the best for classification and CNN for feature extraction from OCT images. The initial lumen segmentation is performed using a deformable model and, further, two prominent layers such as the intima and media and their boundaries are identified. The objective of the paper was to create a ground truth based on peak information and image quantization. Compared to the work presented by Ughi *et al* [43], it was found that, with the same RF as a classifier, CNN finds more prominent features when compared to those obtained with optical and texture features. Tissue characterization has not been presented by the author in the reported study, and identification of plaque components can help in performing risk assessment. However, this was the first paper of its kind to assess the coronary artery layers using a deep-learning approach and hence remains important.

3.4 A link between carotid and coronary artery disease

Carotid artery and coronary artery disease are interlinked with each other and hence the biomarkers used in carotid artery disease (CAD) have been widely used to predict the risk associated with coronary artery disease. Autopsy studies also verified a strong correlation between both of these diseases [59, 60]. Carotid intima–media thickness (cIMT), carotid plaque and coronary artery calcification are the major predictors that provide a strong association between both types of atherosclerosis and have been used widely for prediction of CVD risk.

3.4.1 Carotid intima–media thickness and CAD

cIMT is a well-known and widely accepted biomarker for the prediction of atherosclerosis and coronary artery diseases. For the first time, in 1986 Pignoli *et al* [61] showed the relation between aortic wall thickness measured by ultrasound and cardiovascular disease. Later this study was extended to visualize the role of cIMT in patients with high cholesterol levels. From then onwards cIMT has been independently recognized as the gold standard for CVD risk prediction. Salonen *et al* [62] reported an 11% increase in the risk of myocardial infarction for every 0.1 mm increase in cIMT. Furthermore, as cIMT measurements is a non-invasive technique computed using B-mode ultrasound, it has become the first choice for researchers when dealing with both carotid and coronary artery disease. Ziembicka *et al* [63] investigated the relation between cIMT and coronary artery disease and reported a CAD risk of 94% for patients with an IMT > 1.15 mm. Kao *et al* reported that an IMT > 0.8 mm leads to CV events, whereas a study presented by Ikeda *et al* [64] shows that a cIMT > 0.9 mm has a strong association with CAD. Various studies, including meta-analyses, multi-ethnic studies and clinical trials, have reported cIMT as a predictor of increased CVD risk [65–68]. Intima–media thickness can be measured from multiple segments in the carotid artery, such as the near and far wall of the common carotid artery (CCA), internal carotid artery (ICA) and external carotid artery (ECA), and carotid bulb or bifurcation [62]. Furthermore, there are some protocols for cIMT measurements such as: (i) mean or maximum of the individual measurement, (ii) mean of the maximum for multiple measurements and (iii) mean of the mean for multiple measurements. Although the CCA-cIMT has been independently considered to predict CV risk, each segment of carotid artery has an independent role in cardiovascular disease prediction [69]. ICA-cIMT has been reported to be statistically significant and has a high association with coronary heart disease [70].

It has been reported that the IMT of both the common carotid artery (CCA) and internal carotid artery (ICA), when combined with the Framingham risk score, can be used as a biomarker for CVD [70]. Furthermore, this study also reported a stronger association between maximum IMT calculated from ICA than the mean CCA.

3.4.2 Carotid plaque and CAD

In addition to cIMT, carotid plaque has also been reported as an important predictor of CV events and its detection was recommended for initial CVD risk assessment [64, 71]. This study suggests the measures for identification of carotid plaque, either as an intrusion into the lumen by more than 0.5 mm compared to the surrounding vessel wall, 50% thicker, or cIMT > 1.5 mm. A recent atherosclerosis risk in communities study indicated that the inclusion of carotid plaque in the study results in an overall 50%–100% increase in CV risk irrespective of IMT [72]. Furthermore, this study also recommended guidelines for measuring carotid plaque with thickness > 1.5 mm between the EEL and IEL. The association between coronary and carotid IMT was investigated using IVUS and it was found that an increase in carotid IMT has a significant correlation with an increase in plaque area in the left main coronary artery [73]. The plaque area was computed using EEL area and lumina area as follows:

$$\%\ \text{plaque area} = \frac{\text{EEM area} - \text{Lumen area}}{\text{EEM area}} \times 100. \tag{3.10}$$

This study reported the correlation between carotid and coronary atherosclerosis as 0.4, which was less than that obtained for autopsy studies, which was improved using a common metric for both the coronary and carotid arteries using IVUS.

The role of carotid ultrasound measurements in assessing the CAD risk in patients suffering from lupus was also analyzed using cIMT and total plaque area (TPA) [74]. This study reported that cIMT has a lower correlation than TPA with CAD risk prediction, and hence recommended the use of TPA as a surrogate biomarker for risk prediction.

3.4.3 Coronary IMT and carotid atheroma for CAD risk detection

A strong correlation between carotid and coronary atherosclerosis exists when the analysis is performed using a common variable, such as carotid IMT and coronary IMT, and using a common imaging modality [75]. This study showed a high correlation between carotid IMT and coronary IMT as well as with the percentage atheroma volume (PAV). The coronary IMT was computed by

$$\text{Coronary IMT} = \left[\sqrt{\text{EEM area} \times \pi^{-1}}\right] - \left[\sqrt{\text{Lumen area} \times \pi^{-1}}\right]. \tag{3.11}$$

The total atheroma volume (TAV) and PAV were calculated as

$$\text{TAV} = \text{Atheroma area} \times \text{pullback length} \tag{3.12}$$

and

$$\text{PAV} = \frac{\sum \text{Atheroma area}}{\sum \text{EEM area}} \times 100, \tag{3.13}$$

where atheroma area was given as [EEM area − Lumen area].

In addition to carotid IMT, brachial–ankle pulse wave velocity and ankle–brachial index were also reported as strong biomarkers of coronary artery disease and can predict the risk of CAD [76].

3.4.4 Femoral and carotid IMT for CAD risk detection

Lekakis *et al* [77] reported that individual carotid IMT computation is a weak marker for CAD risk prediction and hence added another approach by combining carotid and femoral IMT to assess the extent of coronary artery diseases. The authors also reported a high correlation between the IMT computed from the common and internal carotid artery, carotid bifurcation and femoral artery, and concluded that the patients with high IMT values have an increased risk of cardiovascular events.

3.4.5 Coronary calcium and carotid risk factors for risk detection

Carotid plaque has a high association with CV events compared to carotid IMT, but the coronary artery calcium score has been reported as the strongest of these three to predict and stratify coronary artery disease [78]. Another multi-ethnic study on atherosclerosis conducted by Polak *et al* [79] showed a high correlation between CCA and ICA IMT and coronary heart disease. The study integrated a Framingham score and CAC to improve coronary heart disease risk prediction in diverse population. Manual tracing cIMT to assess the CVD leads to inter-observer variability. A multi-ethnic study on atherosclerosis resolved this problem by proposing an automated edge detection method to compute IMT values similar to manual tracings [80].

3.5 Wall quantification

To stratify the risk associated with CVD, information related to plaque composition and its quantification is important. Coronary artery plaque measurement includes the segmentation of the lumen, determination of vulnerable plaque components such as the fibrous cap and macrophages, and quantification of coronary calcium (figure 3.6).

3.5.1 Lumen measurement

The lumen is the region through which blood circulates in the coronary artery. Lumen segmentation is an important step before performing tissue characterization.

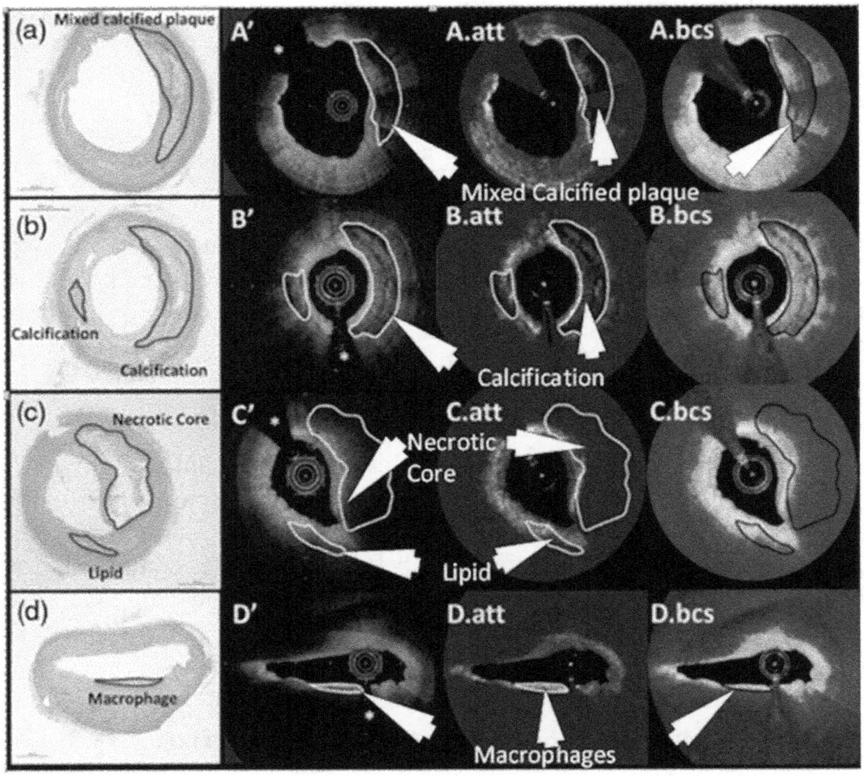

Figure 3.6. (a)–(d) Histological cross sections (column 1) and (A′)–(B′) the corresponding IVOCT cross sections; an asterisk (*) marks the guide wire (column 2). The attenuation coefficients and the backscattering term are labeled as (A.att)–(D.att) (column 3) and (A.bcs)–(B.bcs) (column 4), respectively. The noisy region behind the determined cut-off border is marked with a dark blue overlay in the attenuation and backscatter cross-section. Plaques of interest are labeled in the pathological cross sections, which have been aligned with other-related IVOCT images. (Reprinted with permission from [49]. Copyright SPIE.)

Recently, Roy *et al* [81] proposed an approach to segment the lumen and tunica from the coronary vessel wall using graph theory and a random walks algorithm. The overall segmentation accuracy reported was 97.86%, which was benchmarked against [43]. The performance of the segmented lumen was analyzed manually using Cohen's kappa coefficient, Bhattacharya's distance and the *K–L* distance. The work presented in [82] extracted the lumen contour information using OCT image intensities and their first and second derivatives. The precise measurement of the lumen is possible using OCT with a high degree of reproducibility [83].

3.5.2 Vessel wall measurement

In addition to the lumen, other vessel wall layers also need to be identified for plaque tissue characterization. Recently, Zahnd *et al* [84] proposed a fully automatic method for simultaneously segmenting the intima, media and adventitia contours. A front propagation method is used to segment the intima–media, media–adventitia and adventitia–periadventitia interfaces. Further, the layers are classified into healthy or disease regions using the AdaBoost classifier. A total of 17 features from each column of segmented contours are extracted, which show distinct characteristics of healthy and disease regions. The Boruta method is used for dominant feature selection. The proposed algorithm was validated against manual annotation and evaluated using the Dice similarity coefficient. The proposed method requires the layers to be clearly visible, which cannot always be the case. Moreover, plaque tissue characterization is not possible with the proposed approach.

3.5.3 Fibrous cap measurement

TCFA is considered as the most vulnerable plaque component, which is prone to rupture and leads to thrombotic events. According to pathological information, a fibrous cap thickness $\leqslant 65\ \mu$m is prone to rupture [2]. Hence, for risk assessment, quantification of the fibrous cap is essential. The work presented by Wang *et al* [85] quantified the fibrous cap using a dynamic programming approach. The proposed method is semi-automatic in nature and requires human intervention for identifying the FC in the luminal circumference of the OCT image. Based on region-of-interest (ROI) information provided by three observers, FC boundaries are segmented and quantified using dynamic programming. The paper also benchmarks the work presented by Kume *et al* [16] for the determination of the 'abluminal boundary' of the FC. The abluminal FC boundary is the internal boundary diffused to the necrotic core. The paper also proposed two metrics for FC quantification: (i) mean FC thickness and (ii) FC surface area. The mean FC thickness is computed as the average value of the distance between the points on the FC luminal and abluminal boundaries. The surface area of the FC is computed as the product of the distance between two frames and the arc length of the FC summed over all frames. In addition to these two metrics, FC volume and density have also been computed in this paper. Plaque vulnerability depends upon FC thickness, macrophage infiltration, blood within the lumen and size of the necrotic core, however, this paper only discusses the FC morphology in detail. Moreover, the risk stratification due to plaque rupture has also not been discussed by the authors.

Recently, work presented by Zahnd *et al* [86, 87] used a contour-segmentation-based approach to quantify the fibrous cap thickness. The objective of this paper was to extract the abluminal boundary of the fibrous cap. Similar to the previous work presented in [85], Zahnd used a semi-automatic approach for fibrous cap measurement by extracting the ROI from experts. However, in order to prevent the effects of catheter position, geometrical *a priori* information was also integrated with the dynamic programming approach. *A priori* information was obtained by assuming a little variation of fibrous cap thickness at the adjacent sites. The thickness of the fibrous cap was measured by measuring the distance between two points on the luminal and abluminal boundary, respectively. The results were validated by manual analysis using Bland–Altman plots.

3.5.4 Measurement of calcium

Coronary artery calcification is another dominant factor for stenosis of the artery. The deposition of calcium in the plaque causes the artery to lose its elasticity and become hard. The artery can expand more with the progression of the plaque. The study presented by Mehanna *et al* [88] and that by Mintz in [89] presented the quantification of calcium in the human coronary artery. In an OCT image calcium can be identified as a signal poor region with properly delineated borders. Recently, a group led by Suri [90] proposed a method which can quantify the calcium volume using four segmentation methods (i.e. threshold-based, K-means, FCM and HMRF). In order to improve the speed of the quantification process, five multi-resolution techniques were proposed by this group. The results were validated by ground truth using the Jaccard index (JI) and Dice similarity coefficient (DSC). The paper does not consider the heart–lung motion, although it degrades the quality of IVUS images even at improved speeds. Also, this study assumes that the distance between two adjacent video frames is nearly zero without repetition of an individual frame. Although the paper focuses on calcium volume measurement using IVUS video frames, the techniques proposed by this study can be implemented using OCT images. Even though IVUS has high tissue penetration compared to OCT images, penetration into calcium is higher for OCT than IVUS images. Hence the measurement of calcium area, thickness and volume will be more accurate using OCT images [88].

3.5.5 Quantification of macrophages

Macrophages are another type of vulnerable plaque and their quantification is important for the assessment of CAD risk. The work presented by Tearney *et al* [91] quantified the macrophages within a fibrous cap using backscatter analysis of the OCT image and normalized standard deviation of the OCT signal within the ROI. The study clearly indicates that the fibrous caps containing inflammatory macrophages have strong backscatter reflections in the OCT signal (figure 3.7).

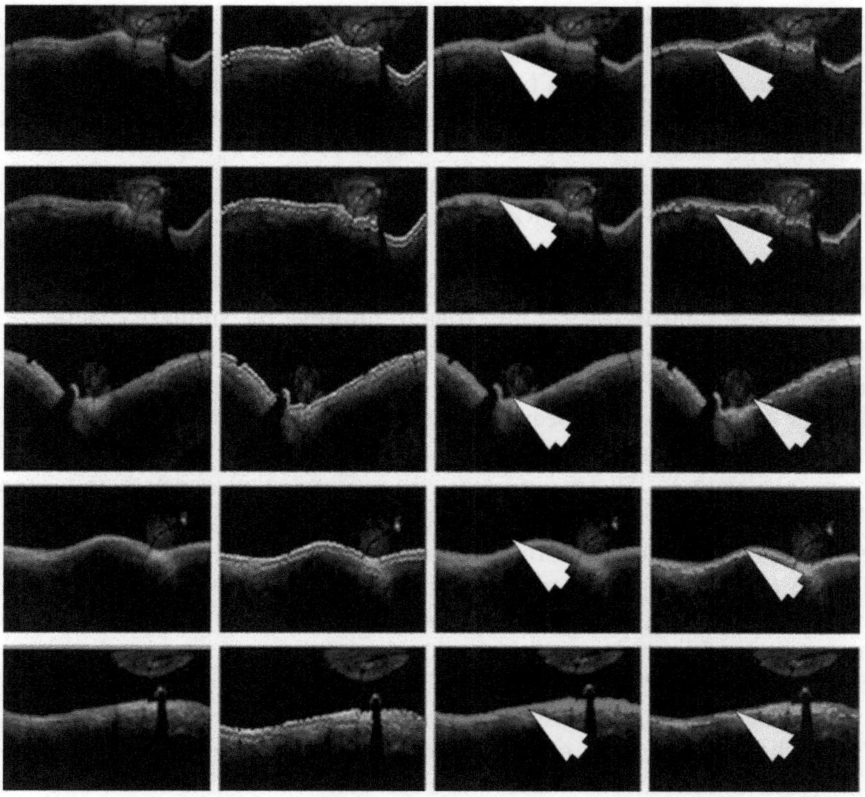

Figure 3.7. Classification results for one frame for five different patients. From left to right for each patient: original image converted to planar representation (column 1), initial segmentation (column 2), intima (red boundary with white arrows, column 3) and media (green boundary with yellow arrows, column 4). (Reproduced with permission from [58]. Copyright Optical Society of America.)

Recently, Vito *et al* [92] proposed a method using the characteristics of the tissue layer (i.e. normalized standard deviation, granulometric index and attenuation) to quantify macrophages. The study was validated with histology and indicated that macrophages were more prominent in the inflammatory region than in the non-inflammatory region. The paper has proposed an inflamed ROI extraction sensitivity equal to 100% and a specificity of 96.8%.

3.5.6 The role of image registration

Registration is a prevalent image processing technique normally used to match images captured from different viewpoints, at different times and locations, or captured using different acquisition systems. It allows images to be aligned and thus obtain a new larger image. Almost every image processing algorithm requires image registration as a preliminary step. In medical imaging, image registration has the additional role of combining two or more images to provide more information. These images can be acquired either from the same or from multiple imaging modalities. This process has also been referred to as medical image fusion in the literature [93]. Combining multi-modality images may add value in diagnosis and aid the doctor to provide an accurate decision in a short time. In general, the image registration process initially has two images, one is fixed and one is movable. Our target is to match the movable image to the fixed one using any of the available registration processes.

Various registration techniques have been reported to date to acquire accurate information from video frames. Registration and fusion have been implemented in [93], which used an intensity-based registration method to register MRI and CT images. Intensity-based registration usually takes two images as the input, compares them based on similarity criteria and tries to improve the similarity value. A rigid transformation is a low-complexity method which uses basic translations and rotations to register the current image with the targeted one. In addition to this, many other techniques have been reported. The recent study by Araki *et al* [94] provided a comparison between four different registration methods to assess coronary calcifications. The paper reported four important registration techniques: rigid, affine, B-spline and demons. In today's era of deep learning, CNNs are also being used for image registration. Recently, Li *et al* [95] proposed a non-rigid registration technique for registering brain MRI images. The technique maximizes the similarity criteria for two images and estimates the spatial transformation using a fully convolutional network, and this technique has been reported to be better than older image registration approaches.

Although medical image registration is widely used as a pre-processing technique, limited work has been done on OCT images, in particular in the domain of the coronary arteries and cardiac tissues [96]. This may be because of the difficulties in visualizing 3D structure from the two-dimensional images of OCT. Angiography was considered as the gold standard for percutaneous coronary intervention (PCI) procedures and hence it is essential to obtain the correct spatial orientation of OCT images with an angiogram. Karanasos *et al* [62] used OCT and angiographic image registration for PCI guidance. The registration of OCT images with angiographic findings enables quick and accurate decision making. The registration process is used to provide accurate information about the target lesion, which requires the implantation of a bioresorbable scaffold and avoids repetitive fluoroscopy. The high-speed acquisition of coronary calcification using IVUS results in blurring or translation of the calcium lesion in successive video frames. Recently, Chiastra *et al* [97] recovered the coronary arteries after stent implantation using OCT. This paper used the iterative closest point (ICP) approach to match the OCT images with the micro-CT images. The ICP technique is based on the rigid transformation discussed above. A similar rigid transformation-based ICP approach has also been used by Ughi *et al* [98] to perform intra-model registration of an IVOCT dataset to assess the stent struts. Prabhu *et al* [99] also proposed a registration algorithm to match and validate an IVOCT pull-back dataset with cryo-images. The recent fusion study presented by Hebsgaard *et al* [100] performed computerized registration of OCT images with coronary angiogram images, and concluded that OCT and coronary angiogram fusion can reduce diagnosis error.

3.6 Risk assessment systems

Risk assessment for CAD can be based on multiple factors such as patient demographics (age, height and weight), consumption of alcohol and smoking. In addition to these, coronary arterial structure, plaque morphology and plaque composition also play a vital role. The deposition of plaque in the periphery of the coronary artery is highly irregular and random in nature. Hence, to identify the risk associated with atherosclerosis, plaque deposition should be analyzed throughout the arterial length, instead of analyzing only a small lesion. A cardiologist needs to perform risk assessment and stratification prior to percutaneous coronary intervention and stent implantation [101, 102]. In general, risk can be classified into three classes: high risk, moderate risk and low risk. In order to perform risk stratification, the morphology of the coronary arterial plaque and its composition play a vital role. The above section on coronary arterial plaque characterization provides a detailed idea of the various plaque components and their morphology. Different plaque components have individual risk factors. The combination and analysis of all of these using OCT or any other imaging modality provides an overall picture of the risk associated with the patient. This is the first review of its kind to characterize, measure and perform risk assessment for the plaque components. The risk assessment criteria for IVUS can be extended to OCT-based studies.

The morphology of the plaque components, provided by texture features and machine-learning approaches, is being used for risk assessment. In OCT each B-scan produces a large number of images, and even more for multiple patients. In order to obtain a reliable and accurate risk assessment system, the system should be trained on huge datasets. Hence, in order to analyze minimal data samples and to perform risk assessment, a transfer learning approach is followed. The features obtained from the trained system are then used for online risk assessment systems with the limited dataset. Recently, a study presented by Araki *et al* [103, 104] performed risk stratification using texture features in a machine-learning paradigm on IVUS images. The risk was stratified based on the link between the carotid artery and coronary artery, as both of them share a common genetic makeup. cIMT is considered a biomarker for risk assessment. Even though the study was carried out on IVUS images, a similar approach can be followed using an IVOCT risk assessment system.

3.7 Discussion

The deposition of plaque on the internal layers of the coronary artery leads to rupture and may result in an occlusive thrombus. As the plaque has multiple clotting agents, it creates blood clotting, reducing the effective diameter of the lumen. This limits the supply of oxygen-rich blood to the heart, resulting in myocardial infarction. IVUS has been widely used to analyze and stratify the risk associated with plaque rupture. But as the microstructural tissue components are outside the resolution, they cannot be detected in IVUS images. Thus this review paper provides a comprehensive understanding of how the plaque components can be characterized using OCT. In addition to characterization and classification, the measurement of vulnerable plaque components along with risk assessment is also presented in the above section.

3.7.1 Benchmarking

Coronary artery diseases are the result of plaque formation in the arterial wall. Section 3.2 discusses plaque morphology and characterization. From the available literature and methodologies discussed in this review, it has been found that TCFA, macrophage infiltration, thrombotic occlusion and formation of the necrotic core within the atheroma region are the vital components which lead to plaque rupture. The dominant methodologies for characterization are (i) depth analysis, (ii) using optical properties of the depth profiles, (iii) machine-learning approaches and (4) deep-learning approaches. Of all the reviewed papers on plaque characterization, the best-of-five benchmarking papers are listed in table 3.5. The table clearly indicates the classification accuracy for the three schools of thought on characterization: (i) based on optical properties, (ii) based on machine-learning approaches and (iii) based on deep-learning approaches. The classification accuracy for deep learning along with the RF classifier are observed to be high compared to other methodologies. The CNN extracts all low-level and high-level features from the image. Instead of using a large dataset for training, transfer learning is used to train and test the offline systems [105].

3.7.2 A note on image acquisition hardware

A catheter-based FD-OCT system (discussed in section 3.2.1) has been widely used for the acquisition of the coronary arterial images for *in vivo* and *ex vivo* applications. The FD-OCT system has a fixed reference mirror, compared to the TD-OCT system. Currently, the C7XR system (LightLab Imaging Inc/Jude Medical, Westford, MA) is widely preferred for performing FD-OCT imaging. The FD-OCT system is capable of performing imaging at the rate of 100 frames/s with a pull-back speed of 20 mm s^{-1} [7]. This will help to visualize the plaque deposition throughout the length of an artery in a short duration.

3.7.3 A note on plaque component quantification

Vulnerable plaque components need to be quantified in order to understand the risk associated with atherosclerosis and MI. Benchmarking over plaque measurement is depicted in table 3.6. A fibrous cap thickness less than 65 μm has been found to be vulnerable. Work presented in [90] proves the calcium detection metrics using IVUS images, which can be extended to OCT images.

3.7.4 Validation of plaque characterization techniques

A coronary arterial plaque characterization, measurement and risk assessment system needs to be properly validated to tackle CVD. At present, only three methods are able to perform the validation: (i) visual analysis by trained

Table 3.5. Promising studies on plaque characterization.

Authors	Technique	Features used	Classifier	*N*	Frames/Video	Accuracy	Validation
Liu *et al* [49]	IVOCT optical properties	Statistical	Attenuation and backscatter coefficients and intensities	135	158	–	Manual analysis
Jimenez *et al* [54]	A-line depth profile estimation	Morphological	LDA	57	9	88%	Histology and manual analysis
Ughi *et al* [43]	A-line and optical properties estimation	Texture and geometry	Random forest	49	100	81.5%	Manual analysis
Gan *et al* [47]	Bayesian framework	Optical, texture and statistical	Relevance vector machine	35	–	80.41%	Histology and manual analysis

Table 3.6. Prospective studies on plaque measurement.

Authors	Technique	Features used	Classifier	N	Frames/video	Accuracy	Validation
Abdolmanafi *et al* [58]	Deep feature learning	CNN features	CNN, RF, SVM	26	–	96%	Manual analysis
Liu *et al* [49]	IVOCT optical properties	Statistical features	Based on attenuation and backscatter coefficients and intensity values	135	158	–	Manual analysis
Jimenez *et al* [54]	A-line depth profile estimation	Morphological features	LDA	57	9	88%	Histology and manual analysis
Ughi *et al* [43]	A-line and optical property estimation	Texture and geom. features	Random forest	49	100	81.5%	Manual analysis
Gan *et al* [47]	Bayesian framework	Optical, texture and statistical	Relevance vector machine	35	–	80.41%	Histology and manual analysis

Table 3.7. Prospective studies on plaque quantification.

Authors	Technique	Features used	Classifier	*N*	Frames/video	Tissue type	ACC*	Validation
Wang *et al* [85]	Dynamic programming	–	–	10	100	Fibrous cap thickness	–	Computer algorithm versus manual analysis
Zahnd *et al* [87]	Contour segmentation with geometrical *a priori*	–	Dynamic programming	21	–	Abluminal interface of fibrous cap	–	Manual analysis
Roy *et al* [81]	Graph representation of the image	CNN features	Random walks for image classification	26	150	Lumen segmentation	97.86%	Lumen annotation by expert
Kume *et al* [16]	Manual comparison using OCT s/w	Statistical	Linear regression, Bland–Altman analysis	135	–	Fibrous cap thickness	90%	Manual and histological analysis
Zahnd *et al* [84]	Front propagation scheme	17 image features	AdaBoost classifier with Boruta feature selection technique	40	100	IM, MA, AP interface	91%	Manual validation with DSC

* ACC: Accuracy

cardiologists, (ii) comparison of the results to histological analysis and, finally, (iii) performing multi-modality analysis for each patient (table 3.7).

3.7.5 A note on the future of OCT

In this review, it has been found that the overall characterization and classification accuracies depend on some dominant factors, such as characterization technique, classifier, number of patients, type of feature vector and number of features. Most of the available studies are lacking in one or all of the mentioned factors. A combination of deep and machine learning was found to be the method with the highest possible accuracy for layer segmentation [58]. Atherosclerotic plaque tissue characterization, detection of TCFA, quantification of macrophages and calcification of coronary arteries using OCT are the potential issues that are yet to be solved.

3.8 Conclusion

To the best of our knowledge, this is the first review of its kind which provides coronary arterial plaque characterization, measurement and risk stratification. This review is the outcome of 70 papers selected out of 150. The main objective was to understand the various vulnerable plaque components which lead to CVD. TCFA is the vital plaque component which leads to rupture and thus myocardial infarction. It has been found that the deep feature learning approach can be best suited for tissue characterization and classification.

Appendix A

A.1 Clinical trials

Various OCT-based trials have been reported in the literature for tissue characterization as well as for PCI guidance. This appendix discusses the evidence of widespread utilization of OCT using clinical trials, cohort studies and meta-analyses (figures A1 and A2).

A.1.1 Tissue characterization trials

The noticeable attributes of lipid-rich plaque are detected by OCT because of its high resolution. Lipid-rich plaque is the common cause of major adverse coronary events (MACEs) [1]. The recent clinical trial reported by Xing *et al* [106] showed that lipid-rich plaque is prevalent in non-culprit lesions of the coronary vessel wall and may lead to an MACE. The study was performed considering a cohort of 1474 patients from six countries. One third of the total patients were found to have a lipid-rich plaque in non-culprit lesion using OCT. The results also showed that lipid-rich plaque in a non-culprit lesion-related MACE has a large lipid length, wide lipid arcs and small luminal area. The risk associated with future MACEs has also been provided by quantifying the lipid-rich plaque in the non-culprit region using lipid arc, lipid length, lipid index and stenosis area. The clinical study presented by Shin *et al* [107] used OCT to characterize thrombus and plaque erosion in patients with vasospastic angina (VSA). The study was carried out on 183 patients with VSA.

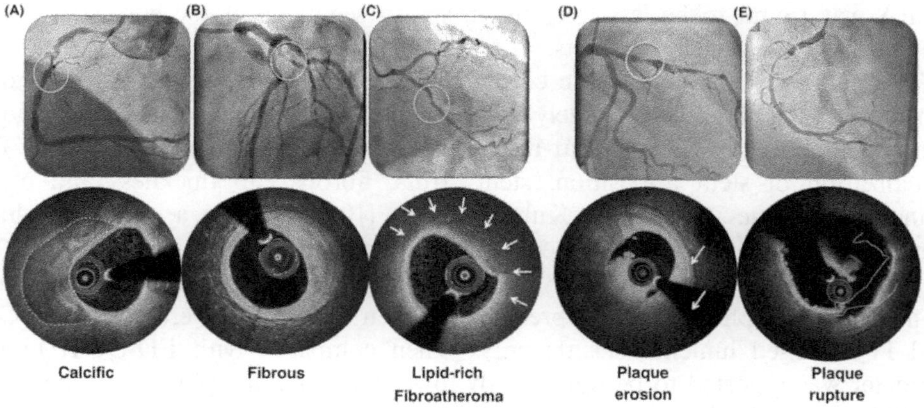

Figure A1. Plaque characterization by OCT. The upper panels show the coronary angiogram and the lower panels show the OCT appearance of the corresponding lesions. (A) Eccentric calcific plaque. Calcium appears in the OCT image as a dark, well-delineated structure from the 6 o'clock to 12 o'clock position. (B) Eccentric fibrous plaque. Fibrous tissue appears signal intense, a homogeneous structure at the 12 o'clock to 5 o'clock position, while in the other circumference a thin, three-layered, normal vessel wall is discernible (corresponding to the intima (signal intense), media (signal poor) and adventitia (signal rich)). (C) Lipid-rich fibroatheroma with a thin cap. The lipid-rich/necrotic core appears as a poor signal, dark with poorly defined borders (arrows) at the 11 o'clock to 3 o'clock position. The media and the adventitia cannot be distinguished due to rapid signal attenuation. (D) Lesion with mural thrombus (dotted line) and without signs of plaque rupture. (E) Lesion with plaque rupture. Remnants of a fibrous cap can be clearly identified as flaps, protruding into the lumen (arrows). (Reproduced with permission from [23]. Copyright Wiley.)

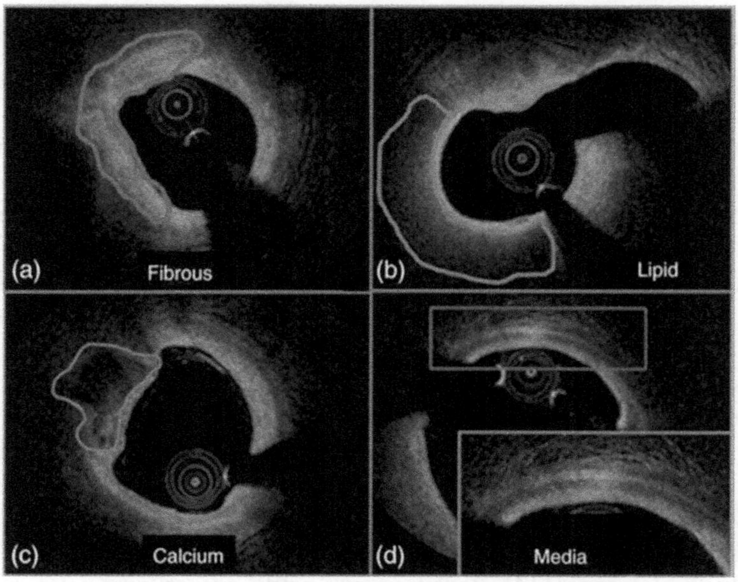

Figure A2. Types of plaque with their OCT appearance. (Reproduced with permission from [30]. Copyright SPIE.)

TCFA was found to be less frequent in spasm segments of the coronary artery compared to non-spasm segments.

The pathophysiology of acute coronary syndrome varies based on the plaque morphology. OCT has been widely used for characterizing microstructural plaque components because of its near-histology resolution. Moreover, it allows the visualization of stent apposition, stent struts, fibrous cap thickness and other vulnerable plaque parameters. Kubo *et al* in [108] reported a study in which coronary assessment was performed using OCT and the analysis was compared to IVUS and angiography. This study included 100 patients from five clinical settings. In addition, five phantoms with predefined lumen dimensions were used for OCT- and IVUS-based lumen measurements. When compared with FD-OCT, lumen diameter was reported to be significantly lower with angiography and higher with IVUS imaging. On the other hand, the phantom study indicated a high similarity of lumen area measured with FD-OCT, whereas IVUS provided overestimated results for the same phantom. Another meta-analysis reported by Iannaccone *et al* [109] verified culprit plaque rupture by OCT and indicated TCFA and smoking as the primary causes of CAD due to plaque rupture. The risk associated with plaque rupture is greater in patients with STEMI than NSTEMI patients. Campos *et al* [110] presented a study to estimate the change in the fibrous cap of coronary artery fibroatheroma within a follow-up period of 6 months using second-generation OCT, the optical frequency domain imaging (OFDI) technique. A cohort of 49 patients with STEMI reported FC volume shrinkage during follow-up. The work presented by Hougaard *et al* [111] analyzed the plaque rupture and healing process after PCI. This indicates the reduction of actual lumen size during the healing process for the follow-up duration. A follow-up duration of 12 months was chosen.

A.1.2 Intravascular PCI trials

PCIs are currently being performed to analyze the culprit lesion in patients with acute coronary artery disease. Angiography has become the first choice for the cardiologist to perform PCIs. However, recent advancements in intravascular imaging, such as OCT and IVUS, have shown the potential to explore more detailed information on the microstructure of the targeted lesion. This helps in understanding the accurate cause of these diseases. A meta-analysis of three random studies has been presented recently by Velagapudi *et al* [112], which compared OCT-guided PCIs with angiography. Death rates are lower using OCT-guided PCIs when compared to angiography-guided PCIs using the hazard ratio (HR). A similar kind of conclusion can be drawn from the study presented by Singh *et al* [113]. This meta-analysis was carried out on 4766 patients from 11 controlled trials. It showed that the risk associated with IVUS-guided coronary interventions is less than that obtained using angiography. OCT has similar risk ratio to IVUS, compared to angiography-guided PCIs. The risk ratios were used for this analysis. Another recent meta-analysis of six clinical trails has been presented by Kuku *et al* [114]. Based on two qualifying criteria, (i) comparison of OCT-guided and angiography-guided PCIs and (ii) a minimum follow-up period of 6 months, a total of 2781 patients (OCT versus angiography guidance 1753 and

OCT versus IVUS guidance 2017 patients) was selected for this analysis. This meta-analysis indicated that PCI with OCT guidance has lower rates of MACEs, cardiac deaths, myocardial infarction and repeat revascularization compared to angiography. However, for MI and revascularization no statistical significance was shown by the study. This study also examined OCT versus IVUS for coronary intervention and concluded that there was no statistically significant difference between the results of both for cardiac events. The estimates of the study have been presented using odds ratios, which measure the relationship between exposure and outcomes [115].

References

[1] Virmani R, Burke A P, Farb A and Kolodgie F D 2006 Pathology of the vulnerable plaque *J. Am. Coll. Cardiol.* **47** C13–8

[2] Virmani R, Burke A P, Kolodgie F D and Farb A 2003 Pathology of the thin-cap fibroatheroma: a type of vulnerable plaque *J. Interv. Cardiol.* **16** 267–72

[3] Ong D S and Jang I- 2015 Fundamentals of optical coherence tomography *Interv. Cardiol. Clin.* **4** 225–37

[4] Roleder T *et al* 2015 The basics of intravascular optical coherence tomography *Postępy Kardiol. Interwencyjnej (Adv. Interv. Cardiol.)* **11** 74

[5] Fujimoto J and Drexler W 2008 Introduction to optical coherence tomography *Optical Coherence Tomography: Technology and Applications* ed W Drexler and J G Fujimoto (Berlin: Springer) pp 1–45

[6] Fujimoto J G and Schmitt J M 2007 Principles of OCT *Optical Coherence Tomography in Cardiovascular Research* ed E Regar, A M G J van Leeuwen and P W Serruys (London: Informa)

[7] Bezerra H G, Costa M A, Guagliumi G, Rollins A M and Simon D I 2009 Intracoronary optical coherence tomography: a comprehensive review: clinical and research applications *JACC: Cardiovasc. Interv.* **2** 1035–46

[8] Terashima M, Kaneda H and Suzuki T 2012 The role of optical coherence tomography in coronary intervention *Korean J. Intern. Med.* **27** 1

[9] Podoleanu A G 2012 Optical coherence tomography *J. Microsc.* **247** 209–19

[10] Hamdan R, Gonzalez R G, Ghostine S and Caussin C 2012 Optical coherence tomography: from physical principles to clinical applications *Arch. Cardiovasc. Dis.* **105** 529–34

[11] Prati F *et al* 2009 Expert review document on methodology, terminology, and clinical applications of optical coherence tomography: physical principles, methodology of image acquisition, and clinical application for assessment of coronary arteries and atherosclerosis *Eur. Heart J.* **31** 401–15

[12] Tearney G J *et al* 1997 *In vivo* endoscopic optical biopsy with optical coherence tomography *Science* **276** 2037–9

[13] Lee J and Saad N 2013 Role of intravascular ultrasound in interventional radiology *Ultrasound Clin.* **8** 185–9

[14] Lee J T and White R A 2004 Basics of intravascular ultrasound: an essential tool for the endovascular surgeon *Seminars in Vascular Surgery: 2004* (Amsterdam: Elsevier) pp 110–8

[15] Finn A V, Chandrashekhar Y and Narula J 2011 IVUS and OCT: either or survivor.... *JACC: Cardiovasc. Imaging* 4 1047–9

[16] Kume T *et al* 2006 Assessment of coronary arterial plaque by optical coherence tomography *Am. J. Cardiol.* **97** 1172–5
[17] Jang I -K 2011 Optical coherence tomography or intravascular ultrasound? *JACC: Cardiovasc. Interv.* **4** 492–4
[18] Jang I-K *et al* 2002 Visualization of coronary atherosclerotic plaques in patients using optical coherence tomography: comparison with intravascular ultrasound *J. Am. Coll. Cardiol.* **39** 604–9
[19] Kume T *et al* 2006 Measurement of the thickness of the fibrous cap by optical coherence tomography *Am. Heart J.* **152** 755.e751–4
[20] Yabushita H *et al* 2002 Characterization of human atherosclerosis by optical coherence tomography *Circulation* **106** 1640–5
[21] Suter M J *et al* 2011 Intravascular optical imaging technology for investigating the coronary artery *JACC: Cardiovasc. Imaging* **4** 1022–39
[22] Tearney G J, Jang I -K and Bouma B E 2006 Optical coherence tomography for imaging the vulnerable plaque *J. Biomed. Opt.* **11** 021002
[23] Regar E *et al* 2016 Identifying stable coronary plaques with OCT technology *Contin. Cardiol. Educ.* **2** 77–88
[24] Matsumoto M *et al* 2012 Morphologic characterization and quantification of superficial calcifications of the coronary artery *in vivo* assessment using optical coherence tomography *Nagoya J. Med. Sci.* **74** 253
[25] Manfrini O *et al* 2006 Sources of error and interpretation of plaque morphology by optical coherence tomography *Am. J. Cardiol.* **98** 156–9
[26] van der Wal A C and Becker A E 1999 Atherosclerotic plaque rupture—pathologic basis of plaque stability and instability *Cardiovasc. Res.* **41** 334–44
[27] Tearney G J *et al* 2012 Consensus standards for acquisition, measurement, and reporting of intravascular optical coherence tomography studies *J. Am. Coll. Cardiol.* **59** 1058–72
[28] Gnanadesigan M *et al* 2016 Automated characterisation of lipid core plaques *in vivo* by quantitative optical coherence tomography tissue type imaging *EuroIntervention* **12** 1490–7
[29] Kim Y, van Soest G, Gnanadesigan M and Johnson T W 2017 A new technique for lipid core plaque detection by optical coherence tomography for prevention of peri-procedural myocardial infarction: a case report *Medicine* **96** e7125
[30] Gargesha M *et al* 2015 Parameter estimation of atherosclerotic tissue optical properties from three-dimensional intravascular optical coherence tomography *J. Med. Imaging* **2** 016001
[31] Riccioni G *et al* 2003 Atherosclerotic plaque formation and risk factors *Int. J. Immunopathol. Pharmacol.* **16** 25–31
[32] Banchhor S K, Londhe N D, Saba L, Radeva P, Laird J R and Suri J S 2017 Relationship between automated coronary calcium volumes and a set of manual coronary lumen volume, vessel volume and atheroma volume in Japanese diabetic cohort *J. Clin. Diagn. Res.* **11** TC09
[33] Kume T *et al* 2006 Assessment of coronary arterial thrombus by optical coherence tomography *Am. J. Cardiol.* **97** 1713–7
[34] Kume T *et al* 2005 Assessment of coronary intima–media thickness by optical coherence tomography *Circ. J.* **69** 903–7
[35] Kume T *et al* 2005 Visualization of neointima formation by optical coherence tomography *Int. Heart J.* **46** 1133–6

[36] Levitz D *et al* 2004 Determination of optical scattering properties of highly-scattering media in optical coherence tomography images *Opt. Express* **12** 249–59
[37] Van Soest G *et al* 2010 Atherosclerotic tissue characterization *in vivo* by optical coherence tomography attenuation imaging *J. Biomed. Opt.* **15** 011105–9
[38] van Leeuwen T G, Faber D J and Aalders M C 2003 Measurement of the axial point spread function in scattering media using single-mode fiber-based optical coherence tomography *IEEE J. Sel. Top. Quantum Electron.* **9** 227–33
[39] Faber D J, Van Der Meer F J, Aalders M C and van Leeuwen T G 2004 Quantitative measurement of attenuation coefficients of weakly scattering media using optical coherence tomography *Opt. Express* **12** 4353–65
[40] Thrane L, Yura H T and Andersen P E 2000 Analysis of optical coherence tomography systems based on the extended Huygens–Fresnel principle *J. Opt. Soc. Am.* A **17** 484–90
[41] van der Meer F J, Faber D J, Sassoon D B, Aalders M C, Pasterkamp G and van Leeuwen T G 2005 Localized measurement of optical attenuation coefficients of atherosclerotic plaque constituents by quantitative optical coherence tomography *IEEE Trans. Med. Imaging* **24** 1369–76
[42] Xu C, Schmitt J M, Carlier S G and Virmani R 2008 Characterization of atherosclerosis plaques by measuring both backscattering and attenuation coefficients in optical coherence tomography *J. Biomed. Opt.* **13** 034003–8
[43] Ughi G J, Adriaenssens T, Sinnaeve P, Desmet W and D'hooge J 2013 Automated tissue characterization of *in vivo* atherosclerotic plaques by intravascular optical coherence tomography images *Biomed. Opt. Express* **4** 1014–30
[44] Vermeer K, Mo J, Weda J, Lemij H and De Boer J 2014 Depth-resolved model-based reconstruction of attenuation coefficients in optical coherence tomography *Biomed. Opt. Express* **5** 322–37
[45] Smith G T *et al* 2015 Automated, depth-resolved estimation of the attenuation coefficient from optical coherence tomography data *IEEE Trans. Med. Imaging* **34** 2592–602
[46] Gnanadesigan M *et al* 2017 Optical coherence tomography attenuation imaging for lipid core detection: an *ex-vivo* validation study *Int. J. Cardiovasc. Imaging* **33** 5–11
[47] Gan Y, Tsay D, Amir S B, Marboe C C and Hendon C P 2016 Automated classification of optical coherence tomography images of human atrial tissue *J. Biomed. Opt.* **21** 101407
[48] Imanaka T *et al* 2013 Analysis of atherosclerosis plaques by measuring attenuation coefficients in optical coherence tomography: thin-cap fibroatheroma or foam cells accumulation without necrotic core? *Eur. Heart J.* **34** P5482
[49] Liu S *et al* 2017 Tissue characterization with depth-resolved attenuation coefficient and backscatter term in intravascular optical coherence tomography images *J. Biomed. Opt.* **22** 096004
[50] Kuppili V *et al* 2017 Extreme learning machine framework for risk stratification of fatty liver disease using ultrasound tissue characterization *J. Med. Syst.* **41** 152
[51] Saba L *et al* 2017 Plaque tissue morphology-based stroke risk stratification using carotid ultrasound: a polling-based PCA learning paradigm *J. Med. Syst.* **41** 98
[52] Sharma A M *et al* 2015 A review on carotid ultrasound atherosclerotic tissue characterization and stroke risk stratification in machine learning framework *Curr. Atheroscler. Rep.* **17** 55

[53] Acharya U R *et al* 2013 Atherosclerotic plaque tissue characterization in 2D ultrasound longitudinal carotid scans for automated classification: a paradigm for stroke risk assessment *Med. Biol. Eng. Comput.* **51** 513–23

[54] Rico-Jimenez J J, Campos-Delgado D U, Villiger M, Otsuka K, Bouma B E and Jo J A 2016 Automatic classification of atherosclerotic plaques imaged with intravascular OCT *Biomed. Opt. Express* **7** 4069–85

[55] Schmitt J, Knüttel A and Bonner R 1993 Measurement of optical properties of biological tissues by low-coherence reflectometry *Appl. Opt.* **32** 6032–42

[56] Schmitt J M, Knuttel A, Yadlowsky M and Eckhaus M 1994 Optical-coherence tomography of a dense tissue: statistics of attenuation and backscattering *Phys. Med. Biol.* **39** 1705

[57] Athanasiou L S *et al* 2014 Methodology for fully automated segmentation and plaque characterization in intracoronary optical coherence tomography images *J. Biomed. Opt.* **19** 026009

[58] Abdolmanafi A, Duong L, Dahdah N and Cheriet F 2017 Deep feature learning for automatic tissue classification of coronary artery using optical coherence tomography *Biomed. Opt. Express* **8** 1203–20

[59] Young W, Gofman J and Tandy R 1960 The quantification of atherosclerosis III. The extent of correlation of degrees of atherosclerosis within and between the coronary and cerebral vascular beds *Am. J. Cardiol.* **6** 300–8

[60] Mitchell J and Schwartz C 1962 Relationship between arterial disease in different sites *Br. Med. J.* **1** 1293

[61] Pignoli P, Tremoli E, Poli A, Oreste P and Paoletti R 1986 Intimal plus medial thickness of the arterial wall: a direct measurement with ultrasound imaging *Circulation* **74** 1399–406

[62] Karanasos A, van der Sijde J N, Ligthart M J, Witberg K and Regar E 2015 *Utility of Optical Coherence Tomography Imaging with Angiographic Co-registration for the Guidance of Percutaneous Coronary Intervention* (Veenendaal: St Jude Medical Nederland)

[63] Maas A and Appelman Y 2010 Gender differences in coronary heart disease *Netherlands Heart J.* **18** 598–603

[64] Towfighi A, Zheng L and Ovbiagele B 2009 Sex-specific trends in midlife coronary heart disease risk and prevalence *Arch. Intern. Med.* **169** 1762–6

[65] ILUMIEN™ OPTIS™ PCI OPTIMIZATION™ SYSTEM www.cardiovascular.abbott/us/en/hcp/resources/product/catalog/ilumien-optis-pci-optimizationsystem.html (Accessed: 1 July 2018)

[66] Greenland P *et al* 2010 ACCF/AHA guideline for assessment of cardiovascular risk in asymptomatic adults: a report of the American College of Cardiology Foundation/American Heart Association task force on practice guidelines developed in collaboration with the American Society of Echocardiography, American Society of Nuclear Cardiology, Society of Atherosclerosis Imaging and Prevention, Society for Cardiovascular Angiography and Interventions, Society of Cardiovascular Computed Tomography, and Society for Cardiovascular Magnetic Resonance *J. Am. Coll. Cardiol.* **56** e50–103

[67] Chambless L E *et al* 1997 Association of coronary heart disease incidence with carotid arterial wall thickness and major risk factors: the Atherosclerosis Risk in Communities (ARIC) Study, 1987–1993 *Am. J. Epidemiol.* **146** 483–94

[68] Saba L, Sanches J M, Pedro L M and Suri J S 2014 *Multi-modality Atherosclerosis Imaging and Diagnosis* (Berlin: Springer)
[69] Laine A, Sanches J M and Suri J S 2012 *Ultrasound Imaging: Advances and Applications* (Berlin: Springer)
[70] Polak J F *et al* 2010 Associations of carotid artery intima–media thickness (IMT) with risk factors and prevalent cardiovascular disease *J. Ultrasound Med.* **29** 1759–68
[71] Boas F E and Fleischmann D 2012 CT artifacts: causes and reduction techniques *Imaging Med.* **4** 229–40
[72] Nambi V *et al* 2010 Carotid intima–media thickness and presence or absence of plaque improves prediction of coronary heart disease risk: the ARIC (Atherosclerosis Risk In Communities) study *J. Am. Coll. Cardiol.* **55** 1600–7
[73] Ogata T, Yasaka M, Yamagishi M, Seguchi O, Nagatsuka K and Minematsu K 2005 Atherosclerosis found on carotid ultrasonography is associated with atherosclerosis on coronary intravascular ultrasonography *J. Ultrasound Med.* **24** 469–74
[74] Eder L, Gladman D, Ibañez D and Urowitz M 2014 The correlation between carotid artery atherosclerosis and clinical ischemic heart disease in lupus patients *Lupus* **23** 1142–8
[75] Amato M *et al* 2017 Carotid plaque-thickness and common carotid IMT show additive value in cardiovascular risk prediction and reclassification *Atherosclerosis* **263** 412–9
[76] Matsushima Y *et al* 2004 Relationship of carotid intima–media thickness, pulse wave velocity, and ankle brachial index to the severity of coronary artery atherosclerosis *Clin. Cardiol.* **27** 629–34
[77] Lekakis J P *et al* 2005 Intima–media thickness score from carotid and femoral arteries predicts the extent of coronary artery disease *Int. J. Cardiovasc. Imaging* **21** 495–501
[78] Gepner A D *et al* 2017 Comparison of carotid plaque score and coronary artery calcium score for predicting cardiovascular disease events: the multi-ethnic study of atherosclerosis *J. Am. Heart Assoc.* **6** e005179
[79] Wasilewski J *et al* 2015 Predominant location of coronary artery atherosclerosis in the left anterior descending artery. The impact of septal perforators and the myocardial bridging effect *Kardiochir. Torakochirurgia Pol. (Pol. J. Cardio-Thorac. Surg.)* **12** 379
[80] Polak J F, Pencina M J, Herrington D and O'leary D H 2011 Associations of edge-detected and manual-traced common carotid intima–media thickness measurements with framingham risk factors *Stroke* **42** 1912–6
[81] Roy A G *et al* 2016 Lumen segmentation in intravascular optical coherence tomography using backscattering tracked and initialized random walks *IEEE J. Biomed. Health Inform.* **20** 606–14
[82] Nam H S, Kim C S, Lee J J, Song J W, Kim J W and Yoo H 2016 Automated detection of vessel lumen and stent struts in intravascular optical coherence tomography to evaluate stent apposition and neointimal coverage *Med. Phys.* **43** 1662–75
[83] Fedele S *et al* 2012 Reproducibility of coronary optical coherence tomography for lumen and length measurements in humans (The CLI-VAR [Centro per la Lotta contro l'Infarto-VARiability] study) *Am. J. Cardiol.* **110** 1106–12
[84] Zahnd G *et al* 2017 Contour segmentation of the intima, media, and adventitia layers in intracoronary OCT images: application to fully automatic detection of healthy wall regions *Int. J. Comput. Assist. Radiol. Surg.* **12** 1923–36

[85] Wang Z *et al* 2012 Volumetric quantification of fibrous caps using intravascular optical coherence tomography *Biomed. Opt. Express* **3** 1413–26
[86] Zahnd G *et al* 2014 Semi-automated quantification of fibrous cap thickness in intracoronary optical coherence tomography *Int. Conf. on Information Processing in Computer-assisted Interventions: 2014* (Berlin: Springer) pp 78–89
[87] Zahnd G *et al* 2015 Quantification of fibrous cap thickness in intracoronary optical coherence tomography with a contour segmentation method based on dynamic programming *Int. J. Comput. Assist. Radiol. Surg.* **10** 1383–94
[88] Mehanna E *et al* 2013 Volumetric characterization of human coronary calcification by frequency-domain optical coherence tomography *Circ. J.* **77** 2334–40
[89] Mintz G S 2015 Intravascular imaging of coronary calcification and its clinical implications *JACC: Cardiovasc. Imaging* **8** 461–71
[90] Banchhor S K *et al* 2016 Five multiresolution-based calcium volume measurement techniques from coronary IVUS videos: a comparative approach *Comput. Methods Programs Biomed.* **134** 237–58
[91] Tearney G J *et al* 2003 Quantification of macrophage content in atherosclerotic plaques by optical coherence tomography *Circulation* **107** 113–9
[92] Di Vito L *et al* 2015 Identification and quantification of macrophage presence in coronary atherosclerotic plaques by optical coherence tomography *Eur. Heart J.-Cardiovasc. Imaging* **16** 807–13
[93] Mohammed H A 2016 The image registration techniques for medical imaging (MRI-CT) *Am. J. Biomed. Eng.* **6** 53–8
[94] Araki T *et al* 2015 A comparative approach of four different image registration techniques for quantitative assessment of coronary artery calcium lesions using intravascular ultrasound *Comput. Methods Programs Biomed.* **118** 158–72
[95] Li H and Fan Y 2017 *Non-rigid image registration using fully convolutional networks with deep self-supervision* (arXiv:170900799).
[96] Makela T *et al* 2002 A review of cardiac image registration methods *IEEE Trans. Med. Imaging* **21** 1011–21
[97] Chiastra C *et al* 2017 Reconstruction of stented coronary arteries from optical coherence tomography images: feasibility, validation, and repeatability of a segmentation method *PLoS One* **12** e0177495
[98] Ughi G J *et al* 2012 Automatic three-dimensional registration of intravascular optical coherence tomography images *J. Biomed. Opt.* **17** 0260051–511
[99] Prabhu D *et al* 2016 Three-dimensional registration of intravascular optical coherence tomography and cryo-image volumes for microscopic-resolution validation *J. Med. Imaging* **3** 026004
[100] Hebsgaard L *et al* 2015 Co-registration of optical coherence tomography and x-ray angiography in percutaneous coronary intervention. The Does Optical Coherence Tomography Optimize Revascularization (DOCTOR) fusion study *Int. J. Cardiol.* **182** 272–8
[101] de Graaf M A *et al* 2013 Automatic quantification and characterization of coronary atherosclerosis with computed tomography coronary angiography: cross-correlation with intravascular ultrasound virtual histology *Int. J. Cardiovasc. Imaging* **29** 1177–90

[102] Honda Y *et al* 2014 Coronary artery calcification as a new predictor of non-target lesion revascularization during the chronic phase after successful percutaneous coronary intervention *Cardiovasc. Interv. Ther.* **29** 315–23

[103] Araki T *et al* 2016 PCA-based polling strategy in machine learning framework for coronary artery disease risk assessment in intravascular ultrasound: a link between carotid and coronary grayscale plaque morphology *Comput. Methods Programs Biomed.* **128** 137–58

[104] Araki T *et al* 2016 A new method for IVUS-based coronary artery disease risk stratification: a link between coronary and carotid ultrasound plaque burdens *Comput. Methods Programs Biomed.* **124** 161–79

[105] Fent G J, Greenwood J P, Plein S and Buch M H 2017 The role of non-invasive cardiovascular imaging in the assessment of cardiovascular risk in rheumatoid arthritis: where we are and where we need to be *Ann. Rheuma. Dis.* **76** 1169

[106] Xing L *et al* 2017 Clinical significance of lipid-rich plaque detected by optical coherence tomography: a 4-year follow-up study *J. Am. Coll. Cardiol.* **69** 2502–13

[107] Shin E-S *et al* 2016 Thrombus and plaque erosion characterized by optical coherence tomography in patients with vasospastic angina *Rev. Esp. Cardiol. (Engl. Ed)* **70** 459–66

[108] Kubo T *et al* 2013 OCT compared with IVUS in a coronary lesion assessment: the OPUS-CLASS study *JACC: Cardiovasc. Imaging* **6** 1095–104

[109] Iannaccone M *et al* 2015 Prevalence and predictors of culprit plaque rupture at OCT in patients with coronary artery disease: a meta-analysis *Eur. Heart J.-Cardiovasc. Imaging* **17** 1128–37

[110] Campos C M *et al* 2018 Serial volumetric assessment of coronary fibroatheroma by optical frequency domain imaging: insights from the TROFI trial *Eur. Heart J.-Cardiovasc. Imaging* **19** 92–100

[111] Hougaard M, Hansen H S, Thayssen P, Antonsen L and Jensen L O 2018 Uncovered culprit plaque ruptures in patients with ST-segment elevation myocardial infarction assessed by optical coherence tomography and intravascular ultrasound with iMap *JACC: Cardiovasc. Imaging* **11** 859–67

[112] Velagapudi P, Turagam M, Aronow H, Khera S, Kolte D and Abbott J 2017 Angiography versus optical coherence tomography guided percutaneous coronary intervention: a meta-analysis of clinical outcomes *J. Am. Coll. Cardiol.* **69** 1113

[113] Singh S, Kovacs D, Parmar S, Arora R and Khosla S 2017 TCT-294 optical coherence tomography or intravascular ultrasound guided PCI leads to superior clinical outcomes in comparison to conventional coronary angiography: a network meta-analysis of randomized controlled trials *J. Am. Coll. Cardiol.* **70** B120

[114] Kuku K O *et al* 2018 Optical coherence tomography-guided percutaneous coronary intervention compared with other imaging guidance: a meta-analysis *Int. J. Cardiovasc. Imaging* **34** 503–13

[115] Szumilas M 2010 Explaining odds ratios *J. Can. Acad. Child Adolesc. Psychiatry* **19** 227

Chapter 4

Rheumatoid arthritis: its link to atherosclerosis imaging and cardiovascular risk assessment using machine-learning-based tissue characterization

Ankush D Jamthikar, Narendra N Khanna, Matteo Piga, Luca Saba, Deep Gupta, Argiris A Giannopoulos, Carlo Carcassi, Andrew Nicolaides, John R Laird, Harman S Suri, Sophie Mavrogeni, George D Kitas and Jasjit S Suri

Rheumatoid arthritis (RA) is a chronic autoimmune disease characterized by synovial joint inflammation. Auto-antibodies and cytokines play a crucial role in the pathogenesis of RA. Extra-articular manifestations include vascular inflammation that may result in a higher risk of cardiovascular (CV) events and stroke. Prevalence of cardiovascular events driven by atherosclerosis is high in RA patients. However, the exact mechanism of atherosclerosis driven by RA is uncertain. Traditional risk factors which are helpful in deriving conventional risk scores using the general population do not provide accurate risk prediction in RA patients. Advancements in medical imaging have facilitated early and accurate risk stratification in vascular diseases compared to traditional risk calculators. Imaging the carotid artery using 2D ultrasound is a non-invasive, economic and efficient imaging approach that provides an atherosclerotic plaque tissue-specific image. Further, it provides support in tissue characterization for cardiovascular disease and stroke risk prediction. Intelligence-based paradigms such as machine-learning and deep-learning techniques not only automate the characterization process but also provide accurate cardiovascular risk stratification for better management of RA diseased patients. This review provides a brief understanding of the pathogenesis of RA and its association with carotid atherosclerosis imaged using B-mode ultrasound techniques. Lacunas in traditional risk scores and the

doi:10.1088/2053-2563/ab0820ch4

role of machine-learning-based tissue characterization algorithms are discussed, which could facilitate cardiovascular risk assessment in RA patients.

Acronym list
RA: rheumatoid arthritis; CUS: carotid ultrasound; OCT: optical coherence tomography; CT: computed tomography; MRI: magnetic resonance imaging; ML: machine learning; DL: deep learning; CNN: convolutional neural network

4.1 Introduction

RA is an autoimmune rheumatic disease causing inflammation of synovial joints. It affects around one percent of the adult population, with females being more likely to be affected than males [1, 2]. With a few exceptions, the prevalence of RA is similar in developed countries and low or middle-income regions of the world [3, 4]. The initial cause of RA remains unclear. Genetic factors (such as the shared epitope of human leukocyte antigen (HLA)-DR), environmental factors (such as smoking) and immune deregulations all play a vital role in RA pathogenesis [5, 6]. This could be in part through the generation of auto-antibodies such as rheumatoid factor (RF) and anti-citrullinated protein antibodies (ACPA) [7, 8], which may have important downstream effects on the immune system.

Risk assessment based on traditional scores has been reported to underestimate cardiovascular disease (CVD) risk in RA patients [9, 10]. One reason behind this inaccurate risk estimation is an exclusive dependence on risk calculators such as the Framingham risk score, which are based on traditional CV risk parameters and derived for the general population. In order to further accurately determine the risk assessment, RA-specific factors may also be required in the risk calculations, as recently attempted by Solomon *et al* [11], who proposed an expanded risk score for RA patients. Other suggestions include the use of a multiplication factor to reflect an increase in risk (for example as per the recently updated EUropean League Against Rheumatism (EULAR) guidance [12] and Cox *et al* [13]), the inclusion of RA as a risk factor in newer versions of established risk calculators (e.g. QRisk II and QRisk III), or inclusion of the inflammatory marker C-reactive protein in the algorithms (e.g. in the Reynold's risk score) [14].

Unfortunately, the evidence suggests that the above approaches lack a comprehensive paradigm in which the image-based characteristics are accounted for, thereby cannot achieve robust prediction of CVD risk in RA patients [15, 16]. The limitations of these scores together with advancements in imaging technology have generated an interest in the utilization of imaging pathways for risk assessment that may provide additional information about the health status of the vascular wall and visualization of the targeted lesion via cross-sectional images.

The recent improvements in multi-modality imaging techniques such as ultrasound (US), computed tomography (CT), magnetic resonance imaging (MRI) and nuclear imaging techniques such as positron emission tomography (PET) have the potential to characterize different biomarkers of CVD [17]. Cardiac CT and MRI have reported changes to the coronary plaque morphology by providing visualization of coronary calcification and vulnerable plaque characteristics [18, 19]. Fluorodeoxyglucose PET (FDG-PET) is a nuclear imaging modality that uses a radioactive material which is normally taken up at sites of active inflammation and, in conjunction with CT, provides possibly an even better visualization of the target lesion [20, 21]. Intravascular optical coherence tomography (IVOCT) is a high-resolution imaging modality (1 pixel = 10–15 μm) that provides the visualization of high-risk microstructural plaque components that are prone to rupture [22]. Carotid ultrasound (CUS) is a non-invasive and cost-efficient imaging modality that provides information about subclinical atherosclerotic carotid plaque and the thickness of the intima and media (cIMT) layers that are associated with CVD risk at an early stage in RA patients [23, 24]. We present a detailed review of various image-based vascular (in particular carotid and coronary) morphology techniques utilizing state-of-the-art machine-learning and deep-learning paradigms for CVD risk stratification in RA patients.

It becomes even more vital to address the CVD risk in RA patients using automated techniques, since manual assessment of image-based characterization for carotid and coronary images leads to significant intra- or inter-observer variability. Thus, intelligence-based strategies are an important pathway and their automation may provide a higher level of comfort and confidence in the eyes of therapy decision makers such as specialty physicians. Keeping the above challenges in mind, the primary objectives of this review are: (i) to understand the role of atherosclerosis driven by RA; (ii) to establish a link between various machine and deep-learning paradigms for morphology-based tissue characterization in CUS and IVOCT in RA patients; and (iii) to envision a future research plan which takes a multi-disciplinary approach to address an important CVD challenge more closely driven by RA rather than standalone conventional CVD.

4.2 Search strategy

The design of this review article is based on rigorous searches on PubMed and Web of Science using the following keywords: tissue characterization in rheumatoid arthritis, pathogenesis of rheumatoid arthritis, angiogenesis in RA, role of anti-bodies in rheumatoid arthritis, CVD risk prediction in RA patients, RA-specific CVD risk factors, non-invasive imaging in rheumatoid arthritis, and machine-learning and deep-learning-based tissue characterization in carotid ultrasound. Furthermore, the bibliographies from research publications of experts in the medical imaging domain have also been shortlisted for this review. Preference was given to research articles published in the last ten years. All the points discussed in this chapter were initially discussed with experts in the field of rheumatology and ultrasound-based tissue characterization.

4.3 Brief description of the pathogensis of rheumatoid arthritis

In order to study vascular tissue characterization in RA patients using intelligence-based learning techniques, it is essential to first understand the mechanism of RA and the sources of pro-inflammatory cytokines that aggressively accelerate atherosclerosis.

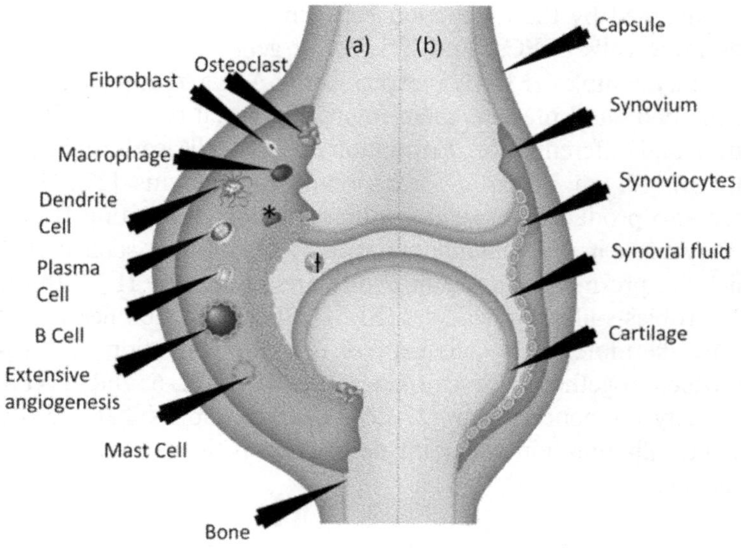

Figure 4.1. Schematic view of (a) a joint affected by RA and (b) a normal joint. (Courtesy of AtheroPoint, Roseville, CA, USA.)

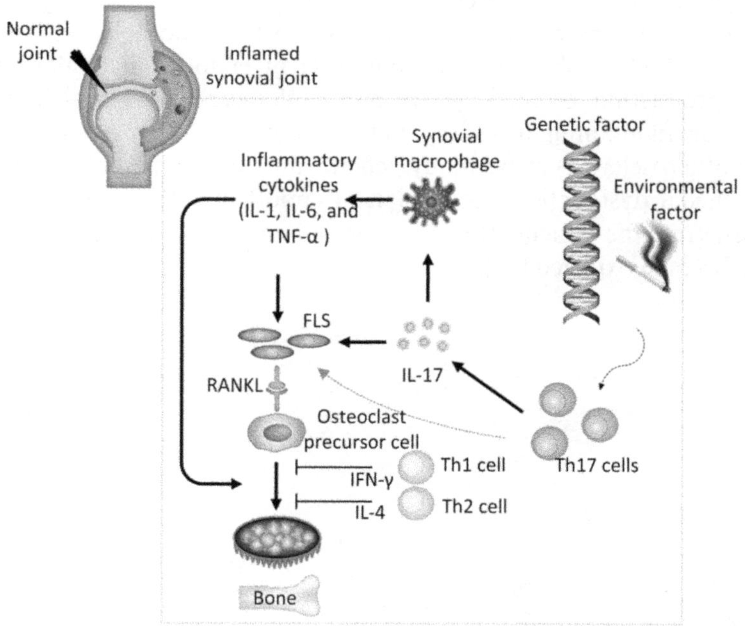

Figure 4.2. Pathogenesis of rheumatoid arthritis. (Courtesy of AtheroPoint, Roseville, CA, USA.)

As shown before in figure 4.1, synovial inflammation is the main attribute of RA [25] and, if not sufficiently well controlled, leads to cartilage and bone destruction. A simplified version of the pathogenesis of RA is shown in figure 4.2. The progression of

RA can be explained by the interaction between genetic or environmental factors, antigen presenting cells (APCs) and T cells. A gene compound known as major histocompatibility complex (MHC) is expressed by APCs, that recognizes the foreign antigens in the body and presents them to specific T cell receptors [26], resulting in their activation and differentiation. Formation of Th17 cells leads to the production of Interleukin (IL)−17, an important contributor to synovitis [27]. Thl and Th2 cytokines are also produced via cellular differentiation, generating IFN-γ and IL-4, which also play a role in synovial inflammation. The cytokines such as IL-1, IL-6 and TNF-α which are pro-inflammatory in nature are induced by IL-17 through macrophages and fibroblasts-like synoviocytes (FLS) [28]. These cytokines activate synovial fibroblasts. Furthermore, FLS cells express receptor activation of NF-κB ligands (RANKL) which together with cytokines (i.e. IL-1, IL-6 and TNF-α) lead to osteoclast activity for bone erosion [27, 29]. Currently, pro-inflammatory cytokine-driven synovial inflammation and joint destruction is an important concept in RA pathogenesis [30].

4.4 Atherosclerosis driven by rheumatoid arthritis

Inflammation is the common link between atherosclerosis and RA [31]. The progression rate of plaque formation in the vessel wall is higher in RA patients [32]. Inflammatory cytokines are confined to the synovium but have systemic effects leading to vascular inflammation and damage to the endothelial cells (figure 4.3) [33]. IL-6 and TNF-α play an important role in endothelial damage since they inhibit the production of nitric oxide and cyclooxygenase-1 which in turn are responsible for maintaining a healthy endothelium.

RA and atherosclerosis share many genetic and environmental factors that may cause endothelial dysfunction. Activation of endothelial cells allows LDL cholesterols to penetrate the lumen–intima border and enter into the subendothelial layer where they become oxidized [34]. This increases the permeability of the endothelium

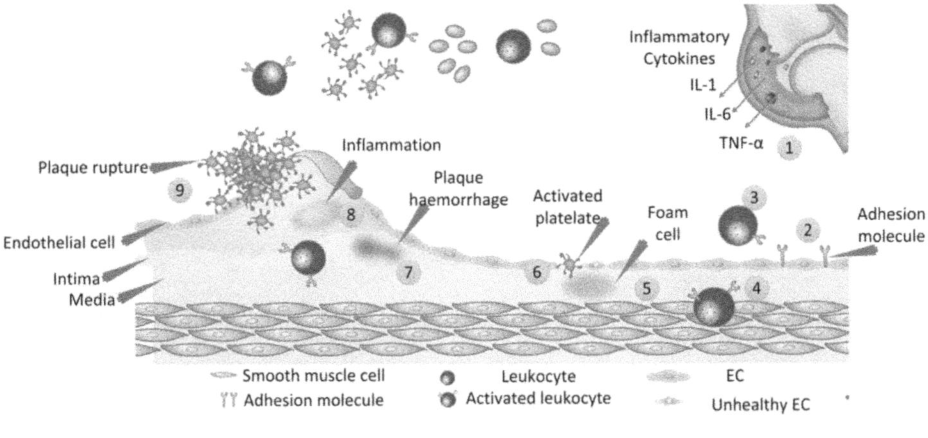

Figure 4.3. Mechanism of atherosclerosis driven by RA. (Courtesy of AtheroPoint, Roseville, CA, USA.)

which results in the growth of immune cells such as T lymphocytes and monocytes in the intimal layer. Upon entering the intimal layer, monocytes are transformed into macrophages and take up oxidized LDL cholesterol, transforming into foam cells [35]. Macrophages further generate pro-inflammatory cytokines such as IL-6 and TNF-α that recruit more monocytes within the intimal layer [36].

Furthermore, macrophages are responsible for the migration of smooth muscle cells (SMCs) and their proliferation into the intima. These SMCs form a thin fibrous protective lining to prevent the encroachment of plaque into the lumen [37]. Pro-inflammatory cytokines, free radicals and enzymes produced by T helper cells and macrophages cause fibrous cap erosion and make the cap vulnerable [38, 39]. Pro-inflammatory TNF-α intensifies the oxidation of LDL cholesterol with levels reported to be higher in RA patients than healthy controls [40, 41]. Furthermore, TNF-α also elevates the expression of adhesion molecules on the endothelial surface and hence, further increases monocyte/macrophage recruitment. Amplification of the inflammatory process leads to plaque formation and rupture resulting in thrombotic events in atherosclerosis.

4.5 The role of platelets in atherothrombosis in RA

Platelets play an important role in vascular inflammation, atherosclerosis and thrombosis in RA patients [42]. The atherothrombosis process involves various stages, starting with endothelial damage, deposition of LDL-C, recruitment of lymphocytes and monocytes/macrophages, formation of plaque and finally plaque rupture [43]. Once the plaque ruptures, platelets start accumulating on the damaged lesion leading to a blood clot or thrombosis [43]. A direct link to platelets in synovial inflammation is not clear, however, some studies have reported an increase of platelets in the synovium and synovial fluid in RA [44, 45]. Pathogenic factors such as inflammatory cytokines (IL-1, IL-6 and TNF-α), oxidized LDL-C, CRP and oxidative stress are linked with the association of platelets that increases the risk of CVD [42].

4.6 The role of amyloidosis in RA

Amyloidosis is a disease of tissue damage due to the deposition of amyloids within the tissues [46]. Amyloids are starch-like proteins of abnormal shapes, primarily produced in bone marrow, and can be deposited in different tissues or body organs. Different types of amyloids affect different body organs such as the heart, liver, kidney, digestive tract, spleen and nervous system. Untreated amyloidosis may lead to the failure of body organs [47]. Amyloid A (AA) amyloidosis is one of the life threatening complications of RA caused by deposition of AA fibrils on different parts of the organ tissues [48]. Amyloid A fibrils are generated from the circulating serum amyloid A (SAA) which is a phase-reactant protein [48], predominantly produced by the liver [49]. SAA is a potential biomarker of inflammation and also contributes in the pathogenesis of RA [49]. Large inflammation in rheumatoid arthritis increases the levels of serum amyloid A (SAA) [50]. Controlling the inflammation using TNF-blocking agents may also aid physicians to prevent the onset of SAA driven amyloidosis. A recent study presented by Majdan and Targońska-Stępniak [51] reported a high association of SAA amyloidosis with RA disease activity, which further results in increased risk of CV and renal diseases in RA patients. Use of SAA may aid a better risk stratification of RA patients using advanced intelligence-based methods such as machine-learning and deep-learning techniques.

4.7 Traditional CV risk factors in rheumatoid arthritis

Prior to envisioning the development of an intelligence-based paradigm to predict the risk of CV comorbidities in RA, the role of traditional and RA-specific risk parameters must be evaluated. Traditional risk factors play an influential role in assessing CV risk. Its contribution leading to CV mortality has been widely assessed [52–54]. Some traditional risk factors have indicated a paradoxical behavior which is specifically observed in inflammatory joint disease such as RA [55, 56]. In addition to the adjustment of traditional risk factors, CV risk does remain elevated in RA. The interplay between the conventional risk factors and RA is discussed below.

4.7.1 Body mass index and physical inactivity

Body mass index (BMI) has a paradoxical behavior in RA patients with a three-fold increase in CV mortality rate for low BMI ($< 20\,\text{kg}\,\text{m}^{-2}$) [55, 57]. However, in non-RA patients, a low BMI reflects a protective role against CV events. Furthermore, a low BMI in RA patients stimulates the systemic inflammation that leads to severe cardiac events. The main reason for this paradoxical effect is the alteration of body mass composition in RA patients, which is uncommon in patients without RA [58]. Patients with RA experience physical inactivity because of joint pain, fatigue, stiffness and a lack of awareness about future CV events [59–61]. Systemic inflammation expresses RANKL proteins, which support the degradation of muscle mass, which is also termed cachexia [62]. Two types of cachexia, classic cachexia and rheumatoid cachexia, indicate the variation in muscle mass and body fat mass. Classic cachexia is rare and indicates a low muscle and body fat mass. Rheumatoid cachexia indicates an increase in body fat mass (as a result of obesity) with a simultaneous decrease in muscle mass (as a result of inflammation) [63]. Both types of cachexia have shown to have an effect on the elevation of CVD risk in RA patients [63]. Elkan *et al* [64] reported an association of rheumatoid cachexia to an elevation of LDL cholesterol level, which further leads to an increase in atherosclerosis. As in non-RA patients, RA patients also represented an elevation in BMI, which further leads to coronary atherosclerosis [65, 66]. Continuation of physical inactivity in RA has shown to increase the BMI resulting in a progression of RA [67], thereby contributing towards CVD risk [68].

4.7.2 Lipids

An increase in both LDL and total cholesterol levels (i.e. hypercholesterolemia) is commonly observed in non-RA patients and is associated with a risk of CV events. However, RA patients demonstrate a paradoxical role of lipids, with a reduction in cholesterol (i.e. for both total and LDL cholesterol levels) being highly associated with an increase in CVD risk [69, 70]. Inflammation in RA not only reduces the lipids, but also modifies the lipid structure and function [70]. For example, HDL cholesterol is highly suppressed in RA, which behaves as a proatherogenic instead of antiatherogenic lipid component. Furthermore, lower values of lipids coincide with an elevation in inflammation, and this further increases the CV risk [56]. In order to perform the CV risk assessment, a ratio of total cholesterol to HDL cholesterol can be an important parameter in an intelligence-based risk assessment and stratification system.

4.7.3 Hypertension

Hypertension is another risk factor which significantly contributes in the development of CVD risk in RA patients [71] and in the general population. There exist two schools of thought pertaining to the role of hypertension in the development of CVD. Some studies indicated a high prevalence of hypertension in RA patients [72, 73], while another study by Boyer *et al* [74] showed a little association of hypertension with RA. Recently, the COMOrbidities in Rheumatoid Arthritis (COMORA) study indicated a prevalence of 40.4% hypertension in RA patients. Another recent study by Balsa *et al* [75] reported a prevalence of 41% hypertension in the recruited RA patients. Several sources such as activation of inflammation, specific genetic polymorphism and the use of antirheumatic drugs have been reported for the initiation of hypertension through an increase in peripheral vascular resistance [76–78].

4.7.4 Smoking

Smoking is a vital CVD risk factor [79], and also plays an important role in RA [80]. Prevalence of smoking is 1.5 times higher in patients with RA compared to controls. Smoking influences the production of RF, ACPA, rheumatoid nodules and rheumatoid cachexia which are the key factors in the development of CVD [81–83]. Although, some studies reported a high prevalence of cigarette smoking in RA patients [84, 85], another study by Gonzalez *et al* has shown a lesser association of smoking with the development of CVD [52].

4.7.5 Insulin resistance and diabetes

The risk of CVD in RA patients has been observed to be similar to that of diabetes mellitus [86]. The CARdiovascular Research in RhEumatoid Arthritis (CARRE) study compared the CVD risk in both RA and diabetes and reported that the proportion of patients with CVD risk in RA is comparable to that in patients with diabetes [87]. Furthermore, a Danish nationwide cohort study presented by Lindhardsen *et al* [88] also found a similar association of myocardial infarction risk in RA patients compared with patients with diabetes. Insulin resistance (IR) is an important diabetic characteristic that is prevalent in RA patients and contributes to the risk of atherosclerosis [89]. Diabetes commonly follows after IR and its mechanism is highly correlated with the levels of inflammatory markers [90, 91].

4.7.6 Ankle–brachial index and arterial stiffness

The ankle–brachial index (ABI) is a ratio of the systolic blood pressure at the ankle and brachial artery and is considered as one of the potential risk factors responsible for CV events. Both high and low values of ABI are associated with a risk of CVD [92, 93]. Some of these studies have reported that a low value of ABI (ABI < 0.9) is associated with an increased risk of CV mortality of 18.7% in males and 12.6% in females, respectively [94]. The risk of coronary heart disease and stroke also increases with ABI < 0.9 [95]. Furthermore, ABI $\geqslant 1.4$ has also been reported to increase the risk of myocardial infarction [93]. ABI is also reported to be linked with an increase in cIMT and plaque in RA patients [96]. In addition to ABI, arterial stiffness is also an important CVD risk factor which is generally elevated in RA patients compared to controls [97]. RA patients experience a high risk of CV events with high arterial stiffness measured using the aortic pulse wave velocity and augmentation index [98].

4.8 RA-specific CV risk factors in rheumatoid arthritis

CVD risk stratification requires inclusion of RA related inflammatory risk markers which are associated with vascular atherosclerosis [99]. Autoimmune RA is commonly characterized by the presence of inflammatory markers (such as erythrocyte sedimentation rate (ESR) and C-reactive protein (CRP)) and antibodies (such as RF and ACPA) [30]. Wallberg-Jonsson *et al* [100, 101] presented a cohort study of 606 RA patients and reported an increase in the CVD mortality rate for seropositive RA. Recently, Majka *et al* [102] presented a study using an African cohort which showed an independent association of RF and ACPA against subclinical atherosclerosis that leads to CVD events. ESR and CRP are vital inflammatory markers and are generally associated with atherosclerotic disease and CVD risk in patients with RA [103]. CRP is a protein generated by the liver and is reported to have a higher concentration in patients with RA. Some studies have reported a direct effect of CRP on the vessel wall which may promote atherosclerosis [104] and coronary heart disease [105]. Higher levels of CRP and ESR are also responsible for an increase in carotid intima–media thickness (cIMT) [106]. Recently, the disease activity score (DAS28) has been reported to have a significant association with cIMT in women [107]. DAS28 is a tool that examines 28 different joints for the analysis of RA disease activity.

4.9 Conventional CV risk algorithms

Risk prediction algorithms should aid the clinician to stratify the patients into either high risk or low risk for CVD risk management. Having RA elevates the risk of atherosclerosis and hence CVD risk algorithms need better intelligence derivatives that reflect the pathogenic changes in the vascular wall. In the last decade, various algorithms have been developed to explain the risk of CVD, such as the Framingham risk score (FRS) [108], systematic coronary risk evaluation (SCORE) [109], modified SCORE (mSCORE) [12], QRISK2 [13], Reynold's risk score [14], pooled cohort equation (PCE) [110] and expanded risk score for rheumatoid arthritis (ERS-RA) [11]. Recent studies have reported that in males, smoking and cardiac history have less association with CV risk in RA patients [52], and hence risk scores developed using these factors do not function well in RA patients. Furthermore, risk indices such as FRS and SCORE use the classical CVD risk parameters and have been shown to underestimate the CVD risk in RA patients [111]. In order to predict the risk in RA patients, the 2009 EULAR task force recommended an index called the modified SCORE (mSCORE), which was 1.5 times the CVD risk determined using traditional risk scores [12]. The use of mSCORE was allowed if the patient satisfied two of three conditions: (i) the duration of RA should be longer than 10 years, (ii) diagnosed with either RF or ACPA and (iii) the presence of extra-articular manifestation. However, the third requirement was excluded by the recent revised EULAR recommendation [112]. The QRISK2 calculator was developed for English and Welsh populations and used RA as an independent factor for CVD risk [13]. An additional multiplication factor of 1.4 was also incorporated in this algorithm. This calculator allowed the patients to be categorized into the high-risk group, but in some cases it overestimated the CV risk and, as a result, patients receiving statins experienced side effects [113]. Reynold's score for the first time included inflammatory markers such as CRP level in its risk prediction algorithm [14]. However, the range of CRP levels was not considered effective for high-grade inflammation [78]. Furthermore, an underestimation of CV risk in RA patients was reported in a comparative study presented recently by Arts *et al* [16]. Several studies have analyzed the conventional risk scores both in the RA population as well as in controls, but have reported no significant accuracy in CVD risk prediction [9, 114, 115]. Furthermore, a recent study by Crowson *et al* [116] reported the failure of CV risk estimation in RA patients with RA-specific risk scores such as QRisk II and ERS-RA. In 2013, the American College of Cardiology and American Heart Association recommended PCE, which gives a ten year risk estimation [117]. Compared to all the previous algorithms, this calculator improved the risk stratification of patients but did not improve the CVD risk prediction.

All the above-mentioned risk algorithms are not robust enough for stratification of CVD risk in RA patients, since they overestimate or underestimate the CV risk [16]. Furthermore, since atherosclerosis is a multifocal disease and plaque can develop randomly throughout the periphery of the arteries, it is not possible, using traditional risk calculators, to provide a clear visualization of such vulnerable cases. Hence there is a strong need to look beyond the scope of these algorithms and use imaging techniques with an automated approach that can train the systems based on available information to perform better risk stratification in both normal and RA patients.

4.10 Cardiovascular imaging in rheumatoid arthritis

4.10.1 Non-invasive imaging techniques

Non-invasive imaging techniques are generally used to assess the CVD risk in rheumatic patients [17]. Assessment of coronary artery calcification is an important tool to perform risk stratification. The coronary artery calcium (CAC) score is a biomarker of atherosclerosis and can be detected reasonably using CT with high histological correlation [118, 119]. Although CAC was widely accepted as a risk predictor in coronary artery disease, it has also been studied in RA. Giles *et al* [120] demonstrated high prevalence of CAC measured using CT in RA patients compared to the controls. Another follow-up study confirmed the elevation of CAC in patients with RA [121]. Although considered as the standard technique for CAC score computation, CT analysis is limited by its inability to characterize the micro-structure of plaque in the blood vessel. Furthermore, CT requires injection of contrast agents and exposure to significant ionizing radiation, which is a leading disadvantage [122]. Non-invasive magnetic resonance imaging has also found its application in RA. Cardiovascular magnetic resonance (CMR) has the ability to perform cardiac anatomy, aortic distensibility and tissue characterization evaluation. Therefore, it represents an excellent tool for early diagnosis of CV involvement, risk stratification and treatment evaluation of patients with cardiovascular disease due to autoimmune rheumatic diseases (ARDs) [123]. Specifically in RA, CMR can accurately diagnose the main pathophysiological phenomena occurring in the myocardium of RA patients, such as myocarditis, vasculitis and macro-/micro-coronary artery disease [123].

CMRA is also capable of differentiating various plaque components based on their biophysical/biochemical properties, such as water–lipid content, molecular diffusion and chemical composition [124]. Recent developments in CMR techniques, such as multi-contrast MR, proton-density (PD)-weighted, time-of-flight (ToF) imaging and parametric imaging helped to characterize fibro-cellular, calcified and lipid-rich regions of atherosclerotic coronary plaques [125–127]. A CMR study investigating the capability of high-resolution, black-blood MR to assess coronary wall thickness, showed a statistical significant difference of cross-sectional area between normal volunteers and patients with atherosclerotic lesions (0.75 ± 0.17 versus 4.38 ± 0.71 mm, $p < 0.0001$).

Similarly, in the last decade, CUS has also been widely studied to characterizes atherosclerotic plaque composition [128, 129]. CUS measurement in RA generally includes assessment of cIMT and atherosclerotic wall plaque morphology [130].

Individual and combined measurement of atherosclerotic plaque and cIMT for CCA, ICA and ECA provides information about plaque severity. Several studies have reported cIMT as the biomarker for CVD risk prediction [131, 132]. In RA patients, cIMT > 0.91 mm was associated with increased CV events (figure 4.4) [133]. Ziembicka *et al* presented a study showing that an increase in cIMT > 1.15 mm is associated with a 94% increase in the chance of having a CV event [132]. PET is an advanced non-invasive nuclear imaging technique that can provide more cellular visualization of body tissues along with 3D representations [134]. Recent

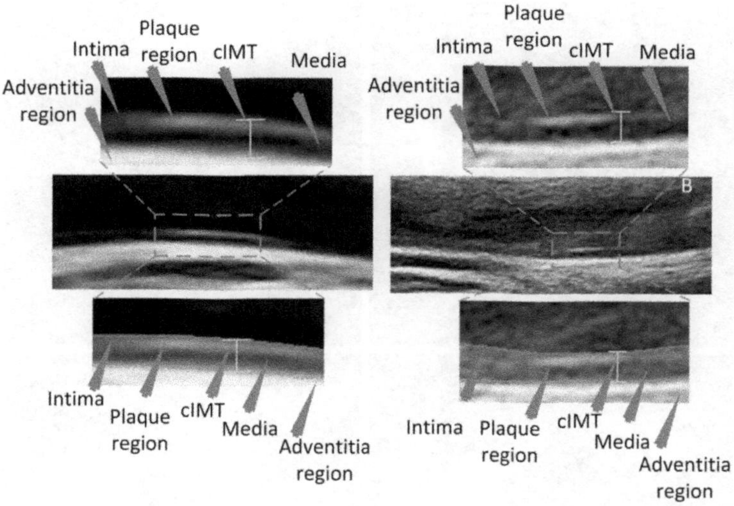

Figure 4.4. Carotid ultrasound image of the common carotid artery (A) for a control patient (cIMT = 0.72 mm) and (B) patients with RA (cIMT = 0.93 mm). (Courtesy of AtheroPoint, Roseville, CA, USA.)

studies have suggested the use of PET in assessing the atherosclerotic plaque and its severity [135, 136].

4.10.2 Invasive imaging techniques: IVUS and OCT

Non-invasive imaging techniques do not provide a detailed visualization of atherosclerotic plaque composition, primarily because of their low spatial resolution [137, 138]. Intravascular ultrasound (IVUS) and OCT are the two advanced techniques that provide cross-sectional visualization of the inner walls of the coronary artery (figures 4.5 and 4.6). OCT has a higher spatial resolution (10–15 μm) compared to IVUS (150–300 μm) [22], and hence it has the potential to characterize different vulnerable plaque components within the vessel wall.

4.11 RA-driven atherosclerotic plaque wall tissue characterization: intelligence paradigm

Atherosclerotic plaque characterization involves identification and classification of different tissue layers for better CVD risk stratification. In RA patients, CUS has shown promise in delineating the three-layered structure of the blood vessel, which is enveloped between the lumen–intima (LI) and media–adventitia (MA) interfaces. Furthermore, the high-risk atherosclerotic plaque tissue components, such as thin-capped fibroatheroma, necrotic core, macrophages and thin fibrous cap, can be characterized using high-resolution imaging techniques such as OCT [139]. CUS and OCT images have been used for tissue characterization in both coronary and carotid arteries [140–142]. Plaque composition, cIMT, echogenicity and echolucency are the

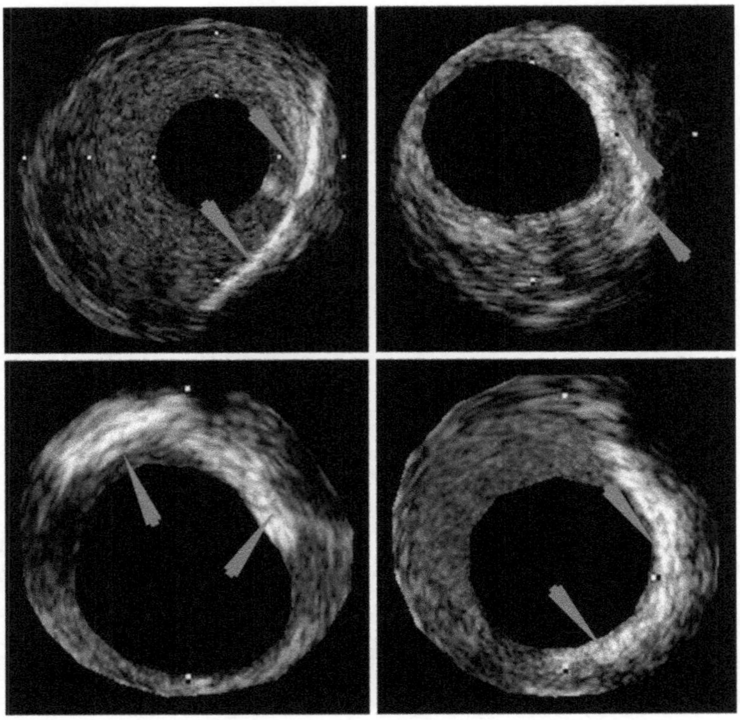

Figure 4.5. IVUS visualization of the atherosclerotic coronary calcified plaque. Red arrows: calcified plaque in a coronary artery. (Courtesy of AtheroPoint, Roseville, CA, USA.)

prevalent biomarkers that provide more information about atherosclerotic lesions [143, 144] and can also be used for future CVD risk estimation in RA patients.

Currently, automation of the tissue characterization process requires an intelligence-based software tool to provide a time-efficient analysis without compromising the risk stratification accuracy. Artificial neural networks (ANNs) are intelligent learning algorithms that extract the information from the input data, process and train their own system and finally provide the targeted output. ANNs are widely used in image analysis applications including tissue characterization in medical images [145–147].

4.11.1 Machine-learning-based tissue characterization

Machine-learning (ML)-based techniques are an extension of neural network algorithms. They are currently used for automation in medical image analysis and computer-aided diagnosis, providing high accuracy in classification and risk stratification tasks [148]. Generally, medical image analysis requires the use of a gold standard as prior information and, based on these prior labels (classes), training coefficients are computed using the training dataset. These training coefficients are then transformed by the test dataset features to predict the output classes (low risk or

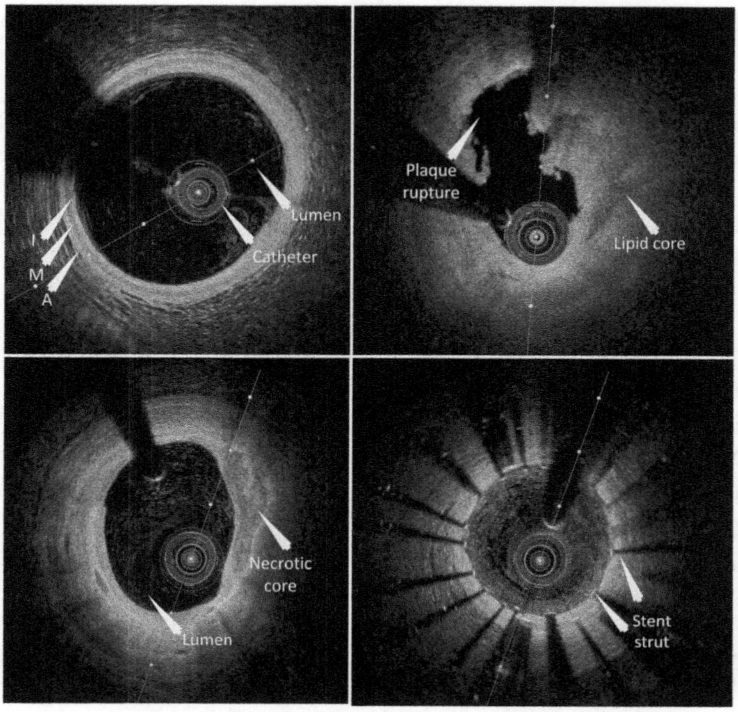

Figure 4.6. Intravascular OCT visualization of atherosclerotic coronary plaque. I: intima; M: media; and A: adventitia. (Courtesy of AtheroPoint, Roseville, CA, USA.)

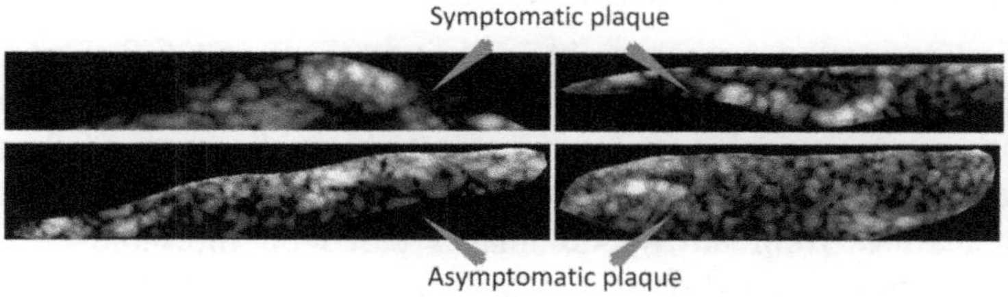

Figure 4.7. Examples of ROIs manually segmented from B-mode ultrasound. (Courtesy of AtheroPoint, Roseville, CA, USA.)

high risk), thereby achieving the objective of the classification. Note that the classes can be binary in nature or have different stages and grades of classification. In characterization or classification tasks, such ML-based techniques collect intelligence from the training images by extracting different features.

Feature extraction is one of the most critical parts of atherosclerotic vascular tissue characterization. The vessel wall of the carotid artery imaged using CUS can be analyzed using a texture-based strategy or morphology-based features which characterize the plaque tissues into either the symptomatic or asymptomatic category (figure 4.7) [149–151]. In general, symptomatic plaque is more severe and inclined towards the occurrence of catastrophic events compared to asymptomatic plaque. The assumption is that both of these have a different appearance in terms of texture.

In addition to texture-based features, wavelet transformed-derived features can also be used to characterize the atherosclerotic carotid plaque tissues using a support vector machine (SVM) [152]. Carotid plaque hypo- and hyper-intensities are useful for tissue characterization and are associated with CVD and stroke risk [153, 154]. Low-intensity (echolucent) plaque is generally comprised of a large lipid content, and a fibrous plaque with or without calcification provides brighter intensity levels (echogenic plaque) [155]. Recent findings suggest that the grayscale median is an important tool to identify this plaque echogenecity or echolucency [156, 157].

CUS images are normally used to perform tissue characterization using ML-based techniques (table A1) [152, 158–160] and can be adapted for RA patients. Furthermore, they have been widely used for risk assessment based on symptomatic and asymptomatic carotid plaque [161, 162]. Similar to CUS, coronary micro-structural plaque tissues can be characterized using the high resolution of OCT [22]. Coronary plaque characterization using OCT is broadly based on three paradigms: (i) analysis of optical coefficients, (ii) ML-based approaches and (iii) DL-based techniques [142, 163–165]. The use of these methods is reported to classify the high-risk plaque components such as thin-cap fibroatheroma, macrophages, calcified lesions and fibrous cap [165–167]. Coronary plaque characterization using OCT has also been presented in various studies (table A2) for CVD risk assessment and can be adapted for RA patients.

4.11.2 Deep-learning-based tissue characterization

The atherosclerotic plaque images obtained from CUS or OCT show an overlap between the different plaque types and hence it becomes difficult to delineate different atherosclerotic plaque components, mainly because of their complex appearance. The conventional feature extraction methods basically detect statistical or imaging features from ultrasound images based on threshold criteria. These features need to be computed prior to classification and, most of the time, they rely on many empirical constants. Furthermore, as the features are extracted from a single isolated pixel intensity value, inaccurate classification may result, as the individual pixels may be affected by noise levels. In recent years, deep learning has become very popular for image feature extraction and classification.

In particular, convolutional neural networks (CNNs) have gained a lot of attention in medical imaging, mainly because of their ability to extract more high-level features compared to handcrafted ones [168]. The general architecture of CNN contains a series of combinations of convolution and pooling layers followed by a fully connected layer (figure 4.8).

The input image at each convolutional layer is convolved with the series of kernels which are responsible for feature extraction. The objective of the pooling layer is to reduce the spatial resolution and converge the extracted features [169].

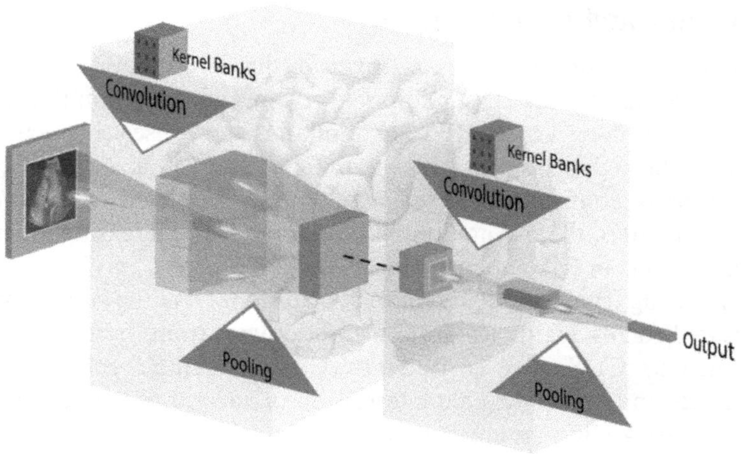

Figure 4.8. A general architecture of a CNN used in medical image analysis applications. (Courtesy of AtheroPoint, Roseville, CA, USA.)

Recently, Lekadir *et al* [170] used deep-learning-based features using a CNN to characterize CUS images into three classes: lipid, fibrous and calcified plaque. Deep-learning features are context-based and do not depend on prior feature information. CNNs can also be used in coronary tissue characterization using OCT images. Various tissue characterization techniques using deep-learning algorithms are depicted in table A2.

4.12 Research agenda

Carotid atherosclerotic plaque tissue characterization in RA patients is largely an unexplored area. cIMT and carotid plaque composition are the important biomarkers that help the physician to estimate future cardiac events. In comparison to the general population, RA patients have been reported to have an increase in cIMT and plaque [171]. Atherosclerotic plaque is a composite of multiple components such as lipids, macrophages and thin-cap fibroatheroma [172]. Plaque characterization and quantification enables one to perform risk assessment at an earlier stage, which is needed in RA patients. The current traditional algorithms are limited by their ability to provide detailed visualization of such 'vulnerable plaque' components. Furthermore, they rely on CVD-specific factors and hence either underestimate or overestimate the CVD risk in RA patients. The use of non-invasive imaging methods has been proposed recently to estimate CVD risk in RA patients [17]. In the last two decades a combination of carotid artery ultrasonography and machine-learning techniques has shown that they have the potential to accurately characterize different plaque components. A combination of CNN and random forest has the ability to extract more robust features from the plaque tissues and classify them into one of the high-risk plaque categories [142]. Hence, these techniques can be helpful to explain the role of atherosclerosis in RA patients.

4.13 Summary and conclusion

Atherosclerosis is no longer a disease driven by lipid formation within the aging vascular walls. Perhaps there is an additional role of inflammation that triggers the plaque build-up in blood vessels. RA is an inflammatory disease which accelerates atherosclerosis through its extra-articular manifestation [32]. Inflammatory cytokines make a major contribution in this acceleration. From the research conducted in the last two decades, there is no doubt about the role of increasing atherosclerosis leading to CV events in RA patients. Hence, there is a need to give special attention to patients suffering from RA. The failure of traditional risk measurements needs to be compensated for by highly accurate methodologies using non-invasive imaging technique such as CUS. The carotid artery has a similar genetic make-up to coronary vessels and hence the characterization of carotid plaque components can be useful in understanding the coronary atherosclerosis process in RA patients. ML-based and DL-based algorithms have been widely implemented in carotid tissue characterization. However, limited efforts have been made to assess the CVD risk using these methodologies in RA patients. At present, limited work has been done to characterize the atherosclerotic plaque tissues in RA and further to predict the cerebrovascular risk. This is the first review of its kind, to the best of our knowledge, to link the risk assessment in RA patients using machine-learning and deep-learning algorithms in carotid atherosclerotic plaque imaging.

Appendix A

Table A1. Machine-learning techniques used for carotid wall tissue characterization.

Authors	Feature types	Feature count	Feature selection	Classifier type	*N*	ACC(%)
Acharya *et al* [152]	2D wavelet transform with wavelet packet	3	*t*-test	SVM	346	83.7
Acharya *et al* [161]	Texture features (deviation, entropy, similarity, run length parameter)	4	*t*-test	AdaBoost + SVM	346	82.4
Tsiaparas *et al* [160]	Texture features (mean, GSM, coarseness, periodicity, roughness, Hurst coefficient)	15	Divergence-based ranking, appropriately threshold by the nonlinear correlation coefficient	SVM and NN	20	—
Acharya *et al* [173]	Grayscale feature extraction (LBP, FGLCM, FRLM, trace transform, HOS features, FS descriptor)	32	*t*-test, Chernoff bound and Bhattacharya distance	SVM	160	90.66
Acharya *et al* [140]	DWT, HOS, texture features	—	—	SVM	146	91.7
Pazinato *et al* [174]	Statistical moments, texture-based, gradient-based, local binary patterns	—	—	Pixel-based classification	12	73
Huang *et al* [175]	Texture features and morphological characteristics	30	Analysis of covariance, linear discriminant analysis	K-NN	136	72.03 and 88.14
Araki *et al* [158]	Grayscale texture features (GLCM, GLRLM, chaotic features)	16	Statistical analysis	SVM	204	95.08 and 93.47
Qian *et al* [159]	Pixel intensity values	1	—	SVM, AdaBoost, RF	29	—
Lekadir *et al* [170]	Deep-learning features	—	—	CNN	56	—

Table A2. Machine-learning techniques used for coronary wall tissue characterization (partially taken from [142]).

Authors	Technique	Feature type	Classifier type	*N*	Frames/ Video	Accuracy (%)	Validation
Abdolmanafi *et al* [165]	Deep-learning features	—	CNN, random forest, SVM	—	26	96	Manual assessment
Liu *et al* [167]	Optical feature characterization	Statistical	Optical features and pixel intensities	135	158	—	Manual assessment
Jimenez *et al* [176]	A-line depth profile estimation	Morphological	Linear discriminant analysis	57	9	88	Histology and manual assessment
Ughi *et al* [164]	Optical feature estimation	Geometrical and texture	Random forest	49	100	81.5	Manual assessment
Gan *et al* [177]	Bayesian framework	Optical, texture and statistical	Relevance vector machine	35	—	80.41	Histology and manual assessment
Wang *et al* [178]	Dynamic programming	—	—	10	100	—	Computer algorithm versus manual analysis
Zahnd *et al* [179]	Contour segmentation with geometrical *a priori*	—	Dynamic programming	21	—	—	Manual assessment
Roy *et al* [180]	Graph representation of the image	CNN features	Random walks for image classification	26	150	97.86	Lumen annotation by an expert
Kume *et al* [181]	Manual comparison using OCT s/w	Statistical features	Linear regression, Bland–Altman analysis	135	—	90	Manual and histological analysis
Zahnd *et al* [182]	Front propagation scheme	17 image features	AdaBoost classifier with Boruta feature selection technique	40	100	91	Manual validation with Dice similarity coefficient

References

[1] Symmons D, Mathers C and Pfleger B 2006 The global burden of rheumatoid arthritis in the year 2000 (Geneva: World Health Organization)

[2] van Vollenhoven R F 2009 Sex differences in rheumatoid arthritis: more than meets the eye *BMC Med.* **7** 12

[3] Rudan I *et al* 2015 Prevalence of rheumatoid arthritis in low–and middle–income countries: a systematic review and analysis *J. Glob. Health* **5** 010409

[4] Hunter T M, Boytsov N N, Zhang X, Schroeder K, Michaud K and Araujo A B 2017 Prevalence of rheumatoid arthritis in the United States adult population in healthcare claims databases, 2004–2014 *Rheumatol. Int.* **37** 1551–7

[5] Padyukov L, Silva C, Stolt P, Alfredsson L and Klareskog L 2004 A gene–environment interaction between smoking and shared epitope genes in HLA–DR provides a high risk of seropositive rheumatoid arthritis *Arthritis Rheumatol.* **50** 3085–92

[6] Klareskog L, Padyukov L, Lorentzen J and Alfredsson L 2006 Mechanisms of disease: genetic susceptibility and environmental triggers in the development of rheumatoid arthritis *Nat. Rev. Rheumatol.* **2** 425

[7] Klareskog L *et al* 2006 A new model for an etiology of rheumatoid arthritis: smoking may trigger HLA–DR (shared epitope)–restricted immune reactions to autoantigens modified by citrullination *Arthritis Rheumatol.* **54** 38–46

[8] Wagner C A *et al* 2015 Identification of anticitrullinated protein antibody reactivities in a subset of anti-CCP-negative rheumatoid arthritis: association with cigarette smoking and HLA-DRB1 'shared epitope' alleles *Ann. Rheum. Dis.* **74** 579–86

[9] Crowson C S, Matteson E L, Roger V L, Therneau T M and Gabriel S E 2012 Usefulness of risk scores to estimate the risk of cardiovascular disease in patients with rheumatoid arthritis *Am. J. Cardiol.* **110** 420–4

[10] Bonek K and Głuszko P 2016 Cardiovascular risk assessment in rheumatoid arthritis—controversies and the new approach *Reumatologia* **54** 128

[11] Solomon D *et al* 2015 Derivation and internal validation of an expanded cardiovascular risk prediction score for rheumatoid arthritis: a Consortium of Rheumatology Researchers of North America Registry Study *Arthritis Rheumatol.* **67** 1995–2003

[12] Peters M J *et al* 2010 EULAR evidence-based recommendations for cardiovascular risk management in patients with rheumatoid arthritis and other forms of inflammatory arthritis *Ann. Rheum. Dis.* **69** 325–31

[13] Hippisley-Cox J *et al* 2008 Predicting cardiovascular risk in England and Wales: prospective derivation and validation of QRISK2 *Br. Med. J.* **336** 1475–82

[14] Ridker P M, Buring J E, Rifai N and Cook N R 2007 Development and validation of improved algorithms for the assessment of global cardiovascular risk in women: the Reynolds Risk Score *J. Am. Med. Assoc.* **297** 611–9

[15] Crowson C S *et al* 2018 Impact of risk factors associated with cardiovascular outcomes in patients with rheumatoid arthritis *Ann. Rheum. Dis.* **77** 48–54

[16] Arts E *et al* 2014 Performance of four current risk algorithms in predicting cardiovascular events in patients with early rheumatoid arthritis *Ann. Rheum. Dis.* **74** 668–74

[17] Fent G J, Greenwood J P, Plein S and Buch M H 2017 The role of non-invasive cardiovascular imaging in the assessment of cardiovascular risk in rheumatoid arthritis: where we are and where we need to be *Ann. Rheum. Dis.* **76** 1169

[18] Maintz D *et al* 2006 Selective coronary artery plaque visualization and differentiation by contrast-enhanced inversion prepared MRI *Eur. Heart J.* **27** 1732–6
[19] Saremi F and Achenbach S 2015 Coronary plaque characterization using CT *Am. J. Roentgenol.* **204** W249–60
[20] Glaudemans A W, de Vries E F, Galli F, Dierckx R A, Slart R H and Signore A 2013 The use of F-FDG-PET/CT for diagnosis and treatment monitoring of inflammatory and infectious diseases *Clin. Dev. Immunol.* **2013** 623036
[21] Furer V, Fayad Z A, Mani V, Calcagno C, Farkouh M E and Greenberg J D 2012 Noninvasive cardiovascular imaging in rheumatoid arthritis: current modalities and the emerging role of magnetic resonance and positron emission tomography imaging *Seminars in Arthritis and Rheumatism* (Amsterdam: Elsevier) pp 676–88
[22] Bezerra H G, Costa M A, Guagliumi G, Rollins A M and Simon D I 2009 Intracoronary optical coherence tomography: a comprehensive review: clinical and research applications *JACC: Cardiovasc. Interv.* **2** 1035–46
[23] Corrales A, González-Juanatey C, Peiró M E, Blanco R, Llorca J and González-Gay M A 2013 Carotid ultrasound is useful for the cardiovascular risk stratification of patients with rheumatoid arthritis: results of a population-based study? *Ann. Rheum. Dis.* **73** 722–7
[24] González-Gay M A, González-Juanatey C and Llorca J 2012 Carotid ultrasound in the cardiovascular risk stratification of patients with rheumatoid arthritis: when and for whom? *Ann. Rheum. Dis.* **71** 796–8
[25] Lee D M and Weinblatt M E 2001 Rheumatoid arthritis *Lancet* **358** 903–11
[26] Taneja V and David C S 2000 Association of MHC and rheumatoid arthritis: regulatory role of HLA class II molecules in animal models of RA—studies on transgenic/knockout mice *Arthritis Res.* **2** 205–7
[27] Choy E 2012 Understanding the dynamics: pathways involved in the pathogenesis of rheumatoid arthritis *Rheumatology* **51** v3–11
[28] Kim K-W, Kim H-R, Kim B-M, Cho M-L and Lee S-H 2015 Th17 cytokines regulate osteoclastogenesis in rheumatoid arthritis *Am. J. Pathol.* **185** 3011–24
[29] McInnes I B and Schett G 2011 The pathogenesis of rheumatoid arthritis *N. Engl. J. Med.* **365** 2205–19
[30] Smolen J S, Aletaha D and McInnes I B Rheumatoid arthritis *Lancet* **388** 2023–38
[31] Libby P 2008 Role of inflammation in atherosclerosis associated with rheumatoid arthritis *Am. J. Med.* **121** S21–31
[32] Skeoch S and Bruce I N 2015 Atherosclerosis in rheumatoid arthritis: is it all about inflammation? *Nat. Rev. Rheumatol.* **11** 390
[33] Sattar N, McCarey D W, Capell H and McInnes I B 2003 Explaining how 'high-grade' systemic inflammation accelerates vascular risk in rheumatoid arthritis *Circulation* **108** 2957–63
[34] Libby P, Ridker P M and Maseri A 2002 Inflammation and atherosclerosis *Circulation* **105** 1135–43
[35] Libby P 2003 Vascular biology of atherosclerosis: overview and state of the art *Am. J. Cardiol.* **91** 3–6
[36] Libby P and Clinton S K 1993 The role of macrophages in atherogenesis *Curr. Opin. Lipidol.* **4** 355–63
[37] Doran A C, Meller N and McNamara C A 2008 Role of smooth muscle cells in the initiation and early progression of atherosclerosis *Arterioscler. Thromb. Vasc. Biol.* **28** 812–9

[38] Hansson G K, Libby P and Tabas I 2015 Inflammation and plaque vulnerability *J. Intern. Med.* **278** 483–93
[39] Virmani R, Burke A P, Farb A and Kolodgie F D 2006 Pathology of the vulnerable plaque *J. Am. Coll. Cardiol.* **47** C13–8
[40] Vuilleumier N, Bratt J, Alizadeh R, Jogestrand T, Hafström I and Frostegård J 2010 Anti-apoA-1 IgG and oxidized LDL are raised in rheumatoid arthritis (RA): potential associations with cardiovascular disease and RA disease activity *Scand. J. Rheumatol.* **39** 447–53
[41] Maziere C, Auclair M and Maziere J-C 1994 Tumor necrosis factor enhances low density lipoprotein oxidative modification by monocytes and endothelial cells *FEBS Lett.* **338** 43–46
[42] Gasparyan A Y, Stavropoulos-Kalinoglou A, Mikhailidis D P, Douglas K M and Kitas G D 2011 Platelet function in rheumatoid arthritis: arthritic and cardiovascular implications *Rheumatol. Int.* **31** 153–64
[43] Vorchheimer D A and Becker R 2006 Platelets in atherothrombosis *Mayo Clinic Proc.* (Amsterdam: Elsevier) pp 59–68
[44] Farr M, Wainwright A, Salmon M, Hollywell C and Bacon P 1984 Platelets in the synovial fluid of patients with rheumatoid arthritis *Rheumatol. Int.* **4** 13–17
[45] Endresen G 1989 Evidence for activation of platelets in the synovial fluid from patients with rheumatoid arthritis *Rheumatol. Int.* **9** 19–24
[46] Migita K *et al* 1996 Increased circulating serum amyloid A protein derivatives in rheumatoid arthritis patients with secondary amyloidosis *Lab. Invest.* **75** 371–5
[47] MayoClinic. Amyloidosis Overview. Patient Care and Health Information www.mayoclinic.org/diseases-conditions/amyloidosis/symptoms-causes/syc-20353178
[48] Nakamura T 2011 Amyloid A amyloidosis secondary to rheumatoid arthritis: pathophysiology and treatments *Clin. Exp. Rheumatol.* **29** 850–7
[49] Eklund K K, Niemi K and Kovanen P T 2012 Immune functions of serum amyloid A *Crit. Rev. Immunol.* **32** 335–48
[50] Obici L, Raimondi S, Lavatelli F, Bellotti V and Merlini G 2009 Susceptibility to AA amyloidosis in rheumatic diseases: a critical overview *Arthritis Care Res.* **61** 1435–40
[51] Targońska-Stępniak B and Majdan M 2014 Serum amyloid A as a marker of persistent inflammation and an indicator of cardiovascular and renal involvement in patients with rheumatoid arthritis *Mediators Inflamm.* **2014** 793628
[52] Gonzalez A *et al* 2008 Do cardiovascular risk factors confer the same risk for cardiovascular outcomes in rheumatoid arthritis patients as in non-rheumatoid arthritis patients? *Ann. Rheum. Dis.* **67** 64–9
[53] Solomon D H, Curhan G C, Rimm E B, Cannuscio C C and Karlson E W 2004 Cardiovascular risk factors in women with and without rheumatoid arthritis *Arthritis Rheumatol.* **50** 3444–9
[54] Semb A *et al* 2010 Lipids, myocardial infarction and ischaemic stroke in patients with rheumatoid arthritis in the Apolipoprotein-related Mortality RISk (AMORIS) Study *Ann. Rheum. Dis.* **69** 1996–2001
[55] Escalante A, Haas R W and del Rincón I 2005 Paradoxical effect of body mass index on survival in rheumatoid arthritis: role of comorbidity and systemic inflammation *Arch. Intern. Med.* **165** 1624–9

[56] Myasoedova E *et al* 2011 Lipid paradox in rheumatoid arthritis: the impact of serum lipid measures and systemic inflammation on the risk of cardiovascular disease *Ann. Rheum. Dis.* **70** 482–7
[57] Kremers H M, Nicola P J, Crowson C S, Ballman K V and Gabriel S E 2004 Prognostic importance of low body mass index in relation to cardiovascular mortality in rheumatoid arthritis *Arthritis Rheumatol.* **50** 3450–7
[58] Roubenoff R *et al* 1994 Rheumatoid cachexia: cytokine-driven hypermetabolism accompanying reduced body cell mass in chronic inflammation *J. Clin. Invest.* **93** 2379–86
[59] Sokka T *et al* 2008 Physical inactivity in patients with rheumatoid arthritis: data from twenty-one countries in a cross-sectional, international study *Arthritis Care Res.* **59** 42–50
[60] Boo S, Oh H, Froelicher E S and Suh C-H 2017 Knowledge and perception of cardiovascular disease risk among patients with rheumatoid arthritis *PLoS One* **12** e0176291
[61] van Zanten J J V *et al* 2015 Perceived barriers, facilitators and benefits for regular physical activity and exercise in patients with rheumatoid arthritis: a review of the literature *Sports Med.* **45** 1401–12
[62] Kitas G D and Gabriel S E 2011 Cardiovascular disease in rheumatoid arthritis: state of the art and future perspectives *Ann. Rheum. Dis.* **70** 8–14
[63] Summers G D, Metsios G S, Stavropoulos-Kalinoglou A and Kitas G D 2010 Rheumatoid cachexia and cardiovascular disease *Nat. Rev. Rheumatol.* **6** 445
[64] Elkan A-C, Håkansson N, Frostegård J, Cederholm T and Hafström I 2009 Rheumatoid cachexia is associated with dyslipidemia and low levels of atheroprotective natural antibodies against phosphorylcholine but not with dietary fat in patients with rheumatoid arthritis: a cross-sectional study *Arthritis Res. Ther.* **11** R37
[65] Qin B *et al* 2015 Body mass index and the risk of rheumatoid arthritis: a systematic review and dose-response meta-analysis *Arthritis Res. Ther.* **17** 86
[66] Kurihara O *et al* 2015 Impact of low body mass index on coronary atherosclerosis: multivessel angioscopic study *Can. J. Cardiol.* **31** S11–12
[67] Stavropoulos-Kalinoglou A, Metsios G S, Koutedakis Y and Kitas G D 2010 Obesity in rheumatoid arthritis *Rheumatology* **50** 450–62
[68] Metsios G S *et al* 2009 Association of physical inactivity with increased cardiovascular risk in patients with rheumatoid arthritis *Eur. J. Cardiovasc. Prev. Rehabil.* **16** 188–94
[69] Myasoedova E, Crowson C S, Kremers H M, Fitz-Gibbon P D, Therneau T M and Gabriel S E 2010 Total cholesterol and LDL levels decrease before rheumatoid arthritis *Ann. Rheum. Dis.* **69** 1310–4
[70] Toms T E *et al* 2011 Are lipid ratios less susceptible to change with systemic inflammation than individual lipid components in patients with rheumatoid arthritis? *Angiology* **62** 167–75
[71] Panoulas V F *et al* 2007 Prevalence and associations of hypertension and its control in patients with rheumatoid arthritis *Rheumatology* **46** 1477–82
[72] Panoulas V F *et al* 2008 Hypertension in rheumatoid arthritis *Rheumatology* **47** 1286–98
[73] Protogerou A D *et al* 2013 Arterial hypertension assessed 'out-of-office' in a contemporary cohort of rheumatoid arthritis patients free of cardiovascular disease is characterized by high prevalence, low awareness, poor control and increased vascular damage-associated 'white coat' phenomenon *Arthritis Res. Ther.* **15** R142
[74] Boyer J-F, Gourraud P-A, Cantagrel A, Davignon J-L and Constantin A 2011 Traditional cardiovascular risk factors in rheumatoid arthritis: a meta-analysis *Joint Bone Spine* **78** 179–83

[75] Balsa A *et al* 2017 Prevalence of comorbidities in rheumatoid arthritis and evaluation of their monitoring in clinical practice: the Spanish cohort of the COMORA study *Reumatol. Clin.* **15** 102–8

[76] Panoulas V F *et al* 2008 Polymorphisms of the endothelin-1 gene associate with hypertension in patients with rheumatoid arthritis *Endothelium* **15** 203–12

[77] Wei Z-H and Du Y-H 2017 Transforming growth factor-β1-509C/T polymorphism might be associated with chronic periodontitis risk *Biomed. Res.* **28** 7763–65

[78] Nurmohamed M T, Heslinga M and Kitas G D 2015 Cardiovascular comorbidity in rheumatic diseases *Nat. Rev. Rheumatol.* **11** 693

[79] Hackshaw A, Morris J K, Boniface S, Tang J-L and Milenković D 2018 Low cigarette consumption and risk of coronary heart disease and stroke: meta-analysis of 141 cohort studies in 55 study reports *Br. Med. J.* **360** j5855

[80] Bergström U, Jacobsson L T H, Nilsson J-Å, Berglund G and Turesson C 2011 Pulmonary dysfunction, smoking, socioeconomic status and the risk of developing rheumatoid arthritis *Rheumatology* **50** 2005–13

[81] Terao C *et al* 2014 Effects of smoking and shared epitope on the production of anti–citrullinated peptide antibody in a Japanese adult population *Arthritis Care Res.* **66** 1818–27

[82] Wolfe F 2000 The effect of smoking on clinical, laboratory, and radiographic status in rheumatoid arthritis *J. Rheumatol.* **27** 630–7

[83] Stavropoulos-Kalinoglou A *et al* 2008 Cigarette smoking associates with body weight and muscle mass of patients with rheumatoid arthritis: a cross-sectional, observational study *Arthritis Res. Ther.* **10** R59

[84] Sugiyama D *et al* 2010 Impact of smoking as a risk factor for developing rheumatoid arthritis: a meta-analysis of observational studies *Ann. Rheum. Dis.* **69** 70–81

[85] Baghdadi L R, Woodman R J, Shanahan E M and Mangoni A A 2015 The impact of traditional cardiovascular risk factors on cardiovascular outcomes in patients with rheumatoid arthritis: a systematic review and meta-analysis *PLoS One* **10** e0117952

[86] Peters M J *et al* 2009 Does rheumatoid arthritis equal diabetes mellitus as an independent risk factor for cardiovascular disease? A prospective study *Arthritis Care Res.* **61** 1571–9

[87] Magda S 2010 Rheumatoid arthritis vs diabetes mellitus as risk factors for cardiovascular disease: the CARRE study *Maedica* **5** 147

[88] Lindhardsen J *et al* 2011 The risk of myocardial infarction in rheumatoid arthritis and diabetes mellitus: a Danish nationwide cohort study *Ann. Rheum. Dis.* **70** 929–34

[89] Montagna G L *et al* 2007 Insulin resistance is an independent risk factor for atherosclerosis in rheumatoid arthritis *Diabetes Vasc. Dis. Res.* **4** 130–5

[90] Chung C P *et al* 2008 Inflammation-associated insulin resistance: differential effects in rheumatoid arthritis and systemic lupus erythematosus define potential mechanisms *Arthritis Rheum.* **58** 2105–12

[91] Giles J T *et al* 2015 Insulin resistance in rheumatoid arthritis: disease-related indicators and associations with the presence and progression of subclinical atherosclerosis *Arthritis Rheumatol.* **67** 626–36

[92] Resnick H E *et al* 2004 Relationship of high and low ankle brachial index to all-cause and cardiovascular disease mortality: the Strong Heart Study *Circulation* **109** 733–9

[93] Hendriks E J *et al* 2016 Association of high ankle brachial index with incident cardiovascular disease and mortality in a high-risk population *Arterioscler. Thromb. Vasc. Biol.* **36** 412–7

[94] Wild S H, Byrne C D, Smith F B, Lee A J and Fowkes F G R 2006 Low ankle-brachial pressure index predicts increased risk of cardiovascular disease independent of the metabolic syndrome and conventional cardiovascular risk factors in the Edinburgh artery study *Diabetes Care* **29** 637

[95] Heald C, Fowkes F, Murray G, Price J and Ankle Brachial Index Collaboration 2006 Risk of mortality and cardiovascular disease associated with the ankle-brachial index: systematic review *Atherosclerosis* **189** 61–9

[96] Kurt T *et al* 2015 Can the ankle brachial pressure index (ABPI) and carotis intima media thickness (CIMT) be new early stage markers of subclinical atherosclerosis in patients with rheumatoid arthritis? *Wien. Klin. Wochenschr.* **127** 529–34

[97] Youssef G, Allam N T, Gaber W, Afifi A and Hesham D 2018 Increased arterial stiffness in rheumatoid arthritis and Its relation to disease activity: a cross sectional study *Egypt. Heart J.* **70** 35–40

[98] Ikdahl E *et al* 2016 Predictive value of arterial stiffness and subclinical carotid atherosclerosis for cardiovascular disease in patients with rheumatoid arthritis *J. Rheumatol.* **43** 1622–30

[99] Ammirati E, Moroni F, Norata G D, Magnoni M and Camici P G 2015 Markers of inflammation associated with plaque progression and instability in patients with carotid atherosclerosis *Mediators Inflamm.* **2015** 718329

[100] Wallberg-Jonsson S, Ohman M and Dahlqvist S R 1997 Cardiovascular morbidity and mortality in patients with seropositive rheumatoid arthritis in Northern Sweden *J. Rheumatol.* **24** 445–51

[101] Wållberg-Jonsson S, Johansson H, Ohman M and Rantapää-Dahlqvist S 1999 Extent of inflammation predicts cardiovascular disease and overall mortality in seropositive rheumatoid arthritis. A retrospective cohort study from disease onset *J. Rheumatol.* **26** 2562–71

[102] Majka D S *et al* 2017 Association of rheumatoid factors with subclinical and clinical atherosclerosis in African American Women: the multiethnic study of atherosclerosis *Arthritis Care Res.* **69** 166–74

[103] Pope J E, Nevskaya T, Barra L and Parraga G 2016 Carotid artery atherosclerosis in patients with active rheumatoid arthritis: predictors of plaque occurrence and progression over 24 weeks *Open Rheumatol. J.* **10** 49

[104] Goodson N J, Symmons D P, Scott D G, Bunn D, Lunt M and Silman A J 2005 Baseline levels of C-reactive protein and prediction of death from cardiovascular disease in patients with inflammatory polyarthritis: a ten-year followup study of a primary care–based inception cohort *Arthritis Rheumatol.* **52** 2293–9

[105] Danesh J *et al* 2004 C-reactive protein and other circulating markers of inflammation in the prediction of coronary heart disease *N. Engl. J. Med.* **350** 1387–97

[106] Gonzalez-Gay M A, Gonzalez-Juanatey C, Piñeiro A, Garcia-Porrua C, Testa A and Llorca J 2005 High-grade C-reactive protein elevation correlates with accelerated atherogenesis in patients with rheumatoid arthritis *J. Rheumatol.* **32** 1219–23

[107] Taverner D, Vallvé J-C, Ferré R, Paredes S, Masana L and Castro A 2018 Variables associated with subclinical atherosclerosis in a cohort of rheumatoid arthritis patients: sex-specific associations and differential effects of disease activity and age *PLoS One* **13** e0193690

[108] D'Agostino R B Sr. *et al* 2008 General cardiovascular risk profile for use in primary care: the Framingham Heart Study *Circulation* **117** 743–53

[109] Conroy R *et al* 2003 Estimation of ten-year risk of fatal cardiovascular disease in Europe: the SCORE project *Eur. Heart J.* **24** 987–1003

[110] Goff D C *et al* 2014 2013 ACC/AHA guideline on the assessment of cardiovascular risk *Circulation* **129** S49

[111] Dessein P H *et al* 2005 Traditional and nontraditional cardiovascular risk factors are associated with atherosclerosis in rheumatoid arthritis *J. Rheumatol.* **32** 435–42

[112] van der Heijde D *et al* 2017 2016 update of the ASAS-EULAR management recommendations for axial spondyloarthritis *Ann. Rheum. Dis.* **76** 978–91

[113] Agca R *et al* 2016 EULAR recommendations for cardiovascular disease risk management in patients with rheumatoid arthritis and other forms of inflammatory joint disorders: 2015/2016 update *Ann. Rheum. Dis.* **76** 17–28

[114] OAMd C *et al* 2016 Assessment of cardiovascular risk in patients with rheumatoid arthritis using the SCORE risk index *Rev. Bras. Reumatol.* **56** 138–44

[115] Arts E *et al* 2016 Prediction of cardiovascular risk in rheumatoid arthritis: performance of original and adapted SCORE algorithms *Ann. Rheum. Dis.* **75** 674–80

[116] Crowson C S *et al* 2017 Rheumatoid arthritis-specific cardiovascular risk scores are not superior to general risk scores: a validation analysis of patients from seven countries *Rheumatology* **56** 1102–10

[117] Stone N J *et al* 2014 2013 ACC/AHA guideline on the treatment of blood cholesterol to reduce atherosclerotic cardiovascular risk in adults: a report of the American College of Cardiology/American Heart Association Task Force on Practice Guidelines *J. Am. Coll. Cardiol.* **63** 2889–934

[118] Hamilton-Craig C *et al* 2017 *Coronary Artery Calcium Scoring–Position Statement.* (Cardiac Society of Australia and New Zealand)

[119] Hou Z-h *et al* 2012 Prognostic value of coronary CT angiography and calcium score for major adverse cardiac events in outpatients *JACC: Cardiovasc. Imaging* **5** 990–9

[120] Giles J T *et al* 2009 Coronary arterial calcification in rheumatoid arthritis: comparison with the Multi-Ethnic Study of Atherosclerosis *Arthritis Res. Ther.* **11** R36

[121] Wahlin B, Meedt T, Jonsson F, Henein M Y and Wållberg-Jonsson S 2016 Coronary artery calcification is related to inflammation in rheumatoid arthritis: a long-term follow-up study *BioMed Res. Int.* **2016** 1261582

[122] de González A B *et al* 2009 Projected cancer risks from computed tomographic scans performed in the United States in 2007 *Arch. Intern. Med.* **169** 2071–7

[123] Mavrogeni S I *et al* 2016 Cardiovascular magnetic resonance in rheumatology: current status and recommendations for use *Int. J. Cardiol.* **217** 135–48

[124] Hamdan A *et al* 2011 A prospective study for comparison of MR and CT imaging for detection of coronary artery stenosis *JACC: Cardiovasc. Imaging* **4** 50–61

[125] Fayad Z A 2003 MR imaging for the noninvasive assessment of atherothrombotic plaques *Magn. Reson. Imaging Clin.* **11** 101–13

[126] Cai J-M, Hatsukami T S, Ferguson M S, Small R, Polissar N L and Yuan C 2002 Classification of human carotid atherosclerotic lesions with *in vivo* multicontrast magnetic resonance imaging *Circulation* **106** 1368–73

[127] Chu B, Phan B A P, Balu N, Yuan C, Brown B G and Zhao X-Q 2006 Reproducibility of carotid atherosclerotic lesion type characterization using high resolution multicontrast weighted cardiovascular magnetic resonance *J. Cardiovasc. Magn. Reson.* **8** 793–99

[128] Park Y B *et al* 2002 Atherosclerosis in rheumatoid arthritis: morphologic evidence obtained by carotid ultrasound *Arthritis Rheumatol.* **46** 1714–9
[129] Del Rincon I 2010 Atherosclerosis in rheumatoid arthritis: what to look for in studies using carotid ultrasound *J. Rheumatol.* **37** 890–3
[130] Patel S *et al* 2017 A study of carotid intimomedial thickness as a primary marker of atherosclerosis in patients with rheumatoid arthritis *Int. Cardiovasc. Forum J.* **9** 31–5
[131] Polak J F *et al* 2010 Associations of carotid artery intima–media thickness (IMT) with risk factors and prevalent cardiovascular disease *J. Ultrasound Med.* **29** 1759–68
[132] Bots M L *et al* 2007 Carotid intima–media thickness and coronary atherosclerosis: weak or strong relations? *Eur. Heart J.* **28** 398–406
[133] Gonzalez-Juanatey C, Llorca J, Martin J and Gonzalez-Gay M A 2009 Carotid intima–media thickness predicts the development of cardiovascular events in patients with rheumatoid arthritis *Seminars in Arthritis and Rheumatism* (Amsterdam: Elsevier) pp 366–71
[134] Narayan N, Owen D R and Taylor P C 2017 Advances in positron emission tomography for the imaging of rheumatoid arthritis *Rheumatology* **56** 1837–46
[135] Evans N R, Tarkin J M, Chowdhury M M, Warburton E A and Rudd J H 2016 PET imaging of atherosclerotic disease: advancing plaque assessment from anatomy to pathophysiology *Curr. Atheroscler. Rep.* **18** 30
[136] Rosenbaum D, Millon A and Fayad Z A 2012 Molecular imaging in atherosclerosis: FDG PET *Curr. Atheroscler. Rep.* **14** 429–37
[137] Lin E and Alessio A 2009 What are the basic concepts of temporal, contrast, and spatial resolution in cardiac CT? *J. Cardiovasc. Comput. Tomogr.* **3** 403–8
[138] Saba L and Suri J S 2013 *Multi-detector CT Imaging: Abdomen, Pelvis, and CAD Applications* (Boca Raton, FL: CRC Press)
[139] Yabushita H *et al* 2002 Characterization of human atherosclerosis by optical coherence tomography *Circulation* **106** 1640–5
[140] Acharya U R *et al* 2013 Atherosclerotic plaque tissue characterization in 2D ultrasound longitudinal carotid scans for automated classification: a paradigm for stroke risk assessment *Med. Biol. Eng. Comput.* **51** 513–23
[141] Nicolaides A, Beach K W, Kyriacou E and Pattichis C S 2011 *Ultrasound and Carotid Bifurcation Atherosclerosis* (Berlin: Springer)
[142] Boi A *et al* 2018 A survey on coronary atherosclerotic plaque tissue characterization in intravascular optical coherence tomography *Curr. Atheroscler. Rep.* **20** 33
[143] Naqvi T Z and Lee M-S 2014 Carotid intima–media thickness and plaque in cardiovascular risk assessment *JACC: Cardiovasc. Imaging* **7** 1025–38
[144] Huang X *et al* 2017 Evaluation of carotid plaque echogenicity based on the integral of the cumulative probability distribution using gray-scale ultrasound images *PLoS One* **12** e0185261
[145] Kalyan K, Jakhia B, Lele R D, Joshi M and Chowdhary A 2014 Artificial neural network application in the diagnosis of disease conditions with liver ultrasound images *Adv. Bioinform.* **2014** 708279
[146] Karimi A, Rahmati S M, Sera T, Kudo S and Navidbakhsh M 2017 A combination of constitutive damage model and artificial neural networks to characterize the mechanical properties of the healthy and atherosclerotic human coronary arteries *Artif. Organs.* **41** E103–17

[147] Kuppili V *et al* 2017 Extreme learning machine framework for risk stratification of fatty liver disease using ultrasound tissue characterization *J. Med. Syst.* **41** 152

[148] El-Baz A, Gimel'farb G and Suzuki K 2017 Machine learning applications in medical image analysis *Comput. Math. Methods Med.* **2017** 2361061

[149] Christodoulou C I, Kyriacou E, Pattichis M S, Pattichis C S and Nicolaides A 2003 A comparative study of morphological and other texture features for the characterization of atherosclerotic carotid plaques *Int. Conf. on Computer Analysis of Images and Patterns* (Berlin: Springer) 503–11

[150] Christodoulou C I, Pattichis C S, Pantziaris M and Nicolaides A 2003 Texture-based classification of atherosclerotic carotid plaques *IEEE Trans. Med. Imaging* **22** 902–12

[151] Acharya U R *et al* 2012 Atherosclerotic risk stratification strategy for carotid arteries using texture-based features *Ultrasound Med. Biol.* **38** 899–915

[152] Acharya U R *et al* 2012 An accurate and generalized approach to plaque characterization in 346 carotid ultrasound scans *IEEE Trans. Instrum. Meas.* **61** 1045–53

[153] Gupta A *et al* 2015 Plaque echolucency and stroke risk in asymptomatic carotid stenosis: a systematic review and meta-analysis *Stroke* **46** 91–7

[154] Mathiesen E B, Bønaa K H and Joakimsen O 2001 Echolucent plaques are associated with high risk of ischemic cerebrovascular events in carotid stenosis: the Tromsø study *Circulation* **103** 2171–5

[155] Arnold J, Modaresi K, Thomas N, Taylor P and Padayachee T 1999 Carotid plaque characterization by duplex scanning: observer error may undermine current clinical trials *Stroke* **30** 61–5

[156] Doonan R *et al* 2013 Association of ultrasonic texture and echodensity features between sides in patients with bilateral carotid atherosclerosis *Eur. J. Vasc. Endovasc. Surg.* **46** 299–305

[157] Irie Y *et al* 2013 The utility of ultrasonic tissue characterization of carotid plaque in the prediction of cardiovascular events in diabetic patients *Atherosclerosis* **230** 399–405

[158] Araki T *et al* 2017 Stroke risk stratification and its validation using ultrasonic Echolucent Carotid Wall plaque morphology: a machine learning paradigm *Comput. Biol. Med.* **80** 77–96

[159] Qian C and Yang X 2018 An integrated method for atherosclerotic carotid plaque segmentation in ultrasound image *Comput. Methods Programs Biomed.* **153** 19–32

[160] Tsiaparas N N, Golemati S, Andreadis I, Stoitsis J S, Valavanis I and Nikita K S 2011 Comparison of multiresolution features for texture classification of carotid atherosclerosis from B-mode ultrasound *IEEE Trans. Inf. Technol. Biomed.* **15** 130–7

[161] Acharya R U *et al* 2012 Symptomatic vs asymptomatic plaque classification in carotid ultrasound *J. Med. Syst.* **36** 1861–71

[162] Acharya U R *et al* 2013 Understanding symptomatology of atherosclerotic plaque by image-based tissue characterization *Comput. Methods Programs Biomed.* **110** 66–75

[163] Xu C, Schmitt J M, Carlier S G and Virmani R 2008 Characterization of atherosclerosis plaques by measuring both backscattering and attenuation coefficients in optical coherence tomography *J. Biomed. Opt.* **13** 034003-1–8

[164] Ughi G J, Adriaenssens T, Sinnaeve P, Desmet W and D'hooge J 2013 Automated tissue characterization of *in vivo* atherosclerotic plaques by intravascular optical coherence tomography images *Biomed. Opt. Express* **4** 1014–30

[165] Abdolmanafi A, Duong L, Dahdah N and Cheriet F 2017 Deep feature learning for automatic tissue classification of coronary artery using optical coherence tomography *Biomed. Opt. Express* **8** 1203–20

[166] Van Soest G *et al* 2010 Atherosclerotic tissue characterization *in vivo* by optical coherence tomography attenuation imaging *J. Biomed. Opt.* **15** 011105-1–9
[167] Liu S *et al* 2017 Tissue characterization with depth-resolved attenuation coefficient and backscatter term in intravascular optical coherence tomography images *J. Biomed. Opt.* **22** 096004
[168] Shen D, Wu G and Suk H-I 2017 Deep learning in medical image analysis *Annu. Rev. Biomed. Eng.* **19** 221–48
[169] Litjens G *et al* 2017 A survey on deep learning in medical image analysis *Med. Image Anal.* **42** 60–88
[170] Lekadir K *et al* 2017 A convolutional neural network for automatic characterization of plaque composition in carotid ultrasound *IEEE J. Biomed. Health Inform.* **21** 48–55
[171] Im C H *et al* 2014 Inflammatory burden interacts with conventional cardiovascular risk factors for carotid plaque formation in rheumatoid arthritis *Rheumatology* **54** 808–15
[172] Libby P, Ridker P M and Hansson G K 2011 Progress and challenges in translating the biology of atherosclerosis *Nature* **473** 317–25
[173] Acharya U R *et al* 2013 Plaque tissue characterization and classification in ultrasound carotid scans: a paradigm for vascular feature amalgamation *IEEE Trans. Instrum. Meas.* **62** 392–400
[174] Pazinato D V *et al* 2016 Pixel-level tissue classification for ultrasound images *IEEE J. Biomed. Health Inform.* **20** 256–67
[175] Huang X *et al* 2016 Classification of carotid plaque echogenicity by combining texture features and morphologic characteristics *J. Ultrasound Med.* **35** 2253–61
[176] Rico-Jimenez J J, Campos-Delgado D U, Villiger M, Otsuka K, Bouma B E and Jo J A 2016 Automatic classification of atherosclerotic plaques imaged with intravascular OCT *Biomed. Opt. Express* **7** 4069–85
[177] Gan Y, Tsay D, Amir S B, Marboe C C and Hendon C P 2016 Automated classification of optical coherence tomography images of human atrial tissue *J. Biomed. Opt.* **21** 101407
[178] Wang Z *et al* 2012 Volumetric quantification of fibrous caps using intravascular optical coherence tomography *Biomed. Opt. Express* **3** 1413–26
[179] Zahnd G *et al* 2015 Quantification of fibrous cap thickness in intracoronary optical coherence tomography with a contour segmentation method based on dynamic programming *Int. J. Comput. Assist. Radiol. Surg.* **10** 1383–94
[180] Roy A G *et al* 2016 Lumen segmentation in intravascular optical coherence tomography using backscattering tracked and initialized random walks *IEEE J. Biomed. Health Inform.* **20** 606–14
[181] Kume T *et al* 2006 Assessment of coronary arterial plaque by optical coherence tomography *Am. J. Cardiol.* **97** 1172–75
[182] Zahnd G *et al* 2017 Contour segmentation of the intima, media, and adventitia layers in intracoronary OCT images: application to fully automatic detection of healthy wall regions *Int. J. Comput. Assist. Radiol. Surg.* **12** 1923–36

Section II

Deep learning strategy for accurate lumen and carotid intima-media thickness measurement

IOP Publishing

Vascular and Intravascular Imaging Trends, Analysis, and Challenges, Volume 2
Plaque characterization
Petia Radeva and Jasjit S Suri

Chapter 5

A deep-learning fully convolutional network for lumen characterization in diabetic patients using carotid ultrasound: a tool for stroke risk

Mainak Biswas, Venkatanareshbabu Kuppili, Luca Saba, Damodar Reddy Edla, Harman S Suri, Aditya Sharma, Elisa Cuadrado-Godia, John R Laird, Andrew Nicolaides and Jasjit S Suri

Manual ultrasound (US)-based methods are used for lumen diameter (LD) measurement to estimate the risk of stroke, but they can be tedious, error prone and subjective, causing variability. We propose an automated deep-learning (DL)-based system for lumen detection. The system consists of a combination of two DL systems: the encoder and the decoder for lumen segmentation. The encoder employs a 13 layer convolution neural-network model (CNN) for rich feature extraction. The decoder employs three up-sample layers of a fully convolutional network (FCN) for lumen segmentation. Three sets of manual tracings were used during the training paradigm, leading to the design of three DL systems. A cross-validation protocol was implemented for all three DL systems. Using the polyline distance metric, the precision-of-merit for the three DL systems over 407 US scans were 99.61%, 97.75% and 99.89%, respectively. The Jaccard index and Dice similarity of the DL lumen segmented region against three ground truth (GT) regions were 0.94, 0.94, 0.93 and 0.97, 0.97, 0.97, respectively. The corresponding area under the curve (AUC) for the three DL systems were 0.95, 0.91 and 0.93. The experimental results demonstrated the superior performance of the proposed deep-learning system over the conventional methods in the literature.

5.1 Introduction

Stroke caused by cardiovascular disease (CVD) is the second most common cause of death [1, 2]. It is also the sixth most common cause of permanent disability [1]. It is

doi:10.1088/2053-2563/ab0820ch5

estimated that every third adult suffers from some form of CVD. There are two kinds of stroke due to CVD: (a) ischemic stroke and (b) hemorrhagic stroke [3]. In ischemic stroke, the blood supply to the brain is blocked due to carotid stenosis. Carotid stenosis occurs due to narrowing of the carotid arteries and thus the inhibition of smoother blood flow. In hemorrhagic stroke, the sudden rupture of a blood vessel within the brain leads to blockages. Stroke causes severe damage to the brain and its cognitive functions [4]. Clinical trials by Bots *et al* [5, 6], Polak *et al* [7, 8] and Nicolaides *et al* [9] have conclusively proven that the increase in stroke risk is related to carotid stenosis, i.e. abnormal narrowing of the lumen walls. The accumulation of plaque (atherosclerosis) along the arterial walls [10] leads to their thickening and ultimately narrowing of the lumen resulting in stenosis. The perpendicular distance between the lumen far wall and lumen near wall, which is also called the lumen diameter (LD), can be used as an effective tool for estimating stenosis [11].

The most popular diagnostic method for fast and accurate evaluation of carotid artery disease is ultrasound (US) imaging. US imaging is low-cost, non-invasive, free from radiation and portable [12], and is therefore very popular. The two most common US imaging techniques used are (a) color Doppler US imaging and (b) B-mode US imaging. Color Doppler US provides knowledge about blood flow within the lumen, enabling the clinician to detect flow reduction, flow turbulence and occlusion in arteries [13, 14]. However, Doppler US is not clinically reliable for stenosis detection, since the blood velocity is variable and the Doppler spectrum is likely to be distorted by the acoustic impedance mismatch between the fluid and the vessel walls [15–17]. Further, pulse-wave velocity measurements are modeled assuming constant arterial wall thickness, which changes over time [18].

The fundamental step for carotid stenosis estimation (which in turn determines the stroke risk [19]) is to first estimate the lumen diameter. This study is focused on developing a novel strategy for detecting the image-based lumen region, which uses cohort intelligence and training neural-network-based models in a deep-learning infrastructure. This requires computing the deep set of grayscale features in learning-based models. The automated measurement of LD from B-mode US images is challenging because of the high variability of grayscale images, the presence of speckle noise, the shape and size of the lumen, and the presence of curvature, plaque, and similar echogenic structures such as the jugular vein [20].

Several techniques have emerged over time to extract lumen boundaries from B-mode US images. Golemati *et al* [21] used a Hough transform to detect four points of the region-of-interest (ROI). However, the algorithm's performance is limited in the case of less bright images where the lumen region may not be detected at all.

Molinari *et al* [22] used an integrated approach for geometric feature extraction, line fitting and classification to extract the CCA. However, the final algorithm outcome is affected by noise and the presence of similar echographic structures, such as the jugular vein, and fails in the classification of final line pairs, i.e. the CCA near wall (also known as LI-near) and CCA far wall (also known as LI-far). Loizou *et al* [23] introduced snake-based CCA segmentation but it suffered from initialization and boundary leakage problems [24, 25]. Araki *et al* [26] combined the scale-space approach with a level set in determining lumen borders. Krishna *et al* [27] used a combination of spatial transformation and scale-space to estimate the LI-far wall and LI-near wall. The major drawback of the methodologies used thus far is that they do not include intelligence in their models. Further, they also lack the accumulated information from the population required for intelligent learning by the system. The earlier systems also lack model-based imaging required for full automation. There is an immediate need for an intelligence-based reliable, accurate and robust method for LD measurement, which in turn is an indicator of atherosclerotic build-up and a risk of stroke. Therefore, we are focused on the design and development of an automated LD measurement using the deep-learning (DL) paradigm, a class of AtheroEdge (AtheroPoint, Roseville, CA) system, in the common carotid artery (CCA), as shown in figure 5.1. We are motivated by training-based learning strategies in the field of classification and segmentation of US images. The extreme learning machine (ELM) and support vector machine (SVM) have been successful in the characterization and stratification of US fatty liver disease (FLD) images [28]. However, they do not produce accurate results in the case of segmentation as they depend on conventional feature extraction techniques. We introduced a deep-learning (DL)-based system [29] for CCA lumen segmentation from US images. Prior to our current study, there has been no DL strategy for the lumen. When we say DL, we are immediately in the learning framework or intelligence-based framework. Prior to this paper, there has been no 'intelligent-based' carotid lumen segmentation work. Further, we want to emphasize that a key

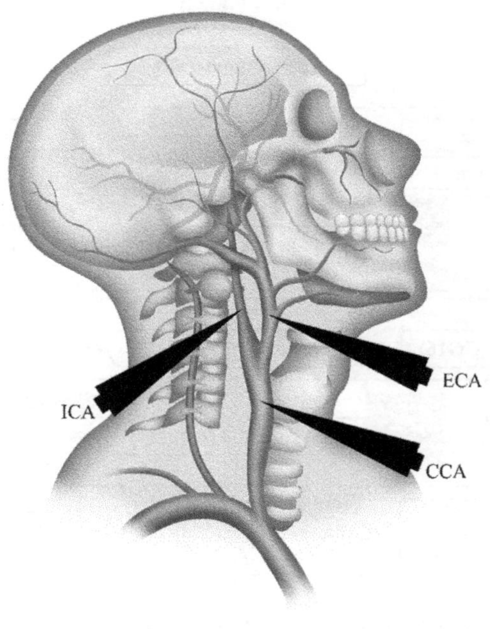

Figure 5.1. Anatomy of CCA (ECA: external carotid artery; ICA: internal carotid artery). (Courtesy of AtheroPoint^{TM}.)

aspect in all measurement techniques is low intra- and inter-observer variability, which should be achieved by the DL paradigm, so this chapter covers variability analysis in a deep-learning environment, which is a unique contribution. In this paper, a comprehensive analysis of the application of DL for lumen segmentation is performed and a comparison of DL to conventional methods for CCA lumen segmentation is provided. One of the major benefits of using DL is that it is independent of conventional feature generation techniques. The DL system generates features internally. In this chapter we apply a three-stage DL-based model for binary lumen segmentation [30, 31], as shown in figure 5.2. The DL-based system uses two stages: the encoder for feature extraction and the decoder for segmentation.

Due to superior learning, the DL-based [32–36] method gives better classification of borders than conventional methods. DL has also been applied for histopathological images [37], the detection and classification of leukocytes [38], and the classification of ulcer tissues [39]. The conventional methods fail in the detection of the LI-near wall because of fuzziness and low contrast. Our DL-based systems, on the other hand, produce accurate detection of the LI-near wall and better segmentation output because of the application of skipping and dense operations. Moreover, the application of a gold standard (GS)/ground truth (GT) in training leads the DL to have better output accuracy. In our study, we have employed three novice tracers for manual CCA delineation of 407 US images. These tracings form the GT for our experiment. The data collected do not have subclinical

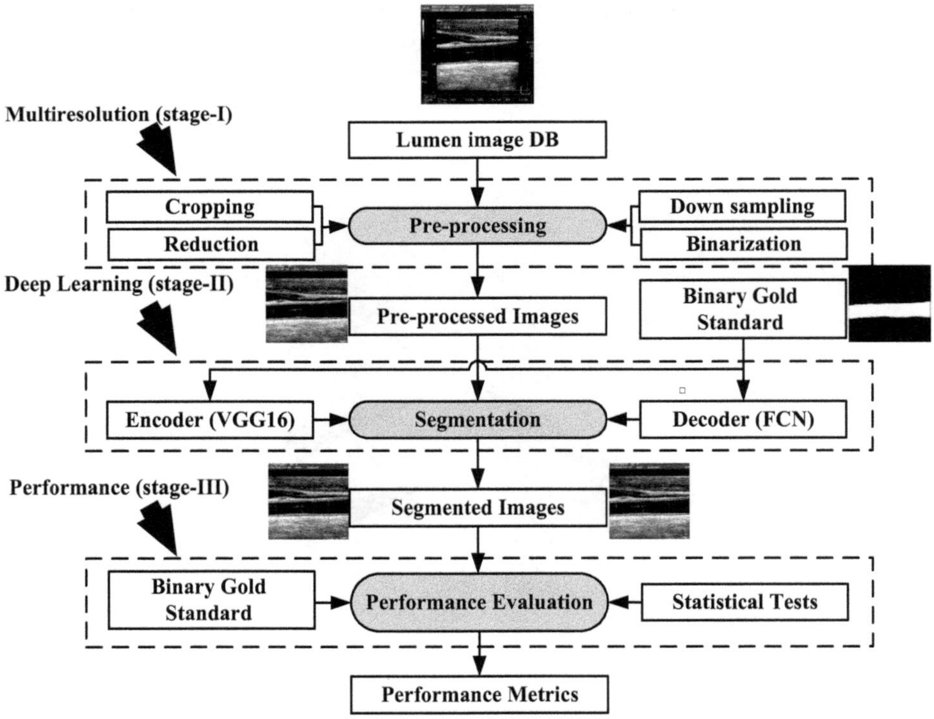

Figure 5.2. Global DL architecture for lumen segmentation.

atherosclerosis disease (SAD) and so we have not computed stenosis. The scope of the study therefore is purely focused on LD measurement. The results from experiments show high accuracy with respect to conventional methods. We have performed inter-operator variability and cross-validation tests for our experiment. We also performed statistical tests. The results of these tests verify the stability and reliability of our system.

5.2 Data demographics

We retrospectively analyzed a database of 407 CCA B-mode sonographic images from 204 patients' left and right carotid arteries: 157 men (76.7%) and 47 women (23.3%), with a mean age of 69 ± 15.9 years. One right carotid artery image for one patient was not available. Ethics approval was granted by the Toho University Institutional Review Board (IRB), Japan. Informed consent was obtained from all the patients. These patients had a mean hemoglobin (HbA1c), glucose, low-density lipoprotein cholesterol, high-density lipoprotein cholesterol and total cholesterol of 5.8 ± 1.0 (mg dl^{-1}), 108 ± 31 (mg dl^{-1}), 99.80 ± 31.30 (mg dl^{-1}), 50.40 ± 15.40 (mg dl^{-1}) and 174.6 ± 37.7 (mg dl^{-1}), respectively. Of the 204 patients, 92 patients were regular smokers. Hypertensive and high cholesterol patients were on adequate medication: statin was prescribed to 93 of them to lower cholesterol levels, while 84 of them received renin–angiotensin system antagonists. Information regarding the blood pressure of these patients was not available. A sonographic scanner (Aplio XV, Aplio XG, Xario, Toshiba, Inc., Tokyo, Japan) equipped with a 7.5 MHz linear array transducer was used to examine the left and right carotid arteries of the recruited patients. All scans were performed under the supervision of an experienced sonographer (with 15 years experience). High-resolution images were acquired according to the recommendations of the American Society of Echocardiography Carotid Intima–Media Thickness Task Force. The mean pixel resolution of the database was 0.05 ± 0.01 mm/pixel.

The manual delineation of the lumen as well as adventitia borders was carried out using ImgTracer™ (AtheroPoint™, USA), a user-friendly commercial software [25]. 15–25 edge points proximal to the bulb were selected in order to delineate the boundaries of the carotid artery. The numbers of points vary depending upon the length of the carotid artery. The observer had the ability to zoom into the image in the wall region for visualization of the wall region. The output of the ImgTracer™ was an ordered set of traced (x, y) coordinates.

5.3 Methodology

In this section, we discuss our model for CCA lumen segmentation in detail. The success of DL in other areas of classification and segmentation of biomedical data inspired us to use the concept for binary lumen segmentation [26–28]. The training/testing paradigm of DL in CCA lumen segmentation is a new field. It gives us an interesting view on DL in the area of lumen segmentation. We apply supervised training for our model. For this, a GT is required. The manual tracings given by the novice examiners form the GT of this experiment. The GT obtained is applied for training of the DL model along with the training data. Figure 5.2 describes the general flow of the data in our model. In this section we discuss in detail each stage in the training/testing paradigms. The training/testing paradigms of the DL model are shown in figure 5.3. As shown in the figure, cross-validation is first applied to the original data. Here, we apply ten-fold cross-validation, i.e. 90% training and 10% testing. The training data along with the GT are fed into the system for training and testing. The training of the three sub-systems was performed in sequence. The first is pre-processing and the two DL systems are the encoder for feature extraction and the decoder for segmentation. The encoder–decoder model for training/testing is shown in figure 5.4. The details of each sub-system are given in the following subsections.

5.3.1 Pre-processing

The pre-processing or multiresolution framework allows us to not just improve the speed, but also to prevent noise from degrading the DL framework. The multi-resolution stage consists of four steps: cropping, binarization, reduction and down-sampling. In the cropping step, all the background information except for the tissue

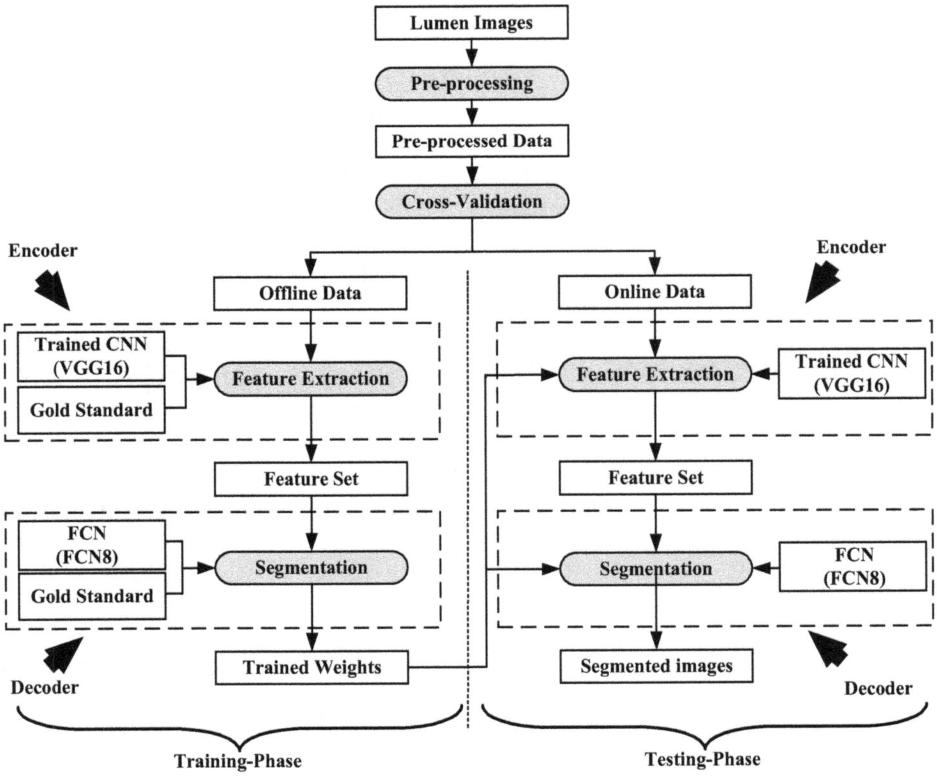

Figure 5.3. Cross-validation paradigm in the DL framework.

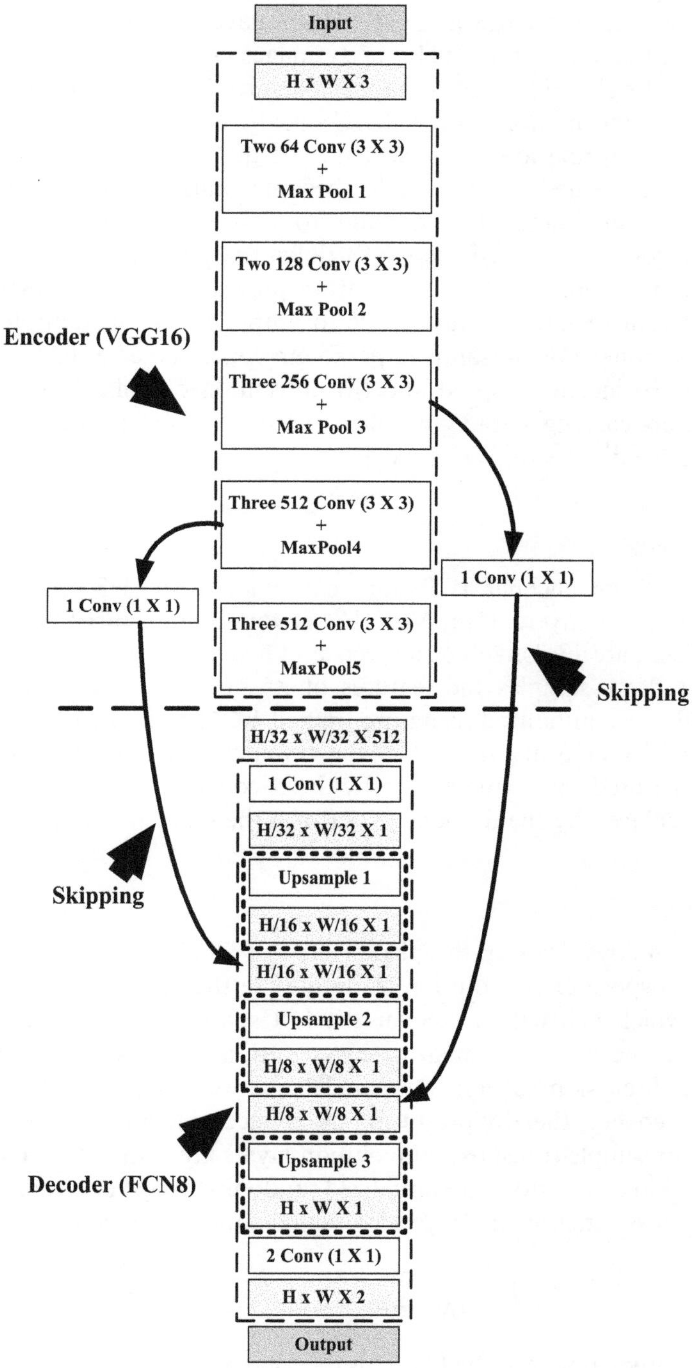

Figure 5.4. Encoder/decoder DL architecture for lumen segmentation.

region is removed from the raw and GT traced images. In the binarization step, the binary GT is extracted from the GT traced images. The tissue information consists of noisy left and right edges, which happens due to the nature of probing the neck region for CCA information. When the US probe is applied to the neck region it obtains the best information from the center of the US probe due to full contact, rather than at the sides, which usually become blurred due to partial contact. Therefore, in the third step, 10% reduction from each side is applied to obtain crisp and sharp images. In the fourth step, 50% down-sampling is applied to each image for faster computation. We have learnt from previous imaging strategies that the global application of filters always leads to amplification of multiplicative noise, therefore we adopted a down-sampled paradigm which served a dual purpose in our infrastructure: (i) increased speed and (ii) prevention of noise from amplification during the deep-learning paradigm, when running filters in more compact dense images. Next, we discuss our DL model.

5.3.2 The encoder

The purpose of the encoder is feature extraction from US CCA images. We employed the first 13 layers of the VGG16 network architecture [26, 27] for feature extraction. There are 13 convolution layers and five max-pooling layers. Each max-pooling layer down-samples the features of its previous convolution layer. The encoder weights are initialized using pre-trained VGG weights on ImageNet. Once the weights of VGG16 are trained in the training phase, high-level features are generated to be used for segmentation by the decoder. The last three layers of the VGG16 are replaced by the decoder to perform the segmentation.

5.3.3 The decoder

The decoder employs three up-sampled layers of FCN [30]. Each up-sampled layer performs a transpose of convolution to up-sample the images. The FCN uses skip connections which helps it to recover the full spatial resolution at the network output, paving the way for semantic segmentation. We discussed skip operation briefly in the discussion section. Our model uses two skip operations for spatial information recovery, thereby producing highly accurate and precise segmentation output. The up-sample/transpose convolution layers are also initialized using VGG weights. The skip connections are initialized randomly using very small weights. The cross-entropy loss function employed for segmentation is given by

$$\theta_{\text{label}}(\alpha, \beta) = \frac{1}{|N|}\sum\nolimits_{n\in N}\sum\nolimits_{c\in C} \beta_n(c) \log \alpha_n(c), \tag{5.1}$$

where α is the prediction, β is the GT, C is the total number of labels/classes and N is the total number of images. The experimental protocol is discussed in the following section.

5.4 Results

5.4.1 Experimental protocol

Our dataset consisted of 407 images. Based on the cross-validation framework in the DL paradigm (as shown in figure 5.3), we conducted our experiment using a ten-fold cross-validation protocol. The dataset is randomly divided into ten partitions. Ten combinations were created from these ten parts. Each combination had two parts: one part for training and the other part for testing. The training part consisted of nine out of ten partitions, constituting 90% of the images. The testing part consisted of the remaining one partition, constituting 10% of the images. The combination was then fed into our model for training and testing. Three kinds of ground truth tracings were considered for variability analysis. The cross-validation protocol was repeated for all the other combinations for each of the three tracers. At the end of the experiment statistical data were accumulated and several performance metrics were computed.

5.4.2 Experimental results

The experiment was carried out on 407 images from 204 patients with one patient missing a right carotid artery image. Plaque was found in almost 40% of images. It is further noted that almost 23% of images had the jugular vein (JV) alongside the carotid artery.

5.4.2.1 Visual LD borders using the DL method

The visual results of our system can be seen in figure 5.5. The figure compares DL with the GT based on four different GT tracings, shown in figure 5.5(a), (b), (c) and (d). It is clearly seen that the DL borders follow the morphology of the lumen.

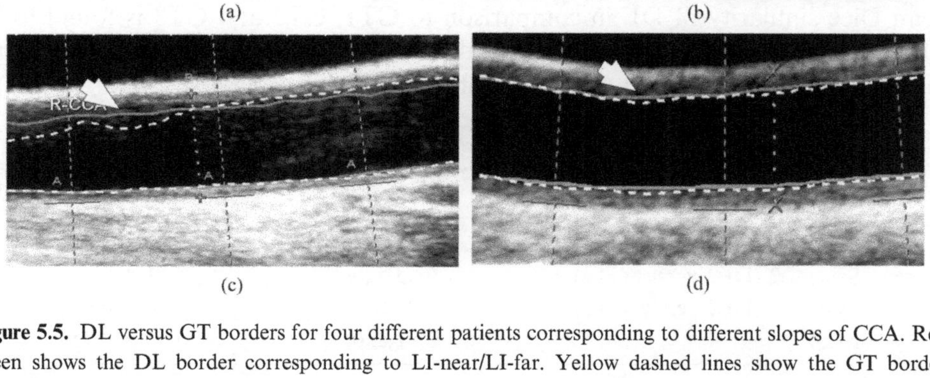

Figure 5.5. DL versus GT borders for four different patients corresponding to different slopes of CCA. Red/green shows the DL border corresponding to LI-near/LI-far. Yellow dashed lines show the GT borders corresponding to LI-near/LI-far.

5.4.2.2 Mean LD values

We have computed the mean LD values for the DL and GT readings using three manual tracings. The mean LD values using DL1 and GT1 were found to be 6.09 ± 0.94 mm and 6.06 ± 0.91 mm, respectively. The corresponding mean LD values using DL2 and GT2 were 6.05 ± 0.91 mm and 5.91 ± 0.88 mm, and finally the mean LD values using DL3 and GT3 were 6.09 ± 0.94 mm and 6.00 ± 0.90 mm, respectively. These values indicate no significant difference between the DL and GT values, with the exception of DL2 and GT2. However, this exception can be attributed to the manual tracer due to inexperience or the hectic schedule of people in medical fields. The LD values for DL along with their GT values are given in table 5.1.

5.4.2.3 Regression correlation plots of DL versus GT for the cohort

The regression plots for LD comparison of DL with GT1, GT2 and GT3 are given in figure 5.6. The correlation coefficient (CC) values obtained for GT1, GT2 and GT3 are given as 0.98, 0.99 and 0.99, respectively. These results validate our theory that DL gives better results due to superior training. The performance evaluation is presented in the next section.

5.4.3 Performance evaluation

Performance evaluation was carried out between DL and their corresponding manual readings. Several metrics for performance evaluation were computed. These included the Jaccard index (JI), Dice similarity (DS), error analysis, precision-of-merit (PoM) and figure-of-merit (F0M) analysis, inter-operator variability studies and receiving operating characteristics (ROCs) analysis. In addition, various statistical tests were carried out to check the reliability and stability of our DL system. They are described in the following sections.

5.4.3.1 Similarity metrics

The Jaccard index and Dice similarity were used to find the similarity between the DL readings and their corresponding GT readings. The mean Jaccard index of DL corresponding to GT1, GT2 and GT3 is observed to be 0.94, 0.94 and 0.93. The mean Dice similarity of DL in comparison to GT1, GT2 and GT3 is found to be 0.97 for all. The results were sorted in decreasing order and their corresponding quartile values were computed. The results are given in table 5.2.

Table 5.1. Mean LD values for DL and corresponding GT using three manual tracings.

	DL (mm)	GT (mm)
System 1 (DL1 versus GT1)	6.09 ± 0.94	6.06 ± 0.91
System 2 (DL2 versus GT2)	6.05 ± 0.91	5.91 ± 0.88
System 3 (DL3 versus GT3)	6.09 ± 0.94	6.00 ± 0.90
Mean	6.07 ± 0.93	5.99 ± 0.95

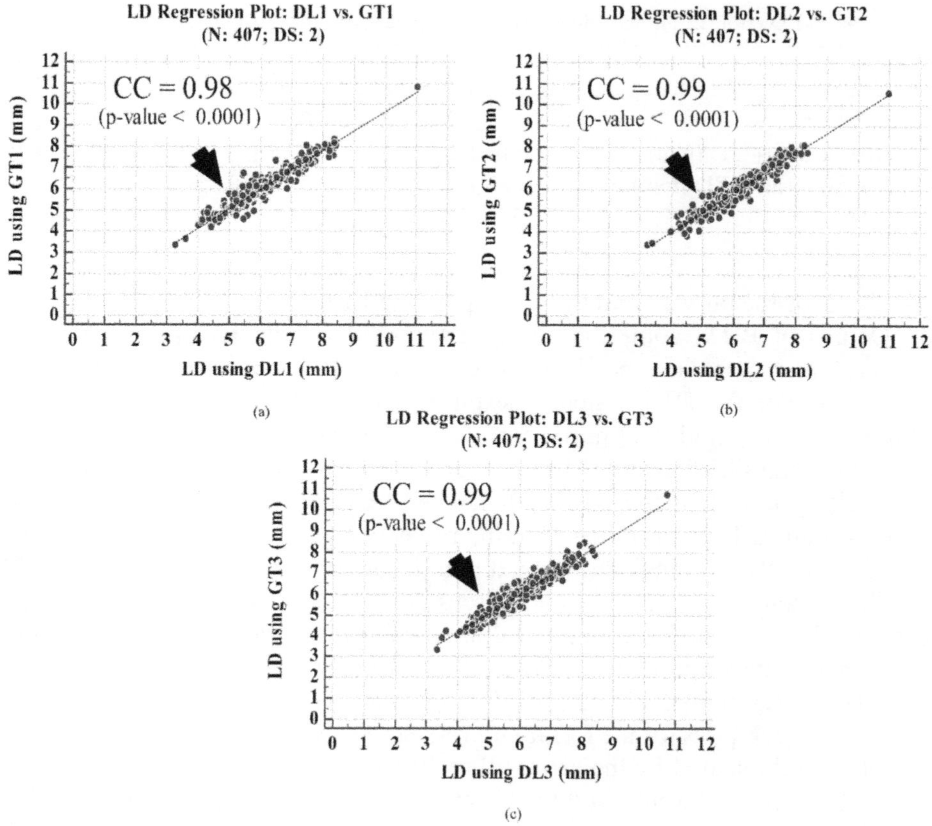

Figure 5.6. Regression plot for (a) DL1 versus GT1, (b) DL2 versus GT2 and (c) DL3 versus GT3. DS denotes double down-sampling since one is applied in pre-processing stage and the second is applied within our model.

Table 5.2. Mean Jaccard index and Dice similarity for three DL systems as per four quartiles.

Quartile	Jaccard index			Dice similarity		
	DL1	DL2	DL3	DL1	DL2	DL3
Q1	0.97 ± 0.004	0.97 ± 0.005	0.97 ± 0.005	0.98 ± 0.002	0.98 ± 0.003	0.98 ± 0.003
Q2	0.95 ± 0.004	0.96 ± 0.003	0.95 ± 0.005	0.98 ± 0.002	0.98 ± 0.002	0.97 ± 0.002
Q3	0.94 ± 0.007	0.94 ± 0.006	0.93 ± 0.005	0.97 ± 0.004	0.97 ± 0.003	0.97 ± 0.003
Q4	0.88 ± 0.052	0.89 ± 0.04	0.89 ± 0.031	0.94 ± 0.034	0.94 ± 0.023	0.94 ± 0.018
Mean	0.94 ± 0.035	0.94 ± 0.035	0.93 ± 0.033	0.97 ± 0.025	0.97 ± 0.025	0.97 ± 0.018
CM*		0.94 ± 0.036			0.97 ± 0.023	

*CM: cumulative mean over all three DL systems for the JI and DS metrics.

5.4.3.2 Error analysis

Here the position error of LI interfaces with respect to the GTs is computed. This is estimated by computing the polyline distance method (PDM) [40] between the LI-far

Table 5.3. Mean LD error, LI-far and LI-near error corresponding to the three DL systems.

Error type	DL1	DL2	DL3
LD error, mean ± SD (mm)	0.19 ± 0.27	0.23 ± 0.23	0.21 ± 0.19
LI-far error, mean ± SD (mm)	0.14 ± 0.07	0.15 ± 0.10	0.14 ± 0.08
LI-near error, mean ± SD (mm)	0.23 ± 0.19	0.20 ± 0.15	0.24 ± 0.15

and GT-far for the far wall and the corresponding LI-near and GT-near for the near wall. Details of PDM computation are given in appendix B. LD and LD error computation are given in appendix C. The LI-far and LI-near errors are discussed in appendix D. For the DL1 system, the three errors LD, LI-far and LI-near were 0.19 mm, 0.14 mm and 0.23 mm, respectively. Similarly, the for DL2 system, the three errors LD, LI-far and LI-near were 0.23 mm, 0.15 mm and 0.20 mm, respectively. Finally, for the DL3 system, the three errors LD, LI-far and LI-near were 0.21 mm, 0.14 mm and 0.24 mm, respectively. These errors are shown in table 5.3. The DL2 readings showed a dip in the error when compared against DL1 and DL3. These results are further discussed and compared with conventional systems in the discussion section. The cumulative distribution function (CDF) plot for LD error is shown in figures 5.7(a)–(c). It is clearly seen that 90% of the images have LD error less than 0.45 mm for DL1, 0.47 mm for DL2 and 0.46 mm for DL3. Similarly, the CDF plots for the LI-far and LI-near errors are given in figures 5.8(a)–(c). It can be seen (indicated by the arrow) that 90% of the images have an LI-far error less than 0.25 mm, 0.28 mm and 0.23 mm corresponding to the three DL systems (DL1, DL2 and DL3). In contrast, the LI-near error shows that 90% of the images have error less than 0.45 mm, 0.38 mm and 0.43 mm, which is slightly higher compared to LI-far. Additionally, the CDF plot for signed LD error is given in figures 5.9(a)–(c). As indicated by the arrows, the signed LD error for DL1 lies between +0.25 and −0.25 mm. In the case of DL2, the signed LD error lies between -0.20 mm and 0.47 mm, and for DL3 the signed LD error lies between −0.30 mm and +0.30 mm. Little variability of LD error is thus observed in DL2 readings. We have previously seen that there is a mismatch of LD mean values for DL2. The reason for this mismatch can be attributed to significant variance of DL2 readings, which is due to inaccurate tracing.

5.4.3.3 FoM and PoM analysis

The figure-of-merit (FoM) and precision-of-merit (PoM) are computed to check the overall system performance. The details of FoM and PoM are given in appendix C. The FoM and PoM values are given in table 5.4. The FoM and PoM values for DL1 and DL3 are nearly 100%. However, DL2 when compared with DL1 and DL3, produces lower FoM and PoM values. The repetition of the anomalies for DL2 values confirms that GT2 was not accurately traced.

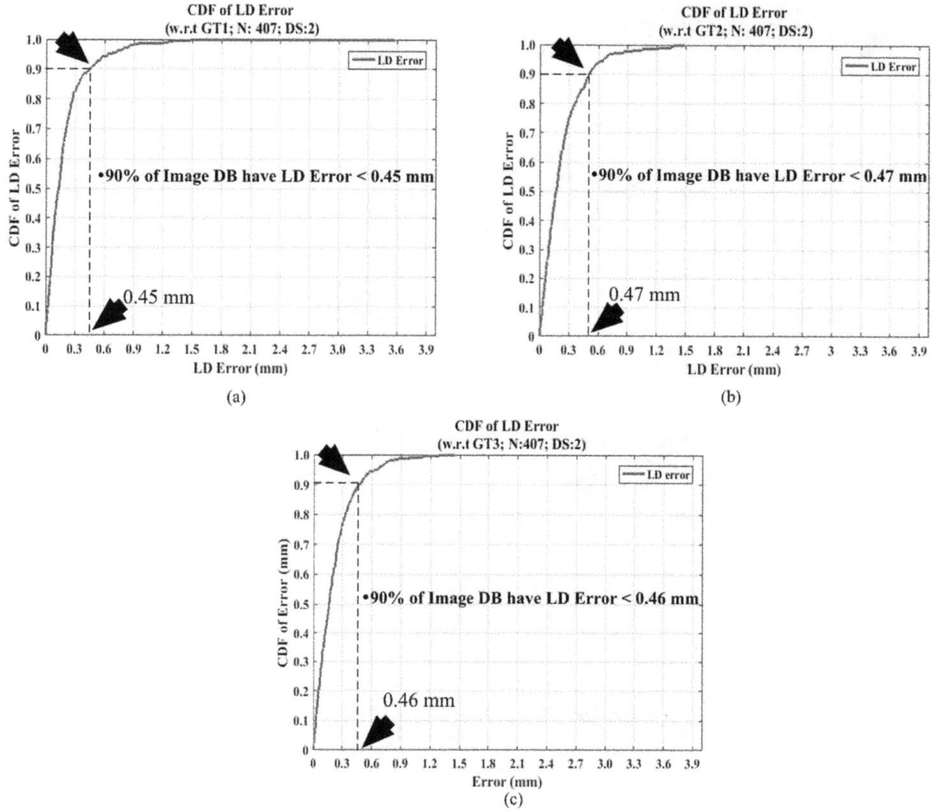

Figure 5.7. CDF plot for LD error for DL1, DL2 and DL3 given in (a), (b) and (c).

5.4.3.4 Inter-operator variability

We test the DL system for inter-operator variability studies. The CC values for DL1 versus DL2, DL1 versus DL3, and DL2 versus DL3 were found to be 0.98, 0.99 and 0.98, respectively. The scatter diagram for regression plots for each is given in figure 5.10. These diagrams clearly show that all the measurements between DL readings are similar.

5.4.3.5 ROC analysis

Stenosis risk assessment was performed in this study. Patients were divided among two groups using a threshold of 0.85 mm: high risk and low risk. Based on the threshold of 0.85 mm, the AUC values for DL1, DL2 and DL3 were found to be 0.95, 0.91 and 0.93, respectively. The low AUC value of DL2 can be attributed to the inaccurate manual tracing which affected the performance of DL2. The ROC curves for all three are given in figure 5.11.

5.4.3.6 Comparison of deep learning against the scale-space method

In this section, performance using the scale-space (SS) method is compared [27] with the same 404 images applied. There is a considerable improvement in the detection

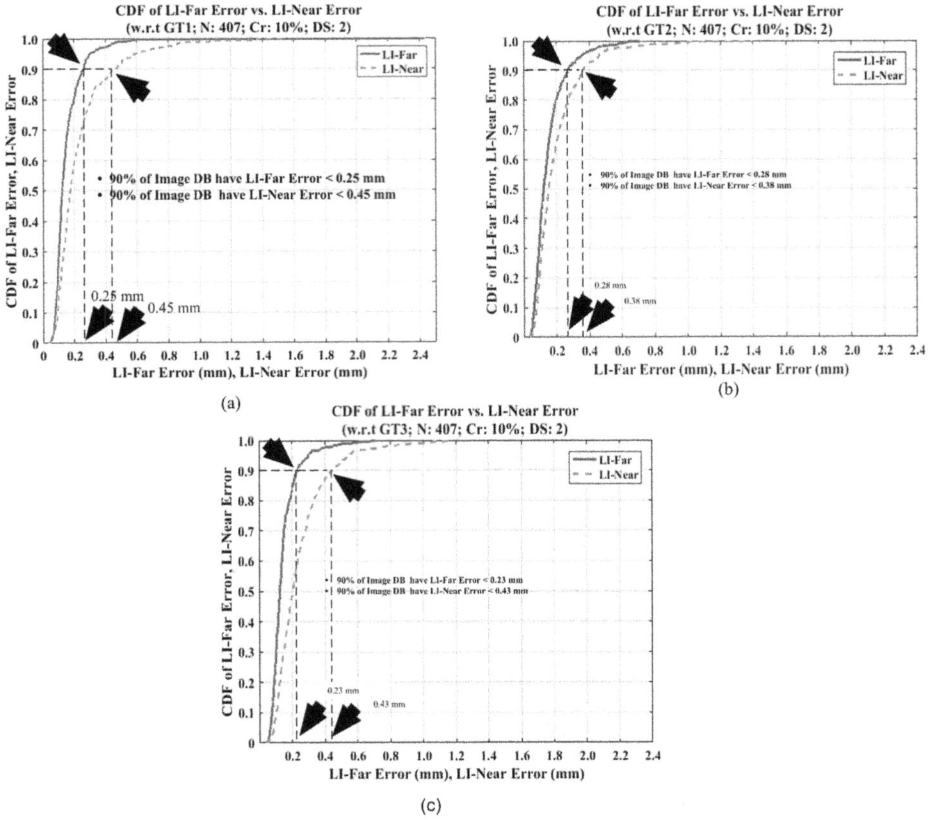

Figure 5.8. CDF error plots for LI-near and LI-far error for DL1, DL2 and DL3 given in (a), (b) and (c).

of LD error by DL w.r.t. GT1 and GT2, i.e. 24% in the case of GT1 and 22.2% in the case of GT2. Improvements are also seen in the case of LI-far error, which is 12.5% in the case of GT1 and 33.3% higher in the case of GT2. In the case of LI-near error, the improvement is only seen in the case of GT2, i.e. 4%. The JI improvement was found to be 5.6% for both GT1 and GT2 and DS improvement was 3.2% for both GT1 and GT2. The FoM improvement was 1.0% for GT1 and 2.2% for GT2. The PoM improvement was 0.8% for GT1 and 1.5% for GT2. The data are given in table 5.5. A comparison between DL and SS is given in figure 5.12.

5.4.3.7 Statistical analysis

The Bland–Altman plot, Wilcoxon test, Mann–Whitney test and paired *t*-test were performed to analyze the relationship between DL and GT readings. The Bland–Altman plots for DL1 versus GT1, DL2 versus GT2, and DL3 versus GT3 are given in figure 5.13. The *p*-values from the Wilcoxon test for DL1, DL2 and DL3 are $p = 0.0071$, $p < 0.0001$ and $p < 0.1548$, respectively. This means that the null hypothesis that data are taken from the same distribution cannot be retained for DL1 and DL2, while it is retained for DL3. The *p*-values from the Mann–Whitney test for DL1, DL2 and DL3 are $p = 0.8192$, $p = 0.0429$ and $p = 0.8846$,

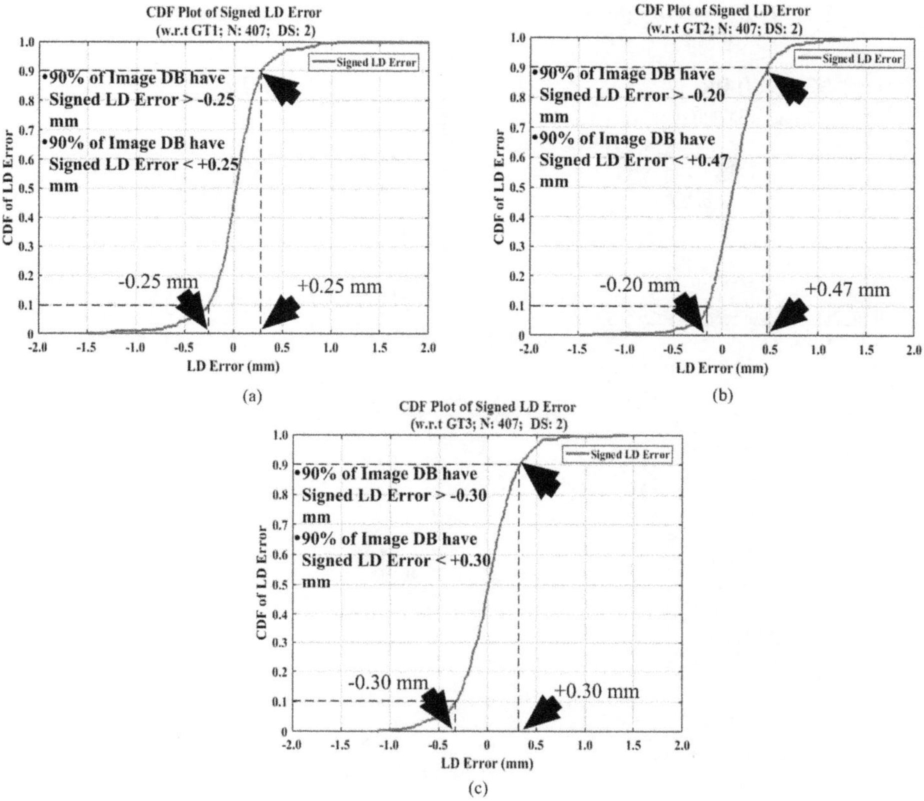

Figure 5.9. CDF plot for signed LD error for DL1, DL2 and DL3 given in (a), (b) and (c).

Table 5.4. FoM and PoM analysis for DL1, DL2 and DL3 for LD error.

Precision analysis	LD error for DL1 (%)	LD error for DL2 (%)	LD error for DL3 (%)
FoM	99.61	97.75	99.89
PoM	96.57	95.94	96.34

respectively. This means that the null hypothesis that data are taken from the same distribution cannot be retained for DL2, while it is retained for DL1 and DL3. The p-values from the paired t-test for DL1, DL2 and DL3 are $p = 0.0595$, $p < 0.0001$ and $p = 0.2730$, respectively. The null hypothesis that data are taken from same distribution cannot be retained for DL2, while it is retained for DL1 and DL3. The corresponding results are given in appendix A, table A1. Corresponding box=plots of the Wilcoxon test, Mann–Whitney test and paired t-test are given in figures A1–A3.

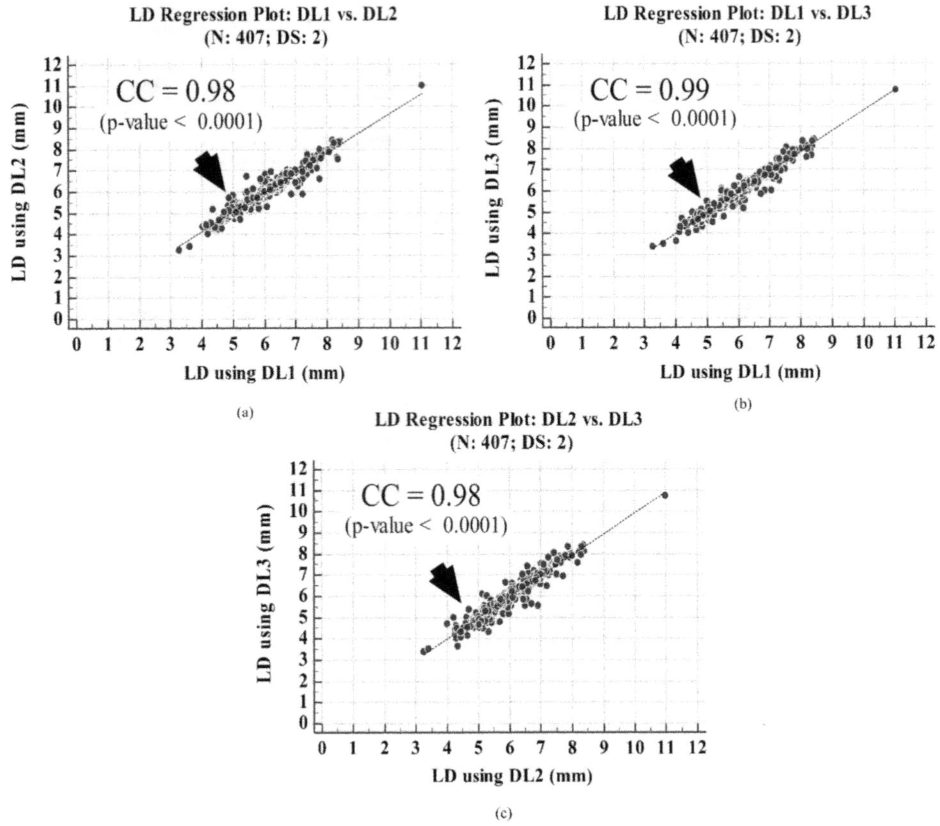

Figure 5.10. Scatter plot of CC values for (a) DL1 versus DL2, (b) DL1 versus DL3 and (c) DL2 versus DL3.

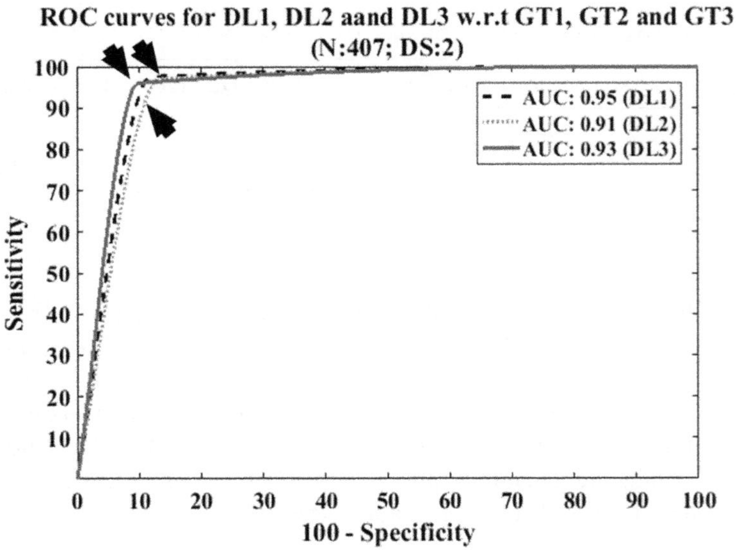

Figure 5.11. ROC curve.

Table 5.5. Comparison of the DL and SS methods for 404 images.

	System 1 w.r.t. GT1			System 2 w.r.t. GT2		
Method and improvement	DL1	SS1	(%) Improv.	DL3	SS2	(%) Improv.
LD error (mm)	0.19 ± 0.25	0.25 ± 0.24	24.0	0.21 ± 0.19	0.27 ± 0.25	22.2
LI-far error (mm)	0.14 ± 0.07	0.16 ± 0.11	12.5	0.14 ± 0.08	0.21 ± 0.34	33.3
LI-near (mm)	0.22 ± 0.14	0.22 ± 0.15	0	0.24 ± 0.15	0.25 ± 0.18	4.0
JI	0.94 ± 0.039	0.89	5.6	0.93 ± 0.03	0.89	5.6
DS	0.97 ± 0.023	0.94	3.2	0.97 ± 0.016	0.94	3.2
FoM (LD) error (%)	99.6	98.7	1.0	99.9	97.7	2.2
PoM (LD) error (%)	96.6	95.9	0.8	96.5	95.1	1.5

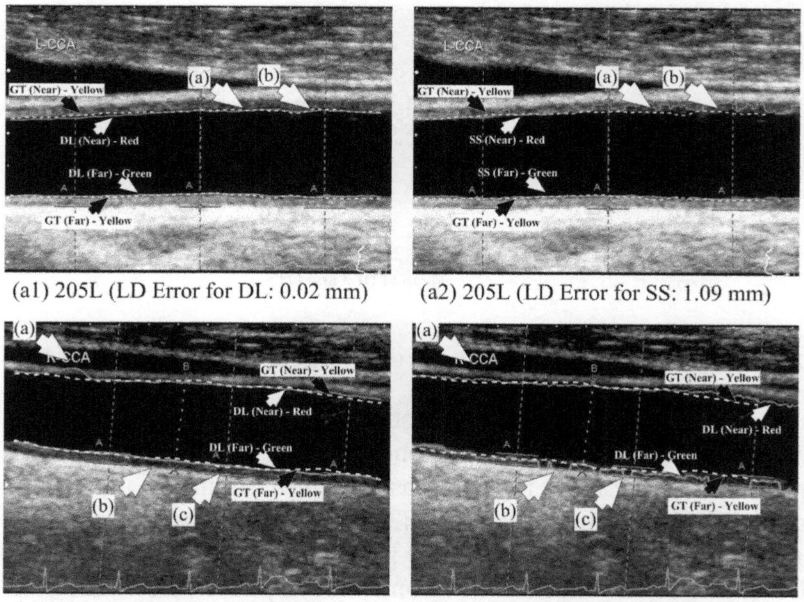

(a1) 205L (LD Error for DL: 0.02 mm) (a2) 205L (LD Error for SS: 1.09 mm)

(b1) 32R (LD Error for DL: 0.26 mm) (b2) 32R (LD Error for SS: 0.37 mm)

Figure 5.12. For patient 205L, the error visible in panel (a2) locations (a) and (b) using the SS method is not present in the corresponding image using the DL method in panel (a1). For patient 32R, in panel (b2) continuous error is noted at locations (b) and (c) using the SS method, while no error is noted in panel (b1) positions (b) and (c) using DL. However, LI-near error is noted in panel (b1) at position (a) using DL, which is not seen in panel (b2) using SS.

5.5 Discussion

The focus of this study has been to develop a fully automated, model-based, intelligent and accurate LD estimation system. In this regard we have developed a DL-based system for LD measurement under the class of AtheroEdge™ systems

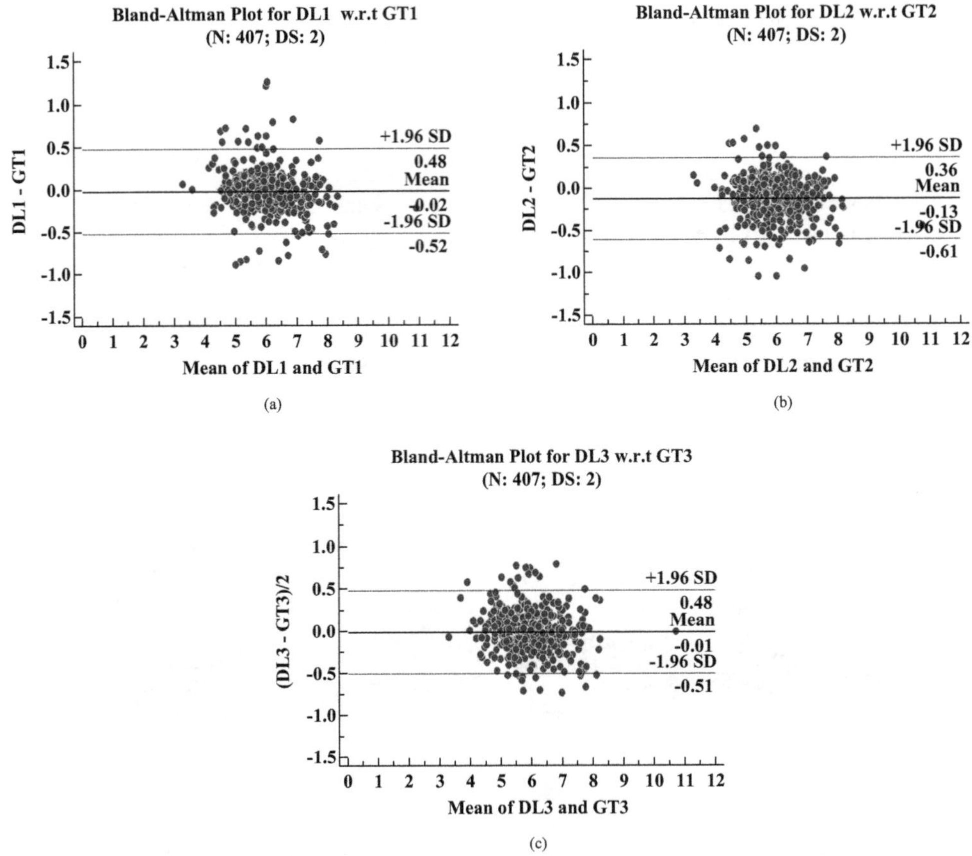

Figure 5.13. Bland–Altman plot for (a) DL1 versus GT1, (b) DL2 versus GT2 and (c) DL3 versus GT3.

from AtheroPoint™. The entire system consists of two stages: the first stage consists of 13-layered DL-based feature extraction, while the second stage consists of three up-sampled layers of FCN for regional-based segmentation. The initialization of the networks used VGG pre-trained weights [36]. The model was trained separately for three sets of GTs that were traced by the physicians, thus producing three kinds of DL systems. Our DL-based model demonstrated superior performance for carotid artery lumen regional segmentation compared to spatial-based techniques due to extraction of the rich features from stage one of the DL combined with three layered FCN network. Comprehensive data analysis was performed yielding the least error of 0.19 ± 0.27 mm, 0.14 ± 0.07 mm and 0.23 ± 0.19 mm (table 5.3) for all 407 images, respectively, corresponding to the DL1 system. Further, cumulative distribution plots showed 90% images had LD error less than 0.25 mm, 0.28 mm and 0.23 mm (figure 5.8), respectively, corresponding to the three DL systems (DL1, DL2 and DL3). Similarly, signed LD error for 90% of the patients had a range from +0.25 to −0.25 mm, −0.20 mm to 0.47 mm, and −0.30 mm to +0.30 mm (figure 5.9) for DL1, DL2 and DL3, respectively. Using the threshold of 0.85 mm, the ROC analysis for high-risk and low-risk patients, showed AUC values of 0.95, 0.91 and

0.93 (figure 5.11) corresponding to DL1, DL2 and DL3, respectively, showing encouraging results for larger cohorts leaning towards clinical trials. The LI-far error was lower than the LI-near error since the highest intensity distribution was noted in the far wall of the CCA. This is very consistent with the results published previously [41].

5.5.1 Benchmarking

The literature on the segmentation of CCA images is limited. Table 5.6 shows the benchmarking table with eight attributes, mainly focusing on the performance of different algorithms, such as the type of technique, data size, LD error, LI-far/LI-near error and PoM. Guiterrez *et al* [42] in 2002 used a combination of a multiresolution-based active contour model for LD estimation. The authors tested using a very limited dataset of 30 images and showed a mean LD error of 0.13 mm, and no information was available on LI-far/LI-near errors. Golemati *et al* [21] applied a Hughes transform on the image data to extract four control points for detection of the ROI in the lumen region. The authors took just ten images demonstrating a mean LD of 0.29 ± 0.30 mm. No information was available on LI-far and LI-near errors. Sahani *et al* [43] applied dynamic threshold-based shape fitting for computing the LD region on only 16 images, and showed an LD error of 0.24 ± 0.13 mm. No information was available on LI-far and LI-near error.

Scale-space paradigms. In 2015, Suri and his team [44] (see table 5.6, row #4) applied Gaussian filter and spectral analysis to extract the binary lumen in 404 images. This can be stated as a major development in the field of automation of a larger cohort of carotid lumen segmentation. In 2016, Suri and his team (Araki *et al* [45]) further applied spectral analysis (see table 5.6, row #5) on 404 images achieving higher FoM and PoM. In the same year, Araki *et al* [26] (see table 5.6, row #(a) and (b)) applied two types of CCA segmentation methods, region-based and boundary-based, on 300 images.

Deep learning versus scale-space. In 2017 again, Suri and his team (Krishna *et al* [27]) applied scale-space for CCA segmentation achieving lower LD, LI-far and LI-near errors (see table 5.6, row #7). The authors took manual tracings (ground truth) for performance evaluation on 404 images. Using the same database, our current study applied deep learning to the CCA segmentation process achieving higher accuracy on 404 images (see table 5.6, row #8), i.e. LD errors: 0.25 ± 0.24 (SS) versus 0.19 ± 0.25 (DL) and 0.27 ± 0.25 (SS) versus 0.21 ± 0.19 (DL). For LI-far 0.16 ± 0.11 (SS) versus 0.14 ± 0.07 (DL) and 0.21 ± 0.34 versus 0.14 ± 0.08. Thus there is an improvement of 24% and 22.2% for LD error and 12.5% and 33.3% for LI-far. This is a considerable improvement using the DL compared to the SS method. A visual representation of DL versus SS is given in figure 5.12.

We took one step ahead and used the entire cohort of 407 images, and the results are shown in table 5.6, row #9. The results are comparable to 404 images, but our study does not omit any patient images, unlike Krishna *et al* [27]. Our results show that the FoM for 407 images (corresponding to DL1 and DL3) using GT1 and GT3 was 99.6% and 99.9%, respectively. The corresponding PoM for 407 images

Table 5.6. Benchmarking table.

SN	Authors	Method	Data size	LD error (mm)	LI-far error (mm)	LI-near error (mm)	FoM (%)	PoM (%)
1	Guiterrez *et al* [42]	Active contour + multiresolution (ACM)	30	ACM: 0.13 ± 0.09	—	—	—	—
2	Golemati *et al* [21]	Hough transform	10	HT: 0.29 ± 0.30	—	—	—	—
3	Sahani *et al* [43]	Dynamic thresholding + shape fitting (DTSF)	16	DTSF: 0.24 ± 0.13	—	—	—	—
4	Saba *et al* [44]	Gaussian filter + spectral analysis (GSA)	404	GSA1: 0.25 ± 0.24 GSA2: 0.27 ± 0.26	GSA1: 0.27 ± 0.15 GSA2: 0.28 ± 0.15	GSA1: 0.23 ± 0.16 GSA2: 0.26 ± 0.22	GSA1: 96.7 GSA2: 97.2	—
5	Araki *et al* [45]	Spectral analysis (SA)	404	SA1: 0.26 ± 0.28 SA2: 0.27 ± 0.26	—	—	SA1: 99.03 SA2: 98.05	—
6(a)	Araki *et al* [26]	Region-based (RB)	300	—	—	—	RB: 99.7	RB: 97.9
6(b)	Araki *et al* [26]	Boundary-based (BB)	300	—	—	—	BB: 87.2	BB: 85.2
7	Krishna *et al* [27]	Scale-space (SS)	404	SS1: 0.25 ± 0.24 SS2: 0.27 ± 0.25	SS1: 0.16 ± 0.11 SS2: 0.21 ± 0.34	SS1: 0.22 ± 0.15 SS2: 0.25 ± 0.18	SS1: 98.7 SS2: 97.7	SS1: 95.9 SS2: 95.1
8	Proposed work	Deep learning	404	DL1: 0.19 ± 0.25 DL3: 0.21 ± 0.19	DL1: 0.14 ± 0.07 DL3: 0.14 ± 0.08	DL1: 0.22 ± 0.14 DL3: 0.24 ± 0.15	DL1: 99.6 DL3: 99.9	DL1: 96.6 DL3: 96.5
9	Proposed work	Deep learning	407	DL1: 0.19 ± 0.27 DL3: 0.21 ± 0.19	DL1: 0.14 ± 0.07 DL3: 0.14 ± 0.08	DL1: 0.23 ± 0.19 DL3: 0.24 ± 0.15	DL1: 99.6 DL3: 99.9	DL1: 96.6 DL3: 96.3

(corresponding to DL1 and DL3) using GT1 and GT3 was 96.6% and 96.3%, respectively. Note that our PoM (using DL: table 5.6, row #9) is slightly lower compared to regional-based LD segmentation (Araki *et al* [26], see table 5.6, row #6 (a)), i.e. 97.9% versus 96.6%. This is mainly due to the difference in data types (407 versus 300) and different sets of ground truth tracings. The hardware configuration for the entire process was an i7, 3.60 GHz processor and 8 GB RAM.

5.5.2 A short note on skip operation in FCN

There is a contraction path and expansion path in an FCN. The contraction path is where the features are down-sampled at an intermediate layer by using convolution and pooling operations. Similarly, in the expansion path, a transpose of convolution is applied to up-sample the features. A skip operation is applied to skip features in the contracting path to intermediate layers in the expansion path, to recover spatial information lost during down-sampling in the expansion path. This is done by merging skipping features from various resolution layers in the contracting path with input features in the expansion path. In this way, highly accurate segmentation output is obtained using three layered FCN. In our model, we have applied two skipping operations. The first skipping feature was extracted from the max-pool layer four of the encoder and merged with the input to the second up-sample layer in decoder. The second skipping layer features were extracted from the max-pool layer three of the encoder and merged with the input to the third up-sample layer in the decoder.

5.5.3 A short note on manual tracings of LI borders

We want to emphasize that ground truth plays an important role in the design of deep-learning strategies. We observed in our data analysis that our mean LDs, LD errors, LI-far and LI-near errors are more consistent when using the GT1 and GT3 manual tracings than when GT2 was used for the development of the DL system. This can be seen in table 5.1, which depicts the behavior and layout of the mean LDs. System 2 (table 5.1, row #2) has a higher difference between the DL and GT computations compared to system 1 and system 3. This is a possible attributable cause for the greater LD, LI-far and LI-near errors in the signed errors (see figure 5.7(c)). This is further supported by the rational that the FoM and PoM are slightly lower for the DL2 system compared to the DL1 and DL3 systems (see table 5.4). Even though the ground truth tracings for GT2 had certain deviations, the deep-learning system was still able to yield a performance close to the DL1 and DL3 systems. Note that all the tracings were performed using ImgTracer™ software which is well established and has been used previously [25]. The possible reason for inaccurate tracing can be attributed to inexperience, fatigue or the busy schedules of people in the medical field.

5.5.4 Strengths and weaknesses

This is the first application of deep learning for lumen segmentation and LD measurement that showed considerable improvement in accuracy and reduction in

LD error. The application of the skip operation where the lower layer and higher layer features were combined and up-sampled gave better segmentation results, thereby reducing wall errors and improving the performance of the system. Further, we demonstrated the stability of the DL system by computing the inter-operator variability.

Even though the technique is novel and being applied for the first time to LD detection with promising and encouraging results, we wanted to share that there is a potential to make this more robust. We note that in the Q4 quartile the JI and DS in table 5.2 showed a slightly degraded performance. Careful evaluation showed that images in the Q4 quartile were noisy and showed larger variability in carotid shapes, which could be a contributing factor in slightly degrading the performance. A greater number of iterations of DL training may be required for better performance along with ML-based calibration strategies. There is a further need to study the application of the model for stenosis detection, and inter-adventitial diameter (IAD) and intima–media thickness (IMT) measurement, which is outside of the scope of the current study.

5.6 Conclusion

Our study showed the application of deep learning for automated lumen detection and LD measurement, using a class of AtheroEdge™ systems from AtheroPoint™. The DL system used a multiresolution strategy in a DL framework. The system consisted of 13 layers for feature extraction in stage one and three layers in stage two of the FCN network. Three expert readings were taken for evaluation. Using the metric of polyline distance, the PoM estimation between the manual and DL system over 407 US scans was 99.61%, 97.75% and 99.89%, respectively. The Jaccard index and Dice similarity of the DL lumen segmented region against three GT regions were 94%, 94% and 93%, and 97%, 97% and 97%, respectively. Using the threshold of 0.85 mm, the AUC obtained from ROC analysis showed an accuracy of 0.95, 0.91 and 0.93, corresponding to three manual readings. Benchmarking the DL system against the previously published literature showed an improvement in LD error, and LI-far and LI-near wall boundary errors of 24%, 33% and 4%, respectively. The new and interesting conclusions can be described briefly as follows. First, the intelligent-based deep-learning strategy provides a more powerful tool of lumen segmentation compared to conventional image processing methods. The capture of global shape using the knowledge derived from the gold standard is a powerful guide during the cross-validation schemes in learning infrastructure. Second, the deep-learning paradigm, a specific class under the intelligent-based infrastructure, when adapted in the multiresolution paradigm can lead to effective feature extraction during deep learning due to noise suppression in the down-sampling mode. Third, since the variability study shows better statistical correlations and successful passing of statistical tests with strong p values (< 0.0001), this clearly demonstrates the power of the deep-learning strategies for lumen risk stratification and characterization. Fourth, our benchmarking evaluation covering the last 15 years of published research shows that deep learning yields the best performance in terms of reliability, accuracy and stability—as demonstrated in this chapter. The deep-learning system is reliable, the best in the class of AtheroEdge™ systems and can be adapted for clinical settings.

Acknowledgment

The authors at the National Institute of Technology, Goa, India, would like to acknowledge MediaLab Asia, the Ministry of Electronics and Information Technology, and the Government of India for their kind support.

Appendix A Statistical test results

Table A1. Statistical tests.

Wilcoxon test (paired samples)			
Readings	DL1	DL2	DL3
Number of positive differences	219	277	207
Number of negative differences	176	118	185
Large sample test statistic Z	−2.692 314	−9.704 749	−1.422 874
Two-tailed probability	$p = 0.0071$	$p < 0.0001$	$p = 0.1548$
Mann–Whitney test (independent samples)			
Average rank of first group	405.6167	390.8108	406.3034
Average rank of second group	409.3833	424.1892	408.6966
Mann–Whitney U	82 058.00	76 032.00	82 337.50
Test statistic Z (corrected for ties)	0.229	2.025	0.145
Two-tailed probability	$p = 0.8192$	$p = 0.0429$	$p = 0.8846$
Paired samples *t*-test			
Mean difference	0.023 89	0.1273	0.013 78
Standard deviation of differences	0.2550	0.2476	0.2533
Standard error of mean difference	0.012 64	0.012 27	0.012 55
95% CI	−0.000 9630 to 0.048 74	0.1032 to 0.1515	−0.010 90 to 0.038 46

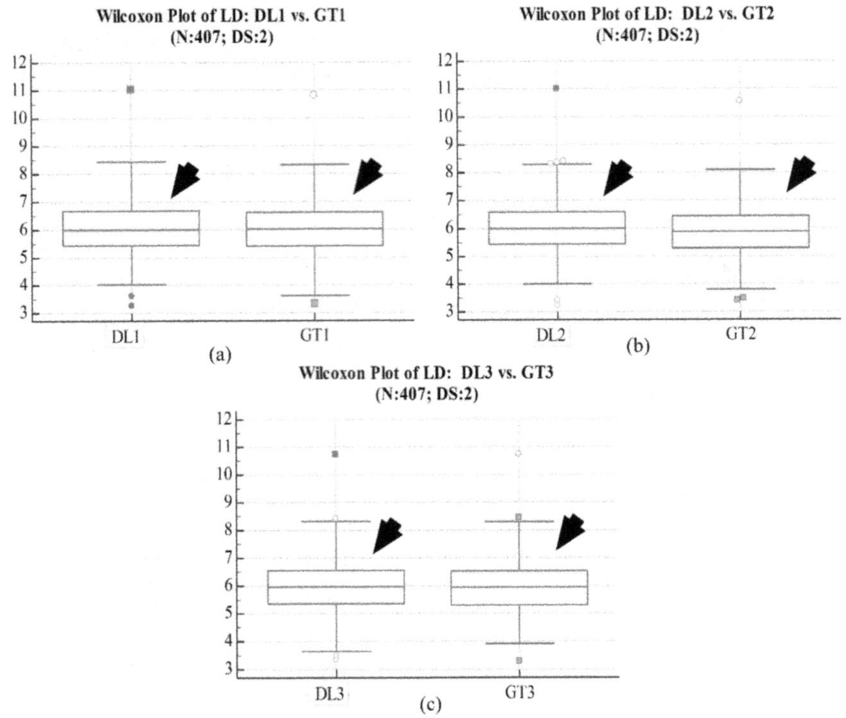

Figure A1. Wilcoxon box plot for (a) DL1 versus GT1, (b) DL2 versus GT2 and (c) DL3 versus GT3.

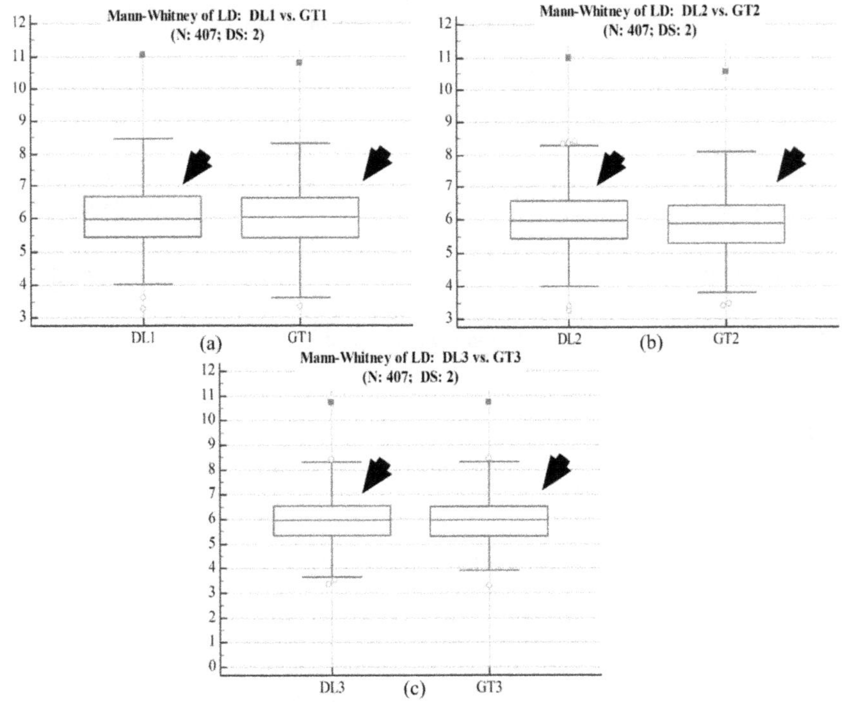

Figure A2. Mann–Whitney box plot for (a) DL1 versus GT1, (b) DL2 versus GT2 and (c) DL3 versus GT3.

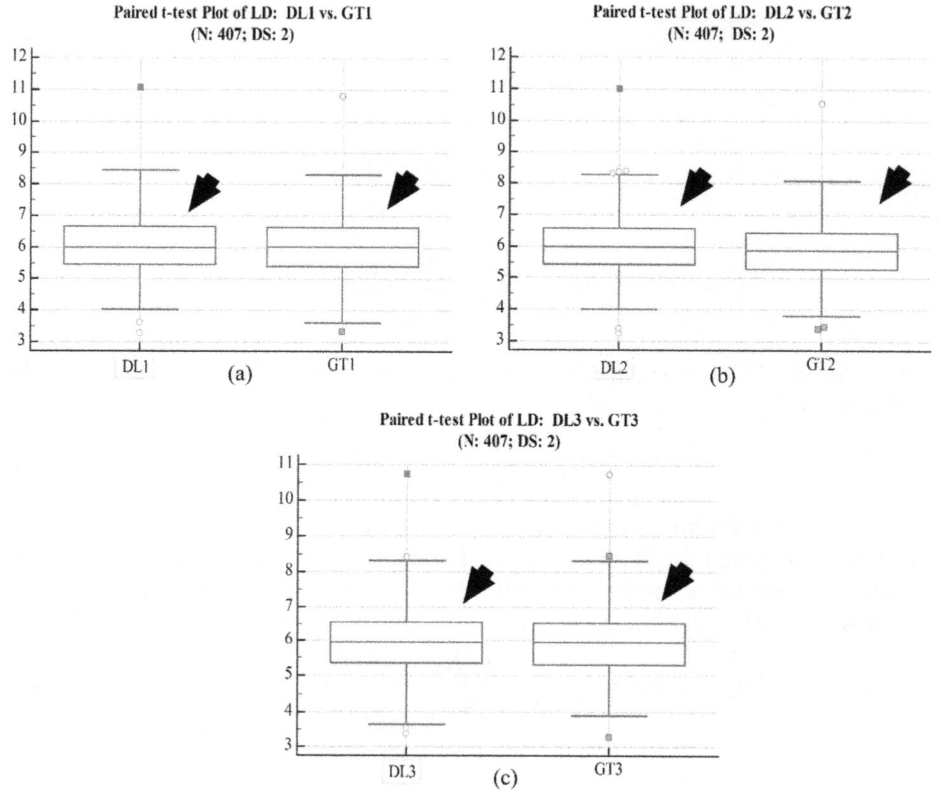

Figure A3. Paired *t*-test box plot for (a) DL1 versus GT1, (b) DL2 versus GT2 and (c) DL3 versus GT3.

Appendix B Polyline distance metric

Polyline distance metric

The polyline distance metric (PDM) [40] is used to measure the LD, LI-far error and LI-near error. Here, we focus on deriving the PDM given two border contours. Let the first and second contour be denoted as I_1 and I_2. Let the reference point on I_1 be vertex A_1 and the segment in I_2 be defined by vertices A_2 and A_3. Let the distance between A_1 and A_2 be d_1 and the distance between A_1 and A_3 be denoted as d_2. Let $D(A_1, L)$ be the polyline distance between vertex A_1: (x_1, y_1) on I_1 and line segment L formed by two points A_2: (x_2, y_2) and A_3: (x_3, y_3). Let phi (φ) be the distance of the reference point A_1 towards the line segment L. The perpendicular distance between the line segment L and the reference point, A_1, is given by d_P. Then, the polyline distance $D(A_1, L)$ can be defined as

$$D(A_1, L) = \begin{cases} |d_P| & 0 < \varphi < 1 \\ \min(d_1, d_2) & \varphi < 0, \varphi > 1, \end{cases} \tag{B.1}$$

where

$$d_1 = \sqrt{(x_1 - x_2)^2 + (y_1 - y_2)^2} \tag{B.2}$$

$$d_2 = \sqrt{(x_1 - x_3)^2 + (y_1 - y_3)^2} \tag{B.3}$$

$$\varphi = \frac{(y_3 - y_2)(y_1 - y_2) + (x_3 - x_2)(x_1 - x_2)}{(x_3 - x_2)^2 + (y_3 - y_2)^2} \tag{B.4}$$

and

$$d_P = \frac{(y_3 - y_2)(x_2 - x_1) + (x_3 - x_2)(y_1 - y_2)}{\sqrt{(x_3 - x_2)^2 + (y_3 - y_2)^2}}. \tag{B.5}$$

The process to obtain $D(A_1, L)$ is repeated for the rest of the points of the contour I_j and is given by

$$D(I_1, I_2) = \sum_{i=1}^{N} D(A_i, S_{I_2}), \tag{B.6}$$

where N is the total number of points on I_1 and S_{I_2} is the segment on contour I_2. This algorithm is repeated in reverse, where I_2 becomes the reference contour and I_1 becomes the segment contour. The reverse is represented as $D(I_2, I_1)$. Finally, by combining both $D(I_1, I_2)$ and $D(I_2, I_1)$, we obtain the PDM which is given by

$$D_{\text{PDM}}(I_1 : I_2) = \frac{D(I_1, I_2) + D(I_2, I_1)}{(\#\text{ points} \in I_1 + \#\text{ points} \in I_2)}. \tag{B.7}$$

Appendix C Figure-of-merit and precision-of-merit

LD and mean LD computation

The LD error is computed as the PDM between the ground truth LD (LD_{gt}) and deep-learning LD (LD_{dl}). The LD for a patient i is computed as the PDM between the LI-far ($\text{LI}_{\text{far}(i)}$) and LI-near ($\text{LI}_{\text{near}(i)}$) wall of the patient. The ground truth LD ($\text{LD}_{\text{gt}(i)}$) for patient i is given as

$$\text{LD}_{\text{gt}(i)} = D_{\text{PDM}}\left(\text{LI}^{\text{gt}}_{\text{far}(i)} : \text{LI}^{\text{gt}}_{\text{near}(i)}\right). \tag{C.1}$$

Similarly, the deep-learning LD ($\text{LD}_{\text{dl}(i)}$) for image i is given as

$$\text{LD}_{\text{dl}(i)} = \text{D}_{\text{PDM}}\left(\text{LI}^{\text{dl}}_{\text{far}(i)} : \text{LI}^{\text{dl}}_{\text{near}(i)}\right). \tag{C.2}$$

The mean LD can therefore be computed as

$$\overline{\text{LD}}_{\text{dl}} = \frac{1}{N}\sum_{i=1}^{N} \text{LD}_{\text{dl}(i)} \tag{C.3}$$

$$\overline{\text{LD}}_{\text{gt}} = \frac{1}{N}\sum_{i=1}^{N} \text{LD}_{\text{gt}(i)}. \tag{C.4}$$

LD error and mean LD error

The LD error ($\varepsilon_{LD(i)}$) for an image i is computed as the absolute difference between $LD_{gt(i)}$ and $LD_{dl(i)}$ and is mathematically represented as

$$\varepsilon_{LD(i)} = |LD_{gt(i)} - LD_{dl(i)}|. \tag{C.5}$$

If $\varepsilon_{LD(i)}$ represents the LD error for an image i, then the mean LD error ($\overline{\varepsilon}_{LD}$) for all N patients is given by

$$\overline{\varepsilon}_{LD} = \frac{\sum_{i=1}^{N} \varepsilon_{LD(i)}}{N}. \tag{C.6}$$

Precision-of-merit (PoM)

Using equations (C.1) and (C.2), one can therefore define mathematically the precision-of-merit (PoM), which is given as

$$PoM_{LD}(\%) = 100 - \left(\frac{\sum_{i=1}^{N} \frac{|LD_{dl(i)} - LD_{gt(i)}|}{LD_{gt(i)}}}{N} \right) \times 100. \tag{C.7}$$

Figure-of-merit (FoM)

The central tendency of the LD distribution can also be used to tell the difference between the DL-based LD and GT-based LD. Using equations (C.3) and (C.4), one can therefore compute the figure-of-merit (FoM), which can be expressed mathematically as

$$FoM_{LD}(\%) = 100 - \left[\left(\frac{|\overline{LD}_{dl} - \overline{LD}_{gt}|}{\overline{LD}_{gt}} \right) \times 100 \right]. \tag{C.8}$$

Appendix D LI-far and LI-near position errors

LI-far error

The LI-far error ($\varepsilon_{far(i)}$) for patient i is computed as the PDM between the GT LI-far wall ($LI^{gt}_{far(i)}$) and DL LI-far ($LI^{dl}_{far(i)}$) wall for the patient, which is given by

$$\varepsilon_{far(i)} = D_{PDM}\left(LI^{gt}_{far(i)}: LI^{dl}_{far(i)}\right). \tag{D.1}$$

If $\varepsilon_{far(i)}$ represents the LI-far error for the patient i, then the mean LI-far error ($\overline{\varepsilon}_{far}$) for all N patients is given by

$$\overline{\varepsilon}_{far} = \frac{\sum_{i=1}^{N} \varepsilon_{far(i)}}{N}. \tag{D.2}$$

LI-near error

Similarly, the LI-near error ($\varepsilon_{\text{near}(i)}$) is computed as the PDM between the GT LI-near wall ($\text{LI}^{\text{gt}}_{\text{near}(i)}$) and DL LI-near ($\text{LI}^{\text{dl}}_{\text{near}(i)}$) wall for patient i, which is given by

$$\varepsilon_{\text{near}(i)} = D_{\text{PDM}}\left(\text{LI}^{\text{gt}}_{\text{near}(i)}:\ \text{LI}^{\text{dl}}_{\text{near}(i)}\right). \tag{D.3}$$

The mean LI-near error ($\bar{\varepsilon}_{\text{near}}$) for all N patients is given by

$$\bar{\varepsilon}_{\text{near}} = \frac{\sum_{i=1}^{N} \varepsilon_{\text{near}(i)}}{N}. \tag{D.4}$$

The corresponding symbol table is given in appendix E, table E1.

Appendix E Symbol table

Table E1. Symbol table.

SN.	Symbol	Abbreviation
1	α	Predicted output
2	β	Ground truth
3	C	Total number of classes
4	N	Total number of patients/images
5	θ	Cross-entropy loss function
6	I_1	First contour
7	I_2	Second contour
8	A_1	Reference point on I_1
9	A_2	Reference point on I_2
10	A_3	Reference point on I_2
11	L	Line segment formed by vertex A_1 and vertex A_2 on I_2
12	d_1	Euclidean distance between vertex A_1 and vertex A_2
13	d_2	Euclidean distance between vertex A_1 and vertex A_3
14	φ	Distance of the reference point A_1 and the line segment L
15	d_P	Perpendicular distance between L and the reference point A_1
16	$D(A_1, L)$	Polyline distance between reference point A_1 and the line segment L
17	$D(I_1, I_2)$	Mean polyline distance between all points on contour I_1 with respect to contour I_2
18	$D(I_2, I_1)$	Mean polyline distance between all points on contour I_2 with respect to contour I_1
19	D_{PDM}	Bidirectional polyline distance metric by combining $D(I_1, I_2)$ and $D(I_2, I_1)$
20	$\text{LI}^{\text{gt}}_{\text{far}(i)}$	LI-far interface or contour taken from ground truth for patient i
21	$\text{LI}^{\text{gt}}_{\text{near}(i)}$	LI-near interface or contour taken from ground truth for patient i
22	$\text{LI}^{\text{dl}}_{\text{far}(i)}$	LI-near interface or contour taken from deep learning for patient i
23	$\text{LI}^{\text{dl}}_{\text{near}(i)}$	LI-near interface or contour taken from deep learning for patient i
24	$\text{LD}_{\text{gt}(i)}$	LD of for patient i taken from ground truth

25	$\overline{\mathrm{LD}}_{\mathrm{gt}}$	Mean LD for N patients taken from ground truth
26	$\mathrm{LD}_{\mathrm{dl}(i)}$	LD of for patient i taken from deep learning
27	$\overline{\mathrm{LD}}_{\mathrm{dl}}$	Mean LD for N patients taken from deep learning
28	$\varepsilon_{\mathrm{LD}(i)}$	Absolute LD error for patient i
29	$\bar{\varepsilon}_{\mathrm{LD}}$	Mean LD error for N patients
30	$\varepsilon_{\mathrm{far}(i)}$	Absolute LI-far error corresponding to patient i
31	$\bar{\varepsilon}_{\mathrm{far}}$	Mean LI-far error computed from N patients
32	$\varepsilon_{\mathrm{near}(i)}$	LI-near error corresponding to patient i
33	$\bar{\varepsilon}_{\mathrm{near}}$	Mean LI-near error computed from N patients
34	$\mathrm{FoM}_{\mathrm{LD}}$	Figure-of-merit
35	$\mathrm{PoM}_{\mathrm{LD}}$	Precision-of-merit computed for individual LD values of DL and GT

References

[1] Ward H *et al* 2012 *Oxford Handbook of Epidemiology for Clinicians* (Oxford: Oxford University Press)

[2] American Heart Association http://www.heart.org/idc/groups/heart-public

[3] Sobieszczyk P and Beckman J 2006 Carotid artery disease *Circulation* 114 e244–7

[4] NIH 2016 What is a stroke? www.nhlbi.nih.gov/health/health-topics/topics/stroke (June 22)

[5] Bots M L, Diederick E G, Hofman A and Witteman J C M 2005 Common carotid intima–media thickness and risk of acute myocardial infarction: the role of lumen diameter *Stroke* **36** 762–7

[6] Bots M L, Arno W H, Koudstaal P J, Hofman A and Grobbee D E 1997 Common carotid intima–media thickness and risk of stroke and myocardial infarction: the Rotterdam Study *Circulation* **96** 1432–7

[7] Tell G S, Joseph F P, Ward B J, Kittner S J, Savage P J and Robbins J 1994 Relation of smoking with carotid artery wall thickness and stenosis in older adults. The Cardiovascular Health Study. The Cardiovascular Health Study (CHS) Collaborative Research Group *Circulation* **90** 2905–8

[8] Polak J F, O'Leary D H, Kronmal R A, Wolfson S K, Bond M G, Tracy R P, Gardin J M, Kittner S J, Price T R and Savage P J 1993 Sonographic evaluation of carotid artery atherosclerosis in the elderly: relationship of disease severity to stroke and transient ischemic attack *Radiology* **188** 363–70

[9] Nicolaides A N *et al* 2010 Asymptomatic internal carotid artery stenosis and cerebrovascular risk stratification *J. Vasc. Surg.* **52** 1486–96

[10] Mann J M and Michael J D 1996 Vulnerable plaque: relation of characteristics to degree of stenosis in human coronary arteries *Circulation* **94** 928–31

[11] Dodge J T, Brown B G, Bolson E L and Dodge H T 1992 Lumen diameter of normal human coronary arteries. Influence of age, sex, anatomic variation, and left ventricular hypertrophy or dilation *Circulation* **86** 232–46

[12] Sakakura K, Nakano M, Otsuka F, Ladich E, Kolodgie F D and Virmani R 2013 Pathophysiology of atherosclerosis plaque progression *Heart Lung Circ.* **22** 399–411

[13] Sarkar S, Ghosh S, Ghosh S K and Collier A 2007 Role of transcranial Doppler ultrasonography in stroke *Postgrad. Med. J.* **83** 683–9

[14] Branas C C, Weingarten M S, Czeredarczuk M and Schafer P F 1994 Examination of carotid arteries with quantitative color Doppler flow imaging *J. Ultrasound Med.* **13** 121–7

[15] Mitchell D G 1990 Color Doppler imaging: principles, limitations, and artifacts *Radiology* **177** 1–10

[16] Mehra S 2010 Role of duplex Doppler sonography in arterial stenoses *J. Indian Acad. Clin. Med.* **11** 294–9

[17] Jones S A *et al* 1996 The influence of acoustic impedance mismatch on poststenotic pulsed-Doppler ultrasound measurements in a coronary artery model *Ultrasound Med. Biol.* **22** 623–34

[18] Hoeks A P G, Brands P J, Willigers J M and Reneman R S 1999 Non-invasive measurement of mechanical properties of arteries in health and disease *Proc. Inst. Mech. Eng.* H **213** 195–202

[19] Wendelhag I, Gustavsson T, Suurküla M, Berglund G and Wikstrand J 1991 Ultrasound measurement of wall thickness in the carotid artery: fundamental principles and description of a computerized analysing system *Clin. Physiol. Funct. Imaging* **11** 565–77

[20] Molinari F, Zeng G and Suri J S 2010 A state of the art review on intima–media thickness (IMT) measurement and wall segmentation techniques for carotid ultrasound *Comput. Methods Programs Biomed.* **100** 201–21

[21] Golemati S, Stoitsis J, Sifakis E G, Balkizas T and Nikita K S 2007 Using the Hough transform to segment ultrasound images of longitudinal and transverse sections of the carotid artery *Ultrasound Med. Biol.* **33** 1918–32

[22] Molinari F, Zeng G and Suri J S 2010 An integrated approach to computer based automated tracing and its validation for 200 common carotid arterial wall ultrasound images *J. Ultrasound Med.* **29** 399–418

[23] Loizou C P, Kasparis T, Spyrou C and Pantziaris M 2013 Integrated system for the complete segmentation of the common carotid artery bifurcation in ultrasound images *Artif. Intell. Appl. Innov.* **412** 292–301

[24] Suri J S *et al* 2002 Shape recovery algorithms using level sets in 2-D/3-D medical imagery: a state-of-the-art review *IEEE Trans. Inf. Technol. Biomed.* **6** 8–28

[25] Suri J S and Laxminarayan S 2002 PDE and level sets (Berlin: Springer)

[26] Araki T *et al* 2016 Two automated techniques for carotid lumen diameter measurement: regional versus boundary approaches *J. Med. Syst.* **40** 1–19

[27] Krishna Kumar P *et al* 2017 Accurate lumen diameter measurement in curved vessels in carotid ultrasound: an iterative scale-space and spatial transformation approach *Med. Biol. Eng. Comput.* **55** 1415

[28] Kuppili V *et al* 2017 Extreme learning machine framework for risk stratification of fatty liver disease using ultrasound tissue characterization *J. Med. Syst.* **41** 152

[29] LeCun Y, Bengio Y and Hinton G 2015 Deep learning *Nature* **521** 436–44

[30] Long J, Shelhamer E and Darrell T 2015 Fully convolutional networks for semantic segmentation *Proc. IEEE Conf. Computer Vision and Pattern Recognition* (Piscataway, NJ: IEEE)

[31] Teichmann M *et al* 2016 *MultiNet: Real-time Joint Semantic Reasoning for Autonomous Driving* arXiv: 1612.07695

[32] Molinari M *et al* 2012 Fully automated dual-snake formulation for carotid intima–media thickness measurement *J. Ultrasound Med.* **31** 1123–36

[33] Iglesias J E and Sabuncu M R 2015 Multi-atlas segmentation of biomedical images: a survey *Med. Image Anal.* **24** 205–19

[34] Ciresan D *et al* 2012 Deep neural networks segment neuronal membranes in electron microscopy images *Advances in Neural Information Processing Systems* (Cambridge, MA: MIT Press) pp 2843–51

[35] Bar Y *et al* 2015 Chest pathology detection using deep learning with non-medical training *2015 IEEE 12th Int. Symp. on IEEE Biomedical Imaging (ISBI)* (Piscataway, NJ: IEEE)

[36] Simonyan K and Zisserman A 2014 *Very deep convolutional networks for large-scale image recognition* arXiv: 1409.1556

[37] Hatipoglu N and Bilgin G 2017 Cell segmentation in histopathological images with deep learning algorithms by utilizing spatial relationships *Med. Biol. Eng. Comput.* **55** 1829–48

[38] Zhao J, Zhang M, Zhou Z, Chu J and Cao F 2017 Automatic detection and classification of leukocytes using convolutional neural networks *Med. Biol. Eng. Comput.* **55** 1287–301

[39] García-Zapirain B, Elmogy M, El-Baz A and Elmaghraby A S 2017 Classification of pressure ulcer tissues with 3D convolutional neural network *Med. Biol. Eng. Comput.* **56** 2245–58

[40] Suri J S, Haralick R M and Sheehan F H 2000 Greedy algorithm for error correction in automatically produced boundaries from low contrast ventriculograms *Pattern Anal. Appl.* **3** 39–60

[41] Molinari F, Ganapathy Krishnamurthi U, Rajendra Acharya S, Vinitha Sree S, Zeng G, Saba L, Nicolaides A and Suri J S 2012 Hypothesis validation of far-wall brightness in carotid-artery ultrasound for feature-based IMT measurement using a combination of level-set segmentation and registration *IEEE Trans. Instrum. Meas.* **61** 1054–63

[42] Gutierrez M A, Pilon P E, Lage S G, Kopel L, Carvalho R T and Furuie S S 2002 Automatic measurement of carotid diameter and wall thickness in ultrasound images *Computers in Cardiology* 359–62

[43] Sahani A K, Joseph J and Sivaprakasam M 2013 Automatic measurement of lumen diameter of carotid artery in A-Mode ultrasound *2013 35th Annu. Int. Conf. of the IEEE Engineering in Medicine and Biology Society (EMBC)* (Piscataway, NJ: IEEE) pp 3873–76

[44] Saba L *et al* 2016 Carotid inter-adventitial diameter is more strongly related to plaque score than lumen diameter: an automated tool for stroke analysis *J. Clin. Ultrasound* **44** 210–20

[45] Araki T *et al* 2016 Ultrasound-based automated carotid lumen diameter/stenosis measurement and its validation system *J. Vasc. Ultrasound* **40** 120–34

Chapter 6

Deep-learning strategy for accurate carotid intima–media thickness measurement: an ultrasound study on a Japanese diabetic cohort

Mainak Biswas, Venkatanareshbabu Kuppili, Tadashi Araki, Damodar Reddy Edla, Elisa Cuadrado Godia, Luca Saba, Harman S Suri, Tomaž Omerzu, John R Laird, Narendra N Khanna, Andrew Nicolai and Jasjit S Suri

Carotid intima–media thickness (cIMT) is an important biomarker for cardiovascular disease and stroke monitoring. This study presents an intelligence-based, novel, robust and clinically strong strategy that uses a combination of deep-learning (DL) and machine-learning (ML) paradigms. A two-stage DL-based system (a class of AtheroEdge™ systems) is proposed for cIMT measurement. Stage 1 consists of a convolution layer-based encoder for feature extraction and a fully convolutional network-based decoder for image segmentation. This stage generates the raw inner lumen borders and raw outer interadventitial borders. To smooth these borders, the DL system uses a cascaded stage 2 that consists of ML-based regression. The final outputs are the far wall lumen–intima (LI) and media–adventitia (MA) borders, which are used for cIMT measurement. There were two sets of gold standards during the DL design, therefore two sets of DL systems (DL1 and DL2) were derived. A total of 396 B-mode ultrasound images of the right and left common carotid artery were used from 203 patients (Institutional Review Board (IRB) approved, Toho University, Japan). For the test set, the cIMT errors for the DL1 and DL2 systems with respect to the gold standard were 0.126 ± 0.134 and 0.124 ± 0.100 mm, respectively. The corresponding LI errors for the DL1 and DL2 systems were 0.077 ± 0.057 and 0.077 ± 0.049 mm, respectively, while the corresponding MA errors for DL1 and DL2 were 0.113 ± 0.105 and 0.109 ± 0.088 mm, respectively. The results showed an up to 20% improvement in cIMT readings for the DL system compared to sonographer readings. Four statistical tests were conducted to evaluate

doi:10.1088/2053-2563/ab0820ch6

reliability, stability and statistical significance. The results showed that the performance of the DL-based approach was superior to the nonintelligence-based conventional methods that use spatial intensities alone. The DL system can be used for stroke risk assessment during routine or clinical trial modes.

6.1 Introduction

Stroke due to cardiovascular disease (CVD) causes the death of approximately five million people and disability among another five million people around the world each year [1]. In the USA, 795 000 people suffered from a stroke in 2010, causing direct medical care costs of approximately USD 33 billion and indirect costs of around USD 20.6 billion [2]. Stroke is generally caused by the blockage or rupturing of the common carotid artery (CCA) or internal carotid artery (ICA) that supply blood to the brain. This blockage or rupture is triggered by the formation of plaque along the arterial walls. Plaque is usually composed of cholesterol, fatty substances, cellular waste products, calcium and fibrin, and is generally formed between the LI and MA interfaces [3].

cIMT is the mean perpendicular distance between the LI and MA interfaces and is an important biomarker for CVD. A comprehensive risk analysis study on 5858 subjects revealed that cIMT values > 1.18 mm led to an increased stroke rate [4]. In 2006, the findings of Bots [5] showed that cIMT was related to the presence of atherosclerosis in the coronary artery. Risk prediction models developed by Nambi *et al* [6] showed an increase in the CVD risk when cIMT and plaque information was added. The study of Meuwese *et al* [7] suggested that an increase in cardiovascular risk was related to an increase in mean cIMT. Ikeda *et al* [8] also confirmed the significant association between cIMT and CVDs. All of the above-mentioned studies also indicate that an increase in cardiovascular events (myocardial infarction) is correlated to an increase in the mean cIMT.

Although Suri *et al* have diligently worked to standardize cIMT measurements [9], there are still challenges regarding accuracy and reproducibility when it comes to the CCA, ICA and bulb regions. Several factors contribute to this, including the variability in studies with regard to nationality, ethnicity, disease, age groups, etc. In this regard, a concerted effort was made to construct a multi-institutional dataset using multiple ethnicities and varying age groups [10]. There are other technical challenges associated with cIMT measurement. For example, the images are obtained through B-mode ultrasound (US) using a linear probe that is manually operated. The CCA extends from the jaw to the shoulder bone; however, the linear probe is unable to cover the entire carotid artery length and imaging has to be performed in sections (i.e. distal, mid and proximal). The CCA image quality also depends on external factors such as speckle noise, probe position, neck position, probe orientation (i.e. anterior, posterior or posterior lateral), probe contact with the skin, linear frequency usage, gain control, dynamic range, and features such as harmonic and compound imaging [11, 12]. The traditional manual segmentation of US images is slow, error-prone, and subject to intra- and interobserver variability. Therefore the measurement of cIMT through automated methods is a growing field of interest. Previously described completely automated methods are briefly discussed here.

Molinari *et al* [9] compared four automated techniques for cIMT measurement. The first method was completely automated layer extraction (CALEX), which is based on the integration of three approaches: feature extraction, line fitting and classification. The second was the completely automated robust edge snapper (CARES) [13], which is based on a combination of feature extraction and edge detection. The third methodology was the completely automated multiresolution edge snapper (CAMES), which is based on a multiresolution approach and uses the concept of scale-space (SS) [14]. Finally, the fourth methodology was the carotid automated double-line extraction system, which is based on edge flow (an edge-detection technique based on US texture and edge energies) [15]. Saba *et al* [16] proposed a fully automated system (AtheroEdge™) for cIMT measurement, while Ikeda *et al* [17] proposed a cIMT measurement system with a classification paradigm that used a combination of global and local strategies involving texture-based entropy and morphology. Saba *et al* [18] later developed a fully automated cloud-based solution called AtheroCloud™ for cIMT measurements. Ikeda *et al* [19] recently proposed an automated segmental cIMT measurement technique that used an automated bulb-edge point as a reference marker. The above-mentioned methodologies use various features such as grayscale median, pixel classification, gradient edges, SS or a combination of these features to predict the cIMT risk assessment. Despite their strong contributions, these external factors make the spatial-based methods prone to variability and a lack of robustness when it comes to completely automated designs.

Another challenge in the segmentation of wall interfaces is the presence of shadows on the far wall due to calcium in the near wall. This causes border position errors in the detection of the LI and MA, even though the average cIMT error is well below the acceptable level. Previous methods took advantage of multiresolution approaches to increase the processing speed; however, the feature extraction at multiple levels was not derived, thus these approaches lack a comprehensive spatial deck of information. Another important point to note is that carotid US cohorts contain shape information that can be learned via neural networks, the intelligence power of which is unsurpassable. This current DL-based study removes all of the above-mentioned challenges to provide reliability and robustness. The spirit of this study was motivated by the work of Suri *et al*, who applied machine intelligence in different fields of medicine including gynecology, urology, dermatology, neurology [20–23] and recently in endocrinology [24].

This study proposed the same intelligence-based paradigm [25, 26] for cIMT measurements. It was hypothesized that by training deep layers of neural networks, the DL-based system could produce more reliable and accurate results when compared to previous methods. Unlike machine learning (ML), DL can generate its own features and thus eliminate the need for less accurate feature-extraction algorithms. The high-level features of DL are more distinctive than the features of conventional methods, thus resulting in a more accurate output. The superior training of the deep layers within the DL system allows the further provision of

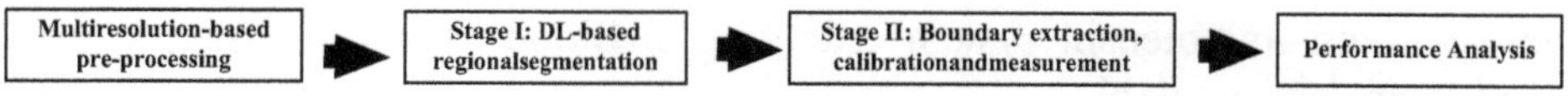

Figure 6.1. Overall concept of the DL-based cIMT measurement system.

better regional segmentation output compared to the conventional methods. The proposed DL-based system is implemented in four phases, as shown in figure 6.1.

Phase I is primarily adapted for data preparation. It removes the nontissue region [27] and prepares the image data using a multiresolution approach to speed up the DL paradigm. This phase is also responsible for generating the cross-validation protocol that splits the cohort into sets of training and testing carotid scans. Phase II is the heart of the DL system that performs the number crunching and consists of encoder and decoder neural networks [28, 29]. In this phase, deep intelligence is derived by externally controlling the number of loops (up to 20 000). The training system uses two kinds of gold standards, namely, lumen regional information and interadventitial regional information, which leads to the design of two DL systems: DL1 and DL2. Phase III performs the boundary extraction that changes regional information to vertex point information (i.e. LI and MA boundaries or the so-called raw DL borders). This phase also ensures smooth boundaries, which attempts to get closer to the ground truth (GT) using an ML-based system, which in turn increases the overall accuracy of the system. The cIMT values are computed from the LI and MA-far walls using the standardized polyline distance metric method (PDM) (discussed in appendix A). The last phase implements performance analysis alongside risk stratification. Four statistical tests were used to assess the statistical significance: a paired *t*-test, Mann–Whitney test, Wilcoxon test and the Kruskal–Wallis test.

In this chapter, section 6.2 discusses the data collection and patient demographics. Sections 6.10 and 6.4 present the methodology and results, respectively. Section 6.5 shows the performance evaluation, and section 6.6 discusses the statistical tests and risk analysis. The discussion and benchmarking are presented in section 6.7 and, finally, the conclusions are presented in section 6.8.

6.2 Data demographics and US acquisition

In this study, 204 patients (157 male and 47 female) with a mean age of 69 ± 11 years were selected. One left carotid image was not available out of the 408 images, therefore the database initially contained 407 images. Eight left carotid and three right carotid US images were rejected due to a lack of grayscale tissue information (including one patient whose left and right carotid images were removed). Thus the final dataset consisted of 396 carotid scans (left and right) from 203 patients. The sonographer's far wall cIMT readings were also available for 193 patients (346 US scans).

Informed consent was obtained from all patients and the IRB, and ethical approval was granted by Toho University, Japan. The mean hemoglobin, glucose, low-density lipoprotein cholesterol, high-density lipoprotein cholesterol and total cholesterol values were 5.8 ± 1.0, 108 ± 31, 99.80 ± 31.30, 50.40 ± 15.40 and 174.6 ± 37.7 mg dL^{-1}, respectively. Of the pool of 203 patients, 92 were regular smokers. Hypertensive and high-cholesterol patients were receiving adequate medication; for example, 93 patients were taking statins to lower their cholesterol levels and 84 were receiving renin–angiotensin system antagonists. Blood pressure statistics for the patients were not available.

A sonographic scanner (Aplio XV, Aplio XG, Xario; Toshiba, Inc., Tokyo, Japan) equipped with a 7.5 MHz linear array transducer was used to examine the left and right carotid arteries. All scans were performed under the supervision of an experienced sonographer (15 years of experience). High-resolution images were acquired as per the recommendations of the American Society of Echocardiography Carotid Intima–Media Thickness Task Force. The mean pixel resolution in the database was 0.05 ± 0.01 mm/pixel.

Manual tracing of the lumen and adventitia borders was performed using ImgTracer™ (AtheroPoint™, Roseville, CA, USA), which is a user-friendly commercial software [30]. The number of points varied with the length of the carotid artery. The software zooms into the image for better visualization of the wall and provides a set of traced (x, y) coordinates.

6.3 Methodology

The heart of the system is an intelligence-based DL platform that supports the extraction of deep features and thereby eliminates the need for algorithms that perform poorly for feature extraction. The platform consists of two DL networks: the encoder, which is used for feature extraction [28], and the decoder, which is used for regional segmentation of the lumen region (LR) or interadventitial region (IAR) [29]. The DL system design allows the LR segmentation to be run in parallel with the IAR segmentation. This is called the regional segmentation block, which is the second phase of the system. Before feeding the binary training images for an LR and an IAR into the DL block, the system design expects the input data to be prepared accordingly for the DL block (the so-called multiresolution block or phase I as shown in figure 6.2). The encoder–decoder is phase II of the DL system. The image processing pipeline is always cascaded with a fine tuner to smooth or refine the outputs, therefore an ML-based system is used to extract LI-far and MA-far borders as part of the phase III subsystem. Finally, performance evaluation is implemented to benchmark the results. This is phase IV of the entire pipeline where the cIMT is measured and undergoes statistical testing. A detailed description of the system is provided in figure 6.2, and the details of these phases and their mathematical representations are discussed below.

6.3.1 Multiresolution as phase I

The objective of phase I was to prepare the data for adaptability to the DL system, which required the grayscale training cohort to be cropped to remove the nontissue information [27]. This automated cropping ensured that the tissue region was retained. The grayscale images were reduced by a further 10% to ensure that very low contrast regions around the edges of the image were eliminated. These grayscale images were down-sampled to improve the processing speed of the DL system under the multiresolution paradigm. In the data preparation block, the binary mapped images were also created which were mapped on a one-to-one basis with the

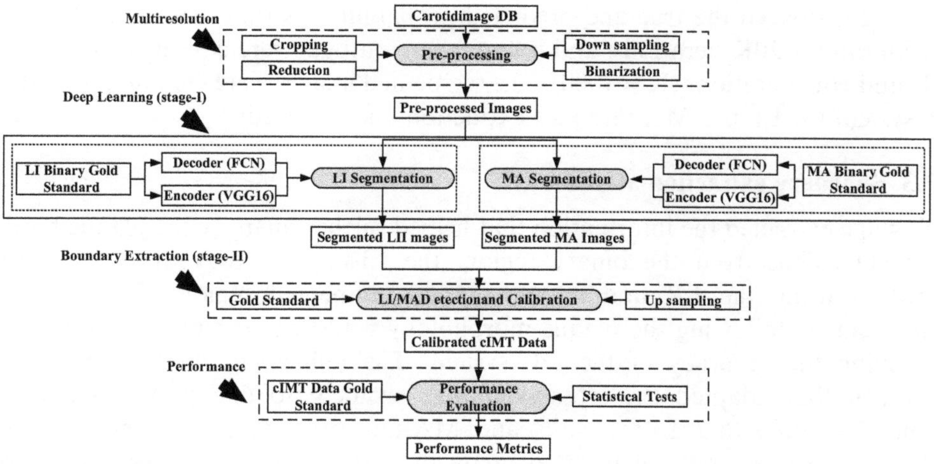

Figure 6.2. The four phases of a DL-based system (a class of AtheroEdge™ system, AtheroPoint™) shown in arrows. Phase I: multiresolution; phase II: the DL-based system; phase III: boundary extraction and calibration; and phase IV: performance analysis.

grayscale down-sampled carotid US scans. If the DL system was prepared for LR extraction, then binary maps corresponded to the LR. Conversely, if the DL system was prepared for IAR extraction, then the binary maps corresponded to the IAR. These LR and IAR binary maps were considered as the gold standard, as their borders were manually traced by experts.

6.3.2 DL as phase II

The DL-based system consisted of two subsystems: the encoder and decoder. The encoder extracted features from the images while the decoder created segmented images from the features. The encoder consisted of 13 convolution layers and five max-pooling layers of the VGG16 network [31]. Details of the decoder network are given in appendix B. The weights were initialized using pretrained visual geometry group (VGG) weights. The convolution layers generated high-level features from the input data, and the max-pooling layers down-sampled the input feature values.

The decoder consisted of three up-sampling layers of the fully convolutional network (FCN) [29]. The up-sampling layers up-sampled the input features but with a twist. They employed two skip operations that helped recover spatial information resulting in highly accurate and crisp segmentation images. Additional information about the skip operation is presented in the discussion section. The up-sampled layers were initialized using VGG weights. The cross-entropy loss function employed for segmentation was

$$\theta_{\text{class}}(\beta_1, \beta_2) = \frac{1}{|N|} \sum_{n \in N} \sum_{l \in L} \beta_{2_n}(l) \log \beta_{1_n}(l), \tag{6.1}$$

where β_1 is the prediction, β_2 is the gold standard or GT, L is the total number of classes and N is the total number of images. The loss function was defined as the difference between the true and predicted probability distributions. The DL-based system ran for 20K iterations, and intermediate outputs were collected for 4K, 8K, 12K and 16K iterations (K = 1000). The segmented images were fed into phase III of the system for LI and MA interface extraction and calibration.

6.3.3 Boundary extraction as phase III

This stage extracted the information that helped further quantify the plaque burden or cIMT. Thus from the binary region, the LI-far and MA-far borders were extracted using the LR and interadventitial segmented regions. This required refinement by following the plaque morphology whilst smoothing the borders and improving the accuracy of the DL system. The refinement used an ML-based approach that adapted the cross-validation protocol to determine accuracy. It should be noted that LI-far walls and MA-far walls were independent of the ML-based system and can be mathematically expressed as a regression or least squares model if GT (or ideal) boundaries are given as

$$\boldsymbol{I}[2N \times P]\colon\ [x_1 \quad y_1 \cdots x_N \quad y_N]^T \tag{6.2}$$

and the raw DL borders extracted using the DL-based method are given as

$$\boldsymbol{D}[2N \times P]\colon\ [a_1 \quad b_1 \ldots a_N \quad b_N]^T, \tag{6.3}$$

where N represents the total number of patients and P represents the total points on the border. In the adaptation of the cross-validation protocol, the DL boundaries were divided into two sets: a training set ($\boldsymbol{D}_{\mathrm{tr}}$) and a test set ($\boldsymbol{D}_{\mathrm{te}}$). Correspondingly, the GT boundaries were also divided into training sets ($\boldsymbol{I}_{\mathrm{tr}}$) and test sets ($\boldsymbol{I}_{\mathrm{te}}$). Using the linear model of least squares presented in [30], one can mathematically express this as a norm equation given as $\|\boldsymbol{I} - \boldsymbol{D}\boldsymbol{\varphi}\|^2$. Letting $\hat{\boldsymbol{\varphi}}_{\mathrm{tr}}$ be the unknown training coefficient matrix of size $[P \times P]$, one can compute it as

$$\hat{\boldsymbol{\varphi}}_{\mathrm{tr}} = \left(\boldsymbol{D}_{\mathrm{tr}}^T \cdot \boldsymbol{D}_{\mathrm{tr}}\right)^{-1} \cdot \boldsymbol{D}_{\mathrm{tr}}^T \cdot \boldsymbol{I}_{\mathrm{tr}}, \tag{6.4}$$

where " · " represents the multiplicative product. These training coefficients were used to estimate the test boundaries ($\hat{\boldsymbol{I}}_{\mathrm{te}}$) as the product of training coefficients and raw test DL borders using

$$\hat{\boldsymbol{I}}_{\mathrm{te}} = \hat{\boldsymbol{\varphi}}_{\mathrm{tr}} \cdot \boldsymbol{D}_{\mathrm{te}}. \tag{6.5}$$

Finally, the DL borders underwent cIMT measurement, as presented in appendix A. The last stage (phase IV) consisted of performance evaluation, as shown in figure 6.2.

6.3.4 Performance evaluation as phase IV

The performance of the DL system required computation of the LI- and MA-far wall position errors. These values were compared against the GT to estimate the precision-of-merit (PoM). These calculations are shown in appendix C. These performance metrics were then compared against other systems for benchmarking (presented in the performance evaluation section).

6.4 Experimental protocol and results

The experimental protocol primarily consisted of the optimization of DL with respect to a number of iterations independent of the LI and MA wall interfaces. As there were two DL systems corresponding to two GTs, the results are presented with respect to GT1 and GT2.

6.4.1 Experimental protocol

In this study, K10 cross-validation (i.e. 90% training dataset and 10% testing dataset) was used for training and testing. In this cross-validation, the dataset was randomly divided into ten parts and ten combinations were formed from these parts. Each combination contained nine parts for training and one part for testing.

The optimization protocol was implemented for 4K, 8K, 12K, 16K and 20K iterations (K = 1000). The iterations were evaluated for LI, MA and cIMT errors to study their effects on the encoder and decoder (shown in figure 6.3) and their ability to smooth out the glitches against the gold standard. The LI, MA and cIMT error values after ML-based calibration were further recorded to show the least error value that smoothed the output borders and improved the accuracy of the entire DL system.

A sample visual output of the DL-based system from phase III is shown in figure 6.4. In addition to the LI, MA and cIMT error evaluations, comprehensive clinical data analyses were also performed (i.e. correlation of age versus cIMT, risk stratification based on the cIMT threshold and receiver operating characteristic (ROC) analysis).

6.4.2 Results

The results were computed for 4K, 8K, 12K, 16K and 20K iterations (K = 1000). The plots for error versus iteration with respect to GT1 and GT2 are shown in figure 6.5(a) and (b), respectively. The LI, MA and cIMT error values for all iterations, including fusion and calibration, corresponding to GT1 and GT2 are presented in tables 6.1 and 6.2, respectively. The term fusion refers to the best result among all iterations. The cIMT values in the fusion rows of tables 6.1 and 6.2 refer to the values obtained from the best optimized LI and MA wall interfaces among all iterations. All values in the calibration rows of tables 6.1 and 6.2 indicate the final values after ML-based calibration was applied (phase II, figure 6.2). The results indicate that the optimized result for LI error with respect to GT1 was obtained at 16K iterations (i.e. 0.135 ± 0.076 mm, which later increased marginally). The optimized result for MA error with respect to GT1 was obtained at 20K iterations (i.e. 0.171 ± 0.153 mm). The best cIMT error with respect to GT1 was computed from the fusion of 16K iterations of LI interface optimization and 20K iterations of MA interface optimization (i.e. 0.128 ± 0.124 mm). After ML-based calibration, the LI, MA and cIMT errors were further reduced to 0.077 ± 0.057, 0.113 ± 0.105 and 0.126 ± 0.134 mm, respectively.

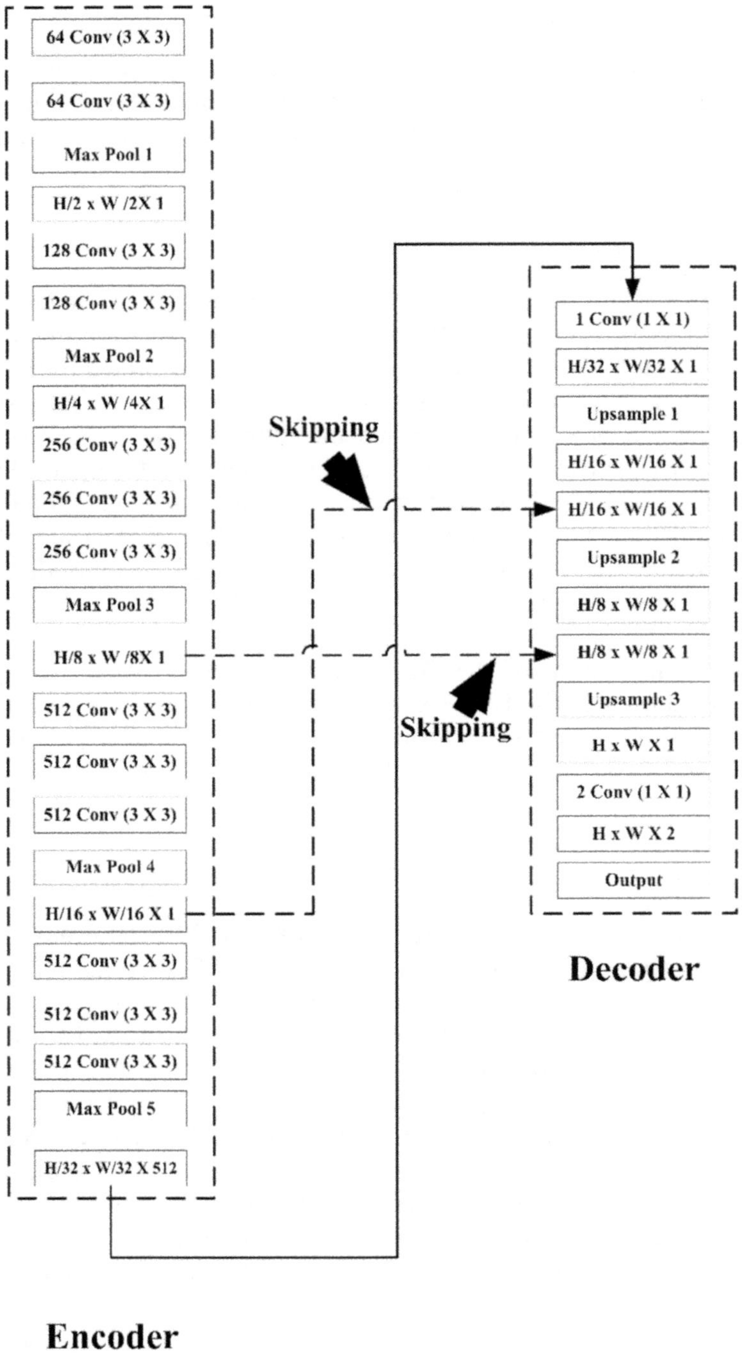

Figure 6.3. The combination of encoder–decoder blocks in the central DL system (a class of AtheroEdge™ system, AtheroPoint).

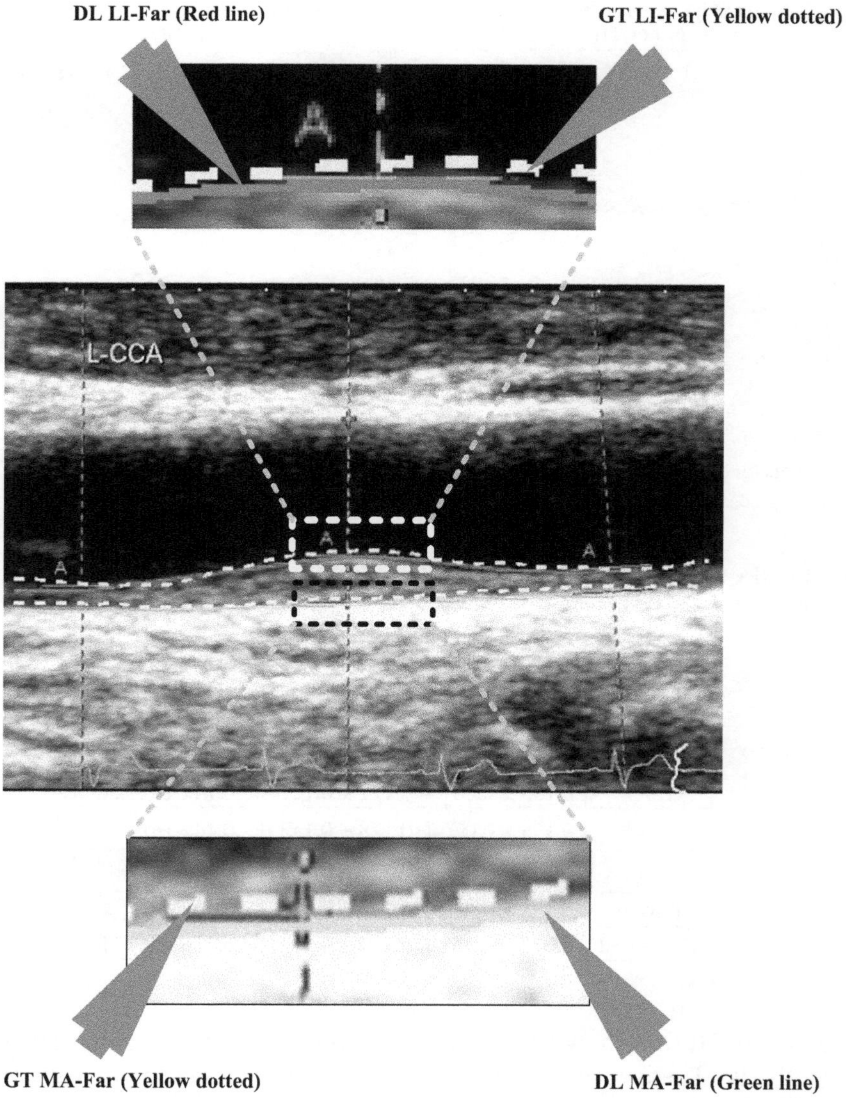

Figure 6.4. The DL-based system showing GT and DL outputs.

Similarly, the best results for LI and MA error optimization using the DL-based system with respect to GT2 were obtained at 20K iterations and were 0.131 ± 0.073 and 0.163 ± 0.132 mm, respectively. The cIMT error for the LI and MA interfaces was 0.124 ± 0.11 mm. After calibration, the LI, MA and cIMT error values were further reduced to 0.077 ± 0.049, 0.109 ± 0.088 and 0.124 ± 0.10 mm, respectively.

The correlation coefficient (CC) for DL1 with respect to GT1 was 0.96 ($p < 0.0001$) and for DL2 with respect to GT2 was 0.95 ($p < 0.0001$). Therefore the CC results show a high degree of correlation between the DL outputs and the corresponding GTs. The correlation plot for DL-based system (DL1 and DL2) output with respect to GT1 and GT2 is shown in figure 6.6. The p-value for both

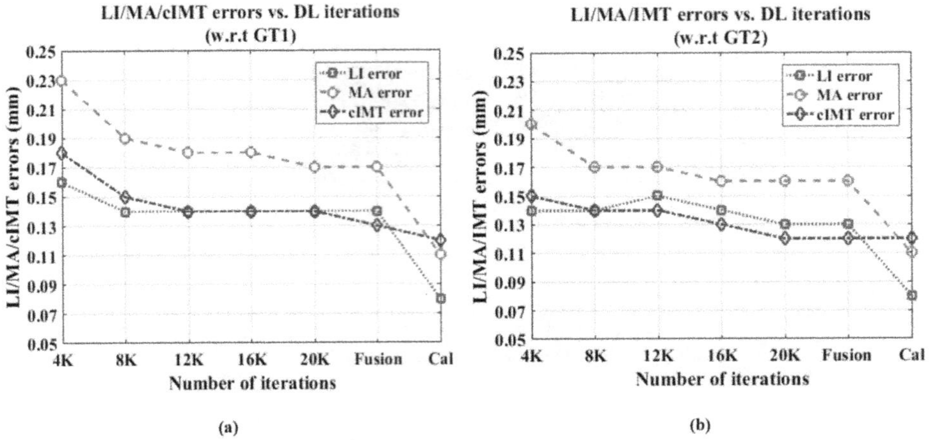

Figure 6.5. Plots for errors versus iterations against (a) GT1 and (b) GT2. Cal: calibration output; cIMT error: blue; LI error: green; MA error; red.

Table 6.1. LI, MA and cIMT error values against GT1. Gray boxes show the optimized results for the corresponding LI, MA and cIMT errors. *K = 1000.

DL Iterations	LI error w.r.t GT1 (mm)	MA error w.r.t GT1 (mm)	cIMT error w.r.t GT1 (mm)
4K*	0.161 ± 0.090	0.230 ± 0.197	0.177 ± 0.179
8K*	0.138 ± 0.078	0.187 ± 0.149	0.146 ± 0.13
12K*	0.135 ± 0.061	0.177 ± 0.122	0.142 ± 0.124
16K*	0.135 ± 0.076	0.178 ± 0.153	0.142 ± 0.132
20K*	0.135 ± 0.078	0.171 ± 0.153	0.140 ± 0.149
Fusion	0.135 ± 0.076	0.171 ± 0.153	0.128 ± 0.124
Calibrated	**0.077 ± 0.057**	**0.113 ± 0.105**	**0.126 ± 0.134**

plots was < 0.0001, thus showing a high correlation and significance that satisfies the null hypothesis. These results prove that the DL-based system is accurate and efficient. The performance of the DL-based system is evaluated in the next section.

The results of the DL-based system with respect to GT1 and GT2 were analyzed using Bland–Altman plots. The corresponding plots with reference to GT1 and GT2 are shown in figure 6.7.

6.5 Performance of the DL systems and variability analysis

Two sets of analyses were performed to evaluate the performance of the DL systems. The first set had four parts: part (i) focused on evaluating DL against manual expert tracers, part (ii) was against the sonographer's readings which were taken in real time in the US vascular laboratory, part (iii) evaluated signed and unsigned cIMT errors of the DL1 and DL2 systems and part (iv) compared the DL system against previously developed methods [33]. The second set had two parts: part (a) consisted

Table 6.2. LI, MA and cIMT error values against GT2. Gray boxes show the optimized results for the corresponding LI, MA and cIMT errors. *K = 1000.

DL Iterations	LI error w.r.t GT2 (mm)	MA error w.r.t GT2 (mm)	cIMT error w.r.t GT2 (mm)
4K*	0.143 ± 0.073	0.198 ± 0.149	0.148 ± 0.134
8K*	0.144 ± 0.088	0.168 ± 0.150	0.136 ± 0.123
12K*	0.149 ± 0.082	0.164 ± 0.137	0.136 ± 0.123
16K*	0.135 ± 0.073	0.164 ± 0.132	0.131 ± 0.121
20K*	0.131 ± 0.062	0.164 ± 0.127	0.124 ± 0.11
Fusion	0.131 ± 0.073	0.163 ± 0.132	0.124 ± 0.11
Calibrated	**0.077 ± 0.049**	**0.109 ± 0.088**	**0.124 ± 0.10**

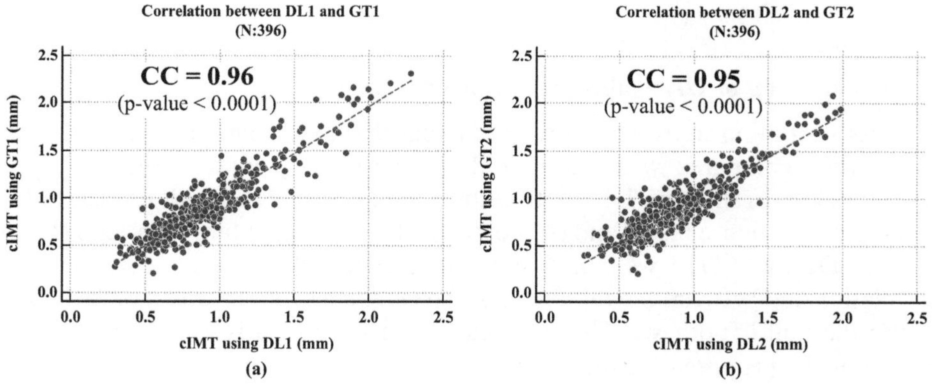

Figure 6.6. Correlation plots of DL-based systems against (a) GT1 and (b) GT2.

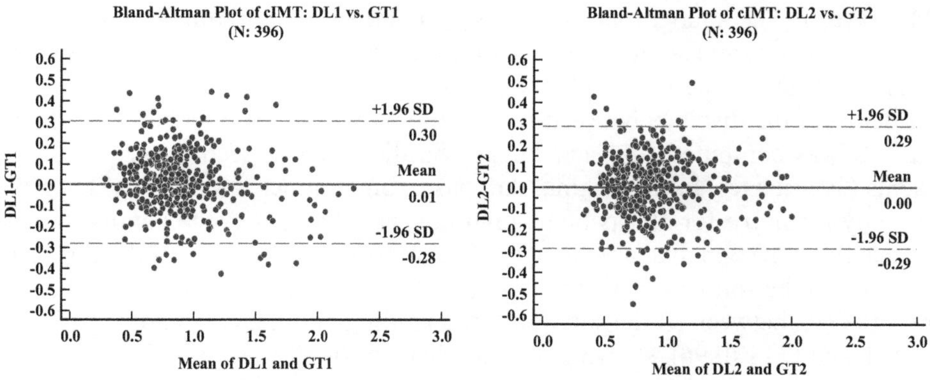

Figure 6.7. Bland–Altman plots of the DL-based system with reference to (a) GT1 and (b) GT2.

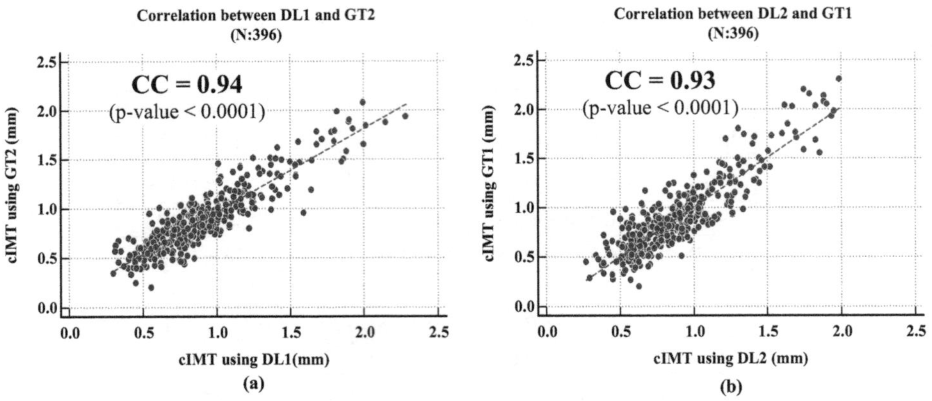

Figure 6.8. Correlation plots of (a) DL1 versus GT2 and (b) DL2 versus GT1.

of the interoperator variability between the two DL systems (DL1 and DL2) and part (b) consisted of interobserver variability between the two GT systems (GT1 and GT2).

6.5.1 Comparison of DL against expert manual tracing

The cross-validation study was performed to check the effectiveness of the DL-based system when compared with other gold standards or ground truths (GTs). The correlation curves showing DL1 with respect to GT2 and DL2 with respect to GT1 are presented in figure 6.8(a) and (b), respectively. The CC values between DL1 and GT2 and DL2 and GT1 were 0.94 and 0.93, respectively, thus showing the strong interrelationship between the DL and GT. The *p*-value for both was < 0.0001, which satisfies the null hypothesis. This also shows the strong statistical significance and stability of the proposed DL-based system.

6.5.2 Comparison of the DL against the sonographer's readings

This study also provided the sonographer's reading of cIMT (symbolized as Sono cIMT (ave.)). The sonographer's reading was taken at one sample point (or one location) or two sample points (two locations) along the CCA. This reading typically consisted of the highest two plaque readings above the baseline but took into consideration the distance between the LI and MA. The mean value from the two locations was computed for each image. As discussed in section 6.2, of the 203 patients (396 images) in the original database, sonographer far wall cIMT readings were only available for 193 patients (346 images). Therefore, the comparison was conducted for the 346 available images. The improvements (in %) in the DL results compared to the sonographer's readings are shown in table 6.3. Row one (R1: CC) shows the CC between (i) the sonographer's reading and the GT reading (0.80) and (ii) DL1 and GT1 (0.96), showing an improvement of 20%. Row 2 (R2: CC) shows the CC between (i) the sonographer's reading and the GT reading (0.83) and (ii) DL2 and GT2 readings (0.95), showing an improvement of 14.5%. The

Table 6.3. Percentage improvement in DL readings compared to the sonographer's readings.

Coefficient of correlation (CC) between three kinds of cIMT (ave.) readings: sonographer (Sono), deep learning (DL1 and DL2 systems) and ground truth (GT1 and GT2)			Percentage Improvement of deep learning (DL) reading over sonographer (Sono) reading
Sono cIMT (ave.) and DL1 cIMT (ave.) against GT1 cIMT (ave.)			
Attribute	Sono vs. GT1	DL1 Vs. GT1	
R1:CC	**0.80**	**0.96**	**20.0%**
Sono cIMT (ave.) and DL2 cIMT (ave.) against GT2 cIMT (ave.)			
	Sono vs. GT2	DL2 vs. GT2	
R2: CC	**0.83**	**0.95**	**14.5%**

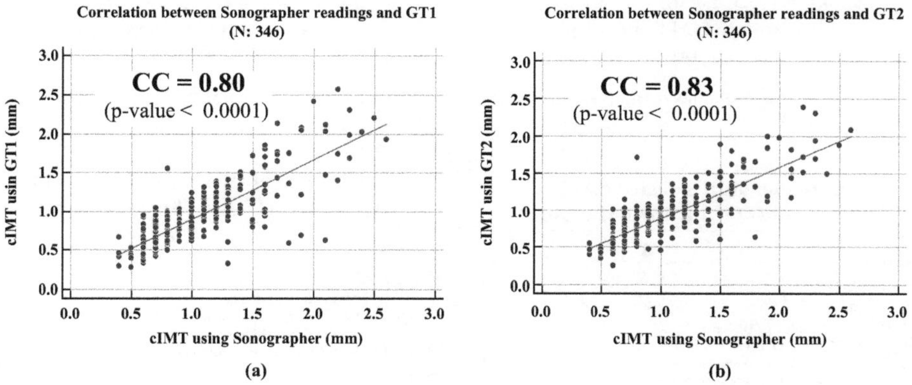

Figure 6.9. Correlation plots of the sonographer's cIMT readings with respect to (a) GT1 and (b) GT2.

correlation plot for the sonographer's cIMT readings with respect to GT1 and GT2 is shown in figure 6.9.

6.5.3 Absolute and signed cIMT error analysis for DL1 and DL2 systems

The cumulative distribution figure plots (CDF) with respect to GT1 and GT2 are shown in figure 6.10. Figure 6.10(a) shows that 90% of patients had an absolute cIMT error < 0.28 mm for GT1. The CDF plot in figure 6.10(b) shows that 90% of patients had an absolute cIMT error < 0.26 mm for GT2. The CDF plots for signed cIMT error are shown in figure 6.11. The CDF plot for signed cIMT error for GT1 indicates that 90% of patients had a signed error > −0.16 mm and 90% had a signed error < 0.18 mm. Similarly for GT2, the signed cIMT error for 90% of patients was > −0.20 mm and for 90% was < 0.19 mm. This further signifies that the DL-based system performs strongly.

6.5.4 DL versus previous methods

The SS method was previously implemented by Suri *et al* [33]. Although the system was clinically stable, it was still compared to the DL-based strategy. A total of 360 attributes were chosen that covered the full spectrum to show the improvement of

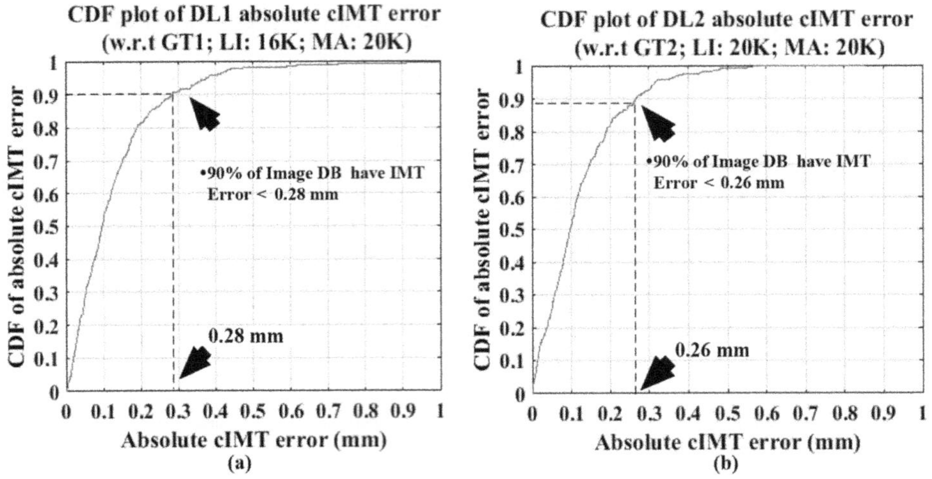

Figure 6.10. Absolute cIMT error for (a) DL1 and (b) DL2.

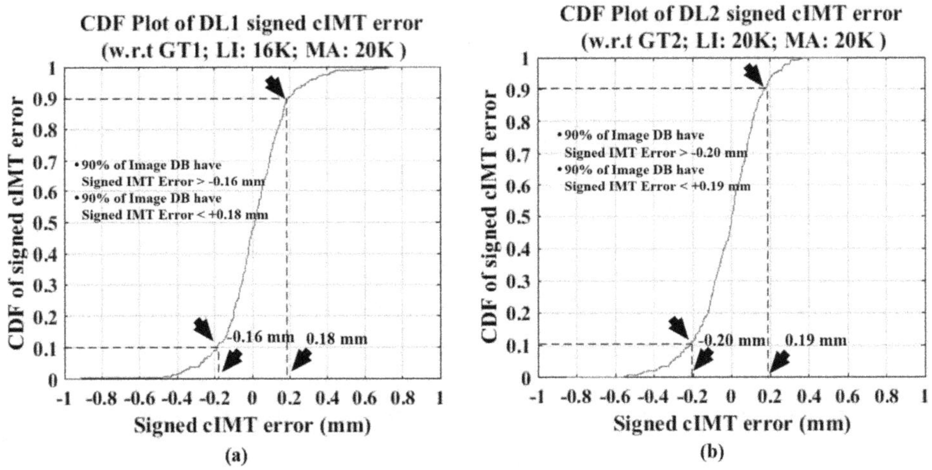

Figure 6.11. Signed cIMT error for (a) DL1 and (b) DL2.

the DL strategy compared to the SS strategy. These are shown in table 6.4 under column one, entitled 'wall characteristics', and seven attributes were considered. The DL method used 396 images, while the SS method used 404 images. For all attributes of the wall characteristics, the percentage improvement is shown in columns C4 and C7 for the DL1 and DL2 systems, respectively. The lumen diameter error improvements for DL1 and DL2 were 33.2% and 39.6%, respectively. The interadventitia diameter error improvements for DL1 and DL2 were 26.7% and 28.7%, respectively. The LI-far error improvements for DL1 and DL2 were 51.9% and 63.3%, respectively. The MA-far error improvements for DL1 and DL2 were 50.9% and 58.1%, respectively. The LI-near error improvements for DL1 and DL2 were 45.5% and 52.4%, respectively. The MA-near error improvements for DL1 and

Table 6.4. Benchmarking of the DL-based system with regards to the SS method.

C0	C1	C2	C3	C4	C5	C6	C7
SN	Wall Characteristics	DL1 w.r.t GT1 (mm)	SS* w.r.t GT1 (mm)	**Improv. (%)**	DL2 w.r.t GT2 (mm)	SS* w.r.t GT2 (mm)	**Improv. (%)**
1	LD error (mm)	0.167 ± 0.181	0.25 ± 0.24	**33.2**	0.163 ± 0.169	0.27 ± 0.25	**39.6**
2	IAD error (mm)	0.176 ± 0.167	0.24 ± 0.24	**26.7**	0.164 ± 0.141	0.23 ± 0.23	**28.7**
3	LI-far error (mm)	0.077 ± 0.057	0.16 ±0.11	**51.9**	0.077 ± 0.049	0.21 ± 0.18	**63.3**
4	MA-far error (mm)	0.113 ± 0.105	0.23 ±0.18	**50.9**	0.109 ± 0.088	0.26 ± 0.15	**58.1**
5	LI-near error (mm)	0.120 ± 0.146	0.22 ± 0.15	**45.5**	0.119 ± 0.179	0.25± 0.18	**52.4**
6	MA-near error(mm)	0.132 ± 0.147	0.23 ± 0.18	**42.6**	0.123 ± 0.137	0.20 ± 0.17	**38.5**
7	JI (lumen region)	0.94 ± 0.03	0.89	**5.6**	0.94 ± 0.03	0.89	**5.6**
8	DS (lumen region)	0.9 ± 0.027	0.94	**3.2**	0.97 ± 0.02	0.94	**3.2**
9	JI (inter-adventitial region)	0.95 ± 0.03	0.91	**4.4**	0.96 ± 0.03	0.91	**5.5**
10	DS (inter-adventitial error)	0.98 ± 0.02	0.95	**3.2**	0.98 ± 0.02	0.95	**3.2**

** computed over 404 images.

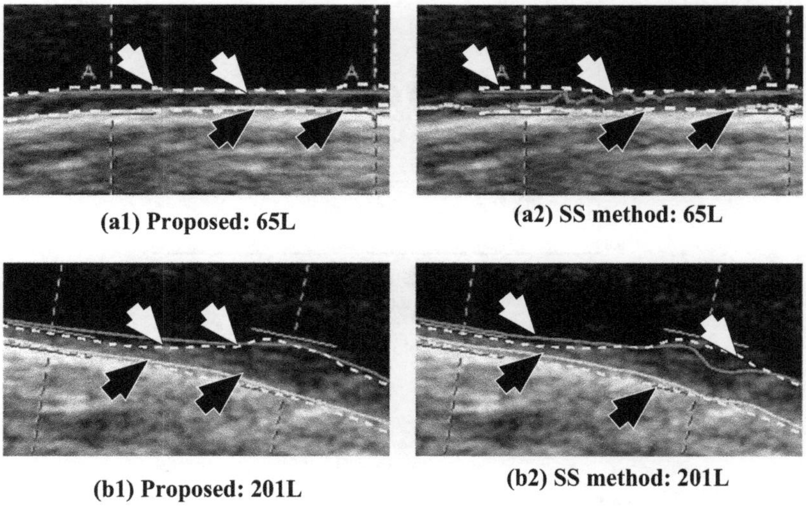

Figure 6.12. Application of the DL-based system and the SS system in patients 65L and 201L. The use of (a1) DL and (a2) SS in patient 65L clearly show that the extracted borders are smoother in the former. For patient 201L, use of the DL-based method (b1) shows better accuracy than the SS system (b2).

DL2 were 42.6% and 38.5%, respectively. The Jaccard index (JI) for the LR improvements for DL1 and DL2 were 5.6% and 5.6%, respectively. The dice similarity (DS) for the LR improvements for DL1 and DL2 were 3.2% and 3.2%, respectively. The JI for the IAR improvements for DL1 and DL2 were 4.4% and 5.5%, respectively. Finally, the DS for the IAR improvements for DL1 and DL2 were 3.2% and 3.2%, respectively. A comparison of two images constructed using both the DL-based system and the SS system is shown in figure 6.12.

6.5.5 Interoperator variability of the DL systems: DL1 and DL2

This study also compared the two DL-based systems to each other to check the reliability of the proposed DL-based system. The correlation between DL1 and DL2 is shown in figure 6.13. The correlation between DL1 and DL2 was 0.95, which indicates a strong interrelationship between DL1 and DL2. The *p*-value was < 0.0001, which further satisfies the null hypothesis and implies that the DL-based system is reliable and stable.

6.5.6 Interobserver variability between the GT systems: GT1 and GT2

The observer readings were also compared to each other to validate that they were compatible. The correlation plot between GT1 and GT2 is shown in figure 6.14. The CC value between GT1 and GT2 was 0.97, which validates that the observer values

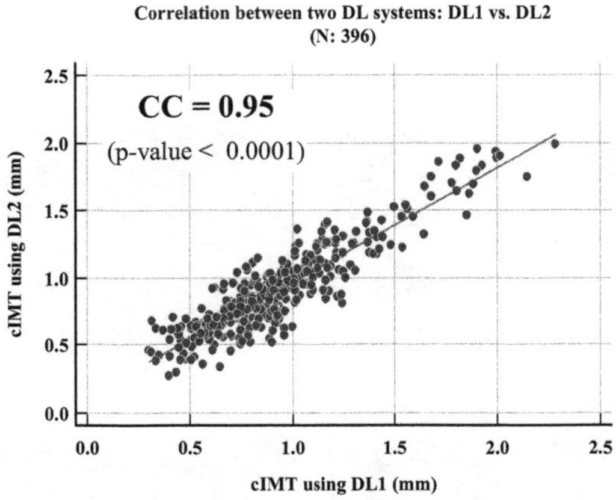

Figure 6.13. Correlation plot between DL1 and DL2.

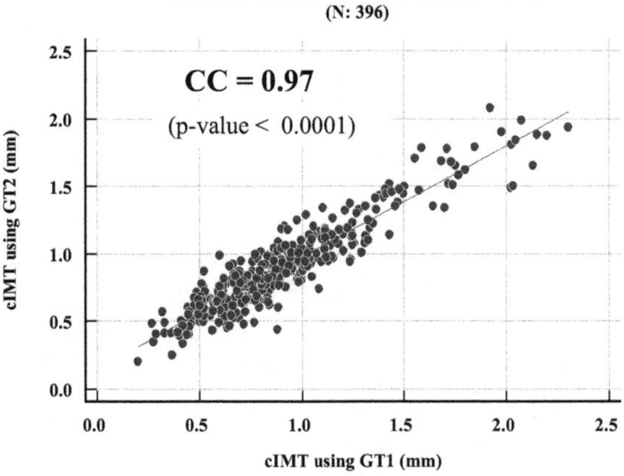

Figure 6.14. Correlation plot between GT1 and GT2.

were compatible. The p-value for the plot was < 0.0001, which further satisfies the null hypothesis and shows that the values were highly correlated.

6.6 Statistical tests and risk analysis

This section presents the four statistical tests that were used to show the significance of the proposed DL system. Risk stratification was also computed using age and risk threshold parameters. This section also presents the ROC curves and area under the curve (AUC) analysis for the DL systems.

6.6.1 Four statistical tests

The outputs of the DL-based system were tested using the paired t-test, Mann–Whitney test and Wilcoxon test, and the corresponding boxplots are shown in figure 6.15. The corresponding p-values for the paired t-tests of DL1 and DL2 with respect to GT1 and GT2 were 0.0105 and 0.0416, respectively. The p-values for the Mann–Whitney tests of DL1 and DL2 with respect to GT1 and GT2 were 0.0320 and 0.0407, respectively. Similarly, the p-values for the Wilcoxon test of DL1 and DL2 with respect to GT1 and GT2 were 0.0488 and 0.0348, respectively. The parameters for the paired t-test, Mann–Whitney test and Wilcoxon test are given in tables 6.5, 6.6 and 6.7, respectively. The p-values from all three tests were statistically significant. The Kruskal–Wallis test was also performed for DL1 and DL2, and the results are given in table 6.8. The p-values with respect to DL1 and DL2 were 0.4905 and 0.4501, respectively. Therefore the null hypothesis that the data were taken from the same distribution was retained for DL1 and DL2.

6.6.2 Risk analysis by age

Several studies showed that cIMT increases with age [18] due to metabolic activity in the arteries [34]. The results obtained in this study were consistent with the previously published literature. cIMT was analyzed against age (years) for the left artery, right artery and the mean of the two carotid arteries. Table 6.9 shows the CC for the left, right and combined cohort using the DL1, DL2, GT1 and GT2 systems. The number of patients in the left, right and combined cohorts was 195, 201 and 203, respectively. Table 6.9 shows the positive correlation between age and cIMT. The right carotid artery showed a higher correlation than the left; however, all patients showed a significant association between age and cIMT ($p < 0.001$).

6.6.3 Risk stratification and ROC curves

This subsection discusses the risk component of the study. Atherosclerosis screening by Bard *et al* [35] suggested that patients with cIMT values > 1.0 mm required more aggressive treatment; however, the population was small (95 patients) and non-diverse. A study of 7983 patients by Bots *et al* [36] suggested that the risk of stroke increased when cIMT values were > 0.9 mm. Other studies also stratified high-risk patients based on cIMT values > 1.0 [37] and 0.80 mm [38]. A study on 100 patients by Saba *et al* [18] recommended a cIMT threshold of 0.9 mm for risk stratification.

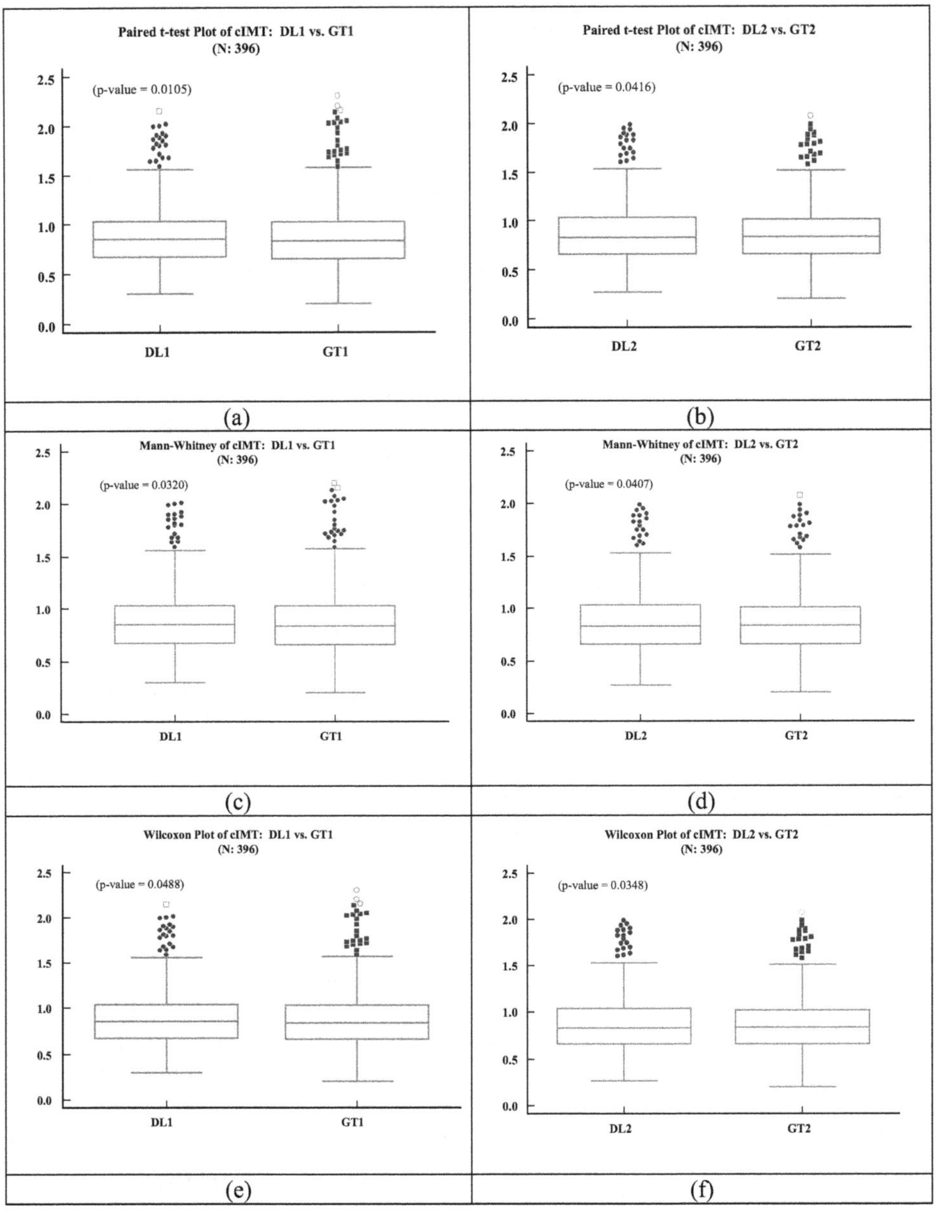

Figure 6.15. Statistical paired *t*-test with respect to (a) GT1 and (b) GT2. Mann–Whitney test with respect to (c) GT1 and (d) GT2. Wilcoxon test with respect to (e) GT1 and (f) GT2.

This dataset contained a diabetic cohort of 201 patients with moderate subclinical atherosclerosis. Although 0.9 mm is recommended as the cutoff for high-risk patients, two sets of cutoffs were actually selected: 0.85 and 0.9 mm. The corresponding ROC curves with respect to these two cutoff values for both DL systems are shown in figure 6.16(a) and (b), respectively. The AUC values for the

Table 6.5. Paired *t*-test.

Parameters	DL1	DL2
Mean difference	−0.012 74	−0.001 553
Standard deviation of differences	0.1490	0.1470
Standard error of mean difference	0.007 489	0.007 385
95% CI	−0.027 47 to 0.001 978	−0.016 07 to 0.012 97
Test statistic t	−1.702	−0.210
Degrees of freedom (DoFs)	395	395
Two-tailed probability	$p = 0.001\ 05\ (< 0.05)$	$p = 0.0416\ (< 0.05)$

Table 6.6. Mann–Whitney test.

Parameters	DL1	DL2
Average rank of first group	404.3687	404.3687
Average rank of second group	388.6313	388.6313
Mann–Whitney U	75 292.00	75 292.00
Large sample test statistic Z	0.968	0.968
Two-tailed probability	$p = 0.032\ 01\ (< 0.05)$	$p = 0.0407\ (< 0.05)$

Table 6.7. Wilcoxon test.

Parameters	DL1	DL2
Number of positive differences	186	188
Number of negative differences	210	208
Large sample test statistic Z	0.733 167	1.852 002
Two-tailed probability	$p = 0.0488\ (< 0.05)$	$p = 0.0348\ (< 0.05)$

Table 6.8. Kruskal–Wallis test.

Parameters	DL1 w.r.t. GT1	DL2 w.r.t. GT2
Test statistic	395.0000	395.0000
Corrected for ties Ht	395.0000	395.0000
Degrees of freedom (DF)	395	395
Significance level	$p = 0.490\ 537\ (> 0.05)$	$p = 0.450\ 0537\ (> 0.05)$

0.85 mm cutoff corresponding to DL1 and DL2 were 0.88 and 0.84. When the cutoff was increased to 0.9 mm, the AUC values for DL1 and DL2 were 0.88 and 0.85, respectively. This shows that 88% of the patients were correctly identified in the low-moderate and high-risk pools.

Table 6.9. Comparative study of age versus cIMT for DL1 and DL2 against GT1 and GT2. The top row shows age versus DL1 and age versus GT1 for the left, right and mean carotid arteries. The bottom row shows age versus DL2 and age versus GT2 for the left, right and mean carotid arteries.

Left cIMT (n = 195)		**Right cIMT (n = 201)**		**Mean of Left and Right cIMT (n = 203)**	
Age Vs DL1 CC (p-value)	Age Vs GT1 CC (p-value)	Age Vs DL1 CC (p-value)	Age Vs GT1 CC (p-value)	Age Vs DL1 CC (p-value)	Age Vs GT1 CC (p-value)
0.20 ($p < 0.001$)	**0.14 ($p < 0.001$)**	**0.19 ($p < 0.001$)**	**0.18 ($p < 0.001$)**	**0.19 ($p < 0.001$)**	**0.14 ($p < 0.001$)**
Age Vs DL2 CC (p-value)	Age Vs GT2 CC (p-value)	Age Vs DL2 CC (p-value)	Age Vs GT2 CC (p-value)	Age Vs DL2 CC (p-value)	Age Vs GT2 CC (p-value)
0.18 ($p < 0.001$)	**0.13 ($p < 0.001$)**	**0.21 ($p < 0.001$)**	**0.16 ($p < 0.001$)**	**0.19 ($p < 0.001$)**	**0.14 ($p < 0.001$)**

*n are the number of patients for left, right and combined carotids.

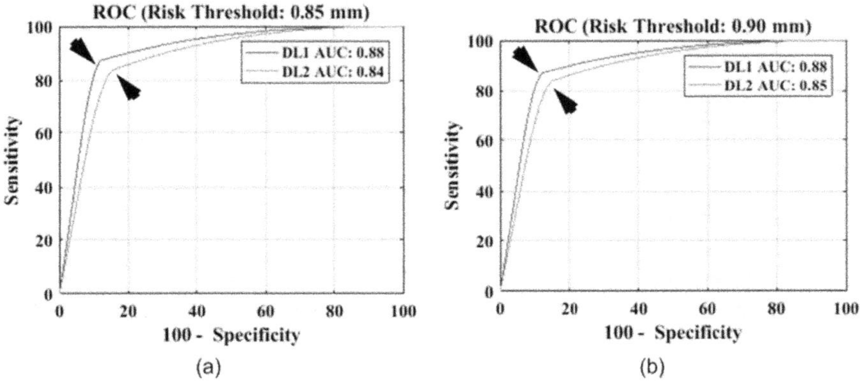

Figure 6.16. ROC curves for two different risk thresholds: (a) 0.85 mm (AUC values of 0.88 and 0.84 corresponding to DL1 and DL2) and (b) 0.90 mm (with AUC values of 0.88 and 0.85).

6.7 Discussion

This paper proposed a two-stage DL-based system implemented serially in four phases to accurately measure LI, MA and cIMT. The DL-based system acquired preprocessed images from the first phase (i.e. multiresolution). The second phase of the entire system was stage I of the DL-based system (the heart of the DL system). The DL-based system was divided into 13 convolution layers (encoder) and three up-sampling layers (decoder). These three up-sampling layers belonged to the FCN. After the images were segmented, they were passed to the third phase (ML-based calibration), which represented the second stage of the DL-based system. In this phase, the LI and MA borders were extracted and calibrated using an ML-based system. The cIMT was computed from the LI and MA borders. Performance analysis was performed in phase IV. The performance results showed that the DL-based system gave better accuracy when compared to contemporary methods and was more robust and efficient. The results for different quartiles are shown in figure 6.17.

6.7.1 Benchmarking table

The algorithms developed for the LI, MA and cIMT measurements are listed in the benchmarking table 6.10. Wendelhag *et al* [39] used dynamic programming for cIMT measurements. Their cIMT error was 0.030 ± 0.032 mm, which was the lowest of all the developed techniques (table 6.10; row #1); however, their dataset was limited to 69 images and the cIMT error varied widely due to different ethnicities, age groups and nationalities. Petroudi *et al* [40] used an active contour model to measure cIMT. The IMT error was 0.080 ± 0.070 mm; however, the dataset was limited to 100 patients (table 6.10; row #2). Suri *et al* [10] developed various

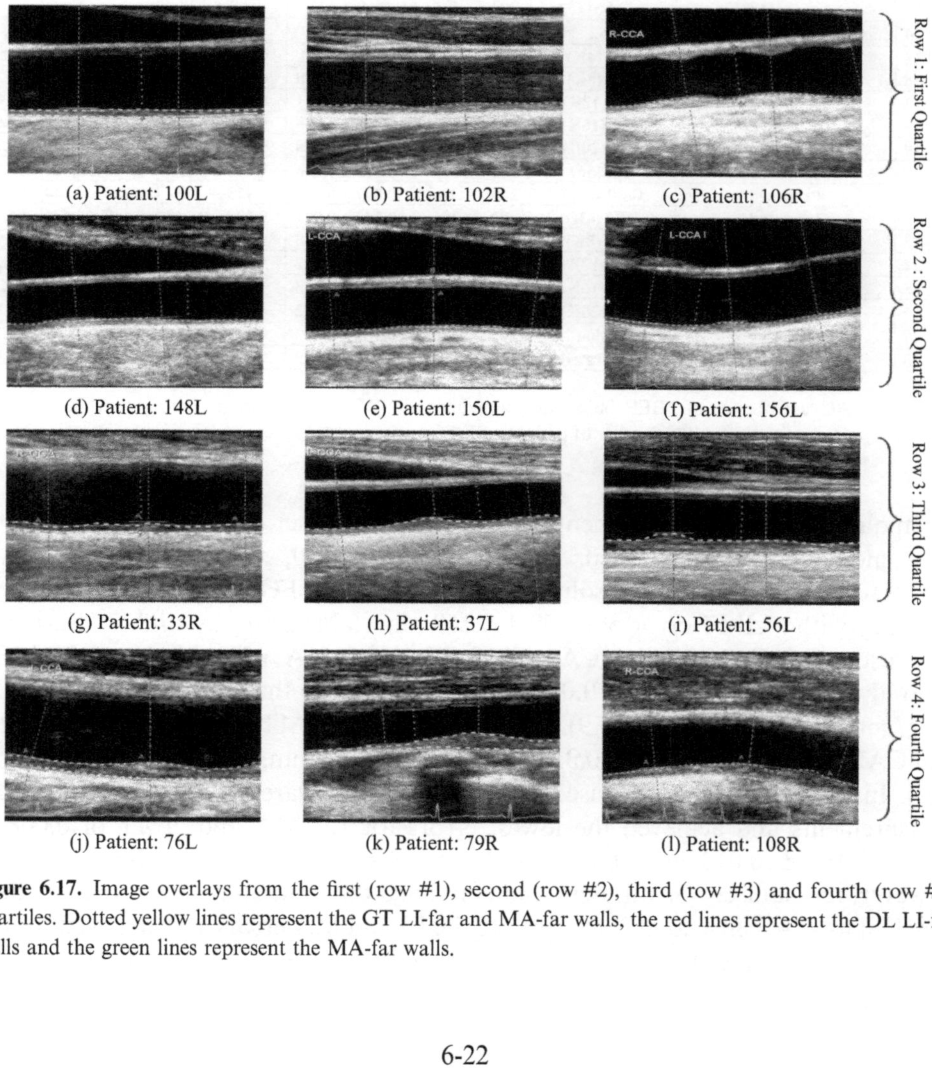

Figure 6.17. Image overlays from the first (row #1), second (row #2), third (row #3) and fourth (row #4) quartiles. Dotted yellow lines represent the GT LI-far and MA-far walls, the red lines represent the DL LI-far walls and the green lines represent the MA-far walls.

Table 6.10. Benchmarking table.

SN	Paper	Method	#P$^+$	Data Size (N)	LI Error (mm)	MA error (mm)	cIMT Error (mm)	PoM
1	Wendelhag *et al.* [39] (1997)	*DP		69	-	-	**0.030 ± 0.032**	-
2	Petroudi *et al.* [40] (2012)	*AC	-	100	-	-	0.080 ± 0.070	-
3	Molinari *et al.* [10] (2012a)	CALEX 1.0	344	665	-	-	0.191 ± 0.217	-
4	Molinari *et al.* [10] (2012a)	CARES	344	647	-	-	0.172 ± 0.222	-
5	Molinari *et al.* [10] (2012a)	CAMES 1.0	344	657	-	-	0.154 ± 0.227	-
6	Molinari *et al.* [10] (2012a)	CAUDLES	344	630	-	-	0.224 ± 0.252	-
7	Molinari *et al.* [10] (2012a)	FOAM	344	665	-	-	0.150 ± 0.169	-
8	Molinari *et al.* [41] (2012b)	CALEX 2.0	365	365	0.088 ± 0.132	0.141 ± 0.201	0.121± 0.334	-
9	Molinari *et al.* [41] (2012b)	CAMES 3.0	365	365	0.081 ± 0.099	0.082 ± 0.197	0.078 ± 0.112	-
10	Ikeda *et al.* [17] (2015)	AtheroEdge™	341	341	**0.008 ± 0.099**	**0.018 ± 0.013**	**0.01 ± 0.01**	-
11	Saba *et al.* [18] (2016)	AtheroCloud™	100	200	0.065 ± 0.037	0.067 ± 0.036	-	-
12	Ikeda *et al.* [19] (2017)	*BEP, SIMT	657	657	0.012 ± 0.012	0.021 ± 0.015	0.165 ± 0.171	-
13	Kumar *et al.* [33] (2017a)	*SS1	202	404	0.16 ± 0.11	0.23 ± 0.18	-	-
14	Kumar *et al.* [33] (2017a)	SS2	202	404	0.21 ± 0.18	0.26 ± 0.15	-	-
15	**Proposed**	**DL1 (4K)**	**203**	**396**	**0.161 ± 0.090**	**0.230 ± 0.197**	**0.177 ± 0.179**	**94.3**
16	**Proposed**	**DL1 (8K)**	**203**	**396**	**0.138 ± 0.078**	**0.187 ± 0.149**	**0.146 ± 0.13**	**94.3**
17	**Proposed**	**DL1 (12K)**	**203**	**396**	**0.135 ± 0.061**	**0.177 ± 0.122**	**0.142 ± 0.124**	**92.0**
18	**Proposed**	**DL1 (16K)**	**203**	**396**	**0.135 ± 0.076**	**0.178 ± 0.153**	**0.142 ± 0.132**	**99.0**
19	**Proposed**	**DL1 (20K)**	**203**	**396**	**0.135 ± 0.078**	**0.171 ± 0.153**	**0.140 ± 0.149**	**98.7**
20	**Proposed**	**Fusion**	**203**	**396**	**0.135 ± 0.076**	**0.171 ± 0.153**	**0.128 ± 0.124**	**97.7**
21	**Proposed**	**Calibrated**	**203**	**396**	**0.077 ± 0.057**	**0.113±0.105**	**0.126 ± 0.134**	**99.9**
22	**Proposed**	**DL2 (4K)**	**203**	**396**	**0.143 ± 0.073**	**0.198 ± 0.149**	**0.148 ± 0.134**	**99.4**
23	**Proposed**	**DL2 (8K)**	**203**	**396**	**0.144 ± 0.088**	**0.168 ± 0.150**	**0.136 ± 0.123**	**99.6**
24	**Proposed**	**DL2 (12K)**	**203**	**396**	**0.149 ± 0.082**	**0.164 ± 0.137**	**0.136 ± 0.123**	**97.2**
25	**Proposed**	**DL2 (16K)**	**203**	**396**	**0.135 ± 0.073**	**0.164 ± 0.132**	**0.131 ± 0.121**	**96.3**
26	**Proposed**	**DL2 (20K)**	**203**	**396**	**0.131 ± 0.062**	**0.164 ± 0.127**	**0.124 ± 0.11**	**99.8**
27	**Proposed**	**Fusion**	**203**	**396**	**0.131 ± 0.073**	**0.163 ± 0.132**	**0.124 ± 0.11**	**98.7**
28	**Proposed**	**Calibrated**	**203**	**396**	**0.077 ± 0.049**	**0.109 ± 0.088**	**0.124 ± 0.10**	**99.9**

*AC, active contours; BEP, bulb-edge point detection; DP, dynamic programming; K, 1000 iterations; P$^+$, number of patients; SIMT, segmental IMT; SS, scale–space.

techniques for IMT measurements using a larger dataset containing 344 patients. Five methods were presented, namely, CALEX 1.0, CARES, CAMES 1.0, CAUDLES and first-order absolute moment (FOAM). FOAM showed the highest accuracy with a cIMT error of 0.150 ± 0.169 mm (table 6.10; rows #3–7). Suri *et al* [41] also used CALEX and CAMES for LI and MA measurements. CAMES showed the lowest LI error at 0.081 ± 0.099 mm, while the MA error was 0.082 ± 0.197 mm (table 6.10; rows #8–9). The corresponding cIMT errors with CALEX 2.0 and CAMES 3.0 were 0.121 ± 0.334 and 0.078 ± 0.112 mm, respectively (table III in [41]). In 2015, Suri *et al* [17] used AtheroEdge™ software for LI, MA and cIMT measurements and achieved the lowest errors for LI, MA and cIMT of 0.008 ± 0.099, 0.018 ± 0.013 and 0.01 ± 0.01 mm, respectively (table 6.10; row #10); however, the dataset was different and contained different ethnicities. In 2016, Suri *et al* [18] used AtheroCloud™ to measure the LI and MA errors and achieved results of 0.065 ± 0.037 and 0.067 ± 0.036 mm, respectively (table 6.10; row #11). In 2017,

Suri *et al* [19] used bulb-edge point detection and segmental cIMT for LI, MA and cIMT error detection and obtained results of 0.012 ± 0.012, 0.021 ± 0.015 and 0.165 ± 0.171 mm, respectively (table 6.10; row #12). This dataset also contained different ethnicities. As discussed in subsection 6.5.4, Kumar *et al* [33] used a diabetic cohort and achieved LI and MA errors of 0.160 ± 0.110 and 0.230 ± 0.180 mm, respectively, for GT1, and 0.210 ± 0.180 and 0.260 ± 0.150 mm, respectively, for GT2 (table 6.10; rows #13–14). The same diabetic cohort was used to assess the novel DL-based system in this study, and the results showed LI and MA errors of 0.077 ± 0.057 and 0.113 ± 0.105 mm, respectively, for GT1, and 0.077 ± 0.049 and 0.109 ± 0.088 mm, respectively, for GT2. This study also reported cIMT errors of 0.126 ± 0.134 and 0.124 ± 0.10 mm for GT1 and GT2, respectively (table 6.10; rows #15–28). The PoM was also computed (described in appendix C) for all experiments (table 6.10; column #9, row #15–28).

6.7.2 A short note on calibration

The ML-based calibration strategy is a regression-based method that was used to fine-tune the raw DL borders to ensure smoothness. It is basically an ML-based cross-validation deformable model to regress DL-based borders from stage I closer to the actual GT borders. An independent coefficient matrix was developed from the training and GT dataset, as shown in equation (6.4). A large number of patients helped to create a more generalized coefficient matrix. The predicted dataset was the product of this training-based coefficient matrix and the online test DL-based matrix. The results showed that LI, MA and cIMT errors were reduced after the use of the ML-based calibration. The best results were obtained when this DL-based pilot study used a jack-knifing strategy for the ML-based paradigm, where all but one instance was used for training and the remaining one was used for testing. Use of the jack-knifing strategy resulted in better accuracy for both stenotic and nonstenotic cases. Thus a strategy where ML-based calibration is cascaded with the core DL-based paradigm is stable, robust and clinically accurate in comparison to previous methods.

6.7.3 A special note on DL optimization

This is the first study to employ a DL strategy for cIMT measurements. Another novelty is the use of both convolution neural network (CNN) and FCN as a combination of LI and MA segmentation. This is also the first time that an ML-based system was introduced to fine-tune the raw DL-based LI and MA borders. The 13 layers of CNN extract high-level features from the CCA US images. These features were up-sampled using up-sampling layers of FCN, and the skipping operation was performed to obtain sharp and crisp segmented images. After extracting the LI and MA borders from these images, ML-based calibration was adapted to smooth any minor glitches in the borders. Finally, the PDM method was adapted to obtain the shortest bidirectional distance.

6.7.4 A special note on skip operation

There are two approaches in FCN: contraction and expansion. In the contraction approach, the features are down-sampled at intermediate layers using convolution and pooling operations. In the expansion approach, the inverse convolution is applied to up-sample the features. Skip operations were applied to extract features (skipping features) from the contracting layers to the intermediate layers to recover spatial information lost during the down-sampling in the contraction path. This was done by merging skipping features from various resolution layers in the contracting path with input features in the expansion path. In this way, a highly accurate segmentation output was obtained from the FCN. Two skipping operations were applied in the model reported here.

6.7.5 Strengths, weaknesses and extensions

The major strength of this DL-based system is its full automation. The accuracy of the system was comparatively higher than contemporary methods and therefore it was clinically stronger. DL is an intelligence-based system that is adapted from neural connections in the brain. This is the first time that a DL-based system was used for cIMT measurements when cascaded with an ML-based calibration, and such a cascade is truly novel. Moreover, once trained, the output from the DL-based system is produced in real time and takes a few milliseconds. However, the dataset used was limited to a Japanese diabetic cohort, and the system has not been tested on a wide variety of datasets. Therefore, the system requires further analysis in a multi-ethnic patient population who have subclinical atherosclerosis with low, moderate and high-risk scenarios. Further analysis also needs to be performed using a different set of original equipment manufacturer (OEM) machines as attempted by Suri *et al* [10]. Finally, this DL desktop version should be extended to a web-based version (previously developed by Suri *et al* [18, 42]) and undergo a reproducibility analysis, which was recently attempted by the same team [43, 44].

6.7.6 Hardware configuration

The system was implemented on central processing unit (CPU)-based hardware (i.e. Intel icore3 2.9 GHz, 8 GB RAM); however, the results were replicated on graphics processing unit (GPU)-based settings (i.e. NVIDIA GeForce GTX with 1280 cores and 5 GB memory).

6.8 Conclusion

This study presents a novel, robust and clinically viable solution to cIMT measurement using an AtheroEdge™ system from AtheroPoint™. The system uses an intelligence-based paradigm for cIMT measurement by employing the DL strategy for the segmentation of the LR and IAR. To fine-tune this, the system adopts an ML-based joint coefficient method for final border extraction for the far wall of the carotid artery. Data are prepared in a multiresolution paradigm which reduces the computational burden. The polyline distance method, which is a standard used in the industry, is adapted for all measurements. The system performs better than previous studies. For example, the LI position error improved by 52% and 63%, and the MA position error improved by 51% and 58%. The cIMT error for DL1 and DL2 was 0.126 ± 0.134 and 0.124 ± 0.10 mm, respectively. The CC between age and cIMT was 0.20, and the AUC had an upper bound close to 90%. The DL-based system can be adapted for clinical settings or multicenter pharmaceutical trial modes, just like AtheroEdge™ or AtheroCloud™.

Acknowledgment

The authors at the National Institute of Technology, Goa, India, would like to acknowledge MediaLab Asia, the Ministry of Electronics and Information Technology and the Government of India for their kind support.

Appendix A Polyline distance method

Polyline distance metric

The polyline distance metric (PDM) [32] is used to measure cIMT between LI and MA interfaces, LI error between deep-learning LI-far and ground truth LI-far interfaces, and MA error between deep-learning MA-far and ground truth MA-far interfaces. The PDM computation is given as follows. Let the first and second interfaces be denoted as C_1 and C_2. Let the reference point on C_1 be vertex P_1 and the segment in C_2 be defined by vertices P_2 and P_3. Let the distance between P_1 and P_2 be d_1 and the distance between P_1 and P_3 be denoted as d_2. Let $D(P_1, L)$ be the polyline distance between vertex P_1: (x_1, y_1) on C_1 and line segment L formed by two points P_2: (x_2, y_2) and P_3: (x_3, y_3). Let delta (δ) be the distance of the reference point, P_1 towards the line segment L. The perpendicular distance between the line segment L and the reference point, P_1, is given by d_p. Then, the polyline distance $D(P_1, L)$ can be defined as

$$D(P_1, L) = \begin{cases} |d_p| \, 0 < \delta < 1 \\ \min(d_1, d_2) \; \delta < 0, \delta > 1 \end{cases}, \tag{A.1}$$

where

$$d_1 = \sqrt{(x_1 - x_2)^2 + (y_1 - y_2)^2} \tag{A.2}$$

$$d_2 = \sqrt{(x_1 - x_3)^2 + (y_1 - y_3)^2} \tag{A.3}$$

$$\delta = \frac{(y_3 - y_2)(y_1 - y_2) + (x_3 - x_2)(x_1 - x_2)}{(x_3 - x_2)^2 + (y_3 - y_2)^2} \tag{A.4}$$

and

$$d_p = \frac{(y_3 - y_2)(x_2 - x_1) + (x_3 - x_2)(y_1 - y_2)}{\sqrt{(x_3 - x_2)^2 + (y_3 - y_2)^2}}. \tag{A.5}$$

The process to obtain $D(P_1, L)$ is repeated for the rest of the points of the contour C_j and is given by

$$D(C_1, C_2) = \sum_{i=1}^{N} D(P_i, S_{C_2}), \tag{A.6}$$

where N is the total number of points on C_1 and S_{C_2} is the segment on contour C_2. This algorithm is repeated in reverse, where C_2 becomes the reference contour and C_1 becomes the segment contour. The reverse is represented as $D(C_2, C_1)$. Finally, by combining both $D(C_1, C_2)$ and $D(C_2, C_1)$, we obtain the PDM which is given by

$$D_{\text{PDM}}(C_1 : C_2) = \frac{D(C_1, C_2) + D(C_2, C_1)}{(\#\text{points} \in C_1 + \#\text{points} \in C_2)}. \tag{A.7}$$

Appendix B Encoder and decoder network

Encoder and decoder

Convolution neural networks have the ability to decompose images into feature maps, generating something like a deck of cards representing the feature maps, which can then be fed into limited layered neural networks for training. Mathematically, a basic convolution can be represented as

$$d(x, y) = I(x, y) \otimes w(x, y) = \sum_{s=-\frac{m}{2}}^{\frac{m}{2}} \sum_{t=-\frac{m}{2}}^{\frac{m}{2}} I(x + s, y + t) \times w(x, y), \tag{B.1}$$

where the image I is convolved with kernel w, yielding an output d, and $\otimes$ represents the convolution operation. The convolution is basically a sum of all products between image I and kernel w, represented by equation (B.1), where the kernel is represented as a vector of size $m \times m$ and is shown for the point locations (x, y), while s and t are the dummy variables. The pooling reduces the dimensionality of each feature map but retaining the most important information, i.e. max-pooling and average pooling. Pooling is done to simplify the output from CNN.

In the architecture given in figure 6.3, for the encoder, we have used 13 convolution layers. Each convolution layer has M (= 64, 128, 256, 512) kernels where each kernel is represented as a vector of size 3 × 3. Small kernels allow large depth without increasing memory requirements. There are intermediate five max-pool layers to down-sample the feature maps which are later concatenated and fed into the next stage. In the decoder, the reverse happens. The input deck is up-sampled to the original size using up-sample layers with the help of skip operations to obtain the segmentation output.

Appendix C LI/MA position errors, cIMT errors and precision-of-merit

LI error

The LI error ($\epsilon_{\mathrm{LI}}(i)$) for patient i is computed as the PDM between the GT LI-far wall ($\mathrm{LI}_{\mathrm{far}}^{\mathrm{gt}}(i)$) and DL LI-far ($\mathrm{LI}_{\mathrm{far}}^{\mathrm{dl}}(i)$) wall for the patient, which is given by

$$\epsilon_{\mathrm{LI}}(i) = D_{\mathrm{PDM}}\left(\mathrm{LI}_{\mathrm{far}}^{\mathrm{gt}}(i):\ \mathrm{LI}_{\mathrm{far}}^{\mathrm{dl}}(i)\right). \tag{C.1}$$

If $\epsilon_{\mathrm{LI}}(i)$ represents the LI error for the patient i, then the mean LI error ($\overline{\epsilon}_{LI}$) for all N patients is given by

$$\overline{\epsilon}_{\mathrm{LI}} = \frac{\sum_{i=1}^{N} \epsilon_{\mathrm{LI}}(i)}{N}. \tag{C.2}$$

MA error

Similarly, the MA error ($\epsilon_{\mathrm{MA}}(i)$) is computed as the PDM between the GT MA-far wall ($\mathrm{MA}_{\mathrm{far}}^{\mathrm{gt}}(i)$) and DL MA-far wall ($\mathrm{MA}_{\mathrm{far}}^{\mathrm{dl}}(i)$) for patient i, given by

$$\epsilon_{\mathrm{MA}}(i) = D_{\mathrm{PDM}}\left(\mathrm{MA}_{\mathrm{far}}^{\mathrm{gt}}(i):\ \mathrm{MA}_{\mathrm{far}}^{\mathrm{dl}}(i)\right). \tag{C.3}$$

The mean MA error ($\overline{\epsilon}_{\mathrm{MA}}$) for all N patients is given by

$$\overline{\epsilon}_{\mathrm{MA}} = \frac{\sum_{i=1}^{N} \epsilon_{\mathrm{MA}}(i)}{N}. \tag{C.4}$$

cIMT error

The cIMT error ($\epsilon_{\mathrm{cIMT}}(i)$) for patient i is computed as the PDM between the ground truth cIMT ($\mathrm{cIMT}_{\mathrm{gt}}(i)$) and deep-learning cIMT ($\mathrm{cIMT}_{\mathrm{dl}}(i)$) for the patient. The $\mathrm{cIMT}_{\mathrm{gt}}(i)$ for patient i is computed as the PDM between the GT LI-far wall ($\mathrm{LI}_{\mathrm{far}}^{\mathrm{gt}}(i)$) and GT MA-far wall ($\mathrm{MA}_{\mathrm{far}}^{\mathrm{gt}}(i)$), which is given as

$$\mathrm{cIMT}_{\mathrm{gt}}(i) = D_{\mathrm{PDM}}\left(\mathrm{LI}_{\mathrm{far}}^{\mathrm{gt}}(i):\ \mathrm{MA}_{\mathrm{far}}^{\mathrm{gt}}(i)\right). \tag{C.5}$$

Similarly, the $\mathrm{cIMT}_{\mathrm{dl}}(i)$ is computed as the PDM between the DL LI-far wall ($\mathrm{LI}_{\mathrm{far}}^{\mathrm{dl}}(i)$) and DL MA-far wall ($\mathrm{MA}_{\mathrm{far}}^{\mathrm{dl}}(i)$), which is given as

$$\mathrm{cIMT}_{\mathrm{dl}}(i) = D_{\mathrm{PDM}}\left(\mathrm{LI}_{\mathrm{far}}^{\mathrm{dl}}(i):\ \mathrm{MA}_{\mathrm{far}}^{\mathrm{dl}}(i)\right). \tag{C.6}$$

Therefore, the cIMT error ($\epsilon_{\mathrm{cIMT}}(i)$) for patient i is computed as the absolute difference between $\mathrm{cIMT}_{\mathrm{gt}}(i)$ and $\mathrm{cIMT}_{\mathrm{dl}}(i)$,

$$\epsilon_{\mathrm{cIMT}}(i) = \left|\mathrm{cIMT}_{\mathrm{gt}}(i) - \mathrm{cIMT}_{\mathrm{dl}}(i)\right|. \tag{C.7}$$

If $\epsilon_{\mathrm{cIMT}}(i)$ signifies the cIMT error for the patient i, then the mean cIMT error ($\overline{\epsilon}_{\mathrm{cIMT}}$) for all N patients is given by

$$\bar{\epsilon}_{\text{cIMT}} = \frac{\sum_{i=1}^{N} \epsilon_{\text{cIMT}}(i)}{N}. \tag{C.8}$$

Precision-of-merit (PoM)

Using equations (C.5) and (C.6), one can, therefore, define the PoM mathematically as

$$\text{PoM}_{\text{cIMT}}(\%) = 100 - \left(\frac{\sum_{i=1}^{N} \frac{\left|\text{cIMT}_{\text{dl}(i)} - \text{cIMT}_{\text{gt}(i)}\right|}{\text{cIMT}_{\text{gt}(i)}}}{N} \right) \times 100. \tag{C.9}$$

All the symbols are discussed in appendix D, table D1.

Appendix D

Table D1. Symbol table.

SN.	Symbol	Abbreviation
1	β_1	Predicted output
2	β_2	Ground truth
3	L	Total number of classes
4	N	Total number of images
5	θ	Loss function
6	$\boldsymbol{I}$	Ground truth boundaries
7	$\boldsymbol{D}$	Predicted DL boundaries
8	m	Total number of boundary points
9	tr	Training symbol
10	te	Testing symbol
11	$\hat{\boldsymbol{\varphi}}_{\text{tr}}$	Estimated coefficient matrix using training data
12	C_1	First interface
13	C_2	Second interface
14	P_1	Reference point on C_1
15	P_2	Reference point on C_2
16	P_3	Reference point on C_2
17	L	Line segment formed by vertex P_1 and vertex P_2 on C_2
18	d_1	Euclidean distance between vertex P_1 and vertex P_2
19	d_2	Euclidean distance between vertex P_1 and vertex P_3
20	δ	Distance of the reference point P_1 and the line segment, L
21	d_p	Perpendicular distance between L and the reference point P_1
22	$D(P_1,L)$	Polyline distance between reference point P_1 and the line segment L
23	$D(C_1, C_2)$	Mean polyline distance between all points on contour C_1 with respect to contour C_2

(*Continued*)

24	$D(C_2, C_1)$	Mean polyline distance between all points on contour C_2 with respect to contour C_1
25	D_{PDM}	Bidirectional polyline distance metric by combining $D(C_1, C_2)$ and $D(C_2, C_1)$
26	$LI^{gt}_{far}(i)$	LI-far interface or contour taken from ground truth for patient i
27	$LI^{dl}_{far}(i)$	LI-near interface or contour taken from deep learning for patient i
28	$MA^{gt}_{far}(i)$	MA-far interface or contour taken from ground truth for patient i
29	$MA^{dl}_{far}(i)$	MA-far interface or contour taken from deep learning for patient i
30	$\epsilon_{LI}(i)$	Absolute LI error for patient i
31	$\bar{\epsilon}_{LI}$	Mean LI error for N patients
32	$\epsilon_{MA}(i)$	Absolute MA error for patient i
33	$\bar{\epsilon}_{MA}$	Mean MA error for N patients
34	$cIMT_{gt}(i)$	PDM between GT LI-far wall and GT MA-far interfaces for patient i
35	$cIMT_{dl}(i)$	PDM between DL LI-far wall and DL MA-far interfaces for patient i
36	$\epsilon_{cIMT}(i)$	Absolute cIMT error for patient i
37	$\bar{\epsilon}_{cIMT}$	Mean absolute cIMT error for N patients
38	PoM_{cIMT}	Precision-of-merit for cIMT

References

[1] Lloyd-Jones D *et al* 2010 Heart disease and stroke statistics—2010 update: a report from the American Heart Association *Circulation* **121** e46

[2] WHO http://who.int/cardiovascular_diseases

[3] Libby P, Geng Y-J, Sukhova G K, Simon D I and Lee R T 1997 Molecular determinants of atherosclerotic plaque vulnerability *Ann. N. Y. Acad. Sci.* **811** 134–45

[4] O'Leary D H, Polak J F, Kronmal R A, Manolio T A, Burke G L and Wolfson S K Jr 1999 Carotid-artery intima and media thickness as a risk factor for myocardial infarction and stroke in older adults *New Engl. J. Med.* **340** 14–22

[5] Bots M L 2006 Carotid intima–media thickness as a surrogate marker for cardiovascular disease in intervention studies *Curr. Med. Res. Opin.* **22** 2181–90

[6] Nambi V, Chambless L, Folsom A R, He M, Hu Y, Mosley T, Volcik K, Boerwinkle E and Ballantyne C M 2010 Carotid intima–media thickness and presence or absence of plaque improves prediction of coronary heart disease risk: the ARIC (Atherosclerosis Risk in Communities) study *J. Am. Coll. Cardiol.* **55** 1600–7

[7] Meuwese M C, de Groot E, Duivenvoorden R, Trip M D, Ose L, Maritz F J and Basart D C G 2009 ACAT inhibition and progression of carotid atherosclerosis in patients with familial hypercholesterolemia: the CAPTIVATE randomized trial *J. Am. Med. Assoc.* **301** 1131–9

[8] Ikeda N *et al* 2013 Automated carotid intima–media thickness and its link for prediction of SYNTAX score in Japanese coronary artery disease patients *Int. Angiol.: J. Int. Union Angiol.* **32** 339–48

[9] Molinari F, Zeng G and Suri J S 2010 Intima–media thickness: setting a standard for a completely automated method of ultrasound measurement *IEEE Trans. Ultrason. Ferroelectr. Freq. Control* **57** 1112–24

[10] Molinari F *et al* 2012 Ultrasound IMT measurement on a multi-ethnic and multi-institutional database: our review and experience using four fully automated and one semi-automated methods *Comput. Methods Programs Biomed.* **108** 946–60
[11] Molinari F, Zeng G and Suri J S 2010 A state of the art review on intima–media thickness (IMT) measurement and wall segmentation techniques for carotid ultrasound *Comput. Methods Programs Biomed.* **100** 201–21
[12] Molinari F, Acharya U R, Zeng G, Meiburger K M and Suri J S 2011 Completely automated robust edge snapper for carotid ultrasound IMT measurement on a multi-institutional database of 300 images *Med. Biol. Eng. Comput.* **49** 935–45
[13] Molinari F, Pattichis C S, Zeng G, Saba L, Acharya U R, Sanfilippo R, Nicolaides A and Suri J S 2012 Completely automated multiresolution edge snapper—a new technique for an accurate carotid ultrasound IMT measurement: clinical validation and benchmarking on a multi-institutional database *IEEE Trans. Image Process.* **21** 1211–22
[14] Molinari F, Meiburger K M, Zeng G, Nicolaides A and Suri J S 2011 CAUDLES-EF: carotid automated ultrasound double line extraction system using edge flow *J. Digit. Imaging* **24** 1059–77
[15] Londhe N D and Suri J S 2016 Superharmonic imaging for medical ultrasound: a review *J. Med. Syst.* **40** 279
[16] Saba L, Molinari F, Meiburger K M, Acharya U R, Nicolaides A and Suri J S 2013 Inter- and intra-observer variability analysis of completely automated cIMT measurement software (AtheroEdge™) and its benchmarking against commercial ultrasound scanner and expert Readers *Comput. Biol. Med.* **43** 1261–72
[17] Ikeda N *et al* 2015 Improved correlation between carotid and coronary atherosclerosis SYNTAX score using automated ultrasound carotid bulb plaque IMT measurement *Ultrasound Med. Biol.* **41** 1247–62
[18] Saba L *et al* 2016 Accurate cloud-based smart IMT measurement, its validation and stroke risk stratification in carotid ultrasound: a web-based point-of-care tool for multicenter clinical trial *Comput. Biol. Med.* **75** 217–34
[19] Ikeda N *et al* 2017 Automated segmental-IMT measurement in thin/thick plaque with bulb presence in carotid ultrasound from multiple scanners: stroke risk assessment *Comput. Methods Programs Biomed.* **141** 73–81
[20] Acharya U R, Mookiah M R K, Sree S V, Yanti R, Martis R J, Saba L, Molinari F, Guerriero S and Suri J S 2014 Evolutionary algorithm-based classifier parameter tuning for automatic ovarian cancer tissue characterization and classification *Ultraschall Med.—Eur. J. Ultrasound* **35** 237–45
[21] Acharya U R, Sree S V, Kulshreshtha S, Molinari F, Koh J E W, Saba L and Suri J S 2014 GyneScan: an improved online paradigm for screening of ovarian cancer via tissue characterization *Technol. Cancer Res. Treat.* **13** 529–39
[22] Pareek G *et al* 2013 Prostate tissue characterization/classification in 144 patient population using wavelet and higher order spectra features from transrectal ultrasound images *Technol. Cancer Res. Treat.* **12** 545–57
[23] Shrivastava V K, Londhe N D, Sonawane R S and Suri J S 2015 Reliable and accurate psoriasis disease classification in dermatology images using comprehensive feature space in machine learning paradigm *Expert Syst. Appl.* **42** 6184–95
[24] Maniruzzaman M, Rahman M J, Al-MehediHasan M, Suri H S, Abedin M M, El-Baz A and Suri J S 2018 Accurate diabetes risk stratification using machine learning: role of missing value and outliers *J. Med. Syst.* **42** 92

[25] LeCun Y, Bengio Y and Hinton G 2015 Deep learning *Nature* **521** 436–44

[26] Teichmann M, Weber M, Zoellner M, Cipolla R and Urtasun R 2016 *MultiNet: real-time joint semantic reasoning for autonomous driving* (arXiv:1612.07695)

[27] Molinari F, Liboni W, Giustetto P, Badalamenti S and Suri J S 2009 Automatic computer-based tracings (ACT) in longitudinal 2-D ultrasound images using different scanners *J. Mech. Med. Biol.* **9** 481–505

[28] Krizhevsky A, Sutskever I and Hinton G E 2012 Imagenet classification with deep convolutional neural networks *Advances in Neural Information Processing Systems* (Cambridge, MA: MIT Press)

[29] Long J, Shelhamer E and Darrell T 2015 Fully convolutional networks for semantic segmentation *Proc. of the IEEE Conf. on Computer Vision and Pattern Recognition* (Piscataway, NJ: IEEE) pp 3431–40

[30] Molinari F, Meiburger K M, Saba L, Zeng G, Acharya U R, Ledda M, Nicolaides A and Suri J S 2012 Fully automated dual-snake formulation for carotid intima–media thickness measurement *J. Ultrasound Med.* **31** 1123–36

[31] Simonyan K and Zisserman A 2014 *Very deep convolutional networks for large-scale image recognition* (arXiv:1409.1556)

[32] Suri J S, Haralick R M and Sheehan F H 2000 Greedy algorithm for error correction in automatically produced boundaries from low contrast ventriculograms *Pattern Anal. Appl.* **3** 39–60

[33] Kumar P K *et al* 2017 Accurate lumen diameter measurement in curved vessels in carotid ultrasound: an iterative scale-space and spatial transformation approach *Med. Biol. Eng. Comput.* **55** 1415–34

[34] Suri J S, Kathuria C and Molinari F (ed) 2010 *Atherosclerosis Disease Management* (Berlin: Springer)

[35] Bard R L, Kalsi H, Rubenfire M, Wakefield T, Fex B, Rajagopalan S and Brook R D 2004 Effect of carotid atherosclerosis screening on risk stratification during primary cardiovascular disease prevention *Am. J. Cardiol.* **93** 1030–32

[36] Bots M L, Hoes A W, Koudstaal P J, Hofman A and Grobbee D E 1997 Common carotid intima–media thickness and risk of stroke and myocardial infarction: the Rotterdam study *Circulation* **96** 1432–37

[37] Araki T, Ikeda N, Dey N, Acharjee S, Molinari F, Saba L, Godia E C, Nicolaides A and Suri J S 2015 Shape-based approach for coronary calcium lesion volume measurement on intravascular ultrasound imaging and its association with carotid intima–media thickness *J. Ultrasound Med.* **34** 469–82

[38] Kao A H *et al* 2013 Relation of carotid intima–media thickness and plaque with incident cardiovascular events in women with systemic lupus erythematosus *Am. J. Cardiol.* **112** 1025–32

[39] Wendelhag I, Liang Q, Gustavsson T and Wikstrand J 1997 A new automated computerized analysing system simplifies reading and reduces the variability in ultrasound measurement of intima media thickness *Stroke* **28** 2195–200

[40] Petroudi S, Loizou C, Pantziaris M and Pattichis C 2012 Segmentation of the common carotid intima–media complex in ultrasound images using active contours *IEEE Trans. Biomed. Eng.* **59** 3060–9

[41] Molinari F, Pattichis C S, Zeng G, Saba L, Acharya U R, Sanfilippo R, Nicolaides A and Suri J S 2012 Completely automated multiresolution edge snapper—a new technique for an

accurate carotid ultrasound IMT measurement: clinical validation and benchmarking on a multi-institutional database *IEEE Trans. Image Process.* **21** 1211–22

[42] Saba L, Banchhor S K, Londhe N D, Araki T, Laird J R, Gupta A, Nicolaides A and Suri J S 2017 Web-based accurate measurements of carotid lumen diameter and stenosis severity: an ultrasound-based clinical tool for stroke risk assessment during multicenter clinical trials *Comput. Biol. Med.* **91** 306–17

[43] Saba L, Banchhor S K, Araki T, Suri H S, Londhe N D, Laird J R, Viskovic K and Suri J S 2018 Intra- and inter-operator reproducibility analysis of automated cloud-based carotid intima–media thickness ultrasound measurement *J. Clin. Diagn. Res.* **12** KC01–11

[44] Saba L, Banchhor S K, Araki T, Viskovic K, Londhe N D, Laird J R, Suri H S and Suri J S 2018 Intra-and inter-operator reproducibility of automated cloud-based carotid lumen diameter ultrasound measurement *Indian Heart J.* **70** 649–64

Section III

Association of morphological and echolucency-based phenotypes with HbA1c

IOP Publishing

Vascular and Intravascular Imaging Trends, Analysis, and Challenges, Volume 2

Plaque characterization

Petia Radeva and Jasjit S Suri

Chapter 7

Echolucency-based phenotype in carotid atherosclerosis disease for risk stratification of diabetes patients

Ankush D Jamthikar, Vasileios Kotsis, Tadashi Araki, Deep Gupta, John R Laird, Argiris A Giannopoulos, Luca Saba, Harman S Suri, Sophie Mavrogeni, George D Kitas, Klaudija Viskovic, Narendra N Khanna, Ajay Gupta, Andrew Nicolaides and Jasjit S Suri

This study investigated the association of a carotid ultrasound echolucent plaque-based biomarker with HbA1c, measured as age-adjusted grayscale median (AAGSM) as a function of chronological age, total plaque area and conventional grayscale median (GSM_{conv}). Two stages were developed: (i) automated measurement of carotid parameters such as total plaque area (TPA) and (ii) computing the AAGSM as a function of GSM_{conv}, age and TPA. Intra-operator (novice and experienced) analysis was conducted. Left/right common carotid artery (CCA) ultrasound scans (408 images; Institutional Review Board (IRB) approved) of 204 patients were collected—mean age: 69 ± 11 years; mean HbA1c: 6.12% ± 1.47%. A moderate inverse correlation was observed between AAGSM and HbA1c (a CC of −0.13, $p = 0.01$), compared to GSM (a CC of −0.06, $p = 0.24$). The left CCA and right CCA showed a CC of −0.18, $p = 0.09$. Using the thresholds for AAGSM and HbA1c as low risk (AAGSM > 100; HbA1c < 5.7%) and moderate risk (40 < AAGSM < 100; 5.7% 6.5%), the area under the curve showed a better performance of AAGSM over GSM_{conv}. A paired *t*-test between the operators and expert gave $p < 0.0001$. Echolucent plaque in patients with diabetes can be more accurately characterized for risk stratification using AAGSM compared to GSM_{conv}.

7.1 Introduction

As per the statistics reported by the World Health Organization (WHO), around 17 million people die due to cardiovascular disease (CVD) every year, mainly because of stroke and heart attack [1]. Moreover, males have been reported to show significantly more events due to both of these diseases compared to females [2]. Vascular plaque build-up, known as atherosclerosis, is the main cause of these diseases [3, 4]. Carotid arteries are the dominant pathways to supply oxygen-rich blood to the brain and can also provide insight into the overall cardiovascular health of a person [5]. Blockages of carotid arteries can lead to an acute compromise of blood flow to the brain resulting in an ischemic stroke.

Carotid atherosclerotic disease has a strong association with the chronological age of a person [6] and its prevalence increases with age [7]. Some studies have also indicated an increase in carotid plaque calcification and lipid core with an increase in age [8]. Furthermore, a high prevalence of softer plaque (so-called morphology) was reported in males compared to females, which increases with the age of a person [9]. In addition to the effect on carotid plaque morphology, the chronological age of a person also contributes to the development of HbA1c levels, which are generally used in the diagnosis and a therapeutic monitoring of diabetes [10]. Both carotid plaque echolucency and higher levels of HbA1c are strongly associated with each other [11]. Furthermore, patients with diabetes have also been identified to have hypoechoic carotid plaque [12]. Some studies have also reported aiming to reduce the prevalence of echolucent plaque in patients with diabetes [13, 14]. It is therefore essential to investigate the simultaneous variations in both the carotid plaque echolucency and HbA1c levels along with the chronological age of a patient for risk stratification and characterization of patients with diabetes.

Higher levels of HbA1c lead to an elevation in the carotid plaque burden which may further lead to stroke and cardiac events [15]. A large number of patients suffering from these diseases have facilitated the risk stratification system by providing additional information about the health status of a person. B-mode ultrasonography is a prevalent imaging technique which is commonly used to assess the morphology of carotid plaque. Compared to other non-invasive imaging modalities, such as computed tomography and magnetic resonance imaging, carotid ultrasound (CUS) is a user-friendly and cost-effective technique and therefore has been adopted in a wide range of practice settings, including the clinician's office. Plaque assessment using CUS images provides information about the morphological characteristics, such as echogenicity and echolucency. The grayscale median (GSM) is a well-known biomarker that is used to measure the plaque echogenicity (hyper-echoic or bright pixels intensity, calcified plaque) or echolucency (hypoechoic or dark pixel intensity, heterogeneous or mixed plaque) [16]. It is normally computed as a median value of all the grayscale intensities that represent the carotid plaque in an US scan. In general, carotid plaque can be of two types based on the intensity appearance in an ultrasound image.

Various studies showed the relationship between echolucent plaque and an increase in the risk of cardiovascular (CV) and stroke events [17–19]. The use of GSM_{conv} in diabetes patients has been reported by some studies to enhance the CVD or stroke risk assessment [12, 20, 21]. Two main findings can be drawn from these studies: (1) echolucent plaques (low GSM_{conv} value) are associated with an elevated risk of stroke and CV events, and (2) patients with diabetes have a high prevalence of echolucent plaque. Although GSM_{conv} is considered as an important biomarker in carotid plaque assessment, it does not consider the chronological age of a person in its formulation and hence cannot explain the age-related variations in plaque morphology.

The main objective of the present study is to investigate the role of chronological age-adjustment in the conventional GSM_{conv} to risk-stratify diabetes patients using a novel age-adjusted GSM (AAGSM) biomarker. Further, we also present the association of AAGSM with HbA1c, which is a potential marker to demarcate the three risk classes, such as low, moderate and high diabetes, and finally we discuss the therapeutic implications of AAGSM. In this chapter we will be using GSM and GSM_{conv} interchangeably since they have the same meaning.

This study has the following layout: the patient demographics and the methodology adopted for AAGSM formulation are discussed in section 7.2, correlation analysis between AAGSM and HbA1c, ROC analysis and visual validation of the three risk classes are presented in section 7.3. The strengths and limitations of this study along with benchmarking are discussed in section 7.4, followed by the conclusion in section 7.5.

7.2 Patient demographics and methodology

7.2.1 Patients demographics

A total of 204 volunteers (157 male and 47 female) with a mean age of 69 ± 11 years were recruited at Toho University Japan (IRB ethics approved and patient consent was obtained). Of the 204 subjects, 49 patients had diabetes and 155 did not. Of 408 B-mode ultrasound scans extracted from the left/right (L/R) CCA, one patient's left CCA artery image was not provided and hence only 407 images were available for analysis in this study. The mean HbA1c levels for all the patients were 6.30% ± 1.1% ranging from 4.8% to 13%. This cohort had a mean total cholesterol of 175.39% ± 38.26 (mg dl^{-1}), high-density lipoprotein (HDL) cholesterol of 50.59 ± 15.13 (mg dl^{-1}) and low-density lipoprotein (LDL) cholesterol of 101.16 ± 31.79 (mg dl^{-1}). Of 204, eighty-three were smokers.

7.2.2 Methodology

The rationale for AAGSM is driven by the role of age in atherosclerosis disease along the morphological-based envelope, which is captured by the lumen–intima (LI) and media–adventitia (MA) interfaces [22, 23]. The conventional GSM is a standardized solution using echolucency analysis; where the probability distribution function is first computed for the far wall region, followed by the cumulative distribution function (CDF) plot whose 50th percentile is chosen [24, 25].

Since the total plaque area (TPA) of the region between the wall interfaces [26, 27] is also age-related, the ratio of age to area is a fraction which plays a role in creating the age-adjusted GSM, by segregating either the symptomatic plaque versus asymptomatic

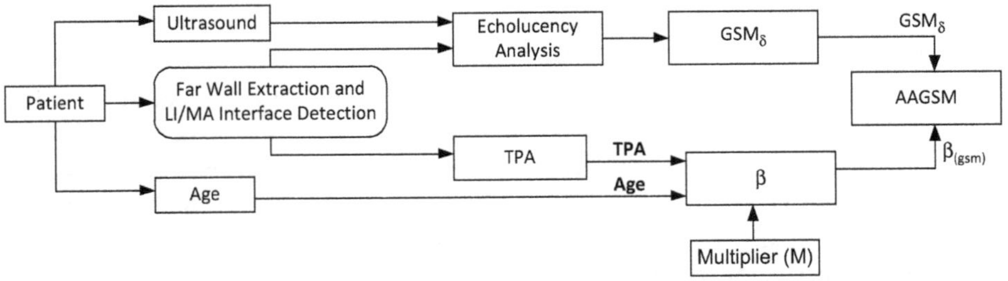

Figure 7.1. AAGSM computation and measurement. TPA was computed using the AtheroEdge™ system and GSM_δ was computed using echolucency analysis.

plaque or controls versus diseased. This can be seen in the simple flow chart of figure 7.1. Thus, the methodology consists of the following steps: (i) automated LI/MA delineation using AtheroEdge™ (a regulatory cleared medical device from AtheroPoint™, Roseville, CA, USA) along with the TPA computation [28, 29] and (ii) computing the AAGSM as a function of current chronological age and wall total plaque area. It consists of correcting the conventional GSM_{conv} to GSM_δ and then scaling GSM_δ by the scale factor β_{gsm}, which is a function of age and area.

This scale factor β_{gsm} is computed by taking the Mth percentage of the age factor divided by the TPA ($\beta_{gsm} = \frac{M * \text{Age}}{\text{TPA}}$), where * represents the product between the pre-multiplier (M) and chronological age. GSM_δ is computed by adjusting the conventional GSM, i.e. GSM_{conv} by scaling and removing the bias. Note that GSM_δ is same as GSM_{conv}, except taken at the δth percentile value, unlike the 50th percentile of CDF, where δth is computed as the mean of CDF shifted by a constant factor ($\alpha \sim 0.375$). This was computed empirically from the training cohort dataset patients whose neurological conditions were known, for example, knowing the symptomatic and asymptomatic condition [30, 31]. A high segregation index, an index which stratifies normal versus abnormal conditions, was used to formulate β_{gsm}. Thus, AAGSM is expressed mathematically as the product of $GSM_\delta \times \beta_{gsm}$, i.e. $\text{AAGSM} = [GSM_\delta \times \frac{M * \text{Age}}{\text{TPA}}]$. Note that TPA was computed by summing all the pixels between the MA and LI borders of the far wall of the carotid artery and multiplying with pixel resolution factors (in millimeters). Step three computes the AAGSM versus HbA1c for each patient independently. The overall system for computation of GSM_δ, β_{gsm} and AAGSM is depicted in figure 7.1.

7.3 Results and statistical analysis

Two operators (one novice and one experienced) were involved in computing the LI/MA interface using AtheroEdge™, which was then used to automatically compute the TPA between the LI/MA and GSM_δ corresponding to the grayscale intensity between LI and MA. The first set of results was used to compute the association between the AAGSM and HbA1c and compared against GSM_{conv} versus HbA1c (figure 7.2). The second set of results is the CC between

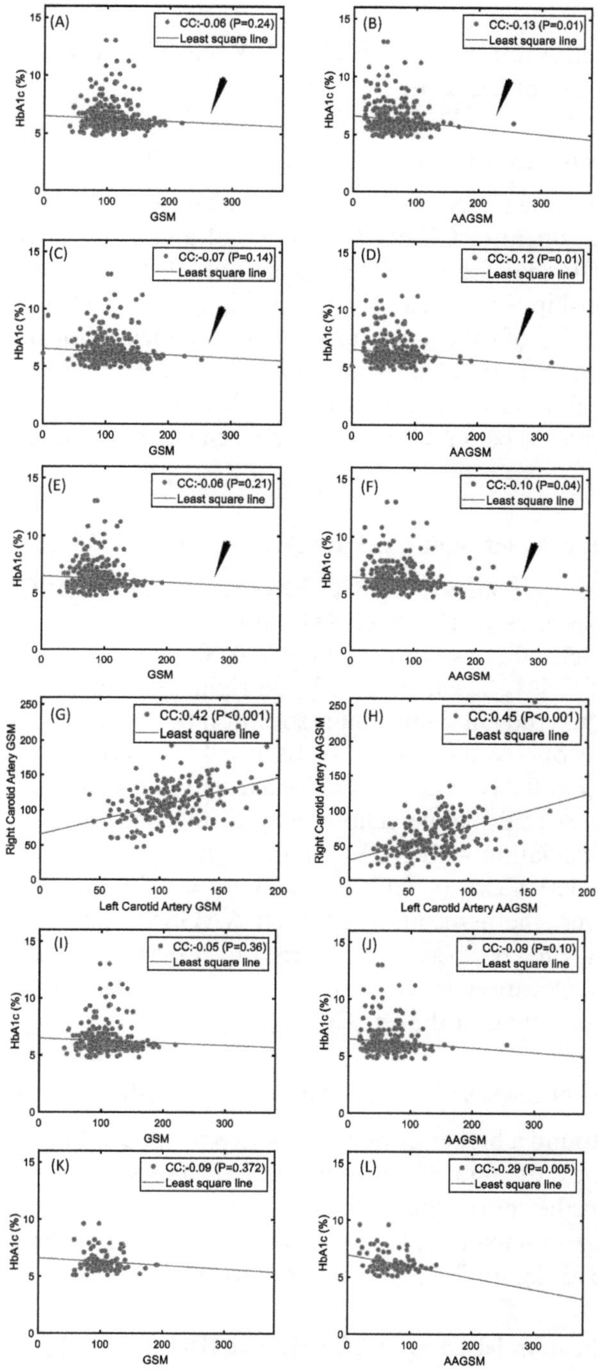

Figure 7.2. Comparison of GSM_{conv} and AAGSM against HbA1c. (A) GSM_{conv} versus HbA1c for the novice operator; (B) AAGSM versus HbA1c for the novice operator; (C) GSM_{conv} versus HbA1c for the experienced operator; (D) AAGSM versus HbA1c for the experienced operator; (E) GSM_{conv} versus HbA1c for the expert; (F) AAGSM versus HbA1c for the expert; (G) GSM_{conv} for the left versus right CCA; (H) AAGSM for the left versus right CCA; (I) GSM_{conv} versus HbA1c for males; (J) AAGSM versus HbA1c for males; (K) GSM_{conv} versus HbA1c for females; and (L) AAGSM versus HbA1c for females. The arrows indicate the regression line for the corresponding CC value.

GSM_{conv} versus HbA1c for the left/right CCA and CC between AAGSM versus HbA1c for the left/right CCA (figure 7.2).

7.3.1 CC analysis of AAGSM and GSM_{conv} against HbA1c

We found a moderate significant inverse CC between AAGSM and HbA1c for both operators (a novice operator CC of −0.13, $p = 0.01$ and an experienced operator CC of −0.12, $p = 0.01$) and for the expert (a CC of −0.10, $p = 0.04$). However, no significant relationship was noted between GSM_{conv} and HbA1c for both operators (a novice operator CC of −0.06, $p = 0.24$ and an experienced operator CC of −0.07, $p = 0.14$) and for the expert (a CC of −0.06, $p = 0.21$). Figures 7.2(A) to (F) represent the CC analysis of AAGSM and GSM_{conv} with HbA1c for two operators and an expert. This indicates a better association between carotid plaque composition and HbA1c levels.

7.3.2 CC between left and right CCA for AAGSM and GSM_{conv}

In this experiment, we compare the CC between left and right CCA for AAGSM and GSM_{conv}, respectively. The AAGSM values of left and right CCAs are more associated with each other (a CC of 0.45, $p < 0.001$, figure 7.2(H)) as compared to the association of GSM_{conv} values of left and right CCA (a CC of 0.42, $p < 0.001$, figure 7.2(I)). In this study, a significant association between AAGSM and HbA1c in the right CCA was observed (a CC of −0.18, $p = 0.01$) as compared to the left CCA (a CC of −0.08, $p = 0.24$), which is attributed to 125% elevation in the CC value indicating higher carotid plaque echolucency in the right neck arteries. A similar trend of higher association with HbA1c in the right CCA (a CC of −0.08, $p = 0.27$) compared to the left CCA (a CC of −0.04, $p = 0.59$) was also observed with GSM_{conv}. However, the association between AAGSM and HbA1c for both L/R CCAs was found to be better compared to GSM_{conv} because of the poor representation of elevation in carotid plaque echolucency (the CC in the right CCA is 100% higher than in the left CCA).

7.3.3 CC analysis of AAGSM–HbA1c and GSM_{conv}–HbA1c in males and females

In our study, we found a higher association between AAGSM and HbA1c in females (CC of −0.29, $p < 0.01$, figure 7.2(L)) compared to males (CC of −0.09, $p = 0.10$, figure 7.2(J)). Furthermore, the correlation between conventional GSM_{conv} and HbA1c was less in both males (CC of −0.05, $p = 0.36$, figure 7.2(I)) and females (CC of −0.09, $p = 0.372$, figure 7.2(K)) compared to AAGSM.

7.3.4 Risk stratification based on AAGSM, and HbA1c and ROC analysis

Based on the two thresholds of AAGSM and HbA1c, patients were classified into three risk classes: low risk (figures 7.3(A), (B)), moderate risk (figures 7.3(C), (D)) and high risk (figures 7.3(E), (F)).

We computed the average values of both conventional and proposed biomarkers of plaque echolucency for the low-risk (GSM_{conv} = 118.70 and 119.32), moderate-

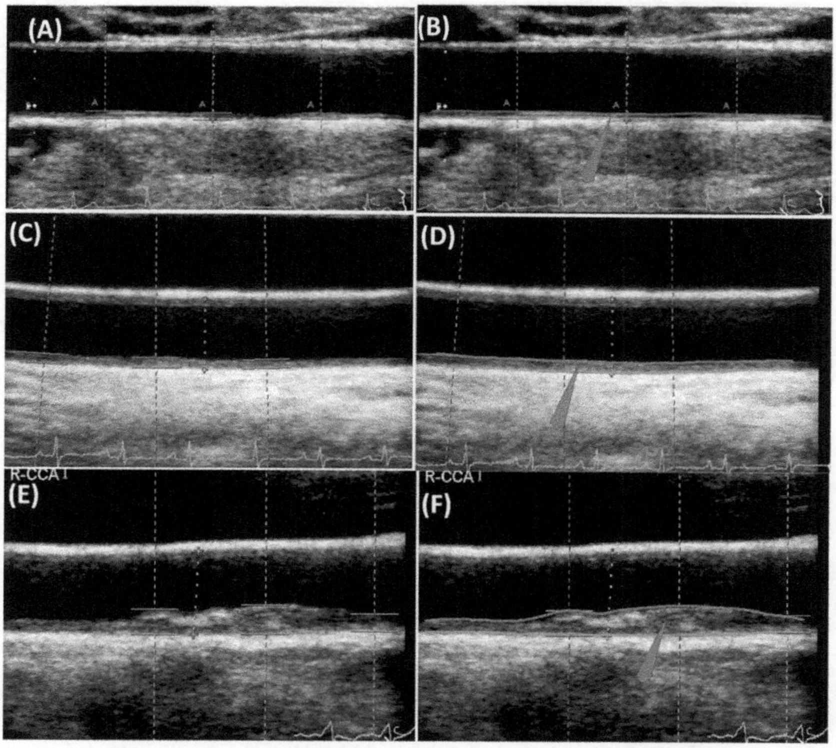

Figure 7.3. Risk stratification based on HbA1c and AAGSM thresholds. Row 1 (low risk)—patient 199R; HbA1c: 6%; AAGSM: 104; (A) original image; (B) processed image using AtheroEdge™ 2.0. Row 2 (moderate risk)—patient 28L; HbA1c: 6.7%; AAGSM: 52.91; (C) original image; (D) processed image using AtheroEdge™ 2.0. Row 3 (high risk)—patient 85R; HbA1c: 6.4%; (E) original image; (F) processed image using AtheroEdge™ 2.0.

risk (GSM_{conv} = 82.57 and AAGSM = 66.20) and high-risk (GSM_{conv} = 44.79 and AAGSM = 30.81) classes. ROC analysis showed a 5% improvement in the area under the curve (AUC) values for AAGSM and HbA1c (AUC: 0.5594) association compared to the relation between conventional GSM_{conv} and HbA1c (AUC: 0.5188).

7.3.5 Statistical tests

All the statistical analysis was carried out using MATLAB 2017b and MedCalc 12.5. We performed a regression analysis between the novice operator and experienced operator. The CC value between both the novice operator and the experienced operator was found to be 0.85 ($p < 0.0001$) (figure 7.4(A)). Furthermore, inter-operator agreement was tested using Bland–Altman analysis (figure 7.4(B)). We found that, except for five outliers, all other AAGSMs lie between the agreement limit of the mean ± 1.96 SD. The possible cause of these outliers can be attributed to the inexperience of the novice operator. We have also performed a paired *t*-test

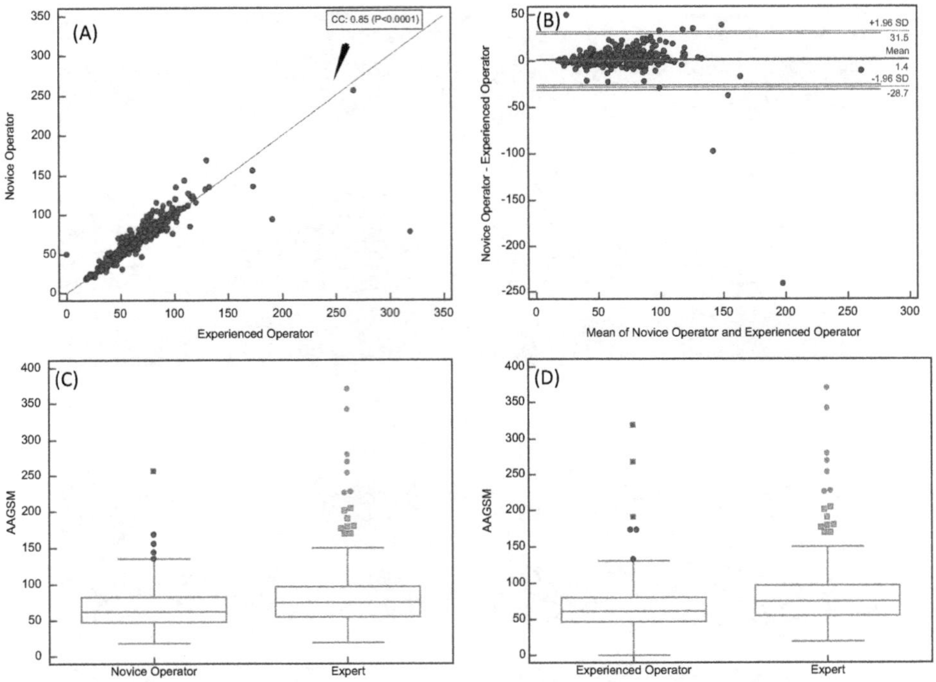

Figure 7.4. Regression and Bland–Altman analysis between AAGSM values of novice and experienced operators. (A) Correlation coefficient: 0.85 ($p < 0.0001$). (B) Agreement between the novice operator and the experienced operator. (C)–(D) Box–whisker plot for the paired t-test between the operators and expert analyst.

between the two operators and the expert analyst. The box–whisker plots in figure 7.4(C), (D) show significant results obtained from the paired t-test.

7.4 Discussion

The grayscale median has been used widely to assess carotid plaque echogenicity, but its specific role in the risk stratification of diabetes patients remains unclear. Furthermore, the conventional GSM_{conv} does not consider the effect of the chronological age of a person, which contributes to the development of the HbA1c levels that are generally used in the diagnosis of diabetes. Chronological age is attributed to an increase in HbA1c levels by 0.05%–1% in 10 years [32]. The Screening for Impaired Glucose Tolerance study indicated an increase in HbA1c levels by 0.085% per decade while the National Health and Nutrition Examination Survey indicated an elevation in HbA1c levels by 0.094% per decade [10]. Hence there is a strong need to explain this age-related variation in HbA1c levels and its association with carotid plaque composition. Our results showed a strong moderate association of AAGSM with diabetes biomarker HbA1c, compared to GSM_{conv}. Thus, our data suggest that AAGSM, as opposed to GSM_{conv}, may be a more effective biomarker for risk stratification of patients with diabetes.

In this study, AAGSM showed a high-moderate negative correlation with HbA1c. This suggests that increasing echolucency of carotid plaques is associated with more advanced diabetes, as measured by HbA1c. The echolucency of the carotid atherosclerotic plaque has also been reported to be associated with a large lipid core without calcification, which can be accurately captured in US scans [33]. Biologically, it has been shown that atherosclerosis may progress rapidly in the context of abnormally elevated HbA1c [34]. Patients with diabetes have been reported to have a large amount of soft carotid atherosclerotic plaque [35]. Furthermore, plaque echolucency and echogenicity in US scans have been tested to assess carotid plaque calcification and to predict cardiovascular events in diabetic patients [36]. The prevalence of mixed plaque (a combination of echolucent and echogenic plaque) has also been reported to be higher in diabetes patients and hence can be assessed using AAGSM. Elevation of HbA1c is also associated with coronary artery calcification indicating a direct association of echogenic plaque with HbA1c [37]. The ultrasound image reconstruction of carotid atherosclerotic disease in diabetic patients tends to have a lower contrast, unlike the presence of higher contrast (hyperechoic or echogenic) due to calcium in the wall. Such a tendency of echolucency leads to lower GSM_δ leading to a stronger negative association. Since AAGSM is derived from GSM_δ, and β_{gsm} is always less than one, therefore AAGSM further decreases with HbA1c.

7.4.1 A note on the HbA1c and AAGSM thresholds for risk stratification

In general, patients with HbA1c $< 5.7\%$ are considered to have a low risk of diabetes and those with HbA1c $\geqslant 6.5\%$ are diagnosed to have high risk of diabetes. An HbA1c range of 5.7%–6.5% has been considered as the intermediate or moderate range and patients in this range have an incremental risk of developing diabetes [38]. However, WHO suggested patients with HbA1c between 6%–6.5% are more likely to develop diabetes and hence should be considered for diabetes preventions treatments [39, 40]. Based on the two HbA1c thresholds, patients' baseline characteristics have been presented in table 7.1. In this study, patients were risk stratified into three classes based on HbA1c levels: low risk (HbA1c $< 5.7\%$ and AAGSM > 100), moderate risk ($5.7 \leqslant$ HbA1c < 6.5 and $40 <$ AAGSM $\leqslant 100$) and high risk ($6.5\% \leqslant$ HbA1c and AAGSM < 40).

In this study, we also performed risk stratification of patients into three classes using AAGSM values. Three risk classes were formed based on two thresholds (AAGSM: 100 and 40). While plotting AAGSM, two cutoff points at which the AAGSM values vary suddenly were considered as the thresholds on both the upper and lower AAGSM curves. It has been found that the AAGSM–HbA1c threshold pair truly stratified a greater number of patients under the correct accurate risk label compared to the conventional HbA1c–GSM_{conv} pair.

7.4.2 Justification of the δth percentile value during GSM_δ measurement

The justification of the δth percentile value for GSM_δ measurement can be attributed to the following reasons. The contribution of each grayscale pixel in a US scan for the carotid plaque region can be visualized using the CDF curve. In general, the mean value of the CDF curve can be considered as representative of all the pixels in

Table 7.1. Patients' baseline characteristics.

Biomarkers	Total patients	Low-risk patients (HbA1c < 5.7%)	Moderate-risk patients (5.7% ⩽ HbA1c < 6.5%)	High-risk patients (HbA1c ⩾ 6.5%)	*p*-value
Total	$n = 204$	$n = 48$	$n = 107$	$n = 49$	—
Age (years)	69 ∓ 11	68 ∓ 12.2	69.13 ∓ 10.51	69.94 ∓ 10.83	0.3974
Male, *n* (%)	158 (77.45)	39 (81.25)	82 (76.63)	36 (73.47)	0.514
Female, *n* (%)	46 (22.54)	9 (18.75)	20 (18.69)	21 (42.86)	0.514
Smokers, *n* (%)	83 (40.68)	21 (39.33)	40 (37.38)	39 (79.59)	0.662
HT, *n* (%)	148 (72.54)	30 (68.67)	78 (72.89)	5 (10.20)	0.142
FH, *n* (%)	25 (12.25)	3 (12.00)	16 (14.95)	2 (4.08)	0.733
HD, *n* (%)	23 (11.27)	12 (12.67)	8 (7.47)	172.60 ∓ 33.78	0.089
TC	175.39 ∓ 38.3	175 ∓ 43.109	177.06 ∓ 38.15	100.37 ∓ 29.43	0.7189
LDL	101.2 ∓ 31.8	97 ∓ 33.26	103.48 ∓ 32.25	49.95 ∓ 13.22	0.7806
HDL	50.6 ∓ 15.1	53 ∓ 16.83	50.02 ∓ 15.24	114.92 ∓ 60.43	0.4588
TG	125.9 ∓ 90.8	117.15 ∓ 82.60	128.89 ∓ 104.04	151.51 ∓ 40.53	0.7220
FBS	121.1 ∓ 35.2	104.71 ∓ 28.93	114.52 ∓ 25.90	$n = 49$	<0.0001

HT: hypertension; FH: family history; TG: triglyceride; FBS: fasting blood sugar; HDL: high-density lipoprotein; LDL: low-density lipoprotein; TC: total cholesterol.

the carotid plaque region. However, as the all the pixels do not represent a normal distribution (a bell-shaped) curve, the mean of the CDF may be affected by darker or brighter intensity levels. In order to avoid this effect, GSM_{conv} considers the 50th percentile of the CDF curve as a grayscale median. However, this method when tried on asymptomatic and symptomatic datasets provided a poor segregation index between these datasets. The segregation index is a tool that provides the status of symptomatic versus asymptomatic plaque. The poor segregation is due to a large number of darker intensities in the plaque region, and this skews the CDF towards the darker intensities (hypoechoic) of the plaque. So, in such cases, considering the median value may not represent the complete central tendency of the carotid plaque CDF. We have considered a higher value compared to the median to obtain a better segregation index. This higher value is denoted by δ in the GSM_{δ} computation.

7.4.3 A special note on age-adjustment pre-multiplier (*M*) selection

The mathematical formulation of AAGSM includes an age-adjustment multiplier (AAM) of the $M \sim$ 25th percentile of chronological age, which was selected by a careful empirical selection and optimization paradigm (figure 7.1). Carotid plaque increases with age. In general, the last 10%–50% of chronological age contributes to the present stage of atherosclerotic plaque in an adult person. Hence we have selected a range of 0.1–0.5 with an increment of 0.05 to classify the patients with diabetes into three risk classes (low, moderate and high). Patients in each risk class were compared against those classified by considering HbA1c and risk stratified into three similar classes. For each AAM, the patients truly and falsely classified into low-risk, moderate-risk and high-risk classes were determined. The multiplier of 0.25 showed the maximum overlap between patients which are truly risk stratified.

7.4.4 A note on the therapeutic implications of AAGSM

Low conventional GSM indicates the prevalence of echolucent plaque which is associated with ischemic stroke [41]. Furthermore, high levels of HbA1c are also associated with the progression of plaque echolucency in patients with diabetes. In this study, we have shown a low moderate to high moderate correlation (up to a CC of −0.29 for females) between plaque echolucency (measured using AAGSM) and HbA1c levels. This leads us to the suggestion that high-risk patients (i.e. with HbA1c $> 6.5\%$ and AAGSM < 40) are more likely to develop the carotid atherosclerotic events which may lead to future ischemic stroke. Hence, it is essential to consider the treatments which can simultaneously lower the lipid-rich echolucent plaque and HbA1c levels. Echolucent plaque determined by using low conventional GSM (GSM_{conv}) has become a therapeutic target to prevent the occurrence of future stroke events [41]. Increasing the echogenicity of carotid plaque is one of the methods discussed in the literature to prevent stroke events [13, 14, 42]. Various types of medication have been discussed in the literature to improve plaque echogenicity (i.e. reducing echolucent plaque). Cesarone *et al* [42] reported the use of a total triterpenic fraction of centella asiatica (TTFCA) to reduce the echolucency of carotid plaque (i.e. to improve the plaque echogenicity) in patients with GSM_{conv}

values less than 18. The authors reported a direct reduction in plaque echolucency and a stabilization of lipid-rich hypoechoic plaque (dark plaque) over a period of 12 months by using TTFCA. Hirano *et al* [13, 14] presented a study of patients with diabetes and reported the use of Pioglitazone and Acarbose to reduce the hypoechoic nature of plaque by improving the carotid plaque echogenicity. Betablockers are also reported to prevent carotid plaque growth. A recent study by Asciutto *et al* [43] reported GSM_{conv} values greater than 30 in patients who were treated with betablockers. Since AAGSM showed a high-moderate association with HbA1c levels, efforts to lower the levels of HbA1c may also contribute to the reduction of carotid plaque echolucency. Hence, it is essential to consider medications that are primarily used in the prevention of the elevation of HbA1c. González-Ortiz *et al* [44] reported the use of metformin glycinate to significantly lower HbA1c levels. We believe a combination of medications to simultaneously lower HbA1c and lower lipid-rich echolucent plaque could possibly be a better scenario, however, more multi-ethnic pilot studies need to be conducted.

7.4.5 Benchmarking against the previous literature

In the literature, GSM_{conv} has been widely used to explain the carotid plaque echogenicity in patients with diabetes. Ostling *et al* [12] presented a study with 98 patients and indicated a high prevalence of echolucent plaque in type-2 diabetes patients. The inverse association between GSM_{conv} and diabetes biomarkers such as triglyceride has been explained by Ostling *et al* [12]. An association between GSM_{conv} and HbA1c was satisfactorily reported by the authors. However, Jorgensen *et al* [11] reported an increasing prevalence of GSM_{conv}-driven plaque echogenicity with HbA1c in a patient without diabetes. GSM_{conv} has also been reported to positively correlate with eicosapentaenoic acid/arachidonic acid ratio which is an important biomarker of CV events in type-2 diabetes patients [20]. In our study, we have found a moderate but significant association between proposed AAGSM and HbA1c. Also, a low negative correlation was observed between GSM_{conv} and HbA1c level.

7.4.6 Strengths, weaknesses and applications of AAGSM

AAGSM is a more specific and finely tuned biomarker compared to GSM_{conv} and appears to offer a better risk stratification of diabetes patients. It is simple to compute AAGSM, since it uses TPA and age, which are adapted as an inverse function and direct functions, respectively, to AAGSM. Another major advantage of AAGSM is its ability to stratify and identify the symptomatic patients based on the ranges of AAGSM. Further, to predict the risk of stroke and CVD, automated systems have been designed in the literature which are based on intelligence-based paradigms, such as machine-learning and deep-learning approaches [31, 45, 46]. The performance of such paradigms is highly dependent on the choice of the features that are extracted from the target atherosclerotic plaque lesion. AAGSM can be integrated as a feature, similar to texture-based or statistical features that will further improve the chances of stroke risk assessment in patients with diabetes.

Recently, Huang *et al* [47] integrated the GSM_{conv} with texture-based features to identify the association with HbA1c. AAGSM can also be used to perform risk stratification in other autoimmune inflammatory diseases such as rheumatoid arthritis using carotid artery plaque morphology [48]. Finally, we want to point out that our AAGSM system was implemented in the C++ language in an MS Window's environment, which offers a better speed compared to MATLAB.

Despite the merits of AAGSM, this pilot study has the limitation that it needs to be extended to a larger database, representing different ethnicities for further fine-tuning the scale factor β_{gsm}, even though the current study was applied to three kinds of cohorts, primarily from Italy [49], the UK [50] and Japan.

7.5 Conclusion

This study has proposed a novel age-adjusted grayscale median which performed better than conventional GSM_{conv} and reported a moderate but significant correlation with HbA1c, which is a potential biomarker of diabetes and useful for risk stratification of diabetes patients. We found a high association of plaque echolucency with HbA1c in the right CCA compared to the left CCA. AAGSM represented this association more significantly compared to conventional GSM_{conv}. A higher association of plaque echolucency was observed in females compared to males using the relation between AAGSM and HbA1c and GSM_{conv} and HbA1c. However, a stronger representation is provided by AAGSM compared to GSM_{conv}. A combination of lipid-lowering and HbA1c lower medications could be a better alternative for diabetic patients with lower AAGSM.

References

[1] World Health Organization The Atlas of Heart Disease and Stroke: Cardiovascular disease http://who.int/cardiovascular_diseases/resources/atlas/en/

[2] Bots S H, Peters S A E and Woodward M 2017 Sex differences in coronary heart disease and stroke mortality: a global assessment of the effect of ageing between 1980 and 2010 *BMJ Glob. Health* **2** e000298

[3] Banerjee C and Chimowitz M I 2017 Stroke caused by atherosclerosis of the major intracranial arteries *Circ. Res.* **120** 502–13

[4] Chen P-C, Jeng J-S, Hsu H-C, Su T-C, Chien K-L and Lee Y-T 2016 Carotid atherosclerosis progression and risk of cardiovascular events in a community in Taiwan *Sci. Rep.* **6** 25733

[5] Lorenz M W *et al* 2012 Carotid intima–media thickness progression to predict cardiovascular events in the general population (the PROG-IMT collaborative project): a meta-analysis of individual participant data *Lancet* **379** 2053–62

[6] Spence J D, Eliasziw M, DiCicco M, Hackam D G, Galil R and Lohmann T 2002 Carotid plaque area: a tool for targeting and evaluating vascular preventive therapy *Stroke* **33** 2916–22

[7] Singh S *et al* 2013 Rapid screening for subclinical atherosclerosis by carotid ultrasound examination: the HAPPY (Heart Attack Prevention Program for You) substudy *Glob. Heart* **8** 83–9

[8] Redgrave J N, Lovett J K and Rothwell P M 2010 Histological features of symptomatic carotid plaques in relation to age and smoking: the Oxford Plaque Study *Stroke* **41** 2288–94

[9] Joakimsen O, Bønaa K H, Stensland-Bugge E and Jacobsen B K 1999 Age and sex differences in the distribution and ultrasound morphology of carotid atherosclerosis: the Tromsø Study *Arterioscler. Thromb. Vasc. Biol.* **19** 3007–13

[10] Dubowitz N *et al* 2014 Aging is associated with increased HbA1c levels, independently of glucose levels and insulin resistance, and also with decreased HbA1c diagnostic specificity *Diabet. Med.* **31** 927–35

[11] Jørgensen L, Jenssen T, Joakimsen O, Heuch I, Ingebretsen O C and Jacobsen B K 2004 Glycated hemoglobin level is strongly related to the prevalence of carotid artery plaques with high echogenicity in nondiabetic individuals: the Tromsø study *Circulation* **110** 466–70

[12] Östling G, Hedblad B, Berglund G and Gonçalves I 2007 Increased echolucency of carotid plaques in patients with type 2 diabetes *Stroke* **38** 2074–8

[13] Hirano M *et al* 2009 Rapid improvement of carotid plaque echogenicity within 1 month of pioglitazone treatment in patients with acute coronary syndrome *Atherosclerosis* **203** 483–8

[14] Hirano M *et al* 2012 Early improvement in carotid plaque echogenicity by acarbose in patients with acute coronary syndromes *Circ. J.* **76** 1452–60

[15] Sun B *et al* 2016 Elevated hemoglobin A1c is associated with carotid plaque vulnerability: novel findings from magnetic resonance imaging study in hypertensive stroke patients *Sci. Rep.* **6** 33246

[16] Griffin M B *et al* 2010 Juxtaluminal hypoechoic area in ultrasonic images of carotid plaques and hemispheric symptoms *J. Vasc. Surg.* **52** 69–76

[17] Gupta A *et al* 2015 Plaque echolucency and stroke risk in asymptomatic carotid stenosis: a systematic review and meta-analysis *Stroke* **46** 91–7

[18] Mathiesen E B, Bønaa K H and Joakimsen O 2001 Echolucent plaques are associated with high risk of ischemic cerebrovascular events in carotid stenosis: the Tromsø study *Circulation* **103** 2171–5

[19] Honda O *et al* 2004 Echolucent carotid plaques predict future coronary events in patients with coronary artery disease *J. Am. Coll. Cardiol.* **43** 1177–84

[20] Ariyoshi K *et al* 2015 Ultrasound analysis of gray-scale median value of carotid plaques is a useful reference index for cerebro-cardiovascular events in patients with type 2 diabetes *J. Diabetes Invest.* **6** 91–7

[21] Irie Y *et al* 2013 The utility of ultrasonic tissue characterization of carotid plaque in the prediction of cardiovascular events in diabetic patients *Atherosclerosis* **230** 399–405

[22] Molinari F, Zeng G and Suri J S 2010 Intima–media thickness: setting a standard for a completely automated method of ultrasound measurement *IEEE Trans. Ultrason. Ferroelectr. Freq. Control* **57** 1112–24

[23] Saba L *et al* 2016 Accurate cloud-based smart IMT measurement, its validation and stroke risk stratification in carotid ultrasound: a web-based point-of-care tool for multicenter clinical trial *Comput. Biol. Med.* **75** 217–34

[24] Huang X *et al* 2017 Evaluation of carotid plaque echogenicity based on the integral of the cumulative probability distribution using gray-scale ultrasound images *PLoS One* **12** e0185261

[25] Shankar P, Forsberg F and Lown L 2003 Statistical modeling of atherosclerotic plaque in carotid B mode images—a feasibility study *Ultrasound Med. Biol.* **29** 1305–9

[26] Spence J D and Solo K 2017 Resistant atherosclerosis: the need for monitoring of plaque burden *Stroke* **48** 1624–9

[27] Alsulaimani S, Gardener H, Elkind M S, Cheung K, Sacco R L and Rundek T 2013 Elevated homocysteine and carotid plaque area and densitometry in the Northern Manhattan Study *Stroke* **44** 457–61

[28] Molinari F *et al* 2012 Automated carotid IMT measurement and its validation in low contrast ultrasound database of 885 patient Indian population epidemiological study: results of AtheroEdge™ Software *Int. Angiol.* **31** 42–53

[29] Ikeda N *et al* 2017 Automated segmental-IMT measurement in thin/thick plaque with bulb presence in carotid ultrasound from multiple scanners: stroke risk assessment *Comput. Methods Programs Biomed.* **141** 73–81

[30] Acharya U R *et al* 2012 Atherosclerotic risk stratification strategy for carotid arteries using texture-based features *Ultrasound Med. Biol.* **38** 899–915

[31] Saba L *et al* 2017 Plaque tissue morphology-based stroke risk stratification using carotid ultrasound: a polling-based PCA learning paradigm *J. Med. Syst.* **41** 98

[32] Loh T P, Ma S, Heng D and Khoo C M 2016 Age-related changes in the cardiometabolic profiles in Singapore resident adult population: findings from the National Health Survey 2010 *PLoS One* **11** e0162102

[33] Grønholdt M-L M *et al* 1997 Lipid-rich carotid artery plaques appear echolucent on ultrasound B-mode images and may be associated with intraplaque haemorrhage *Eur. J. Vasc. Endovasc. Surg.* **14** 439–45

[34] Lee S W *et al* 2017 Association between HbA1c and carotid atherosclerosis among elderly Koreans with normal fasting glucose *PLoS One* **12** e0171761

[35] Zhang Y-H, Gao Y, Mao X, Shang J and Su B-L 2013 Assessment of carotid atherosclerosis in type 2 diabetes mellitus patients with microalbuminuria by high-frequency ultrasonography *Int. J. Endocrinol.* **2013** 819584

[36] de Kreutzenberg S V *et al* 2015 Carotid plaque calcification predicts future cardiovascular events in type 2 diabetes *Diabetes Care* **38** 1937–44

[37] Carson A P *et al* 2015 Hemoglobin a1c and the progression of coronary artery calcification among adults without diabetes *Diabetes Care* **38** 66–71

[38] American Diabetes Association 2010 Diagnosis and classification of diabetes mellitus *Diabetes Care* **33** S62

[39] World Health Organization 2011 *Use of Glycated Haemoglobin (HbA1c) in Diagnosis of Diabetes Mellitus: Abbreviated Report of a WHO consultation*

[40] Day A 2012 HbA1c and diagnosis of diabetes *The Test has Finally Come of Age* (London: SAGE)

[41] Grønholdt M-L M, Nordestgaard B G, Schroeder T V, Vorstrup S and Sillesen H 2001 Ultrasonic echolucent carotid plaques predict future strokes *Circulation* **104** 68–73

[42] Cesarone M, Belcaro G, Nicolaides A and Geroulakos G 2001 Increase in echogenicity of echolucent carotid plaques after treatment with total triterpenic fraction of *Centella asiatica*: a prospective, placebo-controlled, randomized trial *Angiology* **52** S19

[43] Asciutto G, Dias N V, Persson A, Nilsson J and Gonçalves I 2014 Treatment with betablockers is associated with higher grey-scale median in carotid plaques *BMC Cardiovasc. Disord.* **14** 111

[44] Gonzalez-Ortiz M, Martinez-Abundis E, Robles-Cervantes J A, Ramos-Zavala M G, Barrera-Duran C and Gonzalez-Canudas J 2012 Effect of metformin glycinate on glycated hemoglobin A1C concentration and insulin sensitivity in drug-naive adult patients with type 2 diabetes mellitus *Diabetes Technol. Ther.* **14** 1140–44

[45] Araki T *et al* 2017 Stroke risk stratification and its validation using ultrasonic echolucent carotid wall plaque morphology: a machine learning paradigm *Comput. Biol. Med.* **80** 77–96

[46] Biswas M *et al* 2018 Deep learning strategy for accurate carotid intima–media thickness measurement: an ultrasound study on Japanese diabetic cohort *Comput. Biol. Med.* **98** 100–17

[47] Huang X-W *et al* 2016 The relationship between HbA1c and ultrasound plaque textures in atherosclerotic patients *Cardiovasc. Diabetol.* **15** 98

[48] Narendra N *et al* Rheumatoid arthritis: its link to arteriosclerosis imaging and cardiovascular risk assessment using machine learning-based tissue characterization *JACC Cardiovasc. Imaging*

[49] Acharya R U *et al* 2012 Symptomatic vs asymptomatic plaque classification in carotid ultrasound *J. Med. Syst.* **36** 1861–71

[50] Ikeda N *et al* 2015 Improved correlation between carotid and coronary atherosclerosis SYNTAX score using automated ultrasound carotid bulb plaque IMT measurement *Ultrasound Med. Biol.* **41** 1247–62

Chapter 8

Morphologic TPA (mTPA) and composite risk score for moderate carotid atherosclerotic plaque is strongly associated with HbA1c in a diabetes cohort

Elisa Cuadrado-Godia, Md Maniruzzaman, Tadashi Araki, Anudeep Puvvula, Jahanur Rahman, Luca Saba, Harman S Suri, Ajay Gupta, Sumit K Banchhor, Jagjit S Teji, Tomaž Omerzu, Narendra N Khanna, John L Laird, Andrew Nicolaides, Sophie Mavrogeni, George D Kitas and Jasjit S Suri

This study examines the association between six types of carotid artery disease image-based phenotypes and HbA1c in diabetes patients. Six phenotypes (intima–media thickness measurements (cIMT (ave.), cIMT (max.), cIMT (min.)), bidirectional wall variability (cIMTV), morphology-based total plaque area (mTPA) and composite risk score (CRS)) were measured in an automated setting using AtheroEdge™ (AtheroPoint, Roseville, CA, USA). 398 ultrasound (US) scans from 199 patients (157 M, average age 68.96 ∓ 10.98 years) who underwent left/right (L/R) common carotid artery (CCA) US were retrospectively analyzed. Two operators manually calibrated all the US scans. The odds ratio (OR) was computed using logistic regression (LR) and then phenotypes were ranked for the left, right and mean carotids arteries. The fasting blood sugar (FBS) was highly associated with HbA1c ($p < 0.001$) for the left common carotid artery (LCCA), right carotid artery (RCCA), and mean of the left and right common carotid arteries (MCCA). After adjusting the FBS, the OR for mTPA showed a higher risk for LCCA, RCCA and MCCA. The coefficient of correlation (CC) between phenotypes and HbA1c was strong and the inter-CC between cIMT and mTPA/CRS was above 0.9 ($p < 0.001$). The AtheroEdge™ was further tested for inter-operator variability. Among the six carotid phenotypes, all, except for bidirectional wall variability, showed a strong association with HbA1c. mTPA and CRS were equally strong phenotypes as

doi:10.1088/2053-2563/ab0820ch8

cIMT. The CRS phenotype showed the strongest relationship to HbA1c. The AtheroEdge™ system is also noted to be adaptable to clinical settings.

8.1 Introduction

Diabetes mellitus (DM) is a long-term disease that causes high blood sugar levels [1]. On average, there are about 387 million diabetes cases all over the world [2]. This figure is expected to double by the year 2030. Another large cause of mortality worldwide is stroke and cardiovascular disease (CVD), which are diseases of the blood vessels. About 17.7 million people died globally due to CVD in 2015 [3]. Of these deaths, about 7.4 million were due to heart disease and 6.7 million were due to stroke [4]. Years of research have demonstrated a direct link between diabetes and stroke/CVD. An estimated 1.6 million stroke/CVD deaths were directly attributed to diabetes [4]. Observational studies suggest that diabetes patients have a higher risk of coronary artery disease (CAD) compared to their non-diabetes counterparts [5].

Atherosclerosis is an inflammatory disease of blood vessels, a major carotid or coronary artery disorder that promotes lipid metabolism and the development of lesions, so-called plaque [6]. Endothelial cell dysfunction is an initial sign of plaque build-up and occurs due to various risk factors such as high low-density lipoprotein (LDL) cholesterol, smoking, alcohol and tobacco [7–9]. Once entered into the intima, monocytes are transformed into macrophages which take up the oxidized cholesterol and become lipid-laden foam cells. Foam cells, also known as fatty streaks, are an important marker for the atheroma region in atherosclerosis [10, 11]. The atherosclerosis process due to LDL formation is depicted in figure 8.1(a).

In diabetes, the body fails to maintain homeostasis. Diabetes patients are affected by high blood sugar or hyperglycemia. DM not only increases the risk of macrovascular diseases but also causes an increased risk of atherosclerosis and CVD, leading to serious life-threatening problems [12]. Diabetes patients have high blood glucose and endothelial cells take up this glucose. As a result of increased glucose inside the endothelium, the cells produce reactive oxygen species (ROS). The

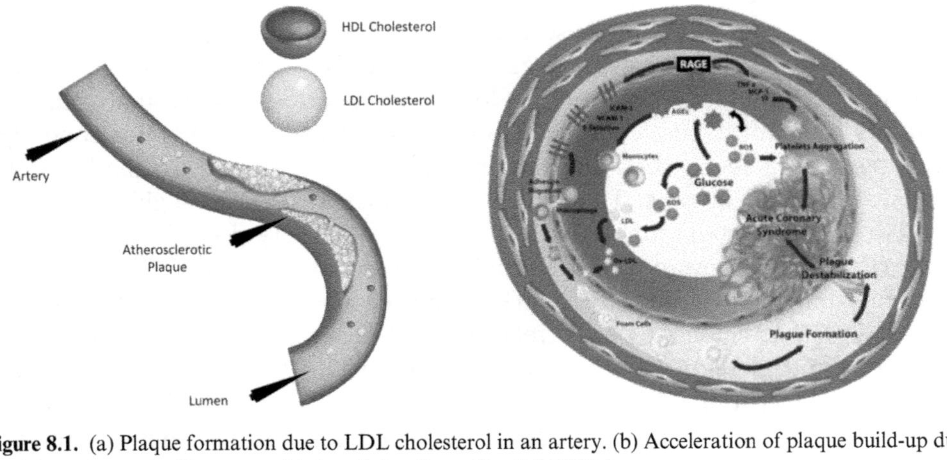

Figure 8.1. (a) Plaque formation due to LDL cholesterol in an artery. (b) Acceleration of plaque build-up due to diabetes in a cell. (Courtesy of AtheroPoint, Roseville, CA, USA.)

increase in ROS leads to the formation of advanced glycolated product (AGP) and protein kinase C (PKC) which leads to endothelial dysfunction. PKC enables the receptors for monocytes over the endothelial cells that increase the vascular permeability of white blood cells, which further leads to plaque build-up and atherosclerosis. Impaired glucose tolerance is a pre-diabetes state of hyperglycemia. It is associated with insulin resistance and CVD. Figure 8.1(b) represents the effect of diabetes on atherosclerosis.

Carotid intima–media thickness (cIMT) has been used as a CAD risk phenotype and can be measured either manually or semi-automatically in the region of interest (ROI) near the far wall of the carotid artery placed by a human operator [13–15]. These methods do not measure the cIMT in a completely automated way. A completely automated method should determine the far wall of the common carotid artery (CCA) all along the longitudinal ultrasound scan [16] without human assistance. Further, the system should detect the bulb edge points as reference markers [17, 18] in the presence of the carotid bulb (or sinuous). Previous techniques also lacked comprehensive inter- and intra-operator variability analysis, an important indicator for the stability of the measurement tool [19]. Finally, previous studies [13–15] did not focus on the association between automated carotid artery disease phenotype measurements and diabetes. The current study focuses on understanding the relationship between six different types of carotid artery disease risk phenotypes and HbA1c in diabetes patients. Previously, Suri and his team showed an association between automated cIMT and HbA1c [19]. The relationship was observed using Pearson correlation coefficients. Olt *et al* [20] also showed the relationship between cIMT and HbA1c using Student's *t*-test and ROC curve analysis, where HbA1c values were not different between the groups ($p > 0.05$). The authors demonstrated that HbA1c values were not associated with subclinical atherosclerosis.

Wall thickness variations are also related to the plaque risk. This is because plaque develops bi-directionally in the walls of the artery (i.e. towards the lumen and towards the adventitia). The wall variations are thus prevalent when measuring cIMT, and hence the variability in cIMT can be computed as a risk index. This phenotype was recently invented by Suri *et al* [21, 22] and termed as cIMTV. In these studies [21, 22], the authors showed that cIMT and cIMTV values were similar between readers (or observers) while using automated AtheroEdge™ software (AtheroPoint™, Roseville, CA, USA) during the study of symptomatic and asymptomatic patients in an Italian cohort. Suri *et al* [23] further studied the association between cIMT and cIMTV using the ankle–brachial index (ABI) for 500 Japanese coronary artery disease patients, where cIMTV and ABI were observed to be highly correlated (based on *p*-values), whereas cIMT was not associated with ABI. Suri *et al* [24] further showed that cIMTV was significantly correlated with leukoaraiosis volume. From the above studies, cIMTV seemed to be a better phenotype compared to cIMT.

The nature of plaque growth in the walls of the arteries has no fixed pattern. Not only is it multi-focal, but due to the complex phenomena of genetic atherosclerosis infiltrated by diabetes metabolic changes, the growth of atherosclerosis disease has a regional spread rather than a plain distance measurement at few locations. Hence, cIMT does not cover all aspects of risk assessment. Rather, the area of the plaque in the far walls of the carotid artery also plays a role in the risk of coronary or carotid artery disease. Recently Spence *et al* [25] studied 4512 patients and showed the effect of LDL-C on TPA. It was observed that neither LDL-C (at follow-up) nor change in LDL-C level (from baseline to follow-up) showed any significant association with a change in plaque burden. Rundek *et al* [26] also used this phenotype to establish the relationship between homocysteine (tHcy) and TPA. A similar observation by Adams *et al* [27] showed the relationship between TPA and CAD. The main challenge with the TPA phenotypes is the involvement of a human operator in its computation, which brings inter- and inter-observer variability. In the current study, TPA is computed automatically and captures the envelope between the lumen–intima and media–adventitia, and this follows the morphology of the plaque along the carotid long axis. We therefore call this TPA morphologic TPA (mTPA). It is important to understand the following: (i) how our automated mTPA and CRS are related; (ii) how the mTPA performance compared to cIMT; (iii) whether mTPA is an important risk indicator; and (iv) whether CRS can be used as a combined risk score representing all the image-based phenotypes. In the current study, a logistic regression tool is applied to understand the relationship between the six phenotypes (cIMT (ave.), cIMT (max.), cIMT (min), cIMTV, mTPA and composite risk score (CRS)) against HbA1c.

As seen in the literature, mTPA has been recommended as a risk phenotype [25–28] but is not automatically computed. Further, there are no studies exploring the association between mTPA and HbA1c. Since both phenotypes, cIMT and mTPA, affect CVD and stroke risk, there is a clear need to understand how these automated phenotype measurements are related to HbA1c in diabetes patients. The main block diagram of the proposed system is shown in figure 8.2. In the first step, two operators (novice and experienced) manually calibrate all the US scans using the AtheroEdge™ system. AtheroEdge™ can detect the far wall interfaces (lumen–intima and media–adventitia) and now six phenotypes (cIMT (ave.), cIMT (max.), cIMT (min.), cIMTV, mTPA and CRS) are automatically computed. In the second step, a logistic regression-based strategy is used to determine the relationship between the six automatically computed phenotypes and HbA1c. Finally, the study analyzes the inter-operator reproducibility of these six phenotypes using AtheroEdge™ software with one novice and one experienced operator. To justify the results of logistic regression, four different types of statistical tests are also performed.

The following are the novelties proposed in our study:

(i) *Six automated measurements*: Measurement of six phenotypes (for each LCCA, RCCA and MCCA) using AtheroEdge™—a standard pharmaceutical batch tool for ultrasound carotid wall analysis.

(ii) *Baseline characterization*: Understanding the significance levels of these six phenotypes with respect to HbA1c.

(iii) *Inter-relationship between phenotypes*: Comprehensive statistical analysis in understanding inter-relationships between the six phenotypes.

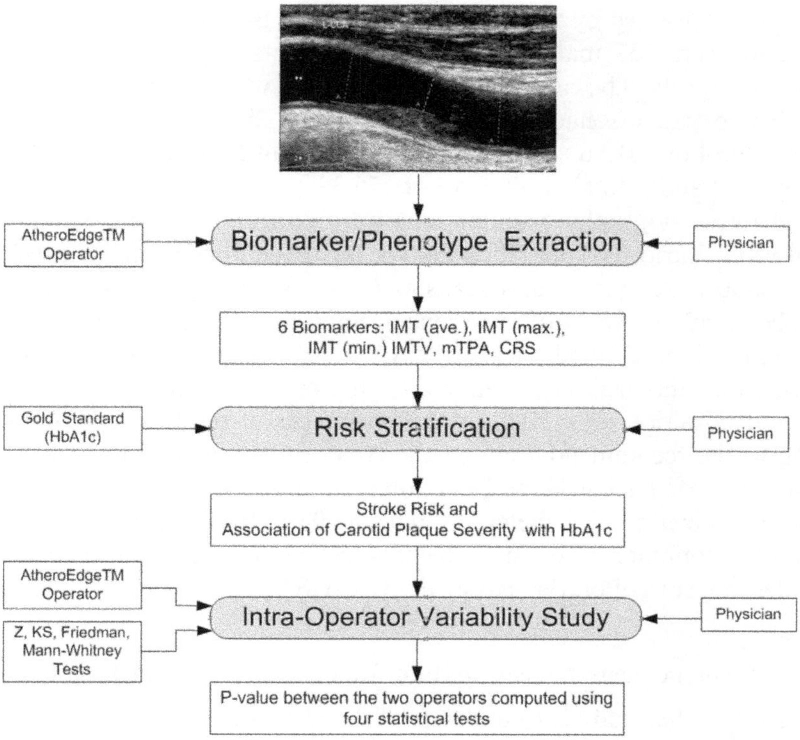

Figure 8.2. The overall system linking carotid disease phenotypes to diabetes (HbA1c). The system consists of three stages: six phenotype measurements, risk stratification and inter-operator variability analysis.

(iv) *Risk assessment using an odds ratio*: Application of logistic regression to identify the order-of-risk in six phenotypes based on an odds ratio (OR).
(v) *Reproducibility*: Inter-operator reproducibility analysis between two operators (with different level of experience: operator 1: a novice; operator 2: experienced).
(vi) *Benchmarking against previous methods*: Benchmarking the proposed model against previously developed methods which link carotid artery disease with diabetes.

The layout of this chapter is as follows. Section 8.2 presents the material and methods. Results are shown in section 8.3. Inter-operator variability and statistical tests are depicted in section 8.4. Benchmarking and discussion are presented in section 8.5. The chapter concludes in section 8.6.

8.2 Materials and methods

8.2.1 Patient demographics

One hundred and ninety-nine (199) patients' left and right common carotid arteries B-mode ultrasound images (398 ultrasound scans) were retrospectively analyzed (ethics approval granted by the Institutional Review Board) from Toho University, Japan. There were 157 male and 42 female patients with mean ages of 67 and 75 years, respectively. The carotid ultrasound images were obtained using a Toshiba scanner. These patients had a mean HbA1c of 6.28 ± 1.1 (mg dl^{-1}), a mean LDL cholesterol of 101.16 ∓ 31.79 (mg dl^{-1}), a mean HDL cholesterol of 50.59 ∓ 15.11 (mg dl^{-1}) and a total cholesterol of 175.39 ∓ 38.26 (mg dl^{-1}).

Carotid ultrasonography examinations were performed with a scanner (Aplio XV, Aplio XG, Xario, Toshiba, Inc., Tokyo, Japan) equipped with a 7.5 MHz linear array transducer. All ultrasound scans were performed by the same experienced sonographer (with 15 years of experience). Subjects were examined in the supine position with the head tilted backward. After the carotid arteries were located by transverse scans, the probe was rotated 90° to acquire a longitudinal image of the anterior and posterior wall. High-resolution images of the far wall were acquired according to the recommendations of the American Society of Echocardiography Carotid Intima–Media Thickness Task Force. In this database, the mean conversion factor of the pixel to millimeters is 0.05 ± 0.01. Two operators (novice and experienced) compute the six phenotypes using the automated system AtheroEdge™ (AtheroPoint™, Roseville, CA, USA).

8.2.2 Six phenotype measurements derived from carotid ultrasound scans

The objective of this study is to understand the role of carotid imaging phenotypes and their association with HbA1c in a Japanese diabetes cohort. As shown in figure 8.2, we collect the six phenotypes from the ultrasound carotid scans using AtheroEdge™ (courtesy of AtheroPoint™, Roseville, CA, USA). Our previous study used part of AtheroEdge™ for epidemiological study [29, 30].

8.2.2.1 Automated carotid imaging phenotypes for the far wall

AtheroEdge™ allows for the automatic computation of the phenotypes from the carotid artery ultrasound scan all along the common carotid artery. The AtheroEdge™ system automatically identifies the far wall [31] of the carotid artery in the ultrasound scans by using an edge detection system [32–34]. The ROI is automatically computed in the carotid artery without the intervention of an operator (a sonographer, sonologist or radiologist). The ROI spans from the carotid bulb edge (distal end) to the proximal end. Since carotid scans are longitudinal in nature, the whole ROI spans from the left edge of the carotid scan (the left edge of the image) to the right edge of the carotid ultrasound scan (the right edge of the image). The AtheroEdge™ system then automatically computes the lumen–intima (LI) and media–adventitia (MA) interfaces in the far wall ROI. The system is fully automated and the operator only has to click a few buttons to measure the six phenotypes. All automated measurements are calibrated to millimeters. cIMT has three kinds of measurements: cIMT (ave.), cIMT (max.) and cIMT (min.). These phenotypes are automatically computed and validated against the gold standard [15, 19] and computed tomography (CT) [34].

8.2.2.2 Automated wall variability

Since the plaque can grow either towards the lumen (i.e. above the baseline) or away from the lumen (protruding towards the adventitia), atherosclerosis disease in the walls is bidirectional in nature [30, 34, 35]. Thus, the special directional wall variability index was developed under the umbrella of AtheroEdge™, so-called cIMTV, and is mathematically expressed as

$$\mathrm{cIMTV} = \sqrt{\frac{\sigma_{\mathrm{LI}}^2 + \sigma_{\mathrm{MA}}^2}{N_{\mathrm{LI}} + N_{\mathrm{MA}}}},$$

where σ_{LI} and σ_{MA} are the standard deviations of the distance of the vertices of LI on segments of MA, and the vertices of MA on segments of LI; N_{LI} and N_{MA} are the number of vertices in the LI and MA borders, respectively. Note that this index has recently been shown to have strong clinical relationships to atherosclerosis disease [22]. The cIMT and cIMTV phenotypes are automatically measured using a standardized distance measurement method using the polyline distance method (PDM) [36].

8.2.2.3 Automated morphologic total plaque area and composite risk score

As discussed earlier, several studies have tried to compute TPA using manual methods, such as Spence *et al* [25, 28]. In these studies, the TPA phenotype envelopes the complete plaque region in the far wall of the carotid artery which shows the biological activity of cell multiplication for the formation of plaque components such as fibrosis, fibro-fatty, calcium and necrosis [37]. Thus, both soft and hard plaque components are part of this envelope. The challenge with the above phenotype is the involvement of the human operator.

In our current study, mTPA was computed automatically and captures the envelope between the lumen–intima and media–adventitia, which follows the entire morphology of the plaque. AtheroEdge™ automatically computes the mTPA (as the area in mm^2) by calculating the area between the LI and MA borders. In this study, the operators perform manual calibration on all US scans. Technically, with a slight calibration change, the proposed AtheroEdge™ software remains insensitive in measuring the six phenotypes [30]. We ensured that the borders are smoothed, interpolated and have common support. Common support allows the envelope to be of the same length for the LI and MA interfaces. The area is computed by summing all pixels in this region and finally calibrated with the resolution factor [11, 25–27]. mTPA not only includes the regions between LI and MA but also any plaque above the baseline. Thus, mTPA measures both the IMT region and the focal thickening region. CRS is computed by giving a score between 1 and 5 for all five phenotypes: cIMT (ave.), cIMT (max.), cIMT (min.), cIMTV and mTPA. These scores are then scaled, summed and then converted to a percentage, the so-called composite risk score.

8.2.2.4 Stratification of diabetes patients into low and high risk

The hemoglobin test indicates the average level of blood sugar over the past 2–3 months. It is also called HbA1c, the glycated hemoglobin test and glycohemoglobin.

Several studies have been conducted where diabetes has been classed into two or three classes that reflect the risk. Cavagnolli *et al* [38] divided HbA1c into three groups: normal (HbA1c < 5.70 mg dl^{-1}), pre-diabetes (5.70 mg dl^{-1} ⩽ HbA1c < 6.40 mg dl^{-1}) and diabetes (HbA1c ⩾ 6.40 mg dl^{-1}). Wang *et al* [39] classified HbA1c into two categories, namely: low-risk patients (HbA1c < 6.5 mg dl^{-1}) and high-risk patients (HbA1c ⩾ 6.50 mg dl^{-1}). Since our database is small, we follow the guidelines of diabetes stratification into two classes: low-risk patients and high-risk patients. There are 150 low-risk patients and 49 high-risk patients. Using this criterion, we compute the demographic and clinic characteristics of the diabetes patients and use logistic regression (LR) to show the effect of the six phenotypes on HbA1c.

8.2.3 Statistical analysis

The baseline characteristics of the study population are presented as mean ∓ SD for continuous variables and numbers (percentages) for the categorical variables, respectively. Differences in variables between low- and high-risk patients are analyzed using an independent t-test for continuous variables and a χ-squared test for categorical variables. LR was used to determine the high-risk phenotype which was responsible for Hb1Ac based on the OR by adjusting the available risk factors of Hb1Ac. A receiver operating curve (ROC) and area under the curve (AUC) were also performed to evaluate the deterministic strength of the phenotypes. Stata version 14 and Ri86 3.4.2 was used for this analysis. All of the tests were two-tailed and p-values < 0.05 and < 0.001 were considered significant.

8.3 Results

8.3.1 Demographics and clinical characteristics of the patients

The patient demographics and study characteristics are shown in table 8.1. A total of 49 of the pool of 199 subjects (25%) are diabetes patients. The study includes 77% male and 40% smokers. There were 36% male diabetes patients with an average age of 69.94 ∓ 10.83 years. Table 8.1 indicates that fasting blood sugar (FBS) is highly associated with HbA1c ($p < 0.001$). The patient's age, sex, smoking, hypertension (HT), family history (FH), hemodialysis (HD), total cholesterol (TC), low/high-density lipoprotein (L/HDL) and triGly (TG) are not associated with HbA1c.

To make the results consistent, we have used two operator readings from the novice operator (operator 1) and experienced operator (operator 2). Table 8.2 shows the mean comparisons of operator 1 and operator 2, which confirm insignificant variation between the operator 1 and operator 2 readings.

8.3.2 Visual display of six phenotypes using AtheroEdge™

This section shows the results of the six phenotypes measured on the entire cohort of 398 images. Two sets of images, low-risk diabetes patients and high-risk diabetes patients, are shown in figures 8.3 and 8.4, respectively. Figure 8.3 shows three rows corresponding to low-risk patients. The left panels show the original image while the

Table 8.1. Demographics and clinical characteristics of diabetes patients.

Phenotypes	Total patients	Low-risk patients (HbA1c < 6.5 mg dl^{-1})	High-risk patients (HbA1c ⩾ 6.5 mg dl^{-1})	*p*-value
Total	*n* = 199	*n* = 150	*n* = 49	—
Age (years)	68.96 ∓ 10.98	68.47 ∓ 10.98	69.94 ∓ 10.83	0.417
Male, *n* (%)	157 (76.96)	117 (78.00)	36 (73.47)	0.514
Smokers, *n* (%)	82 (40.19)	59 (39.33)	21 (42.86)	0.662
HT, *n* (%)	147 (72.06)	103 (68.67)	39 (79.59)	0.142
FH, *n* (%)	24 (11.76)	18 (12.00)	5 (10.20)	0.733
HD, *n* (%)	22 (10.78)	19 (12.67)	2 (4.08)	0.089
TC	175.39 ∓ 38.26	176.53 ∓ 39.48	172.60 ∓ 33.78	0.532
LDL	101.16 ∓ 31.79	101.15 ∓ 32.62	100.37 ∓ 29.43	0.878
HDL	50.59 ∓ 15.11	50.95 ∓ 15.80	49.95 ∓ 13.22	0.692
TG	125.87 ∓ 90.80	128.40 ∓ 96.43	114.92 ∓ 60.43	0.411
FBS*	121.83 ∓ 35.35	112.14 ∓ 27.33	151.51 ∓ 40.53	$p < 0.001$

Categorical and continuous variables are presented as percentages and mean ∓ SD; HT: hypertension; FH: family history: DM, diabetes mellitus; HD: hemodialysis; LDL/HDL: low/high-density lipoprotein; TG: TriGly; FBS: fasting blood sugar. *p*-values are obtained using a χ-squared test for categorical variables and *t*-test for continuous variables.

Table 8.2. Comparison of the means of the six phenotypes between operator 1 and operator 2.

Carotid risk phenotypes	Operator 1 (novice) Mean ∓SD	Operator 2 (experienced) Mean ∓SD
cIMT (ave.) (mm)	0.95 ∓ 0.28	1.15 ∓ 0.39
cIMT (max.) (mm)	1.31 ∓ 0.51	1.38 ∓ 0.52
cIMT (min.) (mm)	0.67 ∓ 0.14	0.65 ∓ 0.18
cIMTV (mm)	0.16 ∓ 0.13	0.18 ∓ 0.12
mTPA (mm^2)	31.27 ∓ 9.75	30.48 ∓ 9.90
CRS (%)	28.94 ∓ 14.74	30.10 ∓ 15.23

right panels show the image processed using AtheroEdge™ 2.0. In row #1 (patient 12L; HbA1c = 5.4 mg dl^{-1}), the measurements of the six imaging phenotypes using AtheroEdge™ 2.0 have the following values: cIMT (ave.) = 0.53 mm; cIMT (max.) = 0.72 mm; cIMT (min.) = 0.43 mm; cIMTV = 0.05 mm; mTPA = 17.75 mm^2; and CRS = 11%. In row #2 (patient 32R; HbA1c = 5.4 mg dl^{-1}), the measurements of the six imaging phenotypes using AtheroEdge™ 2.0 have the following values: cIMT (ave.) = 0.94 mm; cIMT (max.) = 1.14 mm; cIMT (min.) = 0.70 mm; cIMTV = 0.07 mm; mTPA = 32.13 mm^2; and CRS = 26%. In row #3 (patient 112R; HbA1c = 5.5 mg dl^{-1}), the measurements of the six imaging phenotypes have the following values: cIMT (ave.) = 0.73 mm; cIMT (max.) = 0.87; mm cIMT (min.) = 0.53 mm; cIMTV = 0.07 mm; mTPA = 22.23 mm^2; and CRS = 13%. Three set of examples are given for the high-risk diabetes patients. The left panels show the

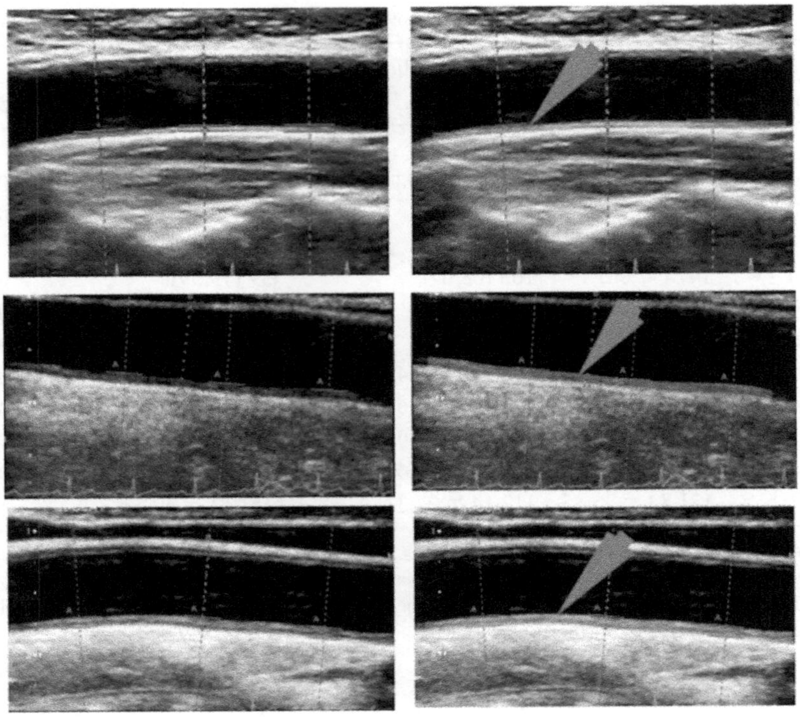

Figure 8.3. Low-risk diabetes patients. Row 1: patient 12L; HbA1c = 5.4 mg dl^{-1}; left: original image; right: image processed using AtheroEdge™ 2.0; measurements of the six imaging phenotypes: cIMT (ave.) = 0.53 mm; cIMT (max.) = 0.72 mm; cIMT (min.) = 0.43 mm; cIMTV = 0.05 mm; mTPA = 17.75 mm^2; and CRS = 11%. Row 2: patient 32R; HbA1c = 5.4 mg dl^{-1}; left: original image; right: image processed using AtheroEdge™ 2.0; measurements of the six imaging phenotypes: cIMT (ave.) = 0.94 mm; cIMT (max.) = 1.14 mm; cIMT (min.) = 0.70 mm; cIMTV = 0.07 mm; mTPA = 32.13 mm^2; and CRS = 26%. Row 3: patient 112R; HbA1c = 5.5 mg dl^{-1}; left: original image; right: image processed using AtheroEdge™ 2.0; measurements of the six imaging phenotypes: cIMT (ave.) = 0.73 mm; cIMT (max.) = 0.87; mm cIMT (min.) = 0.53 mm; cIMTV = 0.07 mm; mTPA = 22.23 mm^2; and CRS = 13%.

original image while the right panels show the image processed using AtheroEdge™ 2.0. In row 1 (patient 63R; HbA1c = 9.6 mg dl^{-1}), the measurements of the six imaging phenotypes have the following values: cIMT (ave.) = 1.89 mm; cIMT (max.) = 3.29 mm; cIMT (min.) = 0.74 mm; cIMTV = 0.82 mm; mTPA = 59.26 mm^2; and CRS = 70%. In row 2 (patient 128R; HbA1c = 9.3 mg dl^{-1}), the measurements of the six imaging phenotypes have the following values: cIMT (ave.) = 1.23 mm; cIMT (max.) = 2.49 mm; cIMT (min.) = 0.65 mm; cIMTV = 0.68 mm; mTPA = 38.51 mm^2; and CRS = 55%. In row 3 (patient 168R; HbA1c = 10.8 mg dl^{-1}), the measurements of the six imaging phenotypes have the following values: cIMT (ave.) = 1.28 mm; cIMT (max.) = 2.49 mm; cIMT (min.) = 0.68 mm; cIMTV = 0.54 mm; mTPA = 47.30 mm^2; and CRS = 55%.

Two AtheroEdge™ (AE) operators measured all six phenotypes for all 407 images. The corresponding coefficient of correlation (CC) along with the *p*-values are shown in table 8.3. The RCCA has stronger correlations between the two

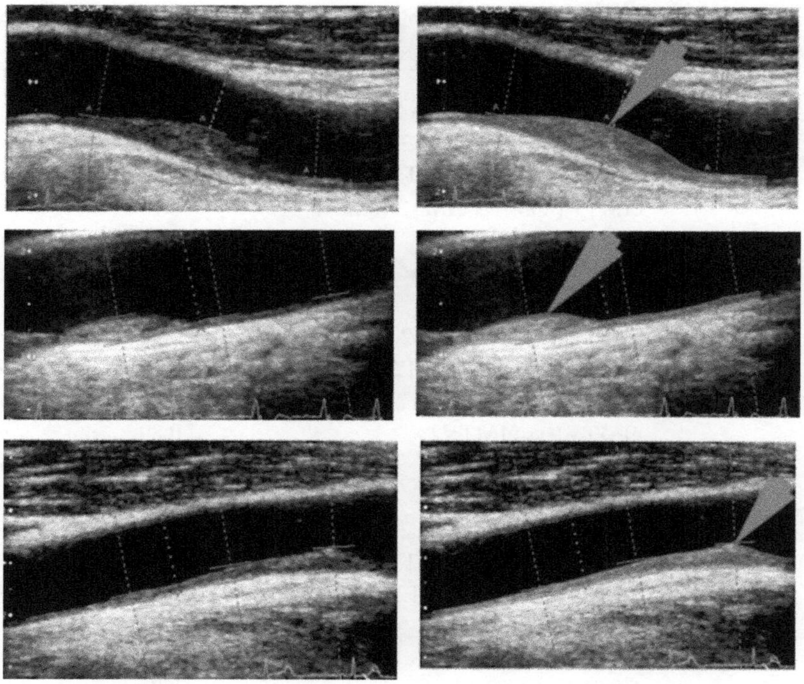

Figure 8.4. High-risk diabetes patients. Row 1: patient 63R; HbA1c = 9.6 mg dl^{-1}; left: original image; right: image processed using AtheroEdge™ 2.0; measurements of the six imaging phenotypes: cIMT (ave.) = 1.89 mm; cIMT (max.) = 3.29 mm; cIMT (min.) = 0.74 mm; cIMTV = 0.82 mm; mTPA = 59.26 mm^2; and CRS = 70%. Row 2: patient 128R; HbA1c = 9.3 mg dl^{-1}; left: original image; right: image processed using AtheroEdge™ 2.0; measurements of the six imaging phenotypes: cIMT (ave.) = 1.23 mm; cIMT (max.) = 2.49 mm; cIMT (min.) = 0.65 mm; cIMTV = 0.68 mm; mTPA = 38.51 mm^2; and CRS = 55%. Row 3: patient 168R; HbA1c = 10.8 mg dl^{-1}; left: original image; right: image processed using AtheroEdge™ 2.0; measurements of the six imaging phenotypes: cIMT(ave.) = 1.28 mm; cIMT(max.) = 2.49 mm; cIMT(min.) = 0.68 mm; cIMTV = 0.54 mm; mTPA = 47.30 mm^2; and CRS = 55%.

readings, unlike the LCCA. Almost all the readings show CC values above 0.90, except for cIMT (min.), which is in the range of 0.90–0.97. Note that mTPA is strongly related to the two operator readings. The same holds for the CRS.

8.3.3 Correlation between operators and correlation between cIMT and mTPA for the left, right, and mean of the left and right carotid arteries

Figures 8.5 and 8.6 show the regression plots of the cross-correlations between the six automated phenotypes. The corresponding tables are shown in appendix B, tables B1, B2 and B3. mTPA is strongly related to cIMT (ave.), cIMT (max.) and cIMT (min.). For MCCA, the CC between mTPA and the three carotid cIMT readings were 0.97 ($p < 0.001$), 0.90 ($p < 0.001$) and 0.78 ($p < 0.001$), respectively, while the CC between CRS and the corresponding three cIMT readings were 0.95 ($p < 0.001$), 0.94 ($p < 0.001$) and 0.72 ($p < 0.001$), respectively. cIMTV was correlated to the rest of the phenotypes (appendix B, table B3). For MCCA, the

Table 8.3. Correlation between the six automated phenotype measurements derived from AtheroEdge™ (AE) using operator 1 and operator 2.

Artery types	AE operator 1 vs AE operator 2	CC	*p*-value
LCCA	cIMT1 (ave.) vs cIMT2 (ave.)	0.97	$p < 0.001$
	cIMT1 (max.) vs cIMT2 (max.)	0.96	$p < 0.001$
	cIMT1 (min.) vs cIMT2 (min.)	0.78	$p < 0.001$
	cIMTV1 vs cIMTV2	0.95	$p < 0.001$
	mTPA1 vs mTPA2	0.95	$p < 0.001$
	CRS1 vs CRS2	0.95	$p < 0.001$
RCCA	cIMT1 (ave.) vs cIMT2 (ave.)	0.94	$p < 0.001$
	cIMT1 (max.) vs cIMT2 (max.)	0.89	$p < 0.001$
	cIMT1 (min.) vs cIMT2 (min.)	0.79	$p < 0.001$
	cIMTV1 vs cIMTV2	0.89	$p < 0.001$
	mTPA1 vs mTPA2	0.92	$p < 0.001$
	CRS1 vs CRS2	0.92	$p < 0.001$
MCCA	cIMT1 (ave.) vs cIMT2 (ave.)	0.90	$p < 0.001$
	cIMT1 (max.) vs cIMT2 (max.)	0.95	$p < 0.001$
	cIMT1 (min.) vs cIMT2 (min.)	0.80	$p < 0.001$
	cIMTV1 vs cIMTV2	0.93	$p < 0.001$
	mTPA1 vs mTPA2	0.95	$p < 0.001$
	CRS1 vs CRS2	0.95	$p < 0.001$

* Index 1 represents AtheroEdge™ measurements using operator 1, while index 2 represents AtheroEdge™ measurements using operator 2; CC: correlation coefficients

values were 0.83, 0.91, 0.51, 0.78 and 0.86 (all $p < 0.001$), clearly demonstrating that cIMTV is strong a phenotype in itself.

8.3.4 Logistic regression for the effect of the six phenotypes on HbA1c for the operator of AtheroEdge™

Tables 8.4 and 8.5 show the logistic regression results for operator 1 and operator 2. The first column represents the artery types, the second column represents the phenotypes, the third and fourth columns represent the odds ratio (OR) and probability value (*p*-value), and the fifth and sixth columns represent the 95% CI for OR and AUC of the ROC, respectively. Table 8.4 indicates that cIMT (ave., max. and min.), mTPA and CRS were highly associated with (higher levels of) HbA1c (low risk versus high risk) after adjusting FBS for LCCA, RCCA and MCCA based on the *p*-values, because their *p*-values almost closed to zero. A box-plot of the six biomarkers for LCCA, RCCA and MCCA between AtheroEdge™ operator 1 and operator 2 is shown in appendix A, figures A1–A3. Spence [25] and Dong [40] showed that TPA was related to stroke and cardiovascular risk, which was consistent with our readings. From the tables, we note that cIMTV is not a significant phenotype for diabetes because the *p*-value is greater than 0.05. Moreover, the CC between cIMTV and HbA1c is very low and the *p*-value is

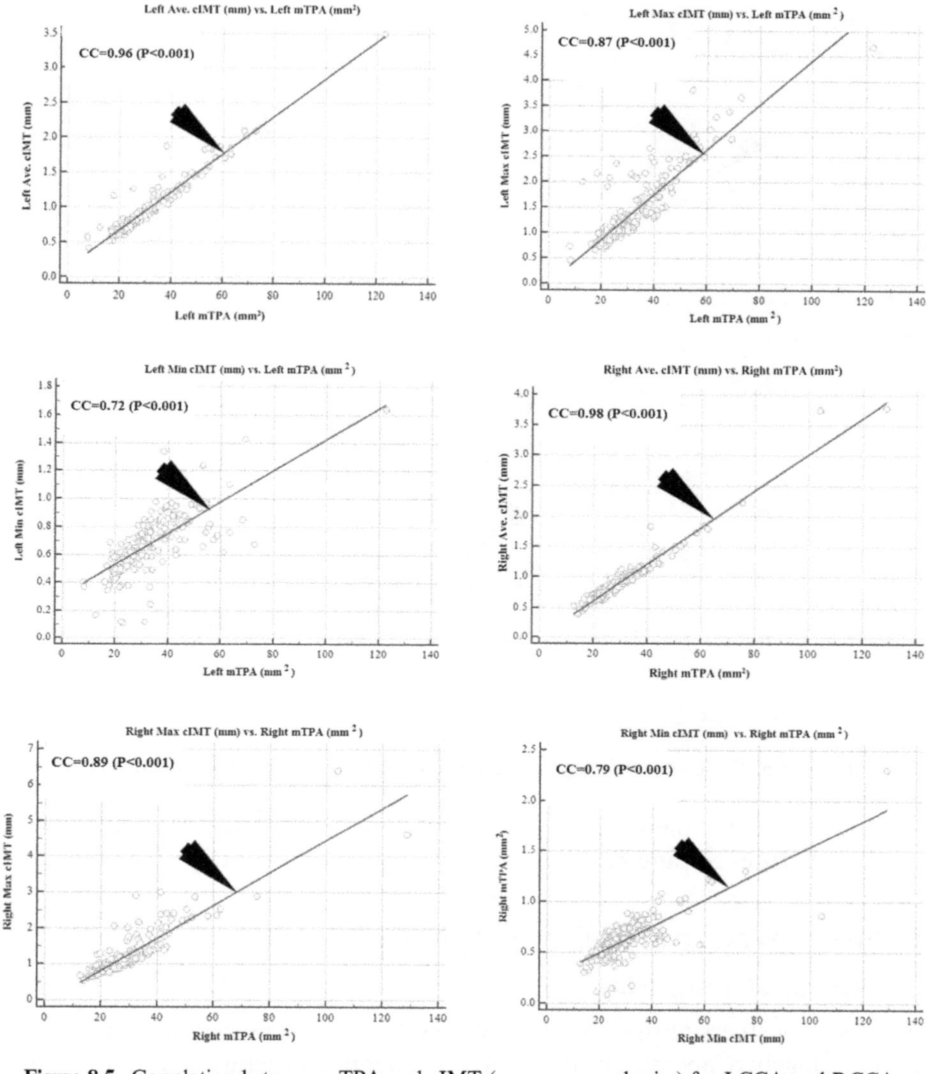

Figure 8.5. Correlation between mTPA and cIMT (ave., max. and min.) for LCCA and RCCA.

greater than 0.05 (see table 8.6). That means cIMTV is not highly related to HbA1c. As a result, the OR is high.

We order the phenotypes for LCCA based on their OR as cIMT (ave.) > cIMT (max.) > cIMT (min.) > mTPA > CRS. The OR of cIMT (ave.) was 1.30 which indicates that the diabetes patients have a 1.30 times (OR = 1.30, 95% CI 0.53–3.21) higher risk compared to non-diabetes patients. mTPA and CRS are stronger phenotypes for diabetes because their *p*-value is almost close to zero. The largest AUC value of 0.82 was computed for cIMT (ave.), cIMT (min.), mTPA, CRS and cIMT (max.). For RCCA, we order the phenotypes as cIMT (max.) > cIMT (ave.) > CRS > mTPA > IMT (min.). A similar result was observed for the MCCA.

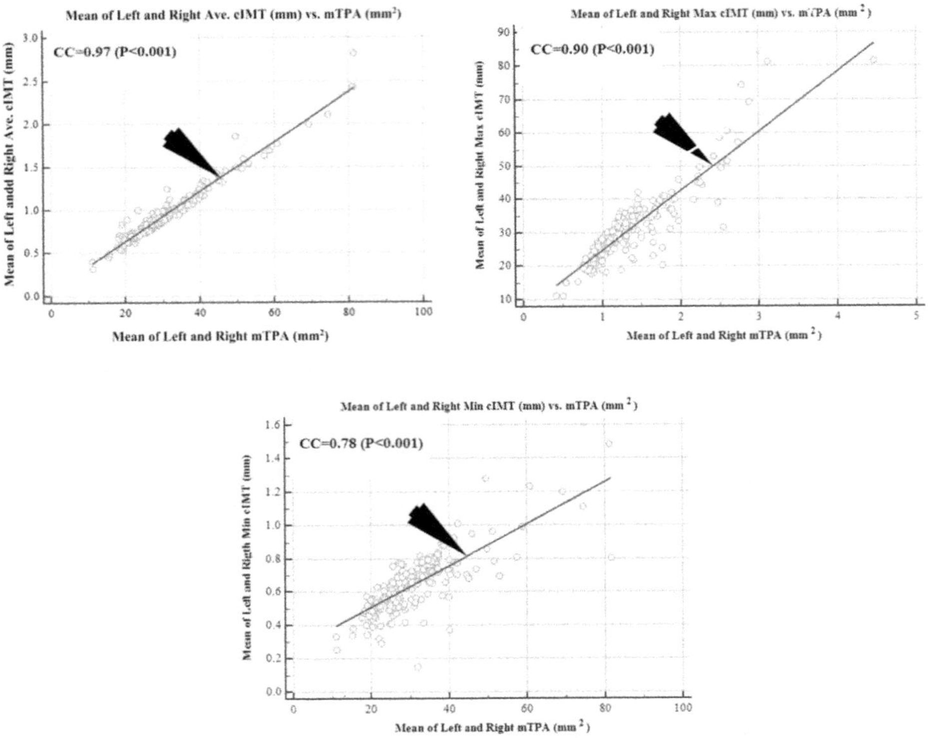

Figure 8.6. Correlation between mTPA and cIMT (ave., max. and min.) for MCCA.

The highest AUC for the CRS was 0.82. Operator 2 also gives the same results as operator 1. Figures 8.7 and 8.8 show the ROC curve for operator 1 and operator 2.

8.4 Inter-operator variability and statistical tests

8.4.1 Inter-operator variability

To test the robustness of the AtheroEdge™ system, we measured the AtheroEdge™ readings using two operators: first a novice operator (operator 1) and second an experienced operator (operator 2). Tables 8.4 and 8.5 show the logistic regression results to determine the effect of the six phenotypes on HbA1c.

8.4.2 Statistical tests

To justify the results of logistic regression, we used four kinds of statistical tests, namely the *z*-test, two sample Kolmogorov–Smirnov test (KS test), Friedman test and Mann–Whitney test, on six phenotype readings using operator 1 and operator 2. According to the power study with the standard normal distribution reflecting a confidence level of 95%, and a desired margin of error less than 5%, our data analysis shows that the sample size needed was 332, while we used over 398 samples. Therefore, this demonstrates that the number of samples was enough for the statistical tests. The significance of the carotid risk phenotype is analyzed on the

Table 8.4. Logistic regression for the effect of the six phenotypes on HbA1c for operator 1.

				95% CI for OR		
Artery type	Phenotype	OR	p-value	Lower	Upper	AUC
LCCA	cIMT (ave.)	1.30	$p < 0.001$	0.53	3.21	0.82
	cIMT (max.)	1.26	$p < 0.001$	0.74	2.15	0.82
	cIMT (min.)	1.20	$p < 0.001$	0.24	6.01	0.81
	cIMTV	2.35	0.543	0.22	25.46	0.82
	mTPA	1.02	$p < 0.001$	0.99	1.05	0.82
	CRS	1.01	$p < 0.001$	0.99	1.03	0.82
RCCA	cIMT(ave.)	1.17	$p < 0.001$	0.33	4.12	0.81
	cIMT (max.)	1.25	$p < 0.001$	0.62	2.50	0.81
	cIMT (min.)	0.62	$p < 0.001$	0.08	4.97	0.81
	cIMTV	2.92	0.242	0.20	43.11	0.81
	mTPA	1.00	$p < 0.001$	0.96	1.04	0.81
	CRS	1.01	$p < 0.001$	0.98	1.03	0.81
MCCA	cIMT (ave.)	1.43	$p < 0.001$	0.54	3.76	0.82
	cIMT (max.)	1.40	$p < 0.001$	0.67	2.93	0.82
	cIMT (min.)	0.90	$p < 0.001$	0.10	7.82	0.81
	cIMTV	4.64	0.524	0.18	116.79	0.82
	mTPA	1.02	$p < 0.001$	0.98	1.06	0.82
	CRS	1.01	$p < 0.001$	0.99	1.04	0.82

cIMT (ave.): carotid intima–media thickness (average); cIMT (max.): carotid intima–media thickness (maximum); cIMT (min.): carotid intima–media thickness (minimum); cIMTV: carotid intima–media thickness variability; mTPA: morphology total plaque area; CRS: composite risk score; OR: odds ratio; CI: confidence interval; AUC: area under the curve. The FBS (a confounding factor) is adjusted in the LR model.

basis of p-values which are less than the cutoff of 0.05. Table 8.7 shows the z-test results using the two different operators.

8.4.2.1 z-tests

For novice operator 1, except for cIMT (min.), all the remaining five phenotypes (cIMT (ave.), cIMT (max.), cIMTV, mTPA and CRS) were statistically significant for LCCA, RCCA and MCCA, while for experienced operator 2, except for cIMT (min.), the remaining five phenotypes (cIMT (ave.), cIMT (max.), cIMTV, mTPA and CRS) were statistically significant for LCCA, RCCA and MCCA. Thus, both operators produced similar results (see table 8.7).

8.4.2.2 Kolmogorov–Smirnov test

For novice operator 1, except for cIMT (min.), the remaining five phenotypes (cIMT (ave.), cIMT (max.), cIMTV, mTPA and CRS) were statistically significant for LCCA, RCCA and MCCA, while for experienced operator 2, except for cIMT (min.), the five remaining phenotypes (cIMT (ave.), cIMT (max.), cIMTV, mTPA,

Table 8.5. Logistic regression for the effect of the six phenotypes on HbA1c for operator 2.

				95% CI for OR		
Artery type	Phenotype	OR	*p*-value	Lower	Upper	AUC
LCCA	cIMT (ave.)	1.40	$p < 0.001$	0.53	3.72	0.82
	cIMT (max.)	1.25	$p < 0.001$	0.72	2.18	0.82
	cIMT (min.)	1.15	$p < 0.001$	0.14	9.48	0.81
	cIMTV	2.73	0.381	0.29	25.77	0.82
	mTPA	1.02	$p < 0.001$	0.99	1.05	0.82
	CRS	1.01	$p < 0.001$	0.99	1.03	0.82
RCCA	cIMT(ave.)	1.46	$p < 0.001$	0.36	5.91	0.81
	cIMT (max.)	1.07	$p < 0.001$	0.52	2.24	0.81
	cIMT (min.)	0.98	$p < 0.001$	0.07	13.17	0.81
	cIMTV	2.07	0.583	0.15	27.69	0.81
	mTPA	1.01	$p < 0.001$	0.97	1.05	0.81
	CRS	1.01	$p < 0.001$	0.99	1.04	0.81
MCCA	cIMT (ave.)	1.64	$p < 0.001$	0.42	6.40	0.82
	cIMT (max.)	1.27	$p < 0.001$	0.60	2.69	0.82
	cIMT (min.)	1.11	$p < 0.001$	0.07	16.72	0.81
	cIMTV	3.68	0.388	0.19	70.57	0.82
	mTPA	1.02	$p < 0.001$	0.98	1.06	0.82
	CRS	1.01	$p < 0.001$	0.99	1.04	0.82

cIMT (ave.): carotid intima–media thickness (average); cIMT (max.): carotid intima–media thickness (maximum); cIMT (min.): carotid intima–media thickness (minimum); cIMTV: intima–media thickness variability; mTPA: morphology total plaque area; CRS: composite risk score; OR: odds ratio; CI: confidence interval; AUC: area under the curve. Only the FBS (a confounding factor) is adjusted in the LR model.

and CRS) were statistically significant for LCCA, RCCA and MCCA. Thus, both operators produced similar results (see appendix C, table C1).

8.4.2.3 Friedman test

Only five phenotypes (cIMT (ave.), cIMT (max.), cIMTV, mTPA and CRS) for operator 1 were statistically significant for LCCA. cIMT (min.) was insignificant because of its *p*-value of 0.064 (i.e. $p > 0.05$). For operator 2, all phenotypes were statistically significant for LCCA (appendix C, table C1). For LCCA, using operator 1, cIMT (max.) and cIMTV were statistically significant ($p < 0.05$). For RCCA, using operator 2, cIMT (ave.) and cIMT (min.) were statistically insignificant ($p > 0.05$). For MCCA, using both operators 1 and 2, all phenotypes were statistically significant ($p < 0.05$).

8.4.2.4 Mann–Whitney test

Except for cIMT (min.) for RCCA, all phenotypes were statistically significant for LCCA, RCCA and MCCA, since they had $p < 0.05$ (appendix C, table C1).

Table 8.6. Correlation between six phenotypes and HbA1c for operator 1 and operator 2.

Artery types	Carotid risk phenotypes against HbA1c	Operator 1		Operator 2	
		CC	p-value	CC	p-value
LCCA	cIMT (ave.) vs HbA1c	0.38	$p < 0.001$	0.37	$p < 0.001$
	cIMT (max.) vs HbA1c	0.39	$p < 0.001$	0.39	$p < 0.001$
	cIMT (min.) vs HbA1c	0.35	$p < 0.001$	0.26	$p < 0.001$
	cIMTV vs HbA1c	0.08	0.253	0.12	0.253
	mTPA vs HbA1c	0.35	$p < 0.001$	0.35	$p < 0.001$
	CRS vs HbA1c	0.38	$p < 0.001$	0.37	$p < 0.001$
RCCA	cIMT (ave.) vs HbA1c	0.37	$p < 0.001$	0.37	$p < 0.001$
	cIMT (max.) vs HbA1c	0.34	$p < 0.001$	0.37	$p < 0.001$
	cIMT (min.) vs HbA1c	0.27	$p < 0.001$	0.29	$p < 0.001$
	cIMTV vs HbA1c	0.14	0.115	0.17	0.238
	mTPA vs HbA1c	0.37	$p < 0.001$	0.38	$p < 0.001$
	CRS vs HbA1c	0.38	$p < 0.001$	0.38	$p < 0.001$
MCCA	cIMT (ave.) vs HbA1c	0.44	$p < 0.001$	0.43	$p < 0.001$
	cIMT (max.) vs HbA1c	0.44	$p < 0.001$	0.45	$p < 0.001$
	cIMT (min.) vs HbA1c	0.38	$p < 0.001$	0.32	$p < 0.001$
	cIMTV vs HbA1c	0.13	0.071	0.02	0.300
	mTPA vs HbA1c	0.42	$p < 0.001$	0.42	$p < 0.001$
	CRS vs HbA1c	0.45	$p < 0.001$	0.43	$p < 0.001$

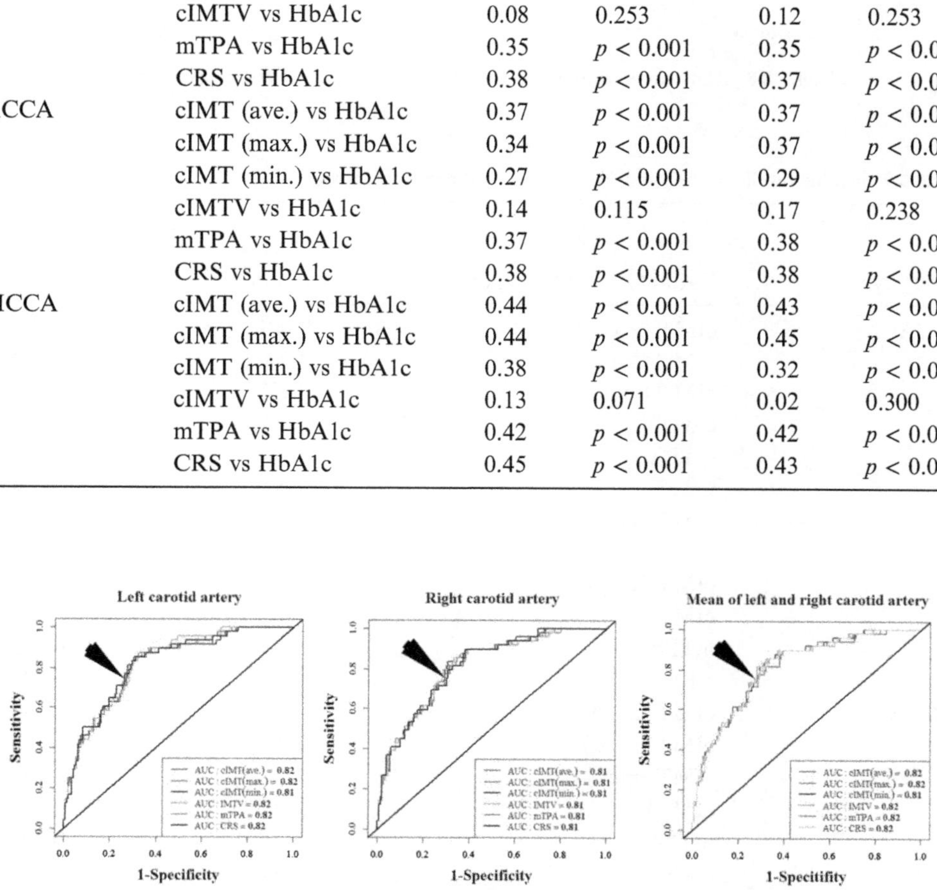

Figure 8.7. ROC curve analysis for LCCA, RCCA and MCCA for operator 1.

In summary, for all the tests, the phenotype cIMT (min.) was insignificant ($p > 0.05$).

8.5 Discussion

Carotid atherosclerosis disease is accelerated by increased HbA1c. As HbA1c increases, the phenotypes show a stronger association. In this study, we have analyzed the association between six automatically measured carotid risk

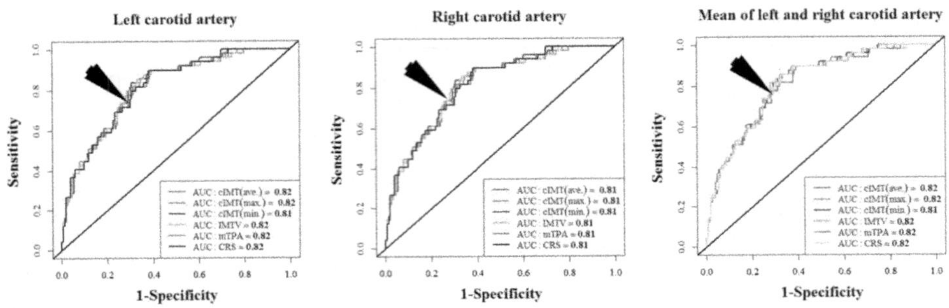

Figure 8.8. ROC curve analysis for LCCA, RCCA and MCCA for operator 2.

Table 8.7. *p*-value results of the *z*-test using the two different operators.

		z-test	
Artery type	Carotid risk phenotypes	Operator 1	Operator 2
LCCA	cIMT (ave.)	0.001	0.001
	cIMT (max.)	0.000	0.000
	cIMT (min.)	0.232	0.137
	cIMTV	0.000	0.000
	mTPA	0.007	0.001
	CRS	0.000	0.000
RCCA	cIMT (ave.)	0.007	0.016
	cIMT (max.)	0.001	0.001
	cIMT (min.)	0.326	0.716
	cIMTV	0.000	0.000
	mTPA	0.008	0.022
	CRS	0.000	0.002
MCCA	cIMT (ave.)	0.000	0.000
	cIMT (max.)	0.000	0.000
	cIMT (min.)	0.142	0.195
	cIMTV	0.000	0.000
	mTPA	0.001	0.001
	CRS	0.000	0.000

phenotypes (cIMT (ave.), cIMT (max.), cIMT (min.), cIMTV, mTPA and CRS) using AtheroEdge™ and HbA1c. It was observed that carotid risk phenotypes have high correlations with HbA1c, ranging from 0.27 to 0.45, except for cIMTV, which was in the range from 0.08 to 0.14 (table 8.6). The behavior of both operators (novice and experienced) was consistent during these correlations (table 8.6). We have also identified the high-risk phenotype by adjusting FBS using logistic regression and the OR. Tables 8.4 and 8.5 demonstrate that all the phenotypes showed an OR higher than 1.00 ($p < 0.001$), except for cIMTV. This was due to the low correlations between cIMTV and HbA1c. For this reason, our findings indicate that all the phenotypes are high risk for HbA1c except for cIMTV. Note that new phenotypes mTPA, IMTV and CRS are strongly related to cIMT (ave.), cIMT (max.), cIMT (min.) for LCCA, RCCA and MCCA (appendix B, tables B1–B3).

8.5.1 A special note on mTPA and CRS

There are several motivations which triggered the evaluation of the image-based phenotypes mTPA and CRS in this study. (a) Since the previous methods for mTPA and CRS were computed using manual methods, which are subjective in nature due to inter- and intra-observer variability, to avoid this issue we developed an automated method for mTPA and CRS measurement. (b) We demonstrated that automated mTPA was highly correlated to automated cIMT (0.96, 0.98 and 0.97, respectively, for the LCCA, RCCA and MCCA), which has never been attempted before, and this provides another reason to know that mTPA is a significant image-based phenotype. There are two basic reasons for this: (i) mTPA has a strong correlation with cIMT because our system computed the complete regional coverage between LI and MA (including focal thickening) and (ii) the number of samples points for cIMT computations were normalized to 100 sample points. (c) We demonstrated that the OR for mTPA was > 1.0 ($p < 0.001$), which further shows that mTPA is highly significant, and this was an important objective of this study. (d) Of all the image-based phenotypes, the CRS was highly related to HbA1c. This is because the CRS is a function of an algebraic addition of all the image-based phenotypes, so the net effect of the association between CRS and HbA1c can be directly adopted for understanding stroke and cardiovascular risk. Thus, due to the above reasons, there was a clear need to understand the role of the image-based phenotypes mTPA and CRS.

8.5.2 Benchmarking

As cIMT is considered as a one candidate phenotype, many studies have tested its significance and tried to observe its relationship with HbA1c in type-2 diabetic patients. The comparison of our current study against previously published studies is summarized in table 8.8. Rema *et al* [41] used 600 type-2 diabetes subjects and applied LR to the dataset. The study observed a significant association of the manually computed cIMT and augmentation index (AI) with diabetes retinopathy. Lee *et al* [42] further studied 133 patients to prove cIMT and plaque score (PS) as risk factors in patients with ischemic stroke in type-2 diabetes. They used LR to determine the risk factors in diabetes patients based on OR. The study observed a correlation between manually computed cIMT and PS for the type-2 diabetes patients. Einarson *et al* [43] studied 15 592 patients to assess the relationship between blood glucose and manually computed cIMT using Pearson correlation coefficients. They showed that there was a small but significant relationship (CC = 0.08,

Table 8.8. Comparisons of our current study against previously published studies.

Authors (year)	Data size	Age (years)	Phenotypes	Study variable	Techniques	Conclusions
Spence *et al* (2004) [28]			TPA	Carotid stenosis	• CC = 0.60 • MLR	Moderately correlated
Rema *et al* (2004) [41]	600	52	cIMT, AL	DM	• LR	cIMT and AI have a significant association with DM
Lee *et al* (2007) [42]	133		cIMT (ave.), PS	D2M	• LR	cIMT (ave.) and PS are risk factors for DM
Einarson *et al* (2010) [43]	15 592		cIMT	Blood glucose	• CC = 0.08 • MLR	A significant relationship between glucose and cIMT
Mathiesen *et al* (2011) [46]	6584	25–84	TPA, IMT	Ischemic stroke	• CPH	TPA is a stronger predictor than IMT for the first-ever ischemic stroke
Rundek *et al* (2012) [26]	2743		TPA, GSM	tHcy	• MLR	tHcy is a modifiable risk factor for carotid plaque
Saba *et al* (2013) [39]	370		cIMT (ave.), PS	HbA1c	• CC = 0.18 • $P < 0.001$	Highly correlated
Ikeda *et al* (2014) [23]	500		cIMT, cIMTV		• ABI	Highly correlated cIMTV and ABI

Lucatelli *et al* (2016) [35]			cIMT, cIMTV	Leukoaraiosis	• CC	cIMTV and LA are significantly associated
Spence *et al* (2017) [25]	4512	62	TPA	LDL-C	• MLR	TPA and LDL-C are highly associated
Kim *et al* (2017) [45]	819	53.3	cIMT, CACS	FLD	• LR	cIMT is a stronger phenotype for FLD than CACS
Proposed study	199		cIMT (ave.), cIMT (max.), cIMT (min.), cIMTV, mTPA and CRS	HbA1c	• CS-test • *t*-test • LR	cIMT (ave.) > cIMT (max.) > cIMT (min.) > mTPA > CRS. All phenotypes are high risk based on OR and *p*-values.

LR: logistic regression; CS: χ-squared test; CACS: coronary artery calcium score; MLR: multiple linear regressions; PS: plaque score; tHcy: total homocysteine; AI: augmentation index; ABI: ankle–brachial index; CPH: Cox proportional hazard model; FLD: fatty liver disease.

$p < 0.001$) between blood glucose levels and cIMT and the results had adverse cardiovascular outcomes. Saba *et al* [44] further utilize 370 patients to indicate the association between automatically computed cIMT, PS and HbA1c using Pearson correlation coefficients. Unlike Lee *et al* [42], the authors observed a moderate correlation (CC = 0.18; p-value < 0.001) between cIMT and HbA1c and no association was observed between PS and HbA1c. Recently, Kim *et al* [45] also investigated a total of 819 patients among which 330 patients had fatty liver disease (FLD). They used LR to show the association between cIMT, coronary artery calcium score (CACS) and fatty liver disease (FLD). The results proved that cIMT is a stronger biomarker for FLD than CACS.

Unlike cIMT, mTPA was recently explored as a phenotype and researchers have tried to observe its relationship with cIMT and HbA1c in type-2 diabetic patients. Spence *et al* [25] studied the relationship between LDL and manually measured TPA for both the near and far wall of the carotid artery. The study observed that in patients with resistant atherosclerosis, LDL follow-up was significantly associated with a change in plaque burden. Mathiesen *et al* [46] studied a total of 6584 patients and showed that the risk of ischemic stroke is associated with the manually computed mTPA and cIMT in the carotid artery. An mTPA cutoff value of 3.9 ± 2.2 mm^2 was considered to categorize the patients into low-risk bins. Cox proportional hazards model are generally used to study the effect of several risk factors on the survival time of patients [47]. They applied Cox's proportional hazard to determine the risk of stroke with mTPA and cIMT based on hazard ratio (HR). The study observed that mTPA was a stronger biomarker than cIMT for the first-ever ischemic stroke. Taking a low-risk cutoff of 9 mm^2, Rundek *et al* [26] applied LR to study the association of homocysteine (tHcy) with manually computed TPA and grayscale median (GSM) by adjusting demographics and vascular risk factors. Based on the p-value and OR, they concluded that tHcy was independently associated with plaque morphology and increased plaque area, which are subclinical markers of stroke risk. In the current study, we examined the effect of six image-based phenotypes (cIMT (ave.), cIMT (max.), cIMT (min.), cIMTV, mTPA and CRS) on HbA1c. Our results demonstrate that mTPA and CRS are high-risk image phenotypes (OR > 1.0, $p < 0.001$). Since HbA1c can give false positives rather than true positives [48], there may be other biomarkers, such as glycated albumin and fructosamine [49], that can be considered in the future.

8.5.3 A special note on the reproducibility of phenotypes

Recently, Suri and his team showed that LDs computed by AtheroCloud™ have high reproducibility. It is very crucial to know that these measurements are reproducible. Inter- and intra-operator reproducibly was also conducted on these phenotypes [34–36]. In our current study, we have performed inter-operator variability analysis and the results are consistent with the previous literature.

8.6 Conclusions

The main objective of this study was to examine the relationship between six automated carotid artery disease risk phenotypes (cIMTave., cIMTmax., cIMTmin., bidirectional wall variability (cIMTV), morphology-based total plaque area (mTPA) and composite risk score (CRS)) and HbA1c in diabetes patients. These risk phenotypes were automatically measured using AtheroEdge™ (AtheroPoint™, Roseville, CA, USA) on carotid ultrasound scans. Fasting blood sugar (FBS) showed a high correlation with HbA1c ($p < 0.001$). Except for bidirectional wall variability (cIMTV), all phenotypes (cIMT (ave.), cIMT (max.), cIMT (min), mTPA and CRS) showed odds ratios > 1.0 ($p < 0.001$) for the LCCA, RCCA and MCCA. After adjusting FBS, the OR for the mTPA phenotype showed a higher risk for diabetes patients for the LCCA, RCCA and MCCA. The automated system was further tested for inter-operator variability (IOV) which showed consistent results using the z-test, KS test, Mann–Whitney test and Friedman test.

Conflict of interest

No conflicting relationship exists for any author.

Contributions

Elisa Cuadrado-Godia: Carotid artery expert advice.
Md Maniruzzaman: Regression analysis.
Tadashi Araki: Data collection and IRB approval.
Anudeep Puvvula: Diabetes expert advisor.
Md Jahanur Rahman: Information technology.
Luca Saba: Radiologist and vascular imaging.
Harman S Suri: Expert writing.
Ajay Gupta: Radiologist and stroke specialist.
Sumit K Banchhor: Manuscript design and rebuttal designs.
Jagjit S Teji: Vascular discussions.
Tomaž Omerzu: Ultrasound vascular radiologist.
Narendra N Khanna: Cardiovascular expert.
John L Laird: Vascular and heart expert.
Andrew Nicolaides: Ultrasound and vascular surgery.
Sophie Mavrogeni: Cardiovascular imaging expert.
George D Kitas: Rheumatology and vascular relation expert.
Jasjit S Suri: Principal Investigator and management.

Financial disclosures

The author(s) received no financial support for the research, authorship and /or publication of this chapter.

Appendix A Box-plots

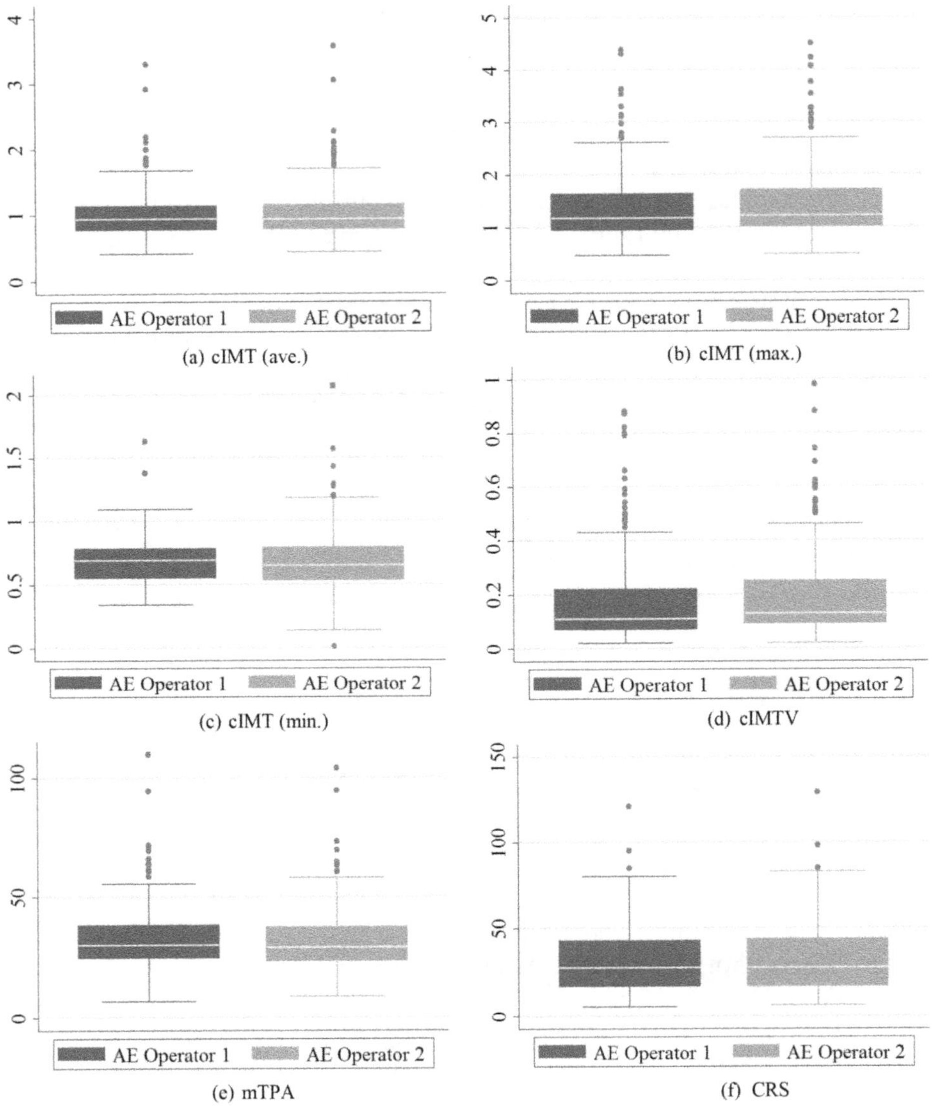

Figure A1. Box-plot of the six phenotypes for the LCCA between operator 1 and operator 2.

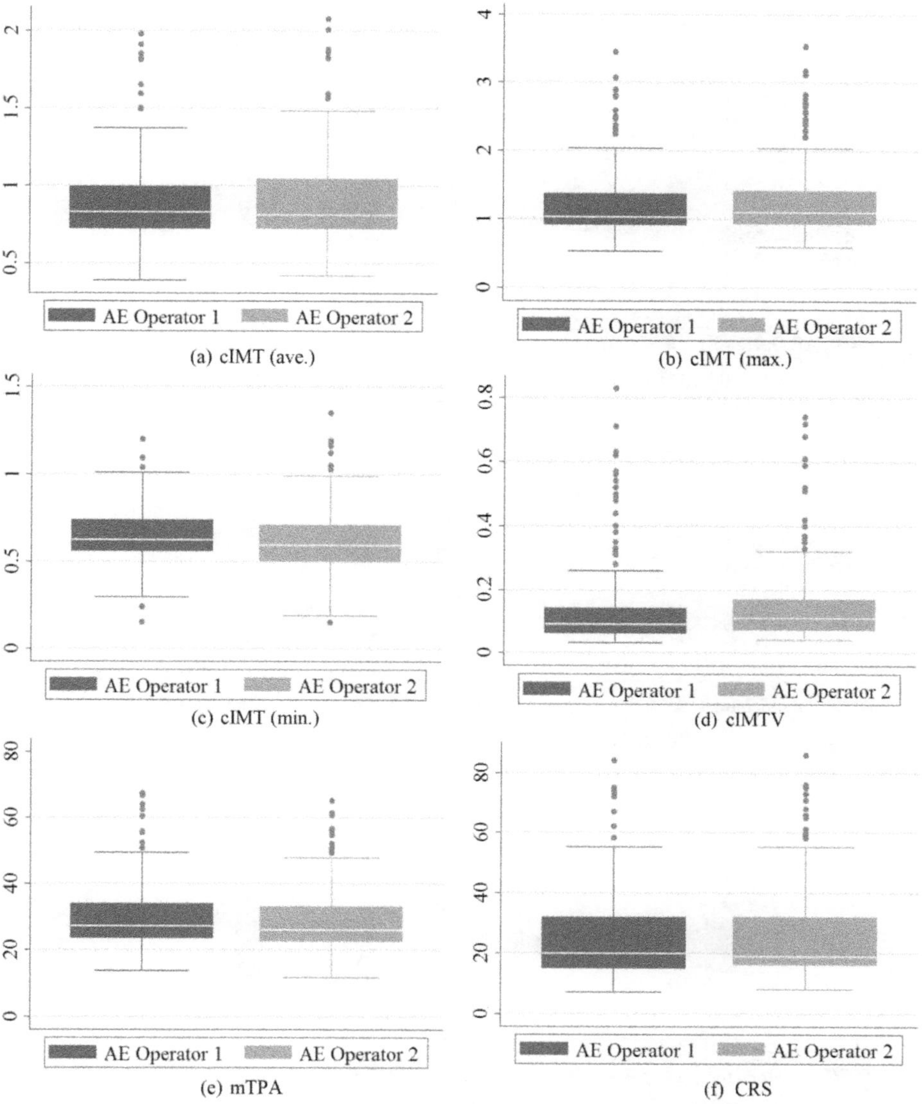

Figure A2. Box-plot of the six phenotypes for the RCCA between operator 1 and operator 2.

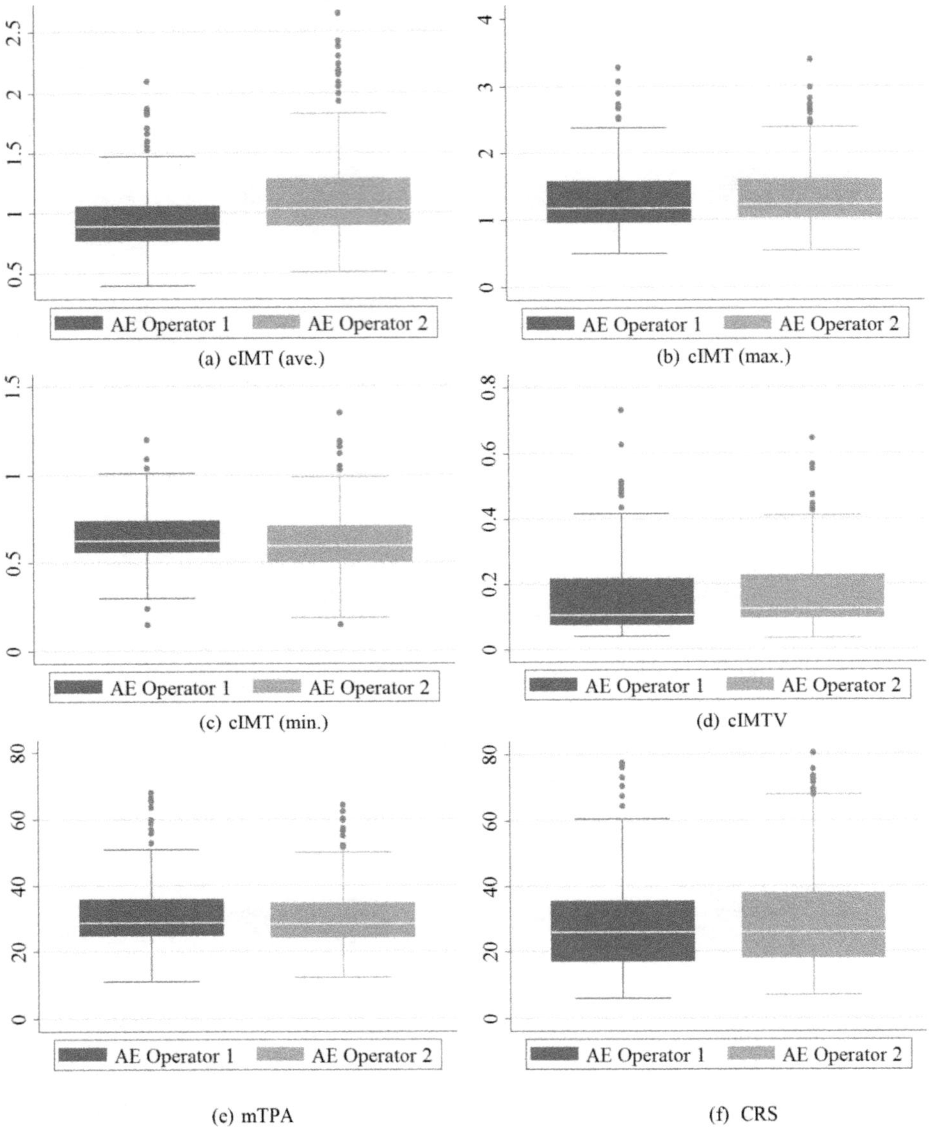

Figure A3. Box-plot of the six phenotypes for the MCCA between operator 1 and operator 2.

Appendix B Correlation tables

Table B1. Correlation between mTPA and cIMT.

Artery types	mTPA vs cIMT relationship	CC	p-value
LCCA	mTPA vs cIMT (ave.)	0.96	$p < 0.001$
	mTPA vs cIMT (max.)	0.87	$p < 0.001$
	mTPA vs cIMT (min.)	0.72	$p < 0.001$
RCCA	mTPA vs cIMT (ave.)	0.98	$p < 0.001$
	mTPA vs cIMT (max.)	0.89	$p < 0.001$
	mTPA vs cIMT (min.)	0.79	$p < 0.001$
MCCA	mTPA vs cIMT (ave.)	0.97	$p < 0.001$
	mTPA vs cIMT (max.)	0.90	$p < 0.001$
	mTPA vs cIMT (min.)	0.78	$p < 0.001$

Table B2. Correlation between CRS and cIMT.

Artery types	CRS vs cIMT	CC	p-value
LCCA	CRS vs cIMT (ave.)	0.93	$p < 0.001$
	CRS vs cIMT (max.)	0.92	$p < 0.001$
	CRS vs cIMT (min.)	0.72	$p < 0.001$
RCCA	CRS vs cIMT (ave.)	0.94	$p < 0.001$
	CRS vs cIMT (max.)	0.95	$p < 0.001$
	CRS vs cIMT (min.)	0.67	$p < 0.001$
MCCA	CRS vs cIMT (ave.)	0.95	$p < 0.001$
	CRS vs cIMT (max.)	0.94	$p < 0.001$
	CRS vs cIMT (min.)	0.72	$p < 0.001$

Table B3. Correlation between cIMTV and the rest of the phenotypes.

Artery type	Relationship	CC	p-value
LCCA	cIMTV vs cIMT (ave.)	0.78	$p < 0.001$
	cIMTV vs cIMT (max.)	0.91	$p < 0.001$
	cIMTV vs cIMT (min.)	0.50	$p < 0.001$
	cIMTV vs mTPA	0.75	$p < 0.001$
	cIMTV vs CRS	0.84	$p < 0.001$
RCCA	cIMTV vs cIMT (ave.)	0.70	$p < 0.001$
	cIMTV vs cIMT (max.)	0.87	$p < 0.001$
	cIMTV vs cIMT (min.)	0.30	$p < 0.001$
	cIMTV vs mTPA	0.70	$p < 0.001$
	cIMTV vs CRS	0.81	$p < 0.001$
MCCA	cIMTV vs cIMT (ave.)	0.83	$p < 0.001$
	cIMTV vs cIMT (max.)	0.91	$p < 0.001$
	cIMTV vs cIMT (min.)	0.51	$p < 0.001$
	cIMTV vs mTPA	0.78	$p < 0.001$
	cIMTV vs CRS	0.86	$p < 0.001$

Appendix C Statistical tests results

Table C1. *p*-value results of statistical tests using two different operators.

Artery type	Carotid risk phenotypes	Operator 1			Operator 2		
		KS test	FD test	MW test	KS test	FD test	MW test
—	—	C1	C2	C3	C4	C5	C6
LCCA	cIMT (ave.)	0.000	0.003	0.001	0.000	0.006	0.001
	cIMT (max.)	0.000	0.002	0.000	0.000	0.002	0.000
	cIMT (min.)	0.129	0.064	0.112	0.001	0.036	0.012
	cIMTV	0.001	0.004	0.000	0.066	0.001	0.000
	mTPA	0.000	0.016	0.007	0.001	0.018	0.003
	CRS	0.000	0.004	0.001	0.000	0.004	0.000
RCCA	cIMT (ave.)	0.053	0.114	0.037	0.158	0.113	0.026
	cIMT (max.)	0.016	0.006	0.001	0.005	0.004	0.000
	cIMT (min.)	0.193	0.279	0.128	0.489	0.490	0.201
	cIMTV	0.006	0.003	0.001	0.003	0.004	0.001
	mTPA	0.053	0.323	0.030	0.041	0.018	0.014
	CRS	0.022	0.069	0.003	0.069	0.011	0.003
MCCA	cIMT (ave.)	0.001	0.002	0.000	0.001	0.000	0.000
	cIMT (max.)	0.000	0.000	0.000	0.000	0.000	0.000
	cIMT (min.)	0.058	0.018	0.076	0.016	0.036	0.050
	cIMTV	0.000	0.000	0.000	0.000	0.000	0.000
	mTPA	0.001	0.002	0.000	0.000	0.000	0.000
	CRS	0.000	0.000	0.000	0.000	0.000	0.000

Note: KS: Kolmogorov–Smirnov; FD: Friedman; MW: Mann–Whitney.

Appendix D Abbreviations

Table D1. List of abbreviations.

Abbreviation	Descriptions
ABI	Ankle–brachial index
AE	AtheroEdge™
AI	Augmentation index
AUC	Area under the curve
CACS	Coronary artery calcium score
CAD	Chronic artery disease
CC	Correlation coefficient
CI	Confidence interval
cIMT (ave.)	Carotid intima–media thickness (average)
cIMT (max.)	Carotid intima–media thickness (maximum)
cIMT (min.)	Carotid intima–media thickness (minimum)
cIMTV	Carotid intima–media thickness variability

CRS	Composite risk score
CVD	Cardiovascular disease
DM	Diabetes mellitus
FBS	Fasting blood sugar
FH	Family history
FLD	Fatty liver disease
HR	Hazard ratio
HT	Hypertension
IOV	Inter-operator variability
KS	Kolmogorov–Smirnov
LDL/HDL	Low/high-density lipoprotein cholesterol
LCCA	Left common carotid artery
LR	Logistic regression
MCCA	Mean of common carotid artery
TPA	Total plaque area
mTPA	Morphological total plaque area
OR	Odds ratio
PS	Plaque score
RCCA	Right common carotid artery
TC	Total cholesterol
TG	TriGly

References

[1] Nathan D M, Genuth S and Lachin J 1993 The effect of intensive treatment of diabetes on the development and progression of long-term complications in insulin-dependent diabetes mellitus *N. Engl. J. Med.* **329** 977–86

[2] Nathan D M, Cleary P A and Backlund J Y 2005 Intensive diabetes treatment and cardiovascular disease in patients with type 1 diabetes *N. Engl. J. Med.* **353** 2643–53

[3] O'Gara P T *et al* 2013 ACCF/AHA guideline for the management of ST-elevation myocardial infarction: executive summary: a report of the American College of Cardiology Foundation/American Heart Association Task Force on Practice Guidelines *Circulation* **127** 362–425

[4] Kishore P, Kim S H and Crandall J P 2012 Glycemic control and cardiovascular disease: what's a doctor to do? *Curr. Diab. Rep.* **12** 255–64

[5] Emerging Risk Factors Collaboration 2010 Diabetes mellitus, fasting blood glucose concentration, and risk of vascular disease: a collaborative meta-analysis of 102 prospective studies *Lancet* **375** 2215–22

[6] Libby P 2003 Vascular biology of atherosclerosis: overview and state of the art *Am. J. Cardiol.* **91** 3–6

[7] Ross R 1999 Atherosclerosis—an inflammatory disease *N. Engl. J. Med.* **340** 115–26

[8] Libby P 2012 History of discovery: inflammation in atherosclerosis *Arterioscler. Thromb. Vasc. Biol.* **32** 2045–51

[9] Patel A K *et al* 2016 A review on atherosclerotic biology, wall stiffness, physics of elasticity, and its ultrasound-based measurement *Curr. Atheroscler. Rep.* **18** 83–95

[10] Riccioni G *et al* 2003 Atherosclerotic plaque formation and risk factors *Int. J. Immunopathol. Pharmacol.* **16** 25–31

[11] Fok P W 2011 Growth of necrotic cores in atherosclerotic plaque *Math. Med. Biol.* **29** 301–27

[12] Zaccardi F *et al* 2015 Pathophysiology of type 1 and type 2 diabetes mellitus: a 90-year perspective *Postgrad. Med. J.* **91** 63–90

[13] O'Leary D H *et al* 1999 Carotid-artery intima and media thickness as a risk factor for myocardial infarction and stroke in older adults *N. Engl. J. Med.* **340** 14–22

[14] Bots M L 2006 Carotid intima–media thickness as a surrogate marker for cardiovascular disease in intervention studies *Curr. Med. Res. Opin.* **22** 2181–90

[15] Nambi V *et al* 2010 Carotid intima–media thickness and presence or absence of plaque improves prediction of coronary heart disease risk: the ARIC (Atherosclerosis Risk In Communities) study *J. Am. Coll. Cardiol.* **55** 1600–7

[16] Molinari F *et al* 2012 Carotid artery recognition system: a comparison of three automated paradigms for ultrasound images *Med. Phys.* **39** 378–91

[17] Ikeda N *et al* 2017 Automated segmental-IMT measurement in thin/thick plaque with bulb presence in carotid ultrasound from multiple scanners: stroke risk assessment *Comput. Methods Programs Biomed.* **141** 73–81

[18] Ikeda N *et al* 2014 Automated and accurate carotid bulb detection, its verification and validation in low quality frozen frames and motion video *Int. Angiol.* **33** 573–89

[19] Saba L *et al* 2013 Inter-and intra-observer variability analysis of completely automated cIMT measurement software (AtheroEdge™) and its benchmarking against commercial ultrasound scanner and expert readers *Comput. Biol. Med.* **43** 1261–72

[20] Olt S *et al* 2016 The relationship between HbA1c and carotid intima–media thickness in type 2 diabetic patients *Pan. Afr. Med. J.* **23** 223–6

[21] Saba L *et al* 2012 Intima Media Thickness Variability (IMTV) and its association with cerebrovascular events: a novel marker of carotid atherosclerosis? *Cardiovasc. Diagn. Ther.* **2** 10–8

[22] Saba L *et al* 2012 Carotid IMT variability (IMTV) and its validation in symptomatic versus asymptomatic Italian population: can this be a useful index for studying symptomaticity? *Echocardiography* **29** 1111–9

[23] Ikeda N *et al* 2014 Ankle–brachial index and its link to automated carotid ultrasound measurement of intima–media thickness variability in 500 Japanese coronary artery disease patients *Curr. Atheroscler. Rep.* **16** 393–410

[24] Lucatelli P *et al* 2016 Relationship between leukoaraiosis, carotid intima–media thickness and intima–media thickness variability: preliminary results *Eur. Radiol.* **26** 4423–31

[25] Spence J D and Solo K 2017 Resistant atherosclerosis: the need for monitoring of plaque burden *Stroke* **48** 1624–9

[26] Alsulaimani S *et al* 2013 Elevated homocysteine and carotid plaque area and densitometry in the Northern Manhattan Study *Stroke* **44** 457–61

[27] Romanens M *et al* 2017 Extensive carotid atherosclerosis and the diagnostic accuracy of coronary risk calculators *Prev. Med. Rep.* **6** 182–6

[28] Spence J D and Hegele R A 2004 Noninvasive phenotypes of atherosclerosis: similar windows but different views *Stroke* **35** 649–53

[29] Molinari F *et al* 2012 Automated carotid IMT measurement and its validation in low contrast ultrasound database of 885 patient Indian population epidemiological studies: results of AtheroEdge™ Software *Int. Angiol.* **31** 42–53

[30] Saba L *et al* 2016 Accurate cloud-based smart IMT measurement, its validation and stroke risk stratification in carotid ultrasound: a web-based point-of-care tool for multicenter clinical trial *Comput. Biol. Med.* **75** 217–34

[31] Rozenberg S G *et al* 1990 Analysis of highly tritium-labeled alanine using 3H-and 1H-NMR *Bioorg. Khim.* **16** 858–61

[32] Molinari F *et al* 2012 Completely automated multiresolution edge snapper—a new technique for an accurate carotid ultrasound IMT measurement: clinical validation and benchmarking on a multi-institutional database *IEEE Trans. Image Process.* **21** 1211–22

[33] Molinari F, Zeng G and Suri J S 2010 Intima–media thickness: setting a standard for a completely automated method of ultrasound measurement *IEEE Trans. Ultrason. Ferroelectr. Freq. Control* **57** 1112–24

[34] Saba L 2011 CT imaging in the carotid artery *Atherosclerosis Disease Management* (Berlin: Springer) pp 353–409

[35] Lucatelli P *et al* 2016 Is there an association between leukoaraiosis volume and diabetes? *J. Neuroradiol.* **43** 273–9

[36] Suri J S, Haralick R M and Sheehan F H 2000 Greedy algorithm for error correction in automatically produced boundaries from low contrast ventriculograms *Pattern Anal. Appl.* **3** 39–60

[37] Libby P, Ridker P M and Hansson G K 2011 Progress and challenges in translating the biology of atherosclerosis *Nature* **473** 317–25

[38] Cavagnolli G, Gross J L and Camargo J L 2012 HbA1c in the diagnosis of diabetes: which cut-off point? *Diabet. Med.* **29** 286–7

[39] Wang B *et al* 2016 Cutoff point of HbA1c for diagnosis of diabetes mellitus in Chinese individuals *PLoS One* **11** e0166597

[40] Dong C *et al* 2015 Sirtuin/uncoupling protein gene variants and carotid plaque area and morphology *Int. J. Stroke* **10** 1247–52

[41] Rema M *et al* 2004 Association of carotid intima–media thickness and arterial stiffness with diabetic retinopathy: the Chennai Urban Rural Epidemiology Study (CURES-2) *Diabetes Care* **27** 1962–7

[42] Lee E J *et al* 2007 Relevance of common carotid intima–media thickness and carotid plaque as risk factors for ischemic stroke in patients with type 2 diabetes mellitus *Am. J. Neuroradiol.* **28** 916–9

[43] Einarson T R, Hunchuck J and Hemels M 2004 Relationship between blood glucose and carotid intima media thickness: a meta-analysis *Cardiovasc. Diabetol.* **9** 37

[44] Saba L *et al* 2013 Association of automated carotid IMT measurement and HbA1c in Japanese patients with coronary artery disease *Diabetes Res. Clin. Pract.* **100** 348–53

[45] Kim H J *et al* 2017 Comparison of carotid intima–media thickness and coronary artery calcium score for estimating subclinical atherosclerosis in patients with fatty liver disease *Cardiovas. J. Afr.* **28** 1–5

[46] Mathiesen E B *et al* 2010 Carotid plaque area and intima–media thickness in prediction of first-ever ischemic stroke: a 10-year follow-up of 6584 men and women: the Tromsø Study *Stroke* **42** 972–80

[47] Cox D R 1992 Regression models and life-tables *Breakthroughs in Statistics* (Berlin: Springer) pp 527–41

[48] Freitas P A, Ehlert L R and Camargo J L 2017 Glycated albumin: a potential biomarker in diabetes *Arch. Endocrinol. Metab.* **61** 296–304

[49] Lee J E 2015 Alternative biomarkers for assessing glycemic control in diabetes: fructosamine, glycated albumin, and 1, 5-anhydroglucitol *Ann. Pediatr. Endocrinol. Metab.* **20** 74–8

Section IV

Risk stratification, speed improvement and cloud-based risk assessment for coronary and carotid arteries

Chapter 9

Plaque tissue morphology-based stroke risk stratification using carotid ultrasound: a polling-based PCA learning paradigm

Luca Saba, Pankaj K Jain, Harman S Suri, Nobutaka Ikeda, Tadashi Araki, Bikesh K Singh, Andrew Nicolaides, Shoaib Shafique, Ajay Gupta, John R Laird and Jasjit S Suri

Severe atherosclerosis disease in carotid arteries causes stenosis which in turn leads to stroke. Machine-learning systems have been previously developed for plaque wall risk assessment using morphology-based characterization. The fundamental assumption in such systems is the extraction of the grayscale features of the plaque region. Even though these systems have the ability to perform risk stratification, they lack the ability to achieve higher performance due their inability to select and retain dominant features.

This chapter introduces a polling-based principal component analysis (PCA) strategy embedded in the machine-learning framework to select and retain dominant features, resulting in superior performance. This leads to more stability and reliability. The automated system uses offline image data along with the ground truth labels to generate the parameters, which are then used to transform the online grayscale features to predict the risk of stroke. A set of sixteen grayscale plaque features is computed. Utilizing the cross-validation protocol ($K = 10$), and the PCA cutoff of 0.995, the machine-learning system is able to achieve accuracies of 98.55% and 98.83% corresponding to the carotid far wall and near wall plaques, respectively. The corresponding reliability of the system was 94.56% and 95.63%, respectively. The automated system was validated against the manual risk assessment system and the precisions-of-merit for the same cross-validation settings and PCA cutoffs are 98.28% and 93.92% for the far and the near wall, respectively.

doi:10.1088/2053-2563/ab0820ch9

PCA-embedded morphology-based plaque characterization is a powerful strategy for risk assessment and can be adopted in clinical settings.

9.1 Introduction

According to the World Health Organization (WHO), 15 million people suffer from stroke every year, of which 5 million die and another 5 million are left permanently disabled [66]. These statistics have also been shown to be significantly higher in women compared to men [71]. A better risk assessment system can significantly improve the diagnosis and subsequent treatment for stroke, and is necessary to curb this global health epidemic.

The main cause of stroke is plaque formation due to the vessel disease called atherosclerosis. The carotid arteries are the main blood supplying arteries to the brain, as shown in figure 9.1(a). Due to the progression of atherosclerotic disease over time, these arteries become blocked, causing stress in the arterial caps [47]. Eventually, the cap of the arterial wall ruptures causing embolization, which results in stroke, ischemic attack and sometimes death [39, 68]. Other attributing factors which accelerate the biology of atherosclerosis are high blood pressure, obesity, smoking, an unhealthy diet, physical inactivity and diabetes mellitus [56]. Secondary risk factors are alcohol, age, gender and heredity [45].

The atherosclerotic arteries can be imaged using ultrasound, magnetic resonance imaging (MRI) and computed tomography (CT) [41, 52, 53]. Ultrasound offers several advantages over MR/CT such as low cost, non-invasive imaging, no risk of radiation, user-friendliness and real-time analysis. Due to the advanced features of ultrasound image reconstruction, such as harmonic and compound imaging [57], we are able to visualize the different kinds of asymptomatic and symptomatic plaques, which in turn incentivizes us to characterize tissue morphologically for stroke risk assessment [44, 58]. The objective of this paper is to

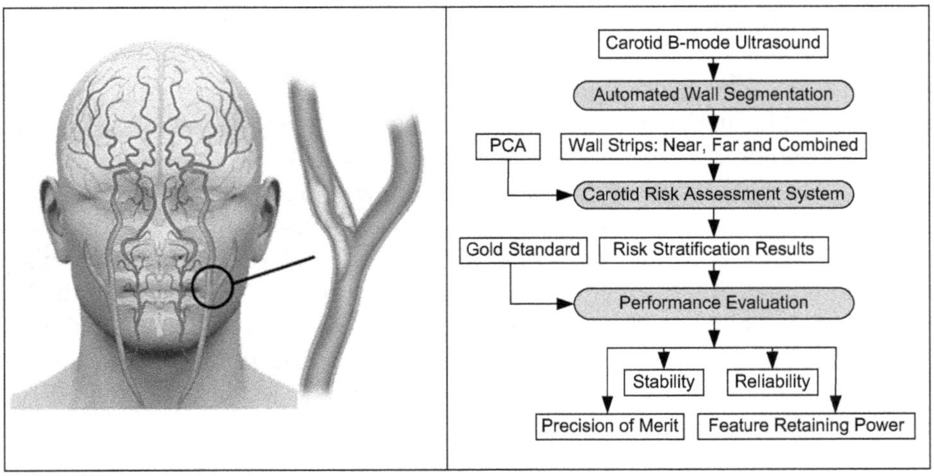

Figure 9.1. (a) Atherosclerotic plaque formation in the far/near walls; (b) Global stroke risk assessment system.

automatically stratify the risk of wall plaque in carotid ultrasound scans based on grayscale tissue characterization utilizing a polling-based PCA strategy in a machine-learning framework.

Plaques seen in ultrasound are traditionally characterized into two kinds: hypo- and hyper-echoic. The hypo-echoic plaques have less echogenicity and are darker than the normal plaques. This kind of plaque is non-homogeneous, has spotty micro-calcifications and a large plaque burden, and has irregular surface characteristics due to the presence of plaque hemorrhage [44]. In contrast, hyper-echoic plaques are brighter and can easily be identified by a radiologist. They are composed of fibrotic or fibro-fatty plaque or calcified plaque with a limited plaque burden, and are homogeneous with a smooth luminescence [44]. However, some asymptomatic patients can have hypo-echoic regions surrounded by hyper-echoic regions and are vulnerable [46]. Further, sometimes calcified plaque results in acoustic shadow (AcS) and poses challenges in visual interpretations of the plaque [27]. It is therefore very important to identify these asymptomatic plaques which are likely to become symptomatic over time. Due to a lack of visual stratification between plaque types, a morphological-based tissue characterization is necessary for stroke risk assessment. One benefit of tissue characterization is that you are then able to compute the risk of rupture. This is because the vulnerable plaque has multiple plaque wall components such as lipids (L), fibro-fatty tissue (FFT), smooth muscle cells (SMCs), calcium (C), necrotic cap thickness (NCT), fibrosis cap (FC), plaque hemorrhage (PH) and thrombus (T) [23, 43]. These plaque patterns can be studied via high-resolution ultrasound scans. Because manual identification of these patterns is prone to error, a machine-learning procedure is better able to predict stroke risk.

One can also characterize the plaque based on the severity of stenosis. In [18, 44] stenosis is defined as 'A focal structure that encroaches into the arterial lumen of at least 0.5 mm or 50% of the surrounding intima–media thickness, or a thickness of greater than or equal to 1.5 mm'. Two major strategies are adopted for stenosis measurement: (a) the North American Symptomatic Carotid Endarterectomy Trial (NASCET) and (b) the European Carotid Surgery Trial (ECST) [27, 40, 41]. Thus several studies have previously adopted stenosis as a risk assessment criterion, but this strategy has a major set-back, in that it does not take into account the tissue-based morphology of the plaque walls.

Lumen diameter (LD) in the common carotid artery (CCA), internal carotid artery (ICA) or external carotid artery (ECA) is the distance between the near wall lumen–intima and far wall lumen–intima. Recently, carotid LD measurement has shown promising signs which can be used globally to present risk. Earlier work has shown a relationship between LD and cardiovascular events (myocardial infarction). In [27, 63] it was suggested that percent cross sectional area reduction is the best method to obtain the severity of stroke risk. Over the years, studies have shown that LD is associated with age, gender and left ventricular mass. Also, studies have shown a relationship between LD and coronary artery disease [15, 17, 26]. We thus hypothesize that LD can play an important role in risk characterization and thus can be used as a biomarker for developing the ground truth labels in the machine-learning (ML) framework. We can therefore build an ML system which can be a combination of grayscale information in the plaque region and the LD.

ML-based tissue characterization for stroke risk has been performed by various authors, including our group [1–3, 5, 6]. A table of comparison between these techniques and the proposed method is shown in appendix A, table A11. One of the bottlenecks in stroke risk stratification based on ML is the choice of feature combination for the best performance. We introduce a polling-based PCA strategy for feature selection and optimization [9]. We adapt the three types of cross-validation (CV) protocols (K5, K10 and JK corresponding to $K = 5$ partitions, $K = 10$ partitions and K = N partitions) for ML accuracy measurements. We will interchangeably use $K = 5$ as K5, $K = 10$ as K10 and K = N as JK. The ML system's performance is evaluated by taking manually segmented walls and computing novel performance indices. Reliability and stability criteria are also established. Figure 9.1(b) shows a complete flow chart of the global risk assessment system.

The layout of this chapter is as follows. The demographics of our patient population are presented in section 9.2, while the methodology adopted is depicted in section 9.3. The experimental protocol and the results are shown in section 9.4. The performance evaluation, consisting of the precision-of-merit, stability and reliability analysis followed by feature retaining power, is presented in section 9.5. The benchmarking, strengths and limitations are elaborated in section 9.6, and the conclusions are presented at the end of section 9.6.

9.2 Demographics, data collection and preparation

9.2.1 Patient demographics

A total of 407 B-mode ultrasound scans corresponding to the left and right CCA were acquired from two hundred and four (204: 157 M/47 F) patients at Toho University, Japan and retrospectively analyzed (patient consent and ethics approval was granted). One patient had only a single image of the right CCA and therefore the total number of images was 407. The mean age of the patients was 69 ± 11 years ranging from 29 to 88 years. Of the 204 patients, 108 patients had a proximal lesion location, 67 a middle location and 29 a distal location in the carotid artery. The database had a mean HbA1c of 6.30 ± 1.1 (mg dl^{-1}), a mean LDL cholesterol of 101.61 ± 31.55 (mg dl^{-1}), a mean HDL cholesterol of 50.66 ± 15.22 (mg dl^{-1}) and total cholesterol of 175.82 ± 37.97 (mg dl^{-1}). Eighty three of the pool of 204 were smokers.

9.2.2 Data acquisition

B-mode ultrasound scans of the carotid arteries were obtained using a scanner (Aplio XV, Aplio XG, Xario, Toshiba, Inc., Tokyo, Japan) equipped with a 7.5 MHz linear array transducer. A sonographer with 15 years' experience performed all the scans. The patients were laid down in a supine position with their heads tilted backwards. As the carotid arteries were located from the transverse view, the probe was rotated by 90° to acquire longitudinal images of the anterior and posterior walls. Adopting the guidelines of the American Society of Echocardiography Carotid

Intima–Media Thickness Task Force 16, the protocol acquired high-resolution images for the far wall and near wall of the CCA. Further, the sonographer acquired high-resolution images of the ICAs and carotid bulbs, as shown in figure 9.2, which was useful for computing the plaque score (PS). The IRB of our institution has provided a full ethics review for this study and written, informed consent was provided by all the patients.

9.2.3 Manual wall region extraction for the manual risk assessment system (mRAS)

For validation of the machine-learning system, one requires manual LD delineation. These LD borders will be used for LD measurement. For manual LD delineation, the neurologists are required to manually trace the LD borders of the CCA. This requires near and far wall tracings of the CCA in carotid ultrasound scans. Figure 9.3 shows sample output images of carotid arteries using ImgTracer®, which correspond to high-risk (left) and low-risk (right) patients. The objective is to extract the grayscale wall region for validation of the stroke risk assessment system (sRAS).

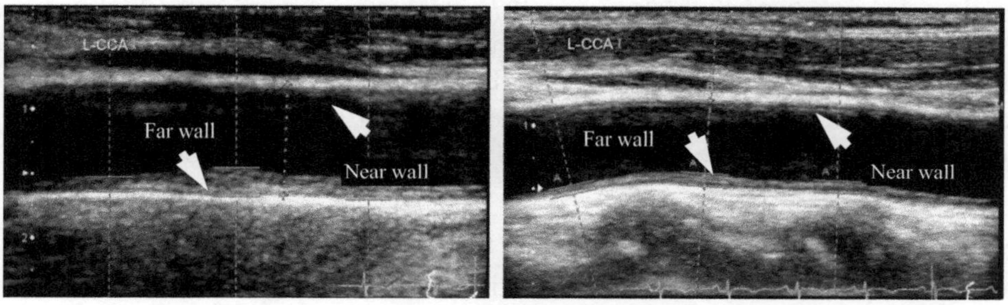

Figure 9.2. Sample raw images of B-mode ultrasound corresponding to high risk and low risk.

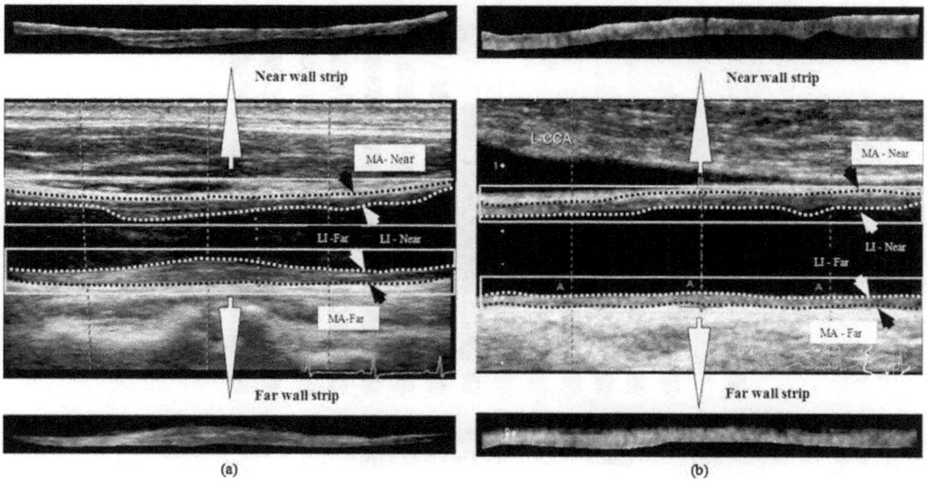

Figure 9.3. Ground truth tracing by an expert on (a) high-risk and (b) low-risk carotid scans.

Thus the manual lumen–intima/media–advetitia (LI/MA) borders are traced for the far and near walls of the carotid arteries. The wall strips for the near and far walls are shown in the top and bottom of the image for a low-risk (right) and high-risk (left) patient. This process is repeated for all 407 patient images. These manual wall strips will be used for validation of the machine-learning system.

9.2.4 Modeling the manual LD into two stratification classes: high risk and low risk

Manual LDs (or ground truth LDs) are required for two reasons: (a) in the automated stroke risk assessment system (sRAS), one needs to model the population into high risk and low risk by knowing the distribution of stenosis; and (b) for the design and validation of the manual risk assessment system (mRAS). To understand the risk distribution, we divide LD measurements into 16 bins starting from 5.0 mm to 8.0 mm with an interval of 0.2 mm. Each of these LD values becomes an LD threshold (LDT) during the sRAS and mRAS system development. Appendix A, table A1 shows the distribution of LDs and the frequency of images using this LDT. For example, considering an LDT of 6.2 mm (row number 7), the number of images which are below 6.2 mm are 234 out of 407, which constitutes 57.49% high risk and 42.51% low risk. The corresponding bar (bar #7) for this LDT is shown in figure 9.4. This is repeated for all the corresponding LDTs and they can be seen in figure 9.4. Note that for each LDT, the total percentage is 100%, which is partitioned into low and high risk based on the images in each bin (high versus low) for that LDT.

9.3 Risk assessment methodology

The fundamental assumption in our stratification modeling is that there is a risk associated with the hyper-echoic and hypo-echoic plaques in the carotid walls. As discussed in the introduction, these components have (a) different contrast levels,

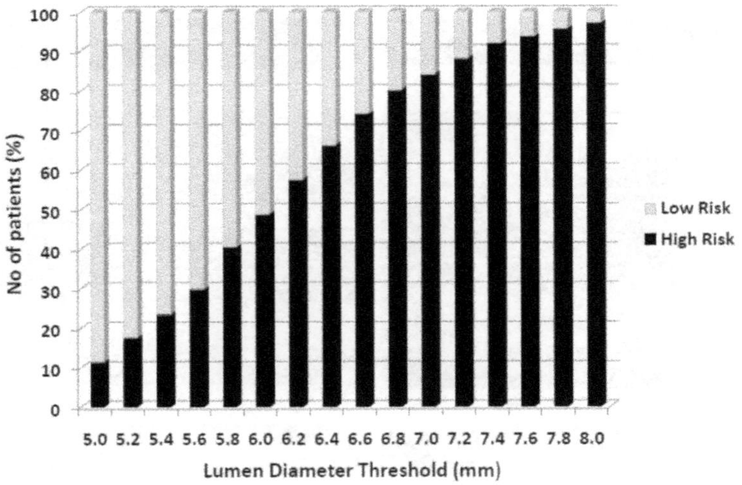

Figure 9.4. Distribution of population into high risk and low risk for modeling stroke risk assessment.

(b) variation in the density of pixels, (c) a chaotic nature of the grayscale intensities and (d) a mixture of dark, gray and bight intensities [67]. As a result, the grayscale texture distribution varies due to the multifocal distribution of plaque. However, the manual risk assessment is very difficult to characterize. Thus, we need an automated system which can morphologically characterize and learn the assessment characteristics to predict the risk of plaque rupture or risk of conversion of plaque from asymptomatic to symptomatic [16].

It is important to note that the growth of plaque is on both sides of the walls in the carotid ultrasound scans. We thus need the wall regions on both sides of the lumen representing the near wall and far wall. We will interchangeably call this wall region the 'IMT wall region' or 'wall strips', since the grayscale information is present between the LI and MA borders. Thus, we can have three kinds of sRAS, corresponding to the near wall, far wall and when we take the near and far wall jointly, the so called 'combined wall'. Thus, the automated segmentation of the wall regions becomes the first step [54] followed by morphologically based risk assessment using LD as a biomarker [58]. This can be seen in figure 9.1(b). Note that for either manual mRAS or automated sRAS, one requires a performance evaluation leading to stability and reliability analysis, which is an integral part of the overall process.

9.3.1 IMT far and near wall strip extraction

One of the important components for automated near and far wall strip extraction requires understanding the intensity distribution in the carotid scans [68]. The biology of atherosclerotic plaque returns the strongest ultrasound signal from the adventitial walls of the carotid artery [28, 31, 32]. This assumption is used to extract the MA borders of the far wall as the starting point. Using the scale-space strategy, which consists of convolution of a higher order Gaussian derivative filter with the carotid scans, helps us extract the MA borders [28–31, 36, 54]. This is the recognition phase of the carotid artery in the ultrasound scans, despite the presence of muscle fascia and the jugular vein [37]. Extending this to the near wall yields the MA borders for the near wall, thus constituting the region-of-interest (ROI) formation [10]. Even though this is independent of the curvature of the carotid artery, one can use image transformation to correct the curvature and then compute the MA borders of the near and far walls [22]. Once the ROI is reconstructed, we can then detect the lumen region based on the assumption that the blood has a constant Reynolds's number [24] and hence represents a unique class of pixels of constant density [11]. Thus, one can apply a pixel-classification scheme to model the lumen region as a binary set of pixel classes. The regional information can be morphologically corrected leading to the formation of the near LI and far LI borders. Thus, the automated delineation of the LI/MA borders can lead to the measurement of IMT [56], LD and inter-adventitial diameter (IAD) [54]. Figure 9.5 shows examples of patients with 'far wall strips' that are high risk (left) and low risk (right). Similarly, figure 9.6 shows examples of patients with 'near wall strips' that are high risk (left) and low risk (right). The block diagram of the LD/IAD detection and measurement

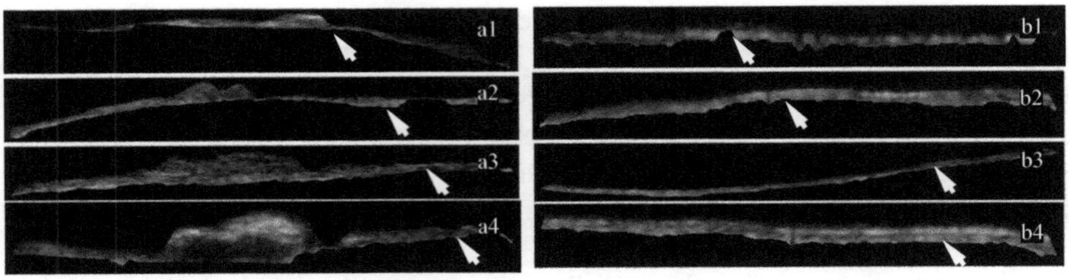

Figure 9.5. Far wall strips of high-risk and low-risk patients. High-risk patients are shown in (a1), (a2), (a3) and (a4). Low-risk patients are shown in (b1), (b2), (b3) and (b4).

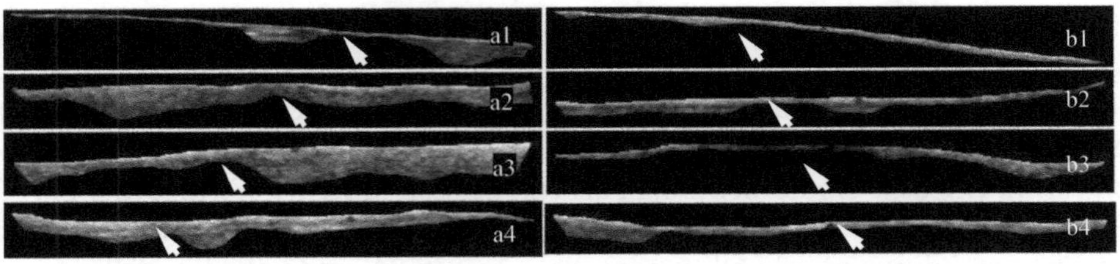

Figure 9.6. Near wall strips of high-risk and low-risk patients. High-risk patients are shown in (a1), (a2), (a3) and (a4). Low-risk patients are shown in (b1), (b2), (b3) and (b4).

is shown in figure 9.1(b). The main advantage of our system is the simultaneous automatic extraction of the LD border, IAD borders and IMT/LD measurements.

9.3.2 Assessment of stroke risk using a machine-learning system

As shown in figure 9.7, the proposed system (a class of AtheroRisk™ systems from AtheroPoint™, Roseville, CA, USA) is divided into two phases, namely offline (the training phase) and online (the testing phase). The system uses a support vector machine (SVM)-based classifier [59–61] for tissue characterization and risk stratification, along with four non-linear kernel functions, called the radial basis function (RBF), Polynomial-1, Polynomial-2 and Polynomial-3. Feature extraction is performed in both the online and offline components, which consists of extraction of three sets of grayscale texture features, namely (i) the gray level co-occurrence matrix (GLCM), (ii) the gray level run length matrix (GLRLM) and (iii) chaotic features, totaling 16 features. Using our second hypothesis, the lumen diameter (LD) is used as the gold standard ground truth to generate the classification labels for high-risk and low-risk (1 and 2, respectively) patients in the training phase [14]. Extracted grayscale features of the plaque images and the class labels derived from the ground truth carotid LD are fed to the offline local processing system that produces the system parameter model as an output. This system parameter model along with the features of the online data are fed to the online classifier to predict the risk of carotid artery disease patients. Thus, we run three experimental protocols using this set-up, which are discussed in section 9.4.

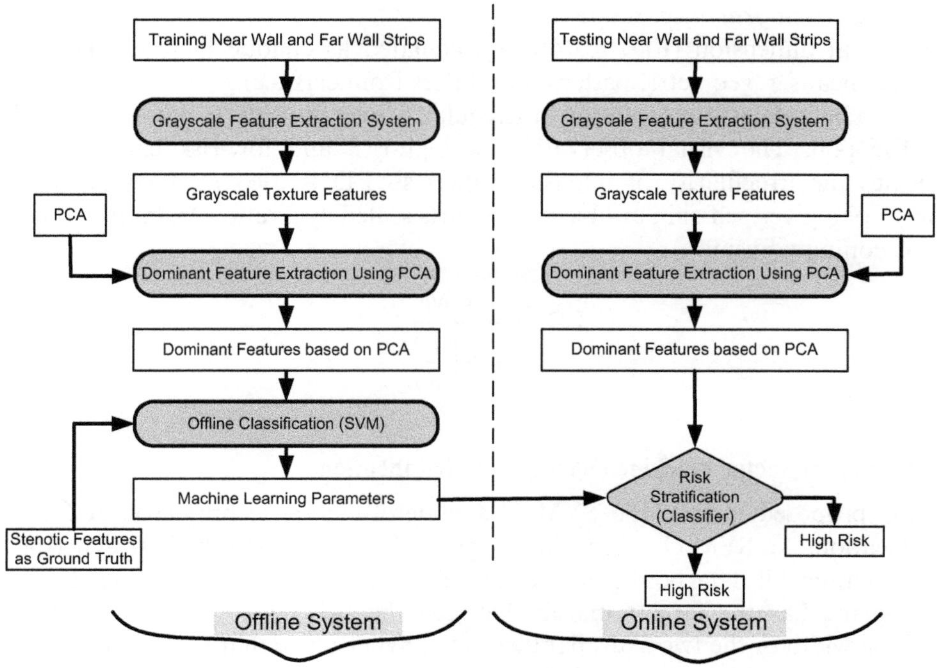

Figure 9.7. Carotid disease risk assessment system (sRAS) for the near wall and far wall—a class of AtheroRisk™ system from AtheroPoint™, Roseville, CA, USA.

9.3.3 Texture features

9.3.3.1 The gray level co-occurrence matrix (GLCM)

The GLCM is a very useful statistical tool for extracting texture information present in an image [20, 64]. Let I be a given grayscale 2D image, containing pixels with gray levels $(0,\ 1,\ldots,\ N_g-1)$. The GLCM is a matrix Mx of order L, where the $P_d(i, j)$th entry of Mx represents the probability of the number of occurrences of a pixel with intensity i being adjacent to a pixel with intensity j. By dividing each element of Mx by the total number of co-occurrence pairs in Mx, the normalized co-occurrence matrix is obtained. The adjacency can be specified in any direction, such as horizontal, vertical, right, left and diagonal, and the texture features can be calculated by taking the average of a specified direction of the co-occurrence matrix. We have extracted four features, shown in appendix B, table B1.

9.3.3.2 The gray level run length matrix (GLRLM)

The GLRLM measures the gray intensity pixel in a particular direction from the reference pixel. Gray intensity pixels can be measured in a particular direction using the GLRLM. Run length is a set of collinear pixels having the same gray level in a particular direction [69]. The GLRLM is a 2D matrix in which element $r(x, y)$ gives the total number of consecutive runs of length y at gray level x. We have extracted 11 features using the GLRLM as shown in appendix B, table B2. Note that M represents the number of gray levels and L represents the maximum run length.

9.3.3.3 Chaotic fractal-based features

The fractal dimension (FD) feature is calculated as chaotic features. The word 'fractal' means a 'geometric pattern' but differs from classical geometry. The degree of fractal boundary fragmentation or irregularity over multiple scales is measured as the FD [25]. The visual observation of high-risk and low-risk carotid images indicates the irregularity of intensity values so FD is used as a feature in this work. For a given self-similar object of N parts scaled by a ratio r from the whole, its FD is computed using

$$\mathrm{FD} = \frac{\log Nr}{\log\left(\frac{1}{r}\right)}. \tag{9.1}$$

9.3.4 Support vector machine (SVM) and classification

In our proposed model, the SVM is a state-of-the-art technique used for risk stratification. An SVM classifies [70] the given data by finding the best hyper-plane that separates all data points of one class from those of the other class. The best hyper-plane has the largest margin between the two classes for an SVM. The maximal width of the two slabs parallel to the hyper-plane that has no interior data points is considered as the margin [21, 38].

9.3.5 Feature reduction technique using polling-based principal component analysis

A polling-based principal component analysis (PCA) technique was used for optimization of feature selection [65]. Our database of 407 images was used to generate 16 grayscale features and, since not all features are weighted for best classification accuracy, we therefore selected the most dominant features which yield the highest accuracy. Although multiple strategies are available for determining the most effect feature set, we adopted the conventional PCA buffered with the polling-based strategy as it is very effective [61, 62]. This is an innovative approach where the polling cutoff is adopted for selecting the best features and retaining them. The major advantage of such a paradigm is that it offers inclusive criteria for bringing new features into the pool while retaining the optimal features and removing redundancy [9]. Thus the linear combination of the feature set keeps improving with an increase in PCA cutoff. The net effect of the polling-based PCA strategy is a reduction in feature space, improving the dominant features and providing a superior stratification between the low- and high-risk plaque classes. The overall algorithmic steps using polling-based PCA is as follows:

- Step 1. Subtract the mean from each of the data dimensions. The mean subtracted is the average across each dimension.
- Step 2. Compute the covariance matrix of the dataset.
- Step 3. Compute the eigenvectors (B_k) and eigenvalues (γ) of the covariance matrix.

- Step 4. Sort the eigenvalues in descending order and arrange the eigenvectors according to the sorted eigenvalues.
- Step 5. Choose the # of principal components (m) to be considered using

$$\sum_{j=1}^{m} \frac{\gamma_j}{\sum_{i=1}^{\eta} \gamma_i} \geqslant R, \tag{9.2}$$

  where R is the cutoff, varying from 0.900 to 0.995 in step sizes of 0.005 (a total of 20 values), and η represents the total number of eigenvalues.
- Step 6. Compute the contribution of each feature C_f using

$$C_f = \sum_{k=1}^{m} | B_k |. \tag{9.3}$$

- Step 7. Sort the features in descending order and select the first m features which will give a reduced number of features (m) (without modifying feature values) with their dominance level from highest to lowest.

Furthermore, the retention of features at different PCA cutoffs is evaluated, termed the feature retaining power (FRP). For each consecutive cutoff, it is observed how many features are retained and FRP is evaluated using equation (9.4):

$$\text{FRP}(\%) = \left[\frac{S_{f(i-j)}}{R_i} \right] \times 100, \tag{9.4}$$

where S_f represents the number of similar features for consecutive cutoffs and R represents the PCA cutoff. Among two consecutive cutoffs, the first and second cutoff are indicated by i and j, respectively.

9.3.6 Kernel optimization based on the machine-learning paradigm

It is important to choose the best kernel during the standardization of the machine-learning paradigm when using classifiers. Adopting the CV protocol on our database, we evaluated four kinds of kernel functions. These are depicted in figure 9.8 and are the radial basis function (RBF) (blue-triangle), Polynomial-1 (green-circle), Polynomial-2 (red-square) and Polynomial-3 (magenta-diamond). As can be seen, there is an increase in stratification accuracy with an increase in PCA cutoff. Even though, all the four kernels are very close, only Polynomial of order 2 (red-square) can be considered as the best kernel. This behavior was seen for all three kinds of walls, far, near and combined, and in both kinds of systems, sRAS and mRAS. Note that, in appendix A, table A2, the values in this table represent the mean values of the entire lumen diameter threshold which includes all 20 PCA cutoffs.

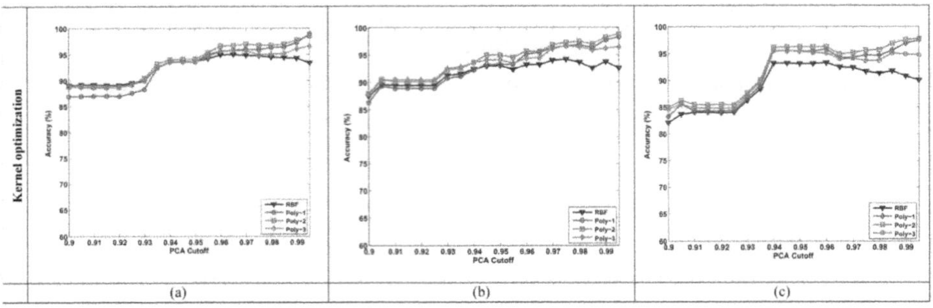

Figure 9.8. Kernel optimization accuracy versus PCA cutoff for (a) the far wall, (b) the near wall and (c) the combined wall.

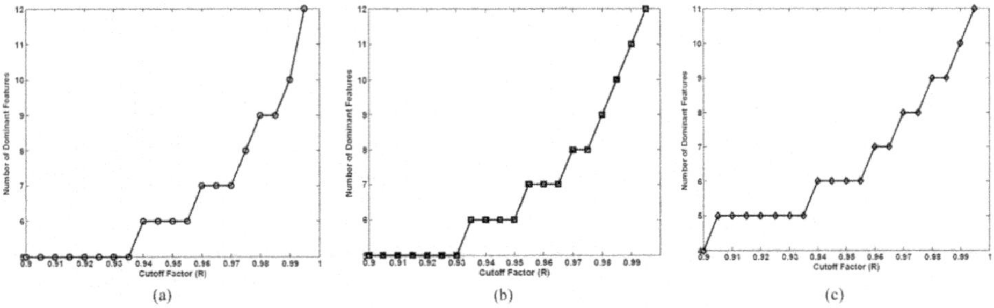

Figure 9.9. Number of dominant features with increasing PCA cutoff: (a) far, (b) near and (c) combined walls (auto), 6.0 mm LD threshold.

9.4 Experimental protocol and results

9.4.1 Experiment 1: dominant feature selection and classification accuracy with changing PCA cutoff

The objective of this experiment to establish the dominant features with changing PCA-cutoff values. This relationship is important to select the best features and retain them during the PCA-cutoff variations. Thus, during this experiment there are two kinds of relationships: (i) dominant features versus changing PCA cutoffs and (ii) classification accuracy versus PCA cutoffs. For this set-up, we demonstrated all three kinds of CV protocols (K5, K10 and JK) while considering the 16 sets of LDT values (ranging from 5.0 mm to 8.0 mm). The PCA cutoffs were changed from 0.900 to 0.995 with an interval of 0.005. This experiment was repeated for all three kinds of carotid walls: the far, near and combined walls.

The results of part (i) can be seen in figure 9.9(a)–(c). This shows the number of dominant features with increasing value of PCA cutoff for the automated far, near and combined walls, respectively. As an example, we took an LD threshold of 6 mm and varied the PCA cutoffs. Note that there is flatness at certain values of the dominant features. This simply indicates that the new incoming features were noisy and did not blend well to become a dominant feature. But with increasing PCA cutoff, the total number of dominant features also increases.

Part (ii) requires computing the mean value of the classification accuracy versus changing PCA cutoff and this can be mathematically expressed as

$$\eta_w^{\text{auto}}(k) = \frac{\sum_l^L \sum_r^R \sum_t^T \eta_w^{\text{auto}}(l,\ r,\ t)}{L \times R \times T} \tag{9.5}$$

$$\eta_w^{\text{manual}}(k) = \frac{\sum_l^L \sum_r^R \sum_t^T \eta_w^{\text{manual}}(l,\ r,\ t)}{L \times R \times T}, \tag{9.6}$$

where $\eta_w^{\text{auto}}(k)$ is the mean classification accuracy for the CV protocol k for the automated method for the wall type w, and $\eta_w^{\text{auto}}(l, r, t)$ is the accuracy using the automated method for wall type w, lumen diameter l, PCA cutoff r and trial number t. Using equation (9.5) the mean accuracies are calculated and are shown in appendix A, table A3 for the automated wall. Similarly, $\eta_w^{\text{manual}}(k)$ is the mean classification accuracy for the CV protocol k for the manual method for the wall type w, and $\eta_w^{\text{manual}}(l, r, t)$ is the accuracy using the manual method for wall type w, lumen diameter l, PCA cutoff r and trial number t. Using equation (9.6) the mean accuracies are calculated and are shown in appendix A, table A4 for manual LD.

The tables are shown in the form of plots. The three kinds of walls corresponding to the automated and manual results and three CV protocols generated a total of 18 plots (3 walls × 2 systems × 3 CV protocols). These 18 plots are arranged in three 2 × 3 matrices, shown in figures 9.10–9.12, respectively, consisting of six plots in each figure. Thus figure 9.10 corresponds to the far wall experiment results for the automated sRAS as (a1), (a2) and (a3) in the first column and the mRAS as (b1), (b2) and (b3) in the second column. Figure 9.11 shows the wall experiment results for the automated sRAS as (c1), (c2) and (c3) in the first column and the mRAS as (d1), (d2) and (d3) in the second column. Similarly, in figure 9.12 the combined wall results are shown for the automated sRAS as (e1), (e2) and (e3) in the first column and the manual sRAS as (f1), (f2) and (f3) in the second column. All 18 plots of figures 9.10–9.12 demonstrate the increase in accuracy with the increase in the PCA cutoffs used for the training and testing of the machine-learning system. Note that in each of the 18 plots, there are 16 curves corresponding to the 16 lumen diameter thresholds that range from 5 mm to 8 mm in intervals of 0.2 mm.

We give the following interpretation. With an increase in PCA cutoffs, there is an increase in the classification accuracy of the system. Irrespective of the wall type and CV protocols, all three walls show generalization of behavior during the machine-learning process. While the classification accuracy is similar in percentage for all three wall types, the patterns in the far wall show more clustering compared to the near wall and combined walls. This is due to the fact that the media regions of the

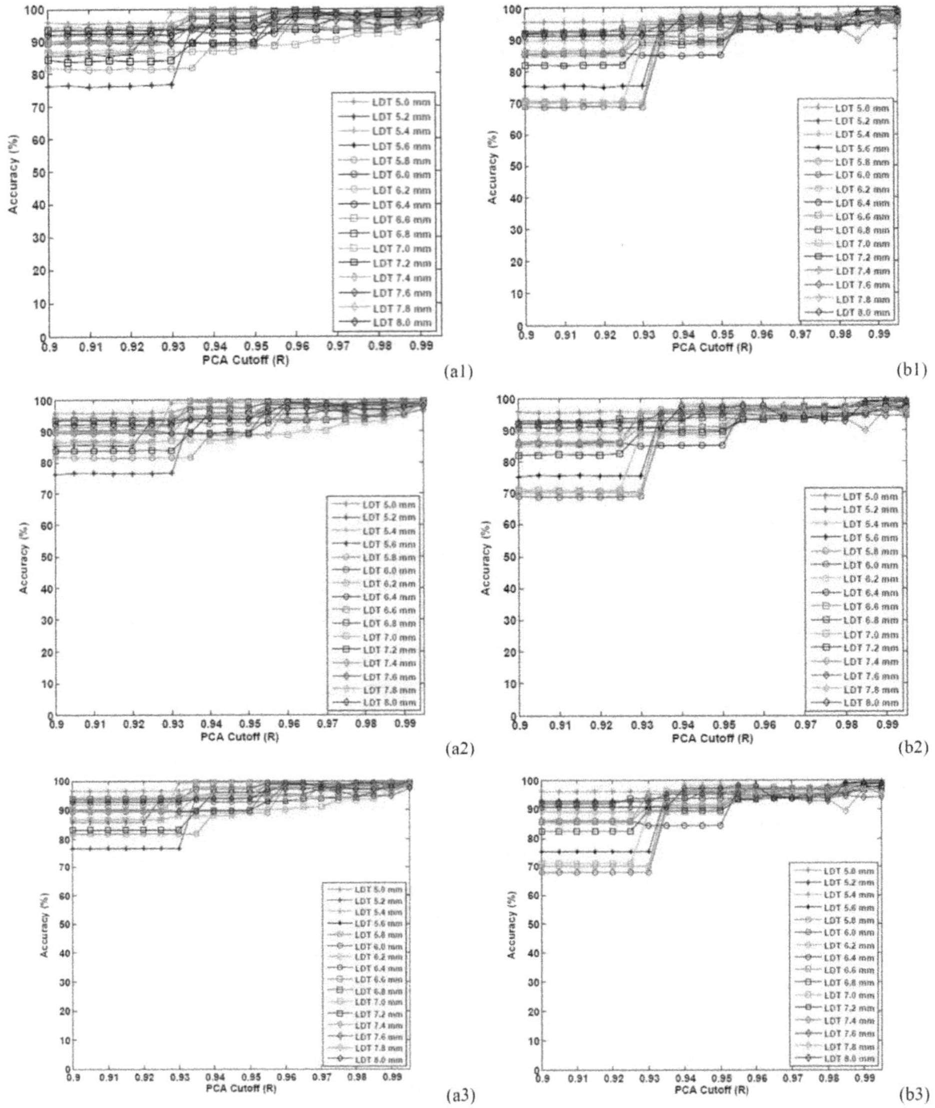

Figure 9.10. Accuracy versus PCA cutoff for the far wall: (a1), (a2) and (a3) automated and (b1), (b2) and (b3) manual.

carotid walls are well detected in the far wall region compared to the near wall region during the image acquisition process. As the PCA cutoff increases beyond 0.95, the classification accuracy stabilizes to nearly 95%, converging to 100% when cutoff (*R*) is 0.995. This behavior is consistent for all three walls, as can be seen in figures 9.10–9.12. Note that the automated (sRAS) systems showed the same behavior as the manual (mRAS) systems, showing stability, reliability and consistency in the machine-learning systems.

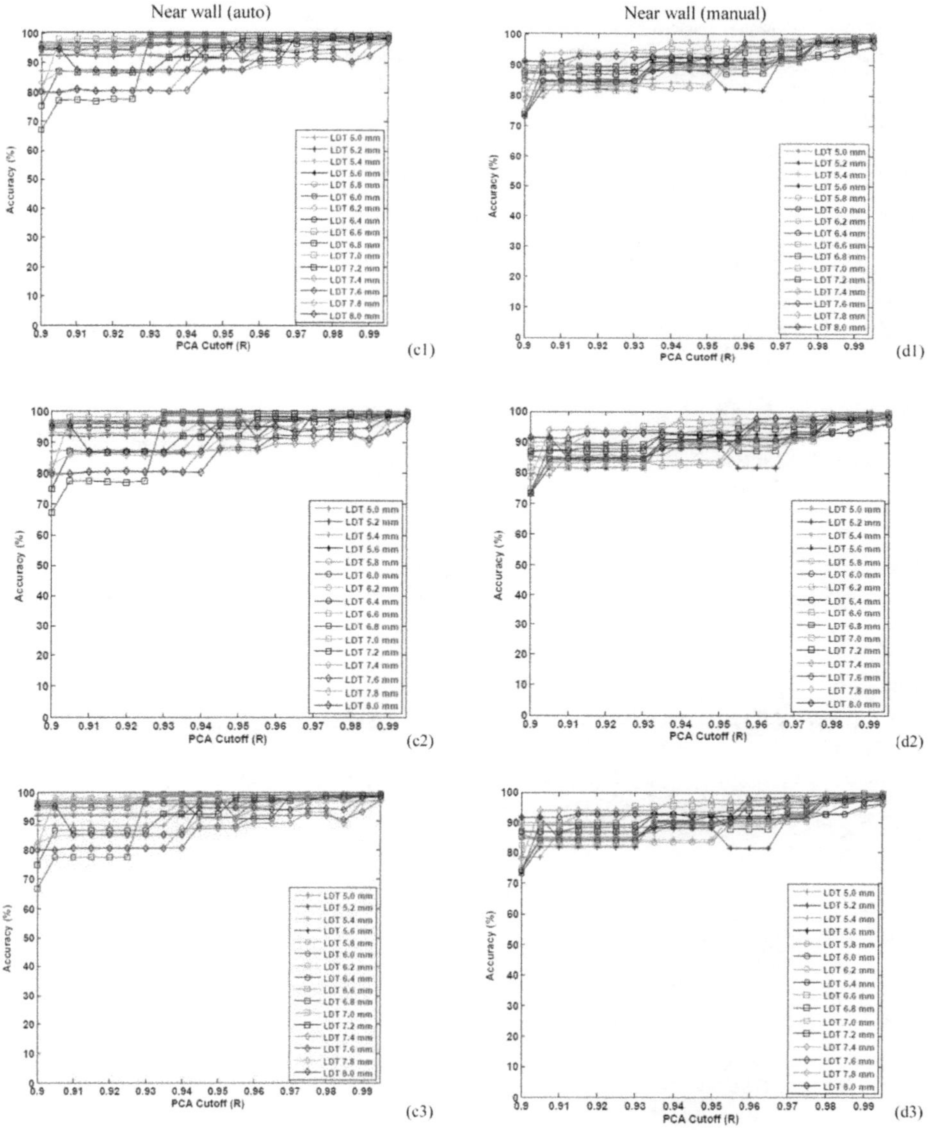

Figure 9.11. Accuracy versus PCA cutoff for the near wall: (c1), (c2) and (c3) automated and (d1), (d2) and (d3) manual.

9.4.2 Experiment 2: the role of data size in the performance of machine-learning

The role of this experiment is to determine what data size for the machine-learning system should be to achieve generalization from the behavior of memorization, given the CV protocol, feature set, PCA cutoffs and wall type. Thus, we partition the dataset into ten datasets with a sampling size increment of 10% starting from a data size of 40 patients approximately equally partitioned into two classes (high risk and

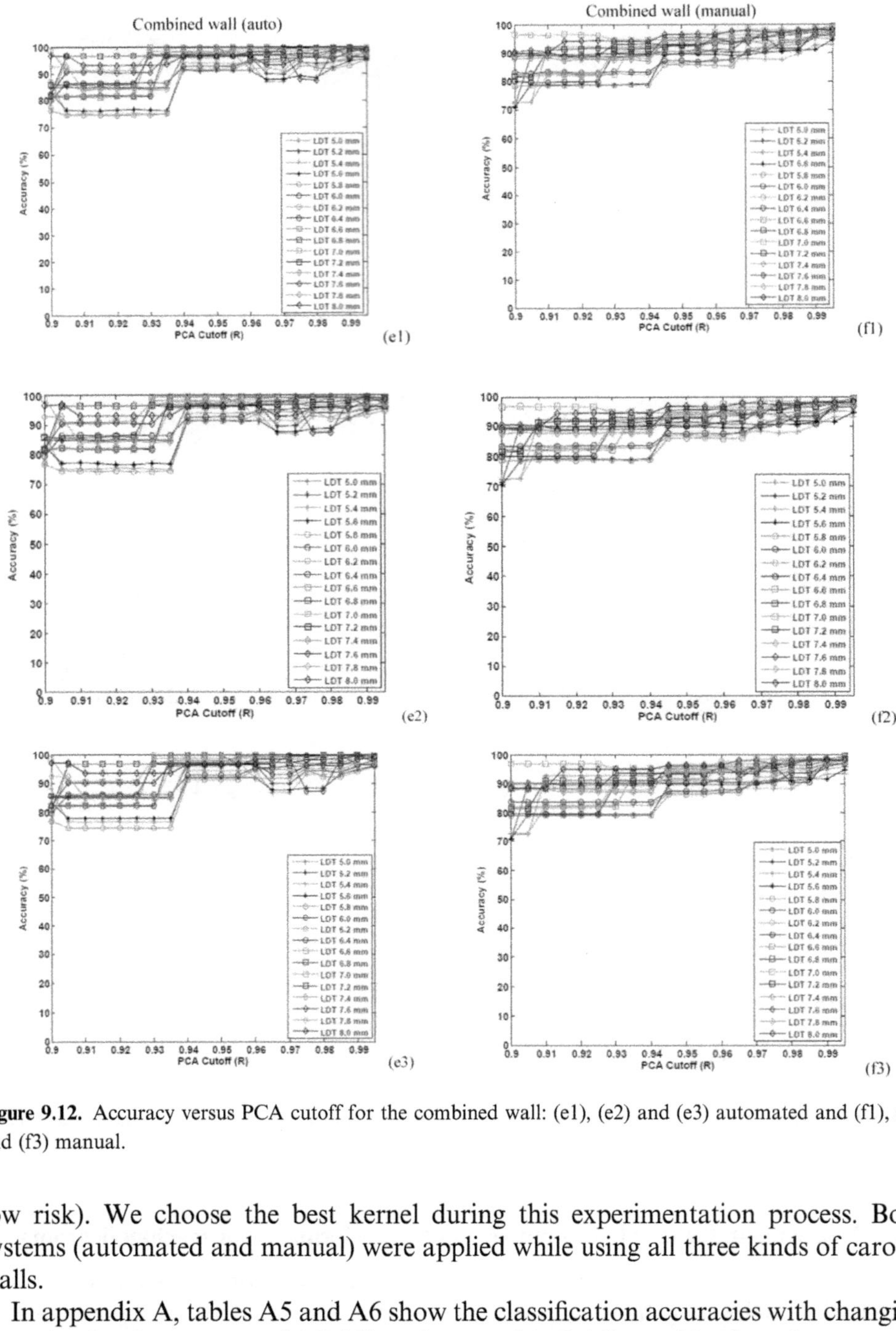

Figure 9.12. Accuracy versus PCA cutoff for the combined wall: (e1), (e2) and (e3) automated and (f1), (f2) and (f3) manual.

low risk). We choose the best kernel during this experimentation process. Both systems (automated and manual) were applied while using all three kinds of carotid walls.

In appendix A, tables A5 and A6 show the classification accuracies with changing data size for the automated (sRAS) and manual (mRAS) machine-learning systems, respectively. The corresponding plots can be seen in figures 9.13–9.15, respectively, reflecting the far wall, near wall and combined walls. Figure 9.13 is divided into two

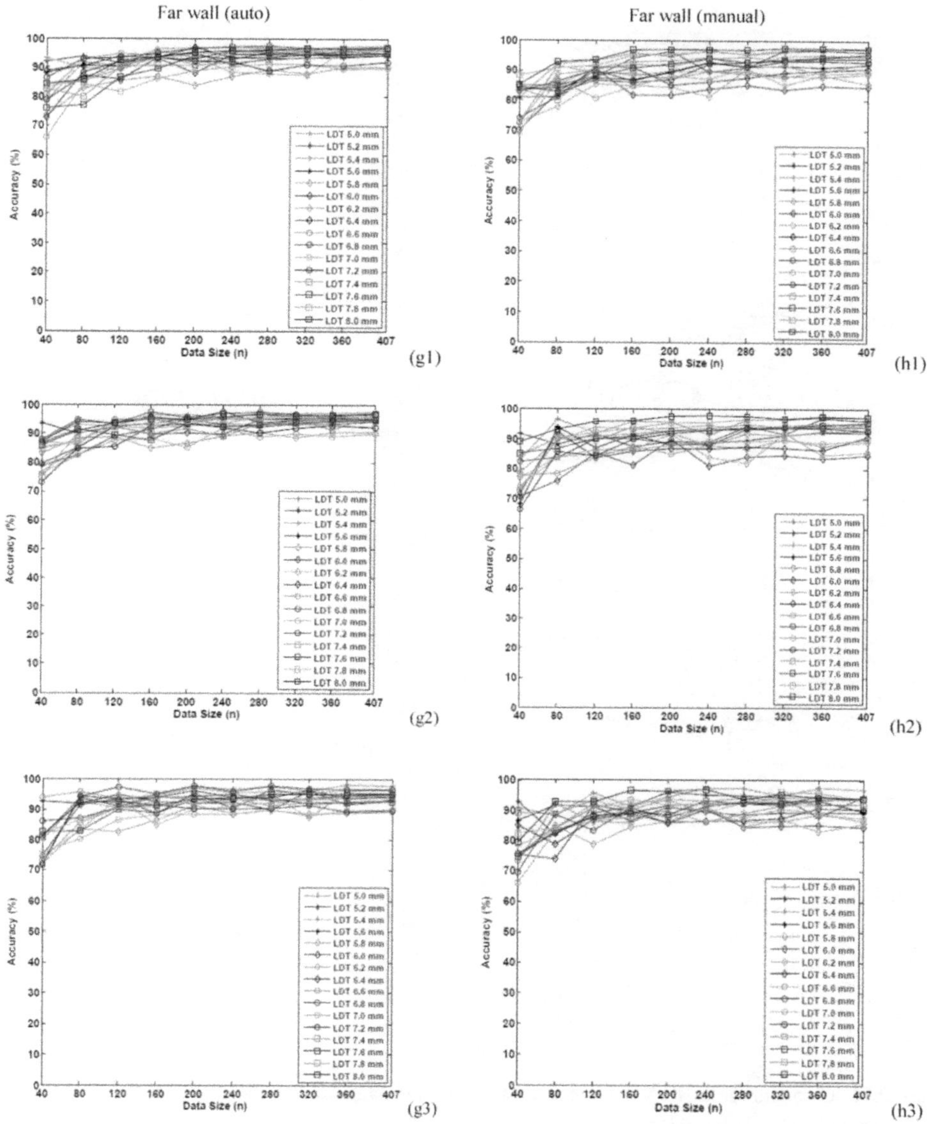

Figure 9.13. Changing data size experiment accuracy versus data size for the far wall: (g1), (g2) and (g3) automated and (h1), (h2) and (h3) manual.

kinds of plots, corresponding to the automated (g1), (g2), (g3) and manual (h1), (h2), (h3) systems. Figure 9.14 is divided into two kinds of plots, corresponding to the automated (i1), (i2), (i3) and manual (j1), (j2), (j3) systems. Figure 9.15 is divided into two kinds of plots, corresponding to automated (k1), (k2), (k3) and manual (l1), (l2), (l3) systems. For the mean accuracies corresponding to all 16 LDs and 20 PCA cutoffs, the changing data size experiment results can be seen in figure 9.16 for the automated (m1), (m2), (m3) and manual (n1), (n2), (n3) systems.

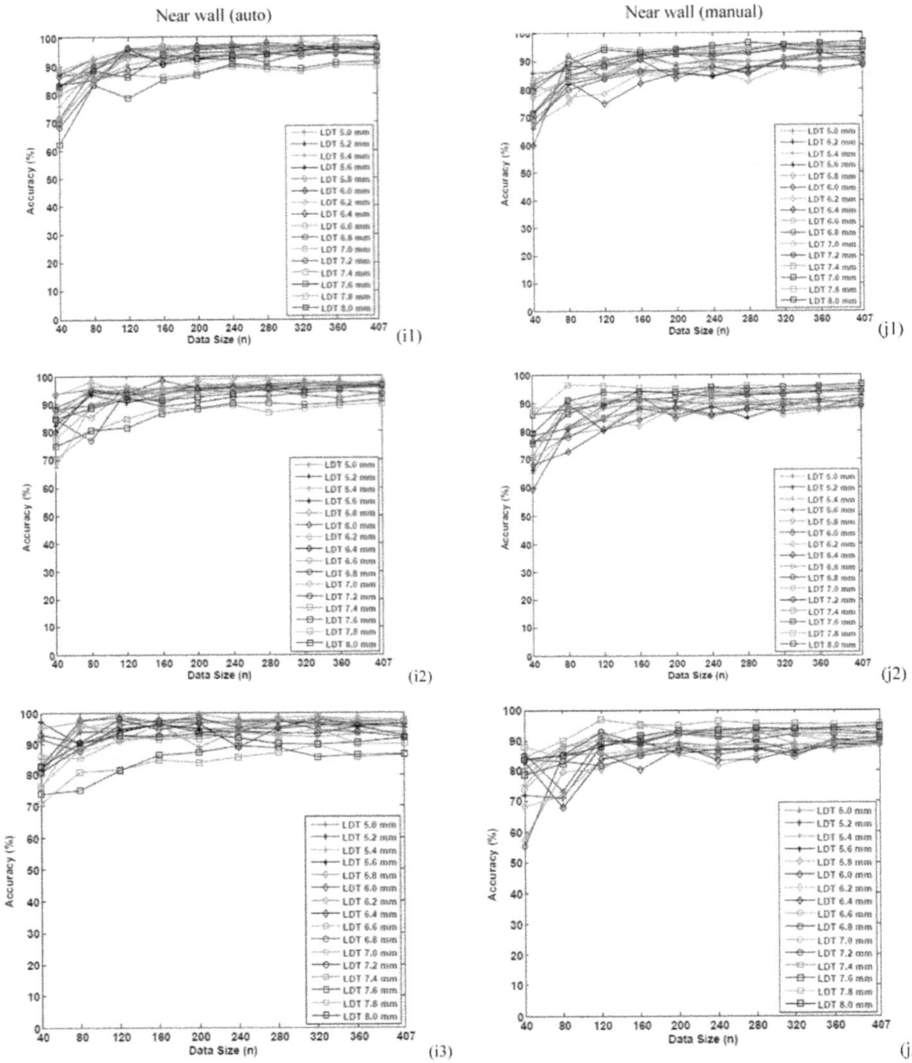

Figure 9.14. Changing data size experiment accuracy versus data size for the near wall: (i1), (i2) and (i3) automated and (j1), (j2) and (j3) manual.

The following interpretations can be given for this experiment. With an increase in data size, the classification accuracy increases gradually and then becomes flat beyond a certain data size. This behavior is seen in all the walls for the automated and manual systems. The point where the accuracy becomes nearly flat is the point where the system is running under the generalization mode. In the far wall this was observed for 240 patients, while in the near and combined walls it was close to 280 patients. This means nearly 60% of the data size is needed for this combination of feature sets, PCA cutoffs and wall types. Note that near wall plaques are noisier

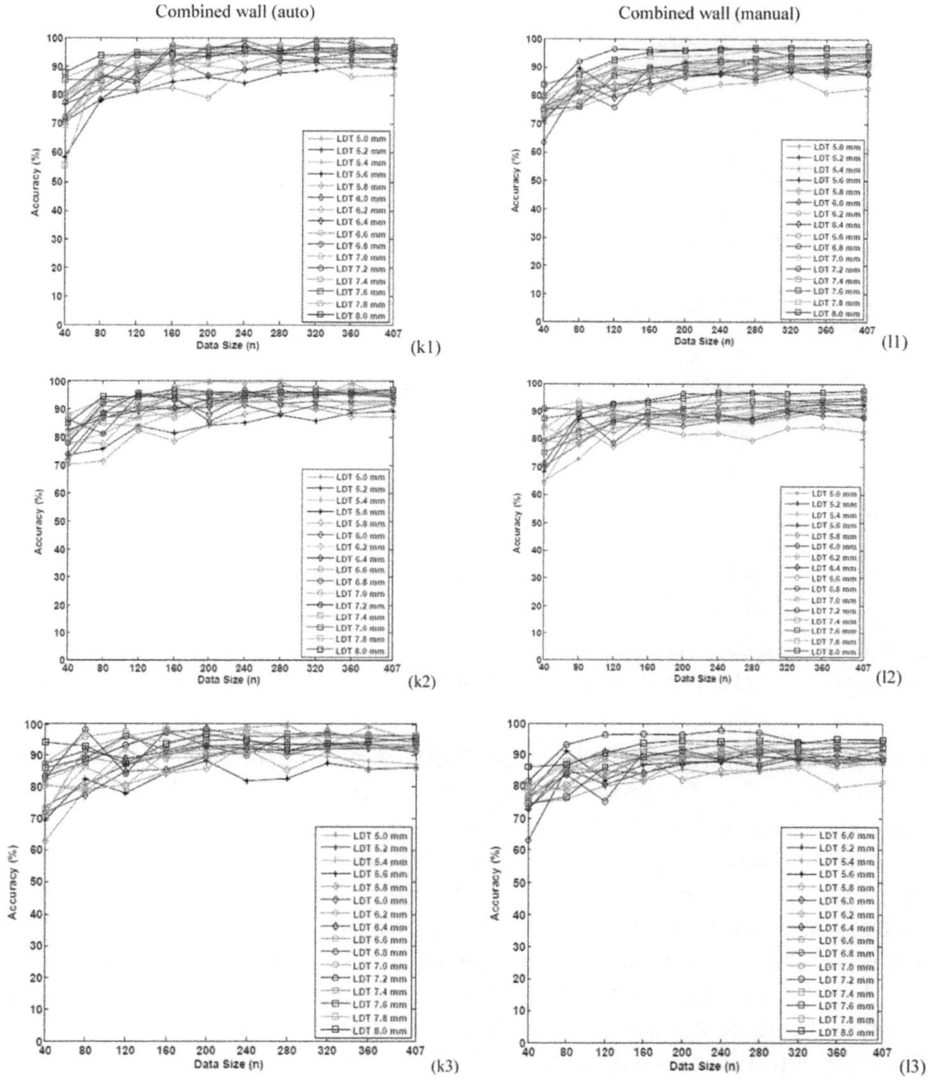

Figure 9.15. Changing data size experiment accuracy versus data size for the combined wall: (k1), (k2) and (k3) automated and (l1), (l2) and (l3) manual.

compared to far wall plaques and hence a greater number of patients are needed to stabilize the generalization during the CV protocol.

9.5 Performance evaluation

The performance of the system can be characterized by looking ay the variations in outputs of the automated system against the manual system considered as the gold standard. These variations can be quantified into four different categories, such as (i) precision-of-merit (PoM) computation; (ii) the reliability of the automated

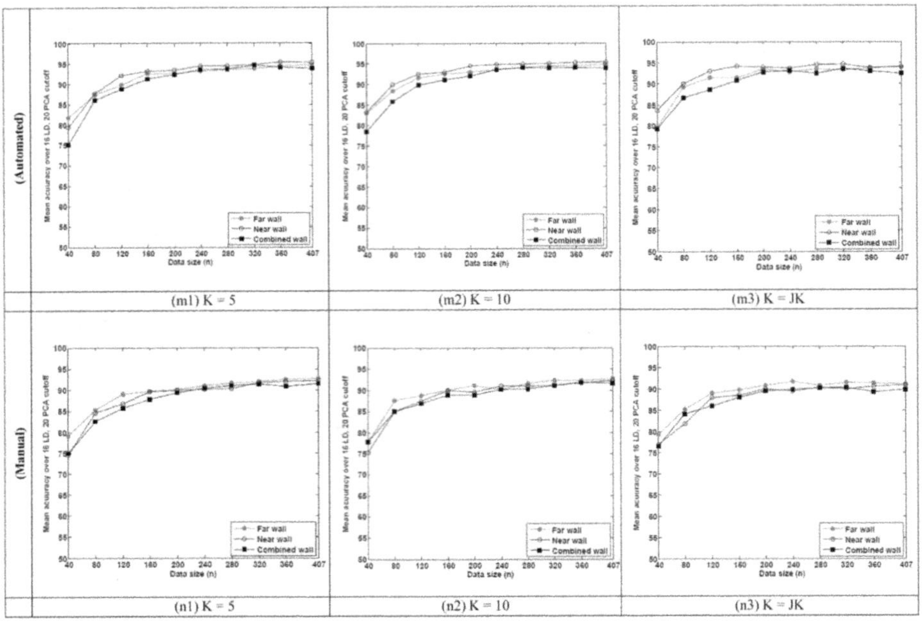

Figure 9.16. Changing data size experiment mean accuracy over all 16 LDs and 20 PCA cutoffs versus data size for the far, near and combined walls: (m1), (m2) and (m3) automated and (n1), (n2) and (n3) manual.

system; (iii) the feature retaining power (FRP) of the machine-learning system and (d) the stability of the machine-learning system. The precision of the system can be calculated by comparing the automated performance against the manual performance, to be mathematically given in section 9.5.1. The reliability of the system can be judged by the response of the system under normal conditions while meeting the hypothesis and assumptions, and will be shown section 9.5.2. Feature retaining power is the ability of the machine-learning system to retain the powerful features as the PCA cutoff increases, and will be discussed in section 9.5.3. Finally, the stability of the system can be calculated if the deviation in mean accuracy corresponding to all the PCA cutoffs for each data size is within the tolerance limit of say 5%, and will be shown in section 9.5.4.

9.5.1 Precision-of-merit (PoM) analysis

PoM analysis of the machine-learning system is performed by replacing the automated grayscale IMT wall strip by the manually segmented IMT wall strip in the training and learning phases. The classification technique (SVM) and the CV protocols (K5, K10 and JK) adapted for the validation of our systems are exactly the same as we applied to the automated set-up. Mathematically, PoM is given by

$$\mathrm{PoM}_w(k) = 100 - \left(\frac{|\eta_w^{\mathrm{auto}}(k) - \eta_w^{\mathrm{manual}}(k)|}{\eta_w^{\mathrm{manual}}(k)} \right) \times 100, \tag{9.7}$$

where $\mathrm{PoM}_w(k)$ is the PoM for the wall type w with partition protocol k, and $\eta_w^{\mathrm{auto}}(l, d, t)$ is the accuracy using the automated method for wall type w, lumen diameter l, selected dominant feature d and trial number t, and is computed using equation (9.5). Correspondingly, $\eta_w^{\mathrm{manual}}(l, d, t)$ is the accuracy using the manual method for wall type w, lumen diameter l, dominant feature d and trial number t, and is computed using equation (9.6).

In appendix A, table A7 shows the PoM values for the far, near and combined walls for the three partition protocols, $K = 5$, $K = 10$ and $K = N$, and total trials ($T = 20$) for two different PCA-cutoff conditions. The mean PoMs for all the walls and all three protocols are computed on the basis of $R < 0.95$ or $R > 0.95$ or all Rs included. As can be seen, when $R > 0.95$ the PoM increases compared to when $R < 0.95$, which is intuitive and consistent with the behavior. The corresponding PoM numbers for the far, near and combined walls with 5% error are 94.43, 88.20 and 90.63 for $K = 5$, 94.55, 88.22 and 90.69 for $K = 10$, and 94.55, 88.17 and 90.74 for $K = \mathrm{JK}$, respectively.

Using the PoM values we generated probability curves for the conditions that the far wall PoM is greater than the near wall PoM and the far wall PoM is greater than the combined wall with a tolerance limit of 5%. These probability curves are shown in figures 9.17(a1)–(a3) for the far wall greater than the near wall and (b1), (b2) and (b3) for the far wall PoM greater than that of the near wall for the three kinds of CV protocols (K5, K10 and JK), respectively. In each curve we stratify the probability curves into three bins: mild (or low), moderate (or medium) and severe (or high) risk. Their computed LD ranges were as follows: mild (low) risk was 7.4 mm–8.0 mm; moderate (or medium) risk was 6.4–7.2 mm; and severe (or high) risk was 5.0–6.2 mm. Further, it is clear from figure 9.17 that a PoM of 90% is achieved when the PCA cutoff R is greater than or equal to 0.95. We therefore selected $R = 0.95$ as a threshold for seeing the overall PoMs.

9.5.2 Reliability analysis of the sRAS

Let the data size set be represented by S given by $S = \{40, 80, 120, \ldots, 360, 407\}$ consisting of i elements, where i takes the values of 1, …, 10. Let N_i be the data size of elements i in set S. The reliability index of the dataset S is mathematically expressed as

$$\mathrm{RI}_N\,(\%) = \left(1 - \frac{\sigma_N}{\mu_{\mathrm{ACC}}}\right) \times 100, \tag{9.8}$$

where σ_N represents the standard deviation and μ_{ACC} represents the mean accuracy for the data size N_i under consideration, and is computed by taking all 20 PCA cutoffs. The objective is to see how reliable the sRAS system is with an increment in PCA cutoff (which in turn increases the optimal feature population) while increasing the data size (N_i). For this analysis we had datasets ranging from 40 to 407 with

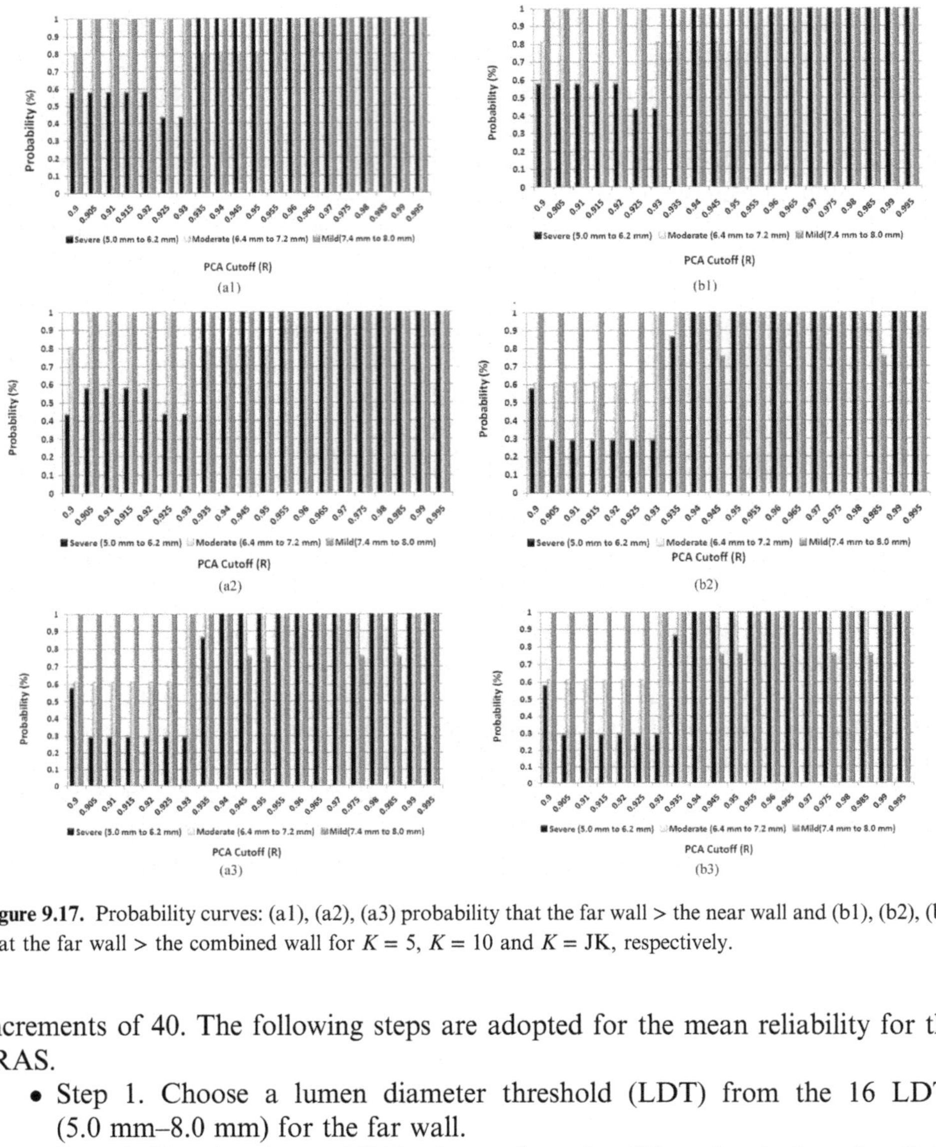

Figure 9.17. Probability curves: (a1), (a2), (a3) probability that the far wall > the near wall and (b1), (b2), (b3) that the far wall > the combined wall for $K = 5$, $K = 10$ and $K = \text{JK}$, respectively.

increments of 40. The following steps are adopted for the mean reliability for the sRAS.

- Step 1. Choose a lumen diameter threshold (LDT) from the 16 LDTs (5.0 mm–8.0 mm) for the far wall.
- Step 2. For the chosen LDT select a data size (N) and calculate the sRAS classification accuracy using PCA-based cutoffs (R), varying from 0.900 to 0.995 with a step size of 0.005, i.e. 20 cutoffs.
- Step 3. Calculate the mean classification accuracy (μ_{ACC}) and standard deviation (σ_N) for the over all 20 cutoffs.
- Step 4. Compute the reliability index (RI_N) using equation (9.8) for all ten data sizes.
- Step 5. Repeat steps 2, 3 and 4 for all ten data sizes (N). Compute the mean reliability index of the carotid risk assessment system ($\overline{\text{RI}}_{\text{sRAS}}$) corresponding to all data sizes using

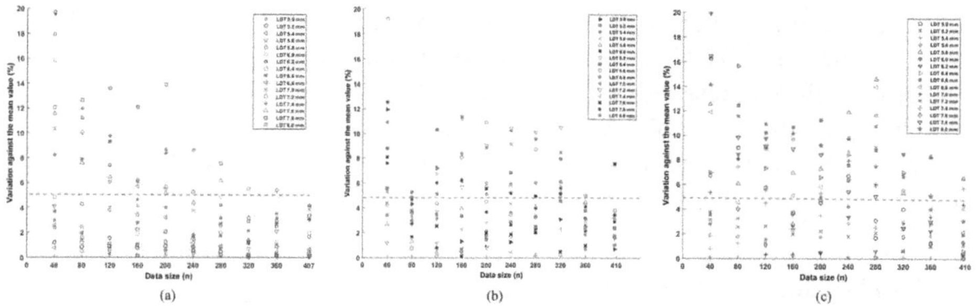

Figure 9.18. Stability analysis. Deviation against mean for 5.0 mm LD: (a) far wall, (b) near wall and (c) combined wall. The red line represents that the deviation against the mean is 5% below the specified value for data size 407.

$$\bar{\mathrm{RI}}_{\mathrm{sRAS}} = \left(\frac{\sum_{N=1}^{N_c} \mathrm{RI}_N}{N_c} \right). \tag{9.9}$$

- Step 6. Repeat step 1 for different LDTs.
- Step 7. Repeat all the steps for the near wall and the combined walls.

Here, N_c represents the cardinality of set S and N_d is the total number of elements in set S. Each element is presented by the data size N. The overall reliability index $\bar{\mathrm{RI}}_{\mathrm{sRAS}}$ of the system for $K = 10$, is 94.56%, 95.63% and 93.88% for the far, near and combined walls, respectively, while the overall reliability index $\bar{\mathrm{RI}}_{\mathrm{mRAS}}$ for the far, near and combined walls is 94.33%, 93.85% and 92.96%, as shown in appendix A, table A8, and figure 9.18.

9.5.3 Feature retaining power of the sRAS

We calculated the feature retaining power (FRP) after applying PCA at different cutoffs for fixed data size (N), i.e. 407 images. The FRP shows how many features are retained for each successive cutoff. FRP is evaluated as follows:

$$\mathrm{FRP}(\%) = \frac{S_{f(i-j)}}{\mathrm{FRP}_i} \times 100, \tag{9.10}$$

where $S_{f(i-j)}$ is the difference in the number of features at two consecutive cutoffs, say FRP_i (lower cutoff) and FRP_j (consecutive cutoff). The denominator FRP_i is the number of features at the lower cutoff. An example is discussed here. FRP is calculated at different cutoff ranges from 0.900 to 0.995 in a step size of 0.005. In FRP, successive features at various cutoffs $S_{f(i-j)}$ are calculated. For example, if the number of features at $\mathrm{FRP}_{0.900}$ is 3 and at $\mathrm{FRP}_{0.905}$ is 3, then the number of similar features in both successive cutoffs $S_{f\,(0.900-0.905)}$ is 3, and thus $S_{f\,(0.900-0.905)}$ is 100%. Likewise, we have calculated $S_{f(i-j)}$ for successive features at different cutoffs. Therefore, we have found average FRPs for the sRAS of 95.39%, 94.66% and 97.63% for the far, near and combined walls, respectively, while for the mRAS these values are 95.08%, 94.61% and 98.07%, respectively, as shown in appendix A, table A9. Also, a list of important features which are selected at each PCA cutoff is given in appendix A, table A10 for the far and near wall.

9.5.4 Stability analysis of the sRAS

The system is said to be stable if the deviation in mean sRAS accuracy corresponding to all the cutoffs for each data size is within the tolerance limit of 5%. The stability of the system is computed in the following way [9, 62]:

- Step 1. Choose any of the 16 lumen diameter thresholds from 5.0 mm to 8.0 mm.
- Step 2. Compute the risk classification accuracy for each data size (N) for corresponding PCA cutoffs (R) varying from 0.900 to 0.995 with a step size of 0.005.
- Step 3. Compute the mean of all risk classification accuracies corresponding to all cutoffs (R).
- Step 4. Compute the deviation of carotid risk classification accuracy from the mean accuracy at every cutoff (R).
- Step 5. Check if the deviation is under the tolerance limit of 5% for the mean accuracy value and declare the system to be stable.
- Step 6. Repeat step 2 to step 5 for each data size (N) and check against the tolerance limit and declare the stability for that LD threshold.
- Step 7. Draw the discrete curve showing the x-axis as increasing data size and y-axis as percentage deviation against the mean accuracy.

Thus, from the result of our proposed system, as shown in figure 9.19, it can be seen that for an LD threshold of 6.2 mm and for each data size, the deviation in mean accuracy corresponding to all the cutoffs is within the tolerance limit of 5%. Since the ability to retain the dominant features, reliability and stability analysis all meet the requirements under the limits of performance evaluation, we therefore conclude that our sRAS yields consistent results while meeting the assumptions. Even though we are able to retain 100% of the features as cutoff advances, at some cutoffs there is a slight drop in FRP due to the inclusion of some noisy features.

Our receiver operating characteristic (ROC) curve (figure 9.20) for an LDT of 6.2 mm using the K10 protocol gave the following areas under the curve (AUCs): (a1) far wall (AUC = 0.993), (a2) near wall (AUC=0.998) and (a3) combined walls (AUC = 0.979) for the automated segmented wall and (b1) far wall (AUC = 0.968), (b2) near wall (AUC = 0.958) and (b3) combined wall (AUC = 0.958) for the manually segmented wall.

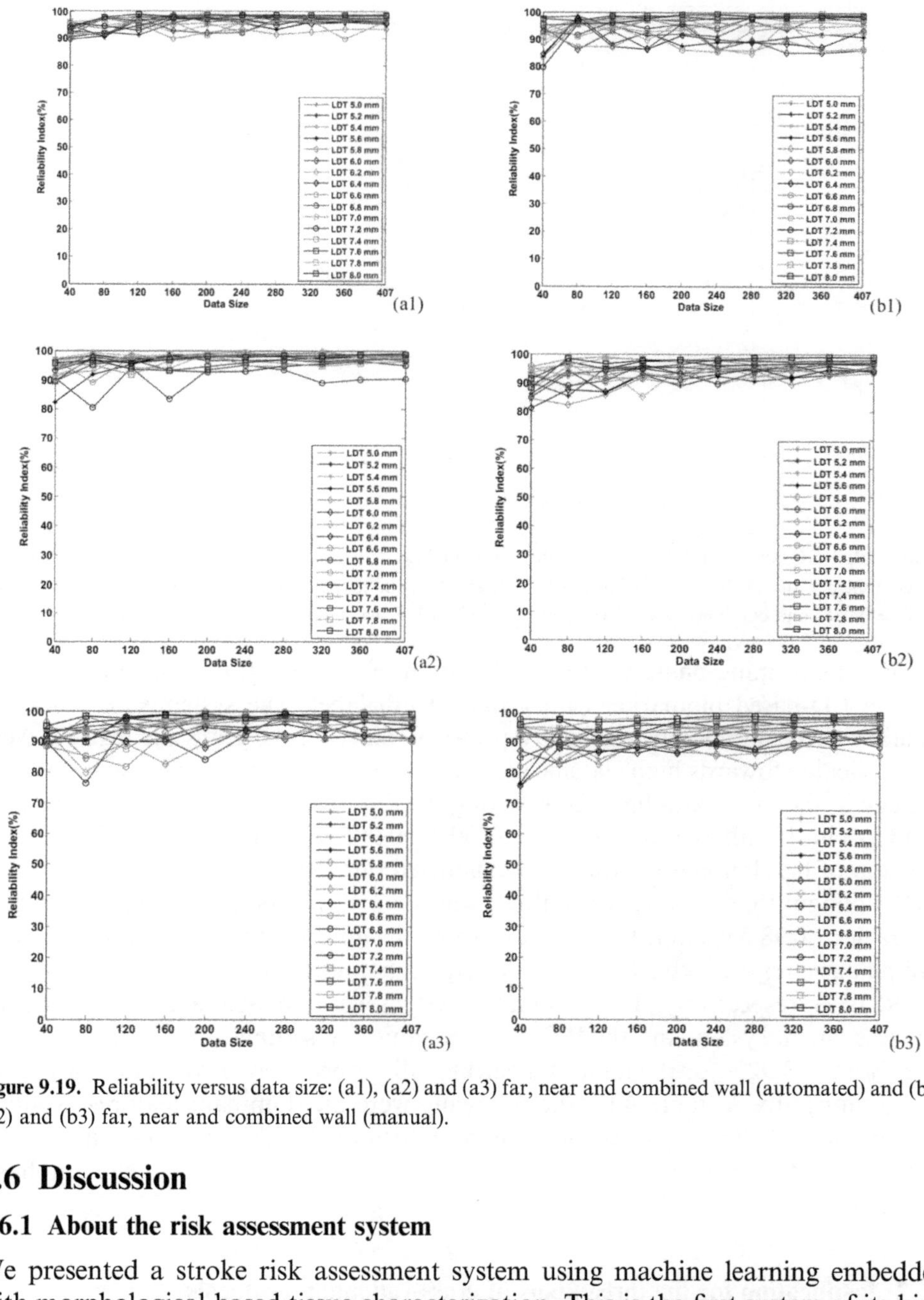

Figure 9.19. Reliability versus data size: (a1), (a2) and (a3) far, near and combined wall (automated) and (b1), (b2) and (b3) far, near and combined wall (manual).

9.6 Discussion

9.6.1 About the risk assessment system

We presented a stroke risk assessment system using machine learning embedded with morphological-based tissue characterization. This is the first system of its kind where the risk assessment can be facilitated independently of the three carotid wall types: (a) far, (b) near and (c) combined far and near walls. The machine-learning system adopted an SVM-based classifier during the training and testing phases,

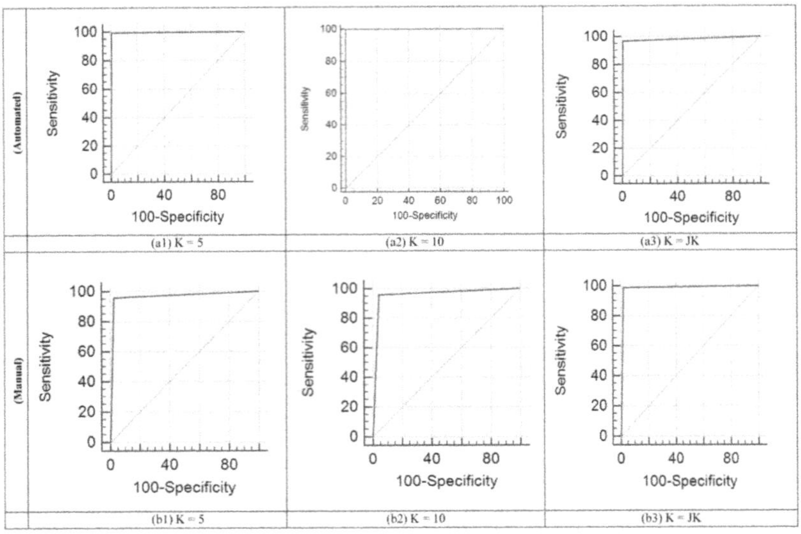

Figure 9.20. ROC curves: (a1), (a2) and (a3) the automated far wall (AUC = 0.993), near wall (AUC = 0.998) and combined walls (AUC = 0.979), and (c1), (c2) and (c3) the manual far wall (AUC = 0.968), near wall (AUC = 0.958) and combined walls (AUC =0.958). The LDT was 6.2 mm using the K10 protocol.

wherein the training phase used a combination of training grayscale features along with an LD-based biomarker as a ground truth label. The system's classification accuracy was computed using the CV protocol. Since all the grayscale features were not applicable towards high classification accuracy, we thus adopted a polling-based strategy in the PCA paradigm for selecting the dominant features. The system used 20 PCA-cutoff values ranging from 0.900 to 0.995 with an interval of 0.005 for selection of the dominant features. We modeled our dataset by using 16 biomarker (LD) configurations along with three sets of CV protocols leading to highest accuracies of 98.55% and 98.83% corresponding to the carotid far wall and near wall plaques, respectively. The corresponding reliabilities of the system were 94.56% and 95.63%, respectively. The automated system was validated against the manual risk assessment system and the PoM for the same CV settings and PCA cutoffs are 98.28% and 93.92% for the far and the near wall, respectively. From our analysis, as seen in appendix A, table A10, the following were the dominant features: 6—long run emphasis (LRE); 7—gray level non-uniformity (GLN); 8—run length non-uniformity (RLN); 9—run percentage (RP); and 10—low gray level run emphasis (LGRE).

9.6.2 Justification for the three kinds of cross-validation protocols

The performance of the machine learning is characterized by the choice of population in each pool: training versus testing. Our population pool had varying degrees of stenosis (1%–10%). In our model, we had 16 types of LD threshold biomarker values ranging from 5 mm to 8 mm in intervals of 0.2 mm. This was stratified into low- and high-risk bins with a limited number of subjects in each bin.

Further, the variability due to the grayscale features and types of SVM-kernels played an equal role in the optimization of stratification accuracy. Thus, the performance of the risk assessment system is governed by the type of CV protocol (K5, K10 or JK) chosen. It is therefore imperative to compare and contrast these protocols to understand the combined effect of the above parameters on the machine-learning paradigm [9, 59, 60].

9.6.3 Choice of biomarker (LD versus cIMT)

LD reflects blockages due to atherosclerotic formation of plaque and has been used previously as an adjunct diagnostic indicator for carotid CT angiography leading to coronary artery bypass grafting (CABG) and cardiovascular events (CVE) [19]. This is an instant carotid biomarker based on Doppler readings. Therefore, it can be a handy biomarker to be used as a the ground truth in the machine-learning paradigm. The weakness of using LD as a biomarker is its inability to access the morphologically based risk components present in walls. For this reason, the machine-learning system uses the grayscale features of the arterial walls rather than (i) between the arterial walls (i.e. lumen region) or (ii) the thickness of the arterial walls (cIMT). Thus, the combination of LD as the biomarker and the grayscale features of the carotid walls are jointly used for developing the learning phase of the ML system.

9.6.4 A note on wall segmentation validation

In the automated risk assessment system, automated LD and IAD borders were utilized. The LD borders were estimated during the segmentation of the LD region. These segmented borders were then validated against the ground truth LD, wherein two observers were used for inter-observer variability studies. The manual tracings were obtained from the trained observer twice, traced over a period of two weeks [55]. Note that the observer was not given access to the previous tracings. ImgTracer™, commercial software from AtheroPoint™, Roseville, CA, USA [12, 13, 42, 48–50], was used for manual tracings of the lumen and adventitia borders [54]. During the manual tracings, the observer was able to zoom in to the wall region for better visualization. The output of ImgTracer™ was an ordered set of traced (x, y) coordinates.

9.6.5 Benchmarking against the current literature

A machine-learning based tissue characterization methodology with only three features was presented in [1]. An SVM-based classification technique was used in this work for learning and testing. The authors obtained a classification accuracy of 83.77%. A manual segmentation method of ROI estimation was adopted, unlike in the current study, which was fully automated. In another study [3], the same authors used 32 texture features obtaining an accuracy of 90.66%. The above studies were performed on plaque cut sections. In [4], the authors adapted both the plaque and far wall models for tissue characterization and risk assessment, leading to accuracies of 83% and 89.5%, respectively. In another model [5], risk stratification was attempted by varying ethnicity by taking data pools from UK and Portugal. Here, the authors adopted 16 features for plaque and 12 features for wall data, leading to accuracies of 85.3% and 93.1%, respectively. All the above models kept the of assessment risk as a primary focus.

Recently, in [6], the authors modeled atherosclerotic disease by identification of asymptomatic and symptomatic classes. They called it the Atheromatic™ system. Seven features, including three texture, two discrete (DWT) and two higher order spectral (HOS) features were used in this study. With the help of a radial basis kernel function and SVM classifier, the study showed an accuracy of 91.77%.

In [46], the authors demonstrated the application of a CADx system which is further being developed and commercialized by AtheroPoint, CA, USA. A total of 16 features were named: the Rayleigh parameter (4th, 5th, 6th), GLCM, wavelet, percentile (10, 50), degree of stenosis, echogenic cap, appearance, mean, skewness, mixture components and plaque disruption were used in this study. They obtained an accuracy of 77% for their CADx system. Under the class of Atheromatic™ systems, the authors of [8] used non-linear HOS features and two wall features with four different classifiers to achieve an accuracy of 99.1%. However CV and performance evaluation was not performed in this research.

9.6.6 Summary of our contribution

Stroke risk assessment models have been approached via two angles: measurement based on stenosis and characterization based on the components of the plaque walls. Our study used a combination of both in a machine-learning framework, leading to a very powerful paradigm for risk assessment. We can summarize different aspects of the novelty of our system: (i) a stroke risk assessment system independent of the wall type or the combined effect of both the near and far walls; (ii) a completely automated risk assessment design which includes automated lumen segmentation, wall region segmentation, tissue characterization during offline learning, and automated wall segmentation and risk prediction during the online phase; (iii) a polling-based PCA strategy for dominant grayscale feature selection during the learning and predicting phases; (iv) comprehensive data analysis for establishing the reliability, stability and robustness of the risk assessment system; and (v) validation of the system against the ground truth-based design of the risk assessment system and performance evaluation of the automated sRAS.

9.6.7 Strengths, weaknesses and extensions

The main strength of the system is its ability to select the dominant features during the characterization protocol leading to higher accuracies. Further, the system is adaptable to all three kinds of walls: near, far and combined. The system was validated via a ground truth segmentation model leading to a higher precision-of-merit. The main shortcoming of this model is a lack in the variability of the data size, and this can be overcome by data acquisition over multiple ethnicities spanning different countries.

Appendix A Experimental results

Table A1. Distribution of high-risk (HR) and low-risk (LR) images in our population-based on the LDT.

SN	LDT (mm)	# of images in HR	# of images in LR	% of images in HR	% of images in LR
1	5.0	46	361	11.30	88.70
2	5.2	71	336	17.44	82.56
3	5.4	95	312	23.34	76.66
4	5.6	121	286	29.73	70.27
5	5.8	165	242	40.54	59.46
6	6.0	198	209	48.65	51.35
7	6.2	234	173	57.49	42.51
8	6.4	269	138	66.09	33.91
9	6.6	302	105	74.20	25.80
10	6.8	326	81	80.10	19.90
11	7.0	342	65	84.03	15.97
12	7.2	358	49	87.96	12.04
13	7.4	374	33	91.89	8.11
14	7.6	381	26	93.61	6.39
15	7.8	389	18	95.58	4.42
16	8.0	395	12	97.05	2.95

Table A2. Kernel optimization results for the automated far, near and combined walls, $K = 10$ and 20 PCA cutoffs.

	Far wall				Near wall				Combined wall			
PCA cutoff	RBF	Poly-1	Poly-2	Poly-3	RBF	Poly-1	Poly-2	Poly-3	RBF	Poly-1	Poly-2	Poly-3
0.900	89.14	87.02	88.80	88.80	87.41	86.29	87.96	87.88	82.12	83.21	85.02	84.64
0.905	89.21	86.99	88.81	88.79	89.69	89.38	90.66	90.31	83.66	85.56	86.28	85.66
0.910	89.16	87.06	88.79	88.69	89.44	88.97	90.16	90.48	84.02	84.33	85.55	84.87
0.915	89.15	87.07	88.79	88.68	89.46	88.95	90.21	90.50	84.04	84.33	85.43	84.80
0.920	89.11	87.01	88.76	88.72	89.44	88.97	90.14	90.46	83.98	84.33	85.52	84.81
0.925	89.57	87.64	89.31	89.17	89.49	88.97	90.21	90.53	84.03	84.38	85.51	84.88
0.930	90.00	88.33	90.40	89.95	91.28	90.78	92.29	92.54	86.29	86.67	87.72	87.36
0.935	92.59	92.40	93.17	92.46	91.59	91.15	92.64	92.94	88.39	88.98	90.07	89.78
0.940	93.56	93.59	94.02	93.64	92.49	92.33	93.70	93.61	93.18	95.48	96.15	95.48
0.945	93.51	93.60	94.06	93.65	92.97	93.30	95.01	94.15	93.19	95.49	96.18	95.43
0.950	93.68	93.57	94.12	93.66	93.05	93.29	95.02	94.10	93.14	95.45	96.18	95.42
0.955	94.20	94.97	95.31	94.53	92.40	93.32	94.57	93.50	93.16	95.46	96.21	95.17
0.960	94.86	95.61	96.58	95.62	93.30	95.23	95.72	94.39	93.28	95.80	96.21	95.07
0.965	94.97	95.80	96.69	95.72	93.30	95.47	95.68	94.46	92.49	94.21	94.93	94.03
0.970	94.82	95.99	96.95	95.72	94.05	96.37	96.88	96.16	92.40	94.29	95.14	94.10
0.975	94.73	95.84	96.76	94.90	94.20	96.67	97.32	96.70	91.70	94.72	95.57	93.64
0.980	94.43	96.29	96.73	95.09	93.70	96.84	97.50	96.52	91.35	94.62	95.67	93.75
0.985	94.40	96.44	96.95	95.14	92.62	96.35	97.04	95.89	91.80	95.64	96.85	95.07
0.990	94.30	97.23	97.70	96.01	93.81	97.75	98.28	96.24	90.78	96.83	97.58	94.88
0.995	93.40	98.87	98.55	96.58	92.65	98.32	98.83	96.43	90.11	97.44	97.73	94.71
Mean	92.44	92.57	93.56	92.78	91.82	92.93	93.99	93.39	89.15	91.36	92.28	91.18

Table A3. Fixed data size results of the automated far, near and combined wall accuracies (mean of all LDs).

	Far wall			Near wall			Combined wall		
PCA cutoff	$K=5$	$K=10$	$K=\text{JK}$	$K=5$	$K=10$	$K=\text{JK}$	$K=5$	$K=10$	$K=\text{JK}$
0.900	88.75	88.80	88.77	87.94	87.96	88.16	85.00	85.02	85.04
0.905	88.74	88.81	88.77	90.60	90.66	90.76	86.11	86.28	86.33
0.910	88.70	88.79	88.77	90.29	90.16	90.20	85.29	85.55	85.63
0.915	88.78	88.79	88.77	90.12	90.21	90.20	85.32	85.43	85.63
0.920	88.78	88.76	88.77	90.20	90.14	90.20	85.26	85.52	85.63
0.925	89.26	89.31	89.28	90.25	90.21	90.20	85.36	85.51	85.63
0.930	90.29	90.40	90.45	92.30	92.29	92.26	87.56	87.72	87.90
0.935	93.07	93.17	93.21	92.58	92.64	92.66	90.07	90.07	90.28
0.940	93.94	94.02	94.13	93.80	93.70	93.77	96.14	96.15	96.18
0.945	93.96	94.06	94.13	94.82	95.01	95.15	96.10	96.18	96.18
0.950	94.07	94.12	94.26	94.92	95.02	95.15	96.10	96.18	96.18
0.955	95.18	95.31	95.42	94.47	94.57	94.69	96.14	96.21	96.22
0.960	96.49	96.58	96.74	95.55	95.72	95.73	96.16	96.21	96.38
0.965	96.59	96.69	96.85	95.56	95.68	95.78	94.87	94.93	95.04
0.970	96.84	96.95	97.11	96.75	96.88	96.90	95.00	95.14	95.22
0.975	96.58	96.76	96.85	97.18	97.32	97.40	95.57	95.57	95.62
0.980	96.58	96.73	96.99	97.37	97.50	97.59	95.51	95.67	95.76
0.985	96.82	96.95	97.24	96.95	97.04	97.10	96.71	96.85	96.90
0.990	97.53	97.70	97.82	98.09	98.28	98.42	97.44	97.58	97.65
0.995	98.46	98.55	98.63	98.71	98.83	98.93	97.62	97.73	97.85
Mean	93.47	93.56	93.65	93.92	93.99	94.06	92.17	92.28	92.36

Table A4. Fixed data size results of the manual far, near and combined wall accuracies (mean of all LDs).

	Far wall			Near wall			Combined wall		
PCA cutoff	$K = 5$	$K = 10$	$K = JK$	$K = 5$	$K = 10$	$K = JK$	$K = 5$	$K = 10$	$K = JK$
0.900	83.28	83.38	83.32	82.80	82.87	82.88	82.45	82.38	82.34
0.905	83.25	83.31	83.32	86.36	86.42	86.46	84.13	84.08	84.09
0.910	83.20	83.34	83.32	86.30	86.40	86.52	86.62	86.60	86.58
0.915	83.26	83.34	83.32	86.64	86.75	86.92	86.62	86.69	86.75
0.920	83.16	83.37	83.32	86.68	86.78	86.92	86.58	86.69	86.75
0.925	83.38	83.46	83.43	86.63	86.75	86.92	86.61	86.73	86.75
0.930	86.44	86.45	86.46	87.00	87.09	87.22	87.94	88.02	88.18
0.935	92.14	92.23	92.28	89.71	89.82	89.97	88.52	88.64	88.87
0.940	92.94	93.19	93.24	90.35	90.36	90.43	88.46	88.69	88.87
0.945	93.32	93.41	93.50	90.39	90.46	90.45	91.67	91.81	92.09
0.950	93.31	93.46	93.50	90.44	90.52	90.59	91.71	91.89	92.09
0.955	95.56	95.65	95.82	90.86	91.02	91.00	91.75	91.88	92.09
0.960	95.45	95.58	95.73	92.05	92.20	92.32	92.63	92.74	92.92
0.965	95.46	95.58	95.72	92.02	92.19	92.32	92.60	92.90	93.01
0.970	95.56	95.65	95.64	94.13	94.32	94.39	93.70	93.87	93.90
0.975	95.46	95.60	95.56	94.19	94.32	94.39	94.32	94.46	94.47
0.980	95.46	95.56	95.59	96.85	97.00	97.02	94.59	94.72	94.82
0.985	96.09	96.29	96.31	97.07	97.18	97.24	94.95	95.08	95.25
0.990	97.14	97.34	97.51	97.57	97.72	97.83	96.48	96.65	96.64
0.995	97.18	97.26	97.27	98.20	98.33	98.42	97.06	97.20	97.42
Mean	91.05	91.17	91.21	90.81	90.92	91.01	90.47	90.59	90.69

Table A5. Changing data size experiment for the automated far, near and combined walls: accuracy for K5, K10 and JK over a mean of 16 LDTs.

	Far wall			Near wall			Combined wall		
Data size	$K = 5$	$K = 10$	$K = JK$	$K = 5$	$K = 10$	$K = JK$	$K = 5$	$K = 10$	$K = JK$
40	81.83	83.03	79.56	79.41	83.24	83.66	75.12	78.52	79.30
80	87.38	88.38	89.25	87.76	89.96	90.07	86.11	85.87	86.63
120	89.88	91.58	91.46	92.18	92.47	93.03	88.83	89.75	88.57
160	92.67	92.50	91.43	93.18	92.96	94.11	91.33	91.07	90.78
200	92.63	93.00	93.42	93.41	94.49	93.88	92.41	91.99	92.72
240	93.22	93.56	92.74	94.52	94.76	93.65	93.60	93.53	93.15
280	93.54	94.00	93.48	94.52	94.96	94.50	93.65	94.03	92.41
320	93.82	94.35	93.20	94.76	95.02	94.63	94.64	93.86	93.59
360	94.40	94.44	93.67	95.41	95.26	93.86	94.11	94.04	92.99
407	94.77	94.86	93.97	95.38	95.45	94.06	93.78	93.86	92.36
Mean	91.41	91.97	91.22	92.05	92.86	92.55	90.36	90.65	90.25

Table A6. Changing data size experiment for the manual far, near and combined wall: accuracy for $K = 5$, 10 and JK over the mean of 16 LDTs.

	Far wall			Near wall			Combined wall		
Data size	$K = 5$	$K = 10$	$K = JK$	$K = 5$	$K = 10$	$K = JK$	$K = 5$	$K = 10$	$K = JK$
40	79.12	77.80	79.27	74.65	75.27	77.03	74.96	77.84	76.61
80	85.39	87.66	85.27	84.65	85.05	81.82	82.63	85.01	84.17
120	89.02	88.80	89.15	86.78	87.55	88.10	85.67	86.88	86.10
160	89.69	90.11	89.86	89.59	89.93	88.56	87.87	88.89	88.09
200	90.10	91.16	90.89	89.98	89.65	90.01	89.34	88.93	89.60
240	91.03	90.18	91.73	90.26	91.07	89.56	90.42	90.24	89.95
280	91.58	91.55	90.82	90.34	91.07	90.34	90.99	90.31	90.30
320	91.98	92.35	91.56	91.80	91.14	90.08	91.39	91.14	90.33
360	92.44	92.30	91.46	92.04	91.73	90.62	90.80	91.80	89.33
407	92.66	92.79	91.21	92.27	92.35	91.01	91.52	91.62	89.92
Mean	89.30	89.47	89.12	88.24	88.48	87.71	87.56	88.27	87.44

Table A7. PoM for the far, near and combined walls and three protocols, $K = 5$, 10 and JK with 5% error.

	$K = 5$			$K = 10$			$K = \text{JK}$		
PCA cutoff	Far	Near	Combined	Far	Near	Combined	Far	Near	Combined
0.900	89.69	83.80	87.68	89.70	84.05	87.56	89.68	83.76	87.33
0.905	89.65	85.29	89.70	89.65	85.20	89.67	89.68	85.04	89.53
0.910	89.66	85.34	90.68	89.69	85.32	90.65	89.68	85.10	90.75
0.915	89.67	84.84	90.76	89.68	84.86	90.88	89.68	84.71	90.96
0.920	89.53	84.96	90.72	89.80	84.99	90.97	89.68	84.71	90.96
0.925	89.13	84.89	90.90	89.15	84.86	90.85	89.13	84.71	90.96
0.930	89.92	84.64	88.68	89.76	84.53	88.80	89.62	84.40	88.95
0.935	95.80	87.54	89.17	95.79	87.41	89.27	95.83	87.32	89.48
0.940	95.87	87.13	86.55	95.89	87.01	86.81	95.97	86.96	86.98
0.945	95.99	88.07	90.36	96.02	88.12	90.41	96.11	88.15	90.73
0.950	96.14	88.14	90.39	96.13	88.15	90.48	96.11	88.20	90.73
0.955	97.31	89.02	90.37	97.32	89.07	90.46	97.34	89.09	90.69
0.960	97.37	89.09	90.90	97.32	89.13	91.02	97.39	89.16	91.07
0.965	97.27	88.89	91.27	97.29	88.94	91.33	97.30	89.01	91.26
0.970	97.78	89.92	92.18	97.80	90.05	92.17	97.65	90.08	92.05
0.975	97.90	90.18	91.87	97.92	90.31	91.86	97.93	90.39	91.72
0.980	97.80	92.45	91.70	97.77	92.47	91.76	97.79	92.48	91.63
0.985	97.93	92.47	92.25	97.93	92.57	92.30	98.02	92.54	92.35
0.990	98.16	93.45	93.08	98.20	93.52	93.21	98.27	93.52	93.27
0.995	98.25	93.86	93.45	98.28	93.92	93.40	98.16	94.00	93.46
Mean	94.54	88.20	90.63	94.55	88.22	90.69	94.55	88.17	90.74
R < 0.95	91.91	85.88	89.60	91.93	85.86	89.67	91.92	85.73	89.76
R > 0.95	97.75	91.04	91.90	97.76	91.11	91.95	97.76	91.14	91.94

Table A8. Reliability versus data size for the far, near and combined walls for the $K = 10$ mean of 16 LDTs.

	Far wall		Near wall		Combined wall	
Data size	Auto	Manual	Auto	Manual	Auto	Manual
40	93.02	90.65	93.15	89.94	91.67	89.16
80	94.54	94.75	94.98	92.85	90.59	90.85
120	96.03	94.18	96.11	92.97	93.31	90.84
160	96.26	94.22	95.80	94.22	94.27	92.04
200	96.25	94.86	97.36	94.13	95.29	92.28
240	96.41	92.89	97.34	95.11	95.69	92.71
280	96.49	94.21	97.35	95.28	96.37	92.53
320	96.99	94.81	96.93	95.01	96.00	93.73
360	96.67	94.27	97.32	95.73	96.09	94.04
407	97.08	94.71	97.23	96.03	95.36	93.79
Mean	95.97	93.96	96.36	94.13	94.46	92.20

Table A9. Feature retaining power for the far, near and combined walls, $K = 10$, LD = 6.2 mm.

Far wall				Near wall				Combined wall			
Auto		Manual		Auto		Manual		Auto		Manual	
Similar features	FRP	Similar features	FRP	Similar features	FRP	Similar features	FRP	Similar features	FRP	Similar features	FRP
5	100	5	100	5	100	2	50	5	100	5	100
5	100	5	100	5	100	5	100	5	100	5	100
5	100	5	100	5	100	5	100	5	100	5	100
5	100	5	100	5	100	5	100	5	100	5	100
5	100	5	100	5	100	5	100	5	100	5	100
5	100	5	100	3	60	5	100	5	100	4	80
5	100	3	60	6	100	3	60	5	100	6	100
3	60	6	100	6	100	6	100	6	100	6	100
6	100	6	100	6	100	6	100	6	100	6	100
6	100	6	100	6	100	6	100	6	100	6	100
6	100	4	66.67	5	83.33	6	100	6	100	6	100
4	66.67	7	100	7	100	7	100	7	100	5	83.33
7	100	7	100	7	100	7	100	7	100	7	100
7	100	7	100	7	100	7	100	7	100	7	100
6	85.71	8	100	8	100	8	100	8	100	8	100
8	100	8	100	7	87.5	7	87.5	6	75	8	100
9	100	8	88.89	7	77.78	9	100	9	100	9	100
9	100	10	100	9	90	9	100	9	100	9	100
10	100	10	90.91	11	100	11	100	8	80	10	100
Mean FRP	95.39	Mean FRP	95.08	Mean FRP	94.66	Mean FRP	94.61	Mean FRP	97.63	Mean FRP	98.07

Table A10. Number of features selected at each PCA cutoff for the far and near wall (automated) 6.2 mm LDT.

PCA cutoff	Total no of features	Name of the features for far wall 6.2 mm LDT	Total no of features	Name of the features for near wall 6.2 mm LDT
0.900	5	6, 15, 7, 9, 10	5	7,14,6,9,10
0.905	5	6, 15, 7, 9, 10	5	7, 14, 6, 9, 10
0.910	5	6, 15, 7, 9, 10	5	7, 14, 6, 9, 10
0.915	5	6, 15, 7, 9, 10	5	7, 14, 6, 9, 10
0.920	5	6, 15, 7, 9, 10	5	7, 14, 6, 9, 10
0.925	5	6, 15, 7, 9, 10	5	7, 14, 6, 9, 10
0.930	5	6, 15, 7, 9, 10	6	6, 14, 7, 12, 16, 8
0.935	5	6, 15, 7, 9, 10	6	6, 14, 7, 12, 16, 8
0.940	6	6, 7, 2, 12, 8, 9	6	6, 14, 7, 12, 16, 8
0.945	6	6, 7, 2, 12, 8, 9	6	6, 14, 7, 12, 16, 8
0.950	6	6, 7, 2, 12, 8, 9	6	6, 14, 7, 12, 16, 8
0.955	6	6, 7, 2, 12, 8, 9	7	16, 7, 6, 14, 12, 2, 9
0.960	7	6, 16, 2, 10, 9, 7, 3	7	16, 7, 6, 14, 12, 2, 9
0.965	7	6, 16, 2, 10, 9, 7, 3	7	16, 7, 6, 14, 12, 2, 9
0.970	7	6, 16, 2, 10, 9, 7, 3	8	16, 7, 2, 9, 14, 8, 12, 6
0.975	8	6, 1, 12, 16, 3, 2, 10, 3	8	16, 7, 2, 9, 14, 8, 12, 6
0.980	9	6, 1, 12, 16, 2, 10, 3, 7, 9	9	16, 2, 6, 7, 12, 10, 9, 5, 14
0.985	9	6, 1, 12, 16, 2, 10, 3, 7, 9	10	16, 2, 12, 6, 7, 10, 1, 4, 8, 9
0.990	10	6, 1, 16, 12, 9, 3, 10, 2, 7, 5	11	2, 12, 16, 6, 7, 1, 10, 14, 8, 5, 9
0.995	11	16, 6, 1, 12, 9, 2, 3, 7, 10, 8, 5	12	7, 12, 2, 6, 10, 16, 9, 1, 14, 3, 5, 8

GLCM features: 1—entropy; 2—energy; 3—contrast; 4—homogeneity; GLRLM features: 5—short run emphasis (SRE); 6—long run emphasis (LRE); 7—gray level non-uniformity (GLN); 8—run length non-uniformity (RLN); 9—run percentage (RP); 10—low gray level run emphasis (LGRE); 11—high gray level run emphasis (HGRE); 12—short run low gray level emphasis (SRLGE); 13—short run high gray level emphasis (SRHGE); 14—long run low gray level emphasis (LRLGE); 15—long run high gray level emphasis (LRHGE); 16—fractal dimension feature.

Table A11. Comparison of various tissue characterization techniques from the literature against our proposed work (ACC = accuracy; Sn: sensitivity; Sp: specificity).

SN	Author	Algorithm	Segmentation technique of wall plaque	Data size and ethnicity of patients	Image features	Machine-learning technique	Cross validation	Validation against GT	Performance evaluation
1	[1] IEEE-TIM	Risk assessment in carotid plaque images (Atheromatic™)	Manual segmentation for plaque region	346 (Cyprus)	Total of 3 features: DWT based average (Dh1), average (Dv1), energy (*E*)	SVM	ACC: 83.7% PPV: 81.8% Sn: 80%, Sp: 86.4%	N/A	N/A
2	[3] IEEE-TIM	Carotid plaque tissue characterization and classification (Atheromatic™)	Manual segmentation for plaque region	160 (Portugal)	Total of 32 features: Texture	SVM	ACC: 90.66% Sn: 83.33% Sp: 95.39%	N/A	N/A
3	[4] UMB)	Risk stratification using texture based features (Atheromatic™)	(a) Manual segmentation for the plaque region. (b) Automated segmentation for the wall using AtheroEdge™.	(a) 346 plaque (b) 342 wall	(a) Total of 8 features: $LBP_{18}Ent$ $LBP_{18}Ene$ $LBP_{216}Ent$ $LBP_{216}Ene$ $LBP_{324}Ene$ LTE_5Ene LTE_6Ene LTE_8Ene (b) Total of 4 features: $LBP_{324}Ene$ LTE_2Ene LTE_8Ene $IMTV_{poly}$	(a) SVM, GMM, DT, KNN, NBC, RBPNN, Fuzzy (b) SVM, KNN, RBPNN	(a) ACC: 83.0% Sn: 87.4% Sp: 79.7% (b) ACC: 89.5% Sn: 89.6% Sp: 88.9%	N/A	N/A

(*Continued*)

4	[5] (MBEC)	Segmentation and risk assessment in carotid plaque images (Atheromatic™)	Manual segmentation for plaque region	346 (UK) 146 (Portugal)	Total features: 16 (UK dataset), 12 (Portugal dataset). FGLCM: *E*, contrast, entropy, correlation, correlation, homogeneity. FRLM: SRE, GLNU, ASM, mean, LRE, RLNU, RP.	SVM, DT, fuzzy	(a) ACC: 85.3% Sn: 84.4% Sp: 85.9% (b) ACC: 93.1% Sn: 99.0% Sp: 80%	N/A	N/A
5	[6] (CMPB)	Classification of carotid plaque images by tissue characterization (Atheromatic™)	Manual segmentation for plaque region	146	Total of 7 features: texture features (3), DWT (2) and HOS features (2)	SVM	ACC: 91.7% Sn: 97% Sp: 80%	N/A	SACI*: 7.04 ± 0.223 and 6.67 ± 0.186
6	[7] (JEM)	Plaque classification in the carotid artery using CT images	Manual segmentation for the plaque region	400 (Italy)	Total of 9 features: standard deviation, DWT, entropy, mean average, LBP	SVM, Adaboost	ACC: 88% Sn: 90.2%, Sp: 86.5%	N/A	Atheromatic index
7	[8] (JCU)	Tissue characterization for far wall (Atheromatic™)	Automated far wall segmentation	118 (Italy)	Total of 8 features: 6 HOS features and 2 wall features (IMT and IMTVpoly)	SVM, KNN, RBPNN, DT	ACC: 91.8% Sn: 83.3% Sp: 95%	N/A	N/A
8	Proposed method	Automated segmentation and risk assessment in carotid far wall and near wall	Automated segmentation for wall region	407 (Japan)	16 texture features: GLCM, GLRLM1. Chaotic features: FDi with feature optimization	SVM, PCA-based feature selection	Far wall: 98.55% Near wall: 98.83% Combined wall: 97.73%	Performed against GT	Stability, PoM, reliability

Appendix B Grayscale features

Table B1. Features of the gray level co-occurrence matrix (GLCM).

Feature name	Equation
Contrast (Con)	$\text{Con} = \sum_{n=0}^{L_g-1} n^2 \left\{ \sum_{i=1}^{L_g} \sum_{j=1}^{L_g} P_d(i, j) \mid \lvert i - j \rvert = n \right\}$
Energy (Eng)	$\text{Eng} = \sum_i \sum_j P_d(i, j)^2$
Entropy (Ent)	$\text{Ent} = -\sum_i \sum_j P_d(i, j) log(P_d(i, j))$
Homogeneity (HOM)	$\text{HOM} = \sum_i \sum_j \frac{1}{1 + (i + j)^2} P_d(i, j)$

Table B2. Features of gray level run length matrix.

Feature Name	Equation
Short run emphasis	$\text{SRE} = \frac{1}{l_t} \sum_{y=1}^{M} \sum_{y=1}^{L} \frac{r(x, y)}{y^2} = \frac{1}{l_t} \sum_{y=1}^{L} \frac{r_t(y)}{y^2}$
Long run emphasis	$\text{LER} = \frac{1}{l_t} \sum_{x=1}^{M} \sum_{y=1}^{L} (r(x, y).\ y^2) = \frac{1}{l_t} \sum_{y=1}^{L} r_t(y).y^2$
Gray level non-uniformity	$\text{GLN} = \frac{1}{l_t} \sum_{x=1}^{M} \left(\sum_{y=1}^{L} (r(x, y)) \right)^2 = \frac{1}{l_t} \sum_{x=1}^{L} r_q(x)^2$
Run length non-uniformity	$\text{RLN} = \frac{1}{l_t} \sum_{x=1}^{M} \left(\sum_{y=1}^{L} (r(x, y)) \right)^2 = \frac{1}{l_t} \sum_{x=1}^{L} r_t(y)^2$
Run percentage	$\text{RP} = \frac{l_t}{l_q}$
Low gray level run emphasis	$\text{LGRE} = \frac{1}{l_t} \sum_{x=1}^{M} \sum_{y=1}^{L} \frac{r(x, y)}{x^2} = \frac{1}{l_t} \sum_{y=1}^{L} \frac{r_q(x)}{x^2}$
High gray level run emphasis	$\text{HGRE} = \frac{1}{l_t} \sum_{x=1}^{M} \sum_{y=1}^{L} (r(x, y).\ x^2) = \frac{1}{l_t} \sum_{y=1}^{L} r_t(y).x^2$
Short run low gray level emphasis	$\text{SRLGE} = \frac{1}{l_t} \sum_{x=1}^{M} \sum_{y=1}^{L} \frac{r(x, y)}{x^2 . y^2}$
Short run high gray level emphasis	$\text{SRHGE} = \frac{1}{l_t} \sum_{x=1}^{M} \sum_{y=1}^{L} \frac{r(x, y) . x^2}{x^2}$
Long run low gray level emphasis	$\text{LRLGE} = \frac{1}{l_t} \sum_{x=1}^{M} \sum_{y=1}^{L} \frac{r(x, y) . y^2}{y^2}$
Long run high gray level emphasis	$\text{LRHGE} = \frac{1}{l_t} \sum_{x=1}^{M} \sum_{y=1}^{L} r(x, y).\ x^2.\ y^2$

References

[1] Acharya U R, Faust O, Sree S V, Molinari F, Saba L, Nicolaides A and Suri J S 2012 An accurate and generalized approach to plaque characterization in 346 carotid ultrasound scans *IEEE Trans. Instrum. Meas.* **61** 1045–53

[2] Acharya U R, Faust O, Alvin A P, Sree S V, Molinari F, Saba L, Nicolaides A and Suri J S 2012 Symptomatic vs asymptomatic plaque classification in carotid ultrasound *J. Med. Syst.* **36** 1861–71

[3] Acharya U R, Rama Krishnan M M, Sree S V, Sanches J, Shafique S, Nicolaides A, Pedro L M and Suri J S 2012 Plaque tissue characterization and classification in

ultrasound carotid scans: a paradigm for vascular feature amalgamation *IEEE Trans. Instrum. Meas.* **62** 392–400

[4] Acharya U R, Sree S V, Rama Krishnan M M, Molinari F, Saba L, Ho S Y S, Ahuja A T, Ho S C, Nicolaides A and Suri J S 2012 Atherosclerotic risk stratification strategy for carotid arteries using texture-based features *Ultrasound Med. Biol.* **38** 899–915

[5] Acharya U R *et al* 2013 Atherosclerotic plaque tissue characterization in 2D ultrasound longitudinal carotid scans for automated classification: a paradigm for stroke risk assessment *Med. Biol. Eng. Comput.* **51** 513–23

[6] Acharya U R, Faust O, Sree S V, Alvin A P C, Krishnamurthi G, José C R, Sanches J and Suri J S 2013 Understanding symptomatology of atherosclerotic plaque by image-based tissue characterization *Comput. Methods Programs Biomed.* **110** 66–75

[7] Acharya U R, Sree S V, Rama Krishnan M M, Saba L, Gao H, Mallarini G and Suri J S 2013 Computed tomography carotid wall plaque characterization using a combination of discrete wavelet transform and texture features: a pilot study *J. Eng. Med.* **227** *643*

[8] Acharya U R, Sree S V, Molinari F, Saba L, Nicolaides A and Suri J S 2015 An automated technique for carotid far wall classification using grayscale features and wall thickness variability *J. Clin. Ultrasound* **43** 302–11

[9] Araki T *et al* 2016 PCA-based polling strategy in machine learning framework for coronary artery disease risk assessment in intravascular ultrasound: a link between carotid and coronary grayscale plaque morphology *Comput. Methods Programs Biomed.* **128** 137–58

[10] Araki T *et al* 2016 Ultrasound-based automated carotid lumen diameter/stenosis measurement and its validation system *J. Vasc. Ultrasound* **40** 120–34

[11] Araki T *et al* 2016 Two automated techniques for carotid lumen diameter measurement: regional versus boundary approaches *J. Med. Syst.* **40** 182

[12] Araki T *et al* 2016 A new method for IVUS-based coronary artery disease risk stratification: a link between coronary and carotid ultrasound plaque burdens *Comput. Methods Programs Biomed.* **124** 161–79

[13] Araki T *et al* 2016 Reliable and accurate calcium volume measurement in coronary artery using intravascular ultrasound videos *J. Med. Syst.* **40** 1–20

[14] Araki T *et al* 2017 Stroke risk stratification and its validation using ultrasonic echolucent carotid wall plaque morphology: a machine learning paradigm *Comput. Biol. Med.* **80** 77–96

[15] Bots M L, Baldassarre D, Simon A, de Groot E, O'Leary D H, Riley W and Grobbee D E 2007 Carotid intima–media thickness and coronary atherosclerosis: weak or strong relations? *Eur. Heart J.* **28** 398–406

[16] Delsanto S, Molinari F, Giustetto P, Liboni W, Badalamenti S and Suri J S 2007 Characterization of a completely user-independent algorithm for carotid artery segmentation in 2-D ultrasound images *IEEE Trans. Instrum. Meas.* **56** 1265–74

[17] Eigenbrodt M L, Sukhija R, Rose K M, Tracy R E, Couper D J, Ewans G W, Bursac Z and Mehta J L 2007 Common carotid artery wall thickness and external diameter as predictors of prevalent and incident cardiac events in a large population study *Cardiovasc. Ultrasound* **5** 1–11

[18] Gupta A *et al* 2015 Plaque echolucency and stroke risk in asymptomatic carotid stenosis: a systematic review and meta-analysis *Stroke* **46** 91–7

[19] Illuminati G, Ricco J B, Caliò F, Pacilè M A, Miraldi F, Frati G, Macrina F and Toscano M 2011 Short-term results of a randomized trial examining timing of carotid endarterectomy in

patients with severe asymptomatic unilateral carotid stenosis undergoing coronary artery bypass grafting *J. Vasc. Surg.* **54** 993–9

[20] Kalyan K, Jakhia B, Lele R D, Joshi M and Chowdhary A 2014 Artificial neural network application in the diagnosis of disease conditions with liver ultrasound images *Adv. Bioinform.* **2014** 7082

[21] Kohavi R 1995 A study of cross-validation and bootstrap for accuracy estimation and model selection *Int. Joint Conf. Artif. Intell.* **14** 1137–43

[22] Krishna Kumar P *et al* 2017 Accurate lumen diameter measurement in curved vessels in carotid ultrasound: an iterative scale-space and spatial transformation approach *Med. Biol. Eng. Comput.* **55** 1415–34

[23] Libby P, Paul M, Ridker P M and Hansson G K 2011 Progress and challenges in translating the biology of atherosclerosis *Nature* **473** 317–25

[24] Mahalingam A, Gawandalkar U U, Kini G, Buradi A, Araki T, Ikeda N, Nicolaides A, Laird J R, Saba L and Suri J S 2016 Numerical analysis of the effect of turbulence transition on the hemodynamic parameters in human coronary arteries *Cardiovasc. Diagn. Ther.* **6** 208

[25] Mandelbrot B B 1983 *The Fractal Geometry of Nature* (New York: Freeman)

[26] Mirek A M and Wolińska-Welcz A 2012 Is the lumen diameter of peripheral arteries a good marker of the extent of coronary atherosclerosis? *Med. Biol. Eng. Comput.* **71** 810–7

[27] Mohebali J, Romero J M, Hannon K M, Jaff M R, Cambria R P and LaMuraglia G M 2015 Acoustic shadowing impairs accurate characterization of stenosis in carotid ultrasound examinations *J. Vasc. Surg.* **62** 1236–44

[28] Molinari F, Zeng G and Suri J S 2010 Intima–media thickness: setting a standard for completely automated method for ultrasound *IEEE Trans. Ultrason. Ferroelectr. Freq. Control (IEEE UFFC)* **57** 1112–24

[29] Molinari F, Zeng G and Suri J S 2010 An integrated approach to computer-based automated tracing and its validation for 200 common carotid arterial wall ultrasound images a new technique *J. Ultrasound Med.* **29** 399–418

[30] Molinari F, Zeng G and Suri J S 2010 A state of the art review on intima–media thickness (IMT) measurement and wall segmentation techniques for carotid ultrasound *Comput. Methods Programs Biomed.* **100** 201–21

[31] Molinari F *et al* 2012 Hypothesis validation of far-wall brightness in carotid-artery ultrasound for feature-based IMT measurement using a combination of level-set segmentation and registration *IEEE Trans. Instrum. Meas.* **61** 1054–63

[32] Molinari F, Constantinos P, Zeng G, Nicolaides A and Suri J S 2012 Completely automated multi-resolution edge snapper ('CAMES')—a new technique for an accurate carotid ultrasound IMT measurement: clinical validation and benchmarking on a multi-institutional database *IEEE Trans.Image Process.* **21** 1211–22

[33] Molinari F *et al* 2012 Ultrasound IMT measurement on a multi-ethnic and multi-institutional database: our review and experience using four fully automated and one semi-automated methods *Comput. Methods Program. Biomed.* **108** 946–60

[34] Molinari F, Kristen M, Meiburger, Saba L, Zeng G, Acharya U R, Ledda M, Nicolaides A and Suri J S 2012 Fully automated dual snake formulation for carotid intima–media thickness measurement: a new approach *J. Ultrasound Med.* **31** 1123–36

[35] Molinari F, Meiburger K M, Saba L, Acharya U R, Ledda M, Nicolaides A and Suri J S 2012 Constrained snake vs conventional snake for carotid ultrasound automated IMT measurements on multi-center data sets *Ultrasonics* **52** 949–61

[36] Molinari F, Meiburger K M, Zeng G, Nicolaides A and Suri J S 2012 CAUDLES-EF: carotid automated ultrasound double line extraction system using edge flow *J. Ultrasound Imaging* **24** 129–62

[37] Molinari F, Meiburger K M, Zeng G, Acharya U R, Liboni W, Nicolaides A and Suri J S 2012 Carotid artery recognition system: a comparison of three automated paradigms for ultrasound images *Med. Phys.* **39** 378–91

[38] Muller K R, Mika S, Ratsch G, Tsuda K and Scholkopf B 2001 An introduction to kernel based learning algorithms *IEEE Trans. Neural Netw.* **12** 181–201

[39] Narula J, Nakano M, Virmani R, Kolodgie F D, Petersen R, Newcomb R, Malik S, Fuster V and Finn A V 2013 Histopathologic characteristics of atherosclerotic coronary disease and implications of the findings for the invasive and noninvasive detection of vulnerable plaques *J. Am. Coll. Cardiol.* **61** 1041–51

[40] Naim C, Cloutier G, Mercure E, Destrempes F, Zid Q, El-Abyad W, Lanthier S, Giroux M F and Soulez G 2013 Characterisation of carotid plaques with ultrasound elastography: feasibility and correlation with high-resolution magnetic resonance imaging *Eur. Soc. Radiol.* **23** 2030–41

[41] Naim C, Douziech M, Therasse E, Robillard P, Giroux M F, Arsenault F, Cloutier G and Soulez G 2014 Vulnerable atherosclerotic carotid plaque evaluation by ultrasound, computed tomography angiography, and magnetic resonance imaging: an overview *Can. Assoc. Radiol. J.* **65** 275–86

[42] Noor N M, Than J C, Rijal O M, Kassim R M, Yunus A, Zeki A A, Anzidei M, Saba L and Suri J S 2015 Automatic lung segmentation using control feedback system: morphology and texture paradigm *J. Med. Syst.* **39** 1–18

[43] Patel A K *et al* 2016 A review on atherosclerotic biology, wall stiffness, physics of elasticity, and its ultrasound-based measurement *Curr. Atheroscler. Rep.* **18** 83

[44] Picano E and Paterni M 2015 Ultrasound tissue characterization of vulnerable atherosclerotic plaque *Int. J. Mol. Sci.* **16** 10121–33

[45] Prabhakaran D, Jeemon P and Roy A 2016 Cardiovascular diseases in India current epidemiology and future directions *Circulation* **133** 1605–20

[46] Pedro L M, Sanches J M, Seabra J, Suri J S, Fernandes E and Fernandes J 2014 Asymptomatic carotid disease—a new tool for assessing neurological risk *Echocardiography* **31** 353–61

[47] Ross R 1995 Cell biology of atherosclerosis *Annu. Rev. Physiol.* **57** 791–804

[48] Saba L, Lippo R S, Tallapally N, Molinari F, Montisci R, Mallarini G and Suri J S 2011 Evaluation of carotid wall thickness by using computed tomography and semiautomated ultrasonographic software *J. Vasc. Ultrasound* **35** 136–42

[49] Saba L, Gao H, Acharya U R, Sannia S, Ledda G and Suri J S 2012 Analysis of carotid artery plaque and wall boundaries on CT images by using a semi-automatic method based on level set model *Neuroradiology* **54** 1207–14

[50] Saba L, Tallapally N, Gao H, Molinari F, Anzidei M, Piga M, Sanfilippo R and Suri J S 2013 Semiautomated and automated algorithms for analysis of the carotid artery wall on computed tomography and sonography: a correlation study *J. Ultrasound Med.* **32** 665–74

[51] Saba L, Molinari F, Meiburger K M, Acharya U R, Nicolaides A and Suri J S 2013 Inter- and intra-observer variability analysis of completely automated cIMT measurement software (AtheroEdge™) and its benchmarking against commercial ultrasound scanner and expert Readers *Comput. Biol. Med.* **43** 1261–72

[52] Saba L and Suri J S (ed) 2014 *Multi-Detector CT Imaging: Principles, Head, Neck, and Vascular Systems* vol 1 (Boca Raton, FL: CRC Press)
[53] Saba L, Sanches J M, Pedro L M and Suri J S 2015 *Multi-Modality Atherosclerosis Imaging and Diagnosis* (New York: Springer)
[54] Saba L *et al* 2016 Carotid inter-adventitial diameter is more strongly related to plaque score compared to lumen diameter: an automated and first ultrasound study in Japanese diabetic cohort *J. Clin. Ultrasound* **44** 210–20
[55] Saba L, Than J C, Noor N M, Rijal O M, Kassim R M, Yunus A, Ng C R and Suri J S 2016 Inter-observer variability analysis of automatic lung delineation in normal and disease patients *J. Med. Syst.* **40** 1–8
[56] Saba L *et al* 2013 Association of automated carotid IMT measurement and HbA1c in Japanese patients with coronary artery disease *Diabetes Res. Clin. Pract.* **100** 348–53
[57] Sanches J M, Laine A F and Suri J S (ed) 2015 *Ultrasound Imaging: Advances and Applications* (Berlin: Springer)
[58] Sharma A M, Gupta A, Kumar P K, Rajan J, Saba L, Nobutaka I, Laird J R, Nicolades A and Suri J S 2015 A review on carotid ultrasound atherosclerotic tissue characterization and stroke risk stratification in machine learning framework *Curr. Atheroscler. Rep.* **17** 55
[59] Shrivastava V K, Londhe N D, Sonawane R S and Suri J S 2015 Reliable and accurate psoriasis disease classification in dermatology images using comprehensive feature space in machine learning paradigm *Expert Syst. Appl.* **42** 6184–95
[60] Shrivastava V K, Londhe N D, Sonawane R S and Suri J S 2015 Exploring the color feature power for psoriasis risk stratification and classification: a data mining paradigm *Comput. Biol. Med.* **65** 54–68
[61] Shrivastava V K, Londhe N D, Sonawane R S and Suri J S 2016 A novel approach to multiclass psoriasis disease risk stratification: machine learning paradigm *Biomed. Sign. Process. Control* **28** 27–40
[62] Shrivastava V K, Londhe N D, Sonawane R S and Suri J S 2016 Reliability analysis of psoriasis decision support system in principal component analysis framework *Data Knowl. Eng.* **106** 1–17
[63] Silver F L *et al* 2011 Safety of stenting and endarterectomy by symptomatic status in the Carotid Revascularization Endarterectomy Versus Stenting Trial (CREST) *Stroke* **42** 675–80
[64] Soh L K and Tsatsoulis C 1999 Texture analysis of SAR sea ice imagery using gray level co-occurrence matrices *IEEE Trans. Geosci. Remote Sens.* **37** 780–95
[65] Song F, Guo Z and Mei D 2010 Feature selection using principal component analysis *IEEE Int. Conf. on System Science, Engineering Design and Manufacturing Informatization (ICSEM)* ((Piscataway, NJ: IEEE) 27–30
[66] The Internet Stroke Center Stroke statistics http://www.strokecenter.org/patients/about-stroke/stroke-statistics/
[67] Suri J S, Yuan C and Wilson D L (ed) 2005 *Plaque Imaging: Pixel to Molecular Level* (Washington, DC: IOS)
[68] Suri J S, Kathuria C and Molinari F (ed) 2011 *Atherosclerosis Disease Management* (Berlin: Springer)
[69] Tang X 1998 Texture information in run-length matrices *IEEE Trans. Image Process.* **7** 1602–9
[70] Vapnik V 1998 *Statistical Learning Theory* (New York: Wiley)
[71] WHO CVD http://www.who.int/mediacentre/factsheets/fs317/en/

Chapter 10

Multiresolution-based coronary calcium volume measurement techniques from intravascular ultrasound videos

Sumit K Banchhor, Tadashi Araki, Narendra D Londhe, Nobutaka Ikeda, Petia Radeva, Ayman El-Baz, Luca Saba, Andrew Nicolaides, Shoaib Shafique, John R Laird and Jasjit S Suri

Fast intravascular ultrasound (IVUS) video processing is required for calcium volume computation during the planning phase of percutaneous coronary interventional (PCI) procedures. Nonlinear multiresolution techniques are generally applied to improve the processing time by down-sampling the video frames. This chapter presents four different segmentation methods for calcium volume measurement, namely threshold-based, fuzzy c-means (FCM) and K-means methods, and a hidden Markov random field (HMRF) method embedded with five different kinds of multiresolution techniques (bilinear, bicubic, wavelet, Lanczos and Gaussian pyramid). This leads to 20 different kinds of combinations. IVUS image datasets consisting of 38 760 IVUS frames taken from 19 patients were collected using a 40 MHz IVUS catheter (Atlantis® SR Pro, Boston Scientific®; pullback speed = 0.5 mm s^{-1}). The performance of these 20 systems is compared with and without multiresolution using the following metrics: (a) computational time; (b) calcium volume; (c) image quality degradation ratio; and (d) quality assessment ratio. The FCM and wavelet experienced the highest percentage mean improvements in computational time, 77.15% and 74.07%, respectively. Wavelet interpolation had the highest mean precisions-of-merit (PoMs) of 94.06% ± 3.64% and 81.34% ± 16.29% compared to other multiresolution techniques for the volume level and frame level, respectively. The wavelet multiresolution technique also presents the highest Jaccard index and Dice similarity of 0.7 and 0.8, respectively. Multiresolution is a nonlinear operation which introduces bias and thus degrades the image. The proposed system also provides a bias correction approach to enrich the system, giving a better mean calcium volume similarity for all the multi-resolution-based segmentation methods. After including the bias correction, bicubic interpolation gives the largest increase in mean calcium volume similarity, 4.13%, compared to the rest of the multiresolution techniques. We demonstrated the time improvement in calcium volume computation without compromising the quality of the IVUS image. Among the 20 different combinations of multiresolution with calcium volume segmentation methods, the FCM embedded with wavelet-based multiresolution gave the best performance. The system is automated and can be adopted in clinical settings.

10.1 Introduction

Cardiovascular disease (CVD) and heart attack are the main causes of death and produce an immense burden in the United States and globally [1]. The most common type of CVD is coronary artery disease (CAD) in which the blood carried by the arteries to the heart becomes blocked (also called stenosis) [2]. The main cause of CAD is calcium deposition, which limits the flow of oxygen-rich blood in these arteries, called atherosclerosis [3]. The progression of atherosclerosis due to calcium blocks the arteries which can lead to myocardial infarction. Clinical symptoms of atherosclerosis appear late in CAD [4] and, therefore, early prediction of calcium volume is crucial in the diagnosis of coronary artery stenosis, which is vital during the planning phase of percutaneous coronary interventional (PCI) procedures [5]. An illustration of the stenotic coronary artery is provided in figure 10.1. Different methods of imaging the coronary artery, such as computed tomography (CT), magnetic resonance imaging (MRI) and optical coherence tomography (OCT), play a vital role in diagnosis and treatment [6, 7]. Intravascular ultrasound (IVUS) is

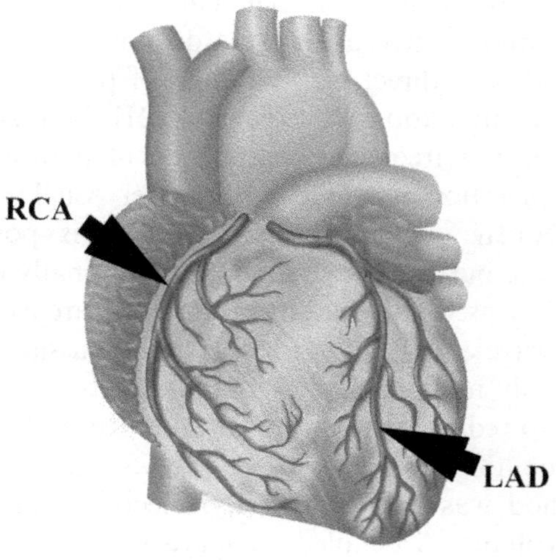

Figure 10.1. LAD: left anterior descending coronary artery; RCA: right coronary artery (courtesy of AtheroPoint™, Roseville, CA, USA).

mostly chosen over other imaging modalities for reasons such as its safety, low acquisition, low cost and ease-of-use with real-time diagnosis [8–10].

Recently, teams led by Suri *et al* [11–14] have been focusing on calcium volume estimation using IVUS. Suri's team [15] compared different software segmentation methods for volume estimation from IVUS videos. This includes methods such as threshold-based [16], K-means [17], FCM [18] and HMRF [19]. This chapter is focused on the application of five multiresolution techniques in the above segmentation framework, primarily for improving the computation time for calcium volume.

The computation of accurate calcium volume is possible only if the calcium is detected all along the coronary artery [15]. Because the pullback speed is 0.5 mm s^{-1} during image acquisition, the video produced by the IVUS scanner consists of a large number of frames (approximately 2040 frames per video). Manual analysis of all the frames is (a) tedious, (b) prone to error [11, 15] and (c) consumes excessive time. To overcome the above limitations, we have developed a set of multiresolution-based automated segmentation methods that can speed up the process without losing accuracy [20].

During the IVUS acquisition, the heart is always moving, so it is important to understand which multiresolution and segmentation method combination is optimal for this set-up. Multiresolution techniques can down-sample large-sized images, improving the speed but storing only the low-frequency components [21]. Storing only the low-frequency components can cause blurring and generates artifacts [22]. This can be removed by introducing proper bias correction in the down-sampled IVUS video frames generated using different multiresolution techniques.

Multiresolution techniques are mainly divided into two categories: adaptive and non-adaptive techniques [11, 23–25]. Adaptive interpolation algorithms extract image features such as texture, intensity value and edge information and use these features as landmarks for the multiresolution method [26]. Various adaptive techniques are context-aware resizing, segment-based, seam carving, wrapping-based, etc [27]. The non-adaptive algorithms do not rely on the image features; instead, they are based on a direct manipulation of pixels, and hence are easy to perform and are less computationally expensive [21]. Non-adaptive interpolation techniques store only the low-frequency components of an original image, leading to blurring of edges or generation of artifacts. For better visual quality, an image must preserve the high-frequency components. This task is possible with adaptive interpolation techniques, but they are more computationally expensive in terms of time [21, 27]. The widely used non-adaptive techniques are as follows: bilinear [28], bicubic [29], discrete wavelet [30], Lanczos [31] and Gaussian pyramid [32]. In this article, we analyzed only non-adaptive techniques embedded with bias correction since our objective is to reduce computational time. Recently, Zhang *et al* [33] had attempted to compute the calcium area in IVUS frames by using contourlet transform. This method was applied to only selective frames and there was no mention of calcium volume. A detailed comparison of the previous techniques is presented in the discussion section (section 10.6).

The novelty of this study is to evaluate the performance of five different multiresolution techniques with the introduction of bias correction in IVUS images using four different segmentation methods for calcium volume measurement. The main contributions of this chapter are as follows: (i) the design and development of four segmentation methods embedded with five kinds of multiresolution techniques, thus a total of 20 different kinds of calcium volume measurement systems; (ii) optimization for the best calcium volume segmentation–multiresolution combination by analyzing computation time with and without multiresolution; (iii) benchmarking the multiresolution-based calcium volumes against non-multiresolution-based calcium volumes using the precision-of-merit (PoM); (iv) benchmarking the multiresolution-based image quality against non-multiresolution techniques using degradation ratio (DR) and quality assessment ratio (QAR) utilizing the signal-to-noise ratio (SNR) and contrast-to-noise ratio (CNR) paradigms; and (v) introduction of bias correction in IVUS images during the multiresolution framework for calcium volume computation. A detailed analysis explaining the key difference between our contributions with respect to our previous work [15] is presented in the discussion section (section 10.6).

The layout of this chapter is as follows: section 10.2 presents the IVUS data acquisition, patient demographics and data preparation. The methods section 10.3 shows the design of five different multiresolution techniques along with four segmentation paradigms. Quantitative and qualitative results are presented in section 10.4. Performance evaluation and comprehensive data analysis are depicted in section 10.5. Benchmarking and discussions on the stability of the system are presented in section 10.6. The chapter concludes in section 10.7.

10.2 Patient demographics and data acquisition

10.2.1 Patient demographics

In this article, IVUS data for nineteen patients were taken from a single-center study [34] of patients who underwent PCIs between July 2009 and December 2010 with stable angina pectoris. Of the 19 patients, 17 were men and 2 women with an age range of 36–81 (average age of 66 years). In this sample, ten patients had a proximal lesion location, five at the middle, and four had distal locations. Ten patients had calcium present on the left anterior descending coronary artery, five on the right coronary artery, three on the left circumflex coronary artery and one on the left main coronary artery. The mean total cholesterol, LDL cholesterol and HDL cholesterol were 165 mg dL^{-1}, 91 mg dL^{-1} and 52 mg dL^{-1}, respectively, and mean hemoglobin was 5.81 g dL^{-1}. Nine patients from the pool of nineteen were smokers. Before performing a coronary intervention procedure, a mixed dose of clopidogrel (75 mg/d) and aspirin (100 mg/d) was given to the patients. Prior to the procedure, intravenous unfractionated heparin was also given to obtain a partial thromboplast in a time of less than 250 s.

10.2.2 IVUS data acquisition

This study underwent a full ethics review by the respective Institutional Review Board and written informed consent was provided by all the patients. A 40 MHz

IVUS catheter (Atlantis® SR Pro, Boston Scientific®) was used for data acquisition and the computer program MATLAB® (MathWorks, Inc., Natick, MA) was used for image analysis. Target lesion imaging was performed during automatic pullback of the catheter at a speed of 0.5 mm s^{-1}. Since the speed of the catheter is constant (0.5 mm s^{-1}), in the current study we have assumed that the gap between the frames and the repetition of frames (vessel sections) is negligible. Standard protocols were used for volume computations using the resolution factor consisting of one pixel as (1/60)th of an millimeter, i.e. 0.0167 mm [15]. Figure 10.2 shows a representative example of five (1–5) frames per patient with calcified plaques, as shown by the arrows. The frames per IVUS video are taken from six patients and arranged in a 5×6 matrix. This is represented as I(1,1)–I(5,4).

10.2.3 Coronary artery data size preparation

Typically, the DICOM format is adopted for image acquisition. DICOM is the proprietary header and the images are 16 bits per pixel. Once the images were converted into AVI QuickTime movies, the resolution of the images was 8 bits per pixel. This affected the image quality compared to the original DICOM format. Recently, Kim *et al* [35] have shown the effect of post-processing on the quality degradation of ultrasound IVUS images. The authors did not use the DICOM video type for this study. Further, the current study is focused on the development of multiresolution-based calcium volume estimation in IVUS videos. Even though the number of patients is only 19, the number of frames processed per patient was 2040. This accounted for a total of 38 760 frames. Here, we have analyzed the complete coronary artery video without eliminating the starting and ending frames, unlike in previous studies [16, 33].

Of these 19 patients, 16 patients had bright calcium (hyperechoic) and the remaining three patients had less bright calcium (hypoechoic). These three patients from the pool of 19 patients were excluded only for the bias correction study. The effect of bias correction on the three patients with hypoechoic calcium is explained later in the discussion (section 10.6).

10.2.4 Region-of-interest estimation

The ImgTracer™ system (courtesy of AtheroPoint™, Roseville, CA, USA) was used for vessel wall region extraction. Since tracing is a tedious process, two expert tracers were used to generate the vessel wall regions. These tracers traced the internal elastic lamina and external elastic lamina borders. Each tracer traced about 19 380 frames, totaling to 38 760 frames acquired from 19 patients. The results of four representative examples are shown in figure 10.3. In figure 10.3(a), the inner yellow curve shows the internal elastic membrane (IEL) and the outer yellow curve shows the external elastic membrane (EEL). An equidistant spline was fitted for smoothing the borders of the vessel wall region. Figure 10.3(b) shows the atherosclerotic wall region used for segmentation of the calcium region.

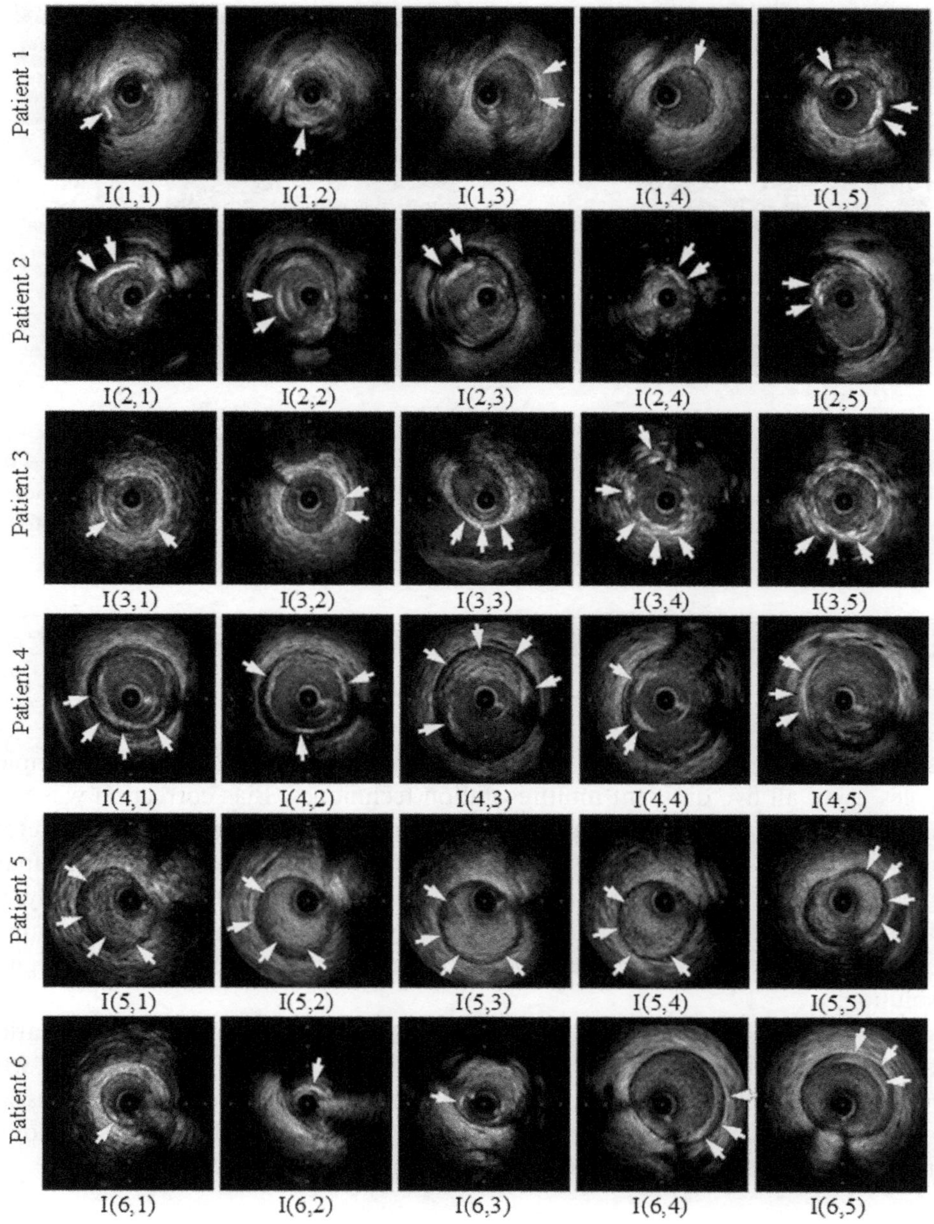

Figure 10.2. Sample frames I(1,1)–I(6,5) with calcified plaques for six patients from their IVUS videos.

10.3 Methodology

10.3.1 Overall system

An image interpolation algorithm can convert an image from one resolution to another while preserving visual content. This feature motivates us to use multi-resolution techniques in our IVUS dataset. The main block diagram of the proposed

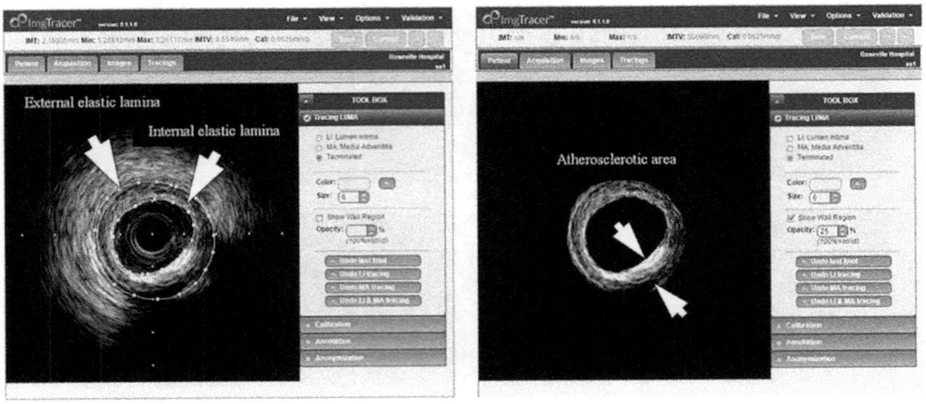

(a) (b)

Figure 10.3. (a) Manually traced IEL and EEL indicated by the inner and outer yellow lines in the vessel wall region, obtained using ImgTracer™. (b) Atherosclerotic grayscale ring image used as the ROI. (Courtesy of AtheroPoint™, Roseville, CA, USA; reproduced with permission from [36]. Copyright 2017 Elsevier.)

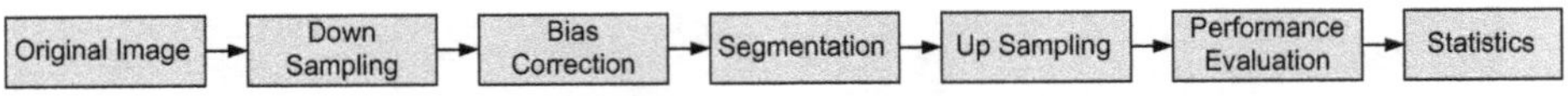

Figure 10.4. Block diagram for the overall system.

system is shown in figure 10.4. We first produced the half-sized down-sampled images using all five different multiresolution techniques. Bias correction was then implemented on the down-sampled IVUS ring images generated from different multiresolution techniques. The bias-corrected sampled images were next binary segmented using four different soft classification segmentation methods. Finally, an up-sampling procedure was applied to the down-sampled binary images obtained from different segmentation methods to restore the image size to the original resolution.

Figure 10.5 shows the decomposed flow chart consisting of four segmentation methods (threshold-based (Santos), FCM, K-means,] and HMRF) embedded with five multiresolution techniques (bilinear, bicubic, wavelet, Lanczos and Gaussian pyramid). This decomposed infrastructure shows the explosion of block 2 (down-sampling) and block 4 (segmentation) of figure 10.4.

10.3.2 Five multiresolution techniques

This section covers the five multiresolution techniques adopted during the segmentation paradigm. Even though the techniques are taken from the literature, the objective is to evaluate the best combination for a calcium volumetric study while removing the bias observed due to the multiresolution paradigm. Further, the aim is to evaluate different metrics for performance evaluation of these combination systems, taking time, volume and image quality into consideration.

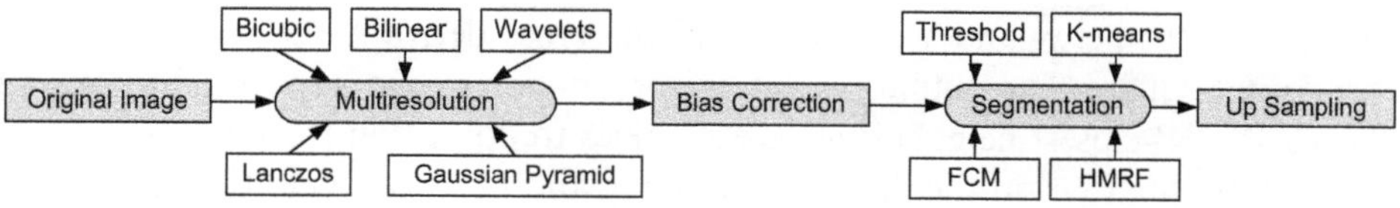

Figure 10.5. Analysis section of the overall system process.

10.3.2.1 Bilinear multiresolution

Bilinear multiresolution is one of the most basic non-adaptive interpolation techniques. Given a point P in the input image, the output pixel in the destination image is the weighted average of four nearest neighboring pixel values to point P.

The intensity value $I(a, b)$ at the interpolated point P at (a, b) in the image can be estimated as

$$\begin{aligned} I(a, b) &= \frac{1}{(a_2 - a_1)(b_2 - b_1)}[(a - a_2)(b - b_2)I(a_1, b_1) \\ &\quad + (a - a_1)(b - b_1)I(a_2, b_2) \\ &\quad - (a - a_1)(b - b_2)I(a_2, b_1) - (a - a_2)(b - b_1)I(a_1, b_2)] \\ &= \frac{(a - a_2)(b - b_2)}{(a_2 - a_1)(b_2 - b_1)}I(a_1, b_1) + \frac{(a - a_1)(b - b_1)}{(a_2 - a_1)(b_2 - b_1)}I(a_2, b_2) \\ &\quad - \frac{(a - a_1)(b - b_2)}{(a_2 - a_1)(b_2 - b_1)}I(a_2, b_1) - \frac{(a - a_2)(b - b_1)}{(a_2 - a_1)(b_2 - b_1)}I(a_1, b_2) \\ &= w_1 I(a_1, b_1) + w_2 I(a_2, b_2) - w_3 I(a_2, b_1) - w_4 I(a_1, b_2), \end{aligned} \tag{10.1}$$

where $I(a_1, b_1)$, $I(a_1, b_2)$, $I(a_2, b_1)$ and $I(a_2, b_2)$ are the intensity values of the four neighboring pixels. Note that, each weight is equivalent to the normalized area between the interpolated point P and the diagonally opposite pixel. For example, the weight multiplied to the intensity $I(a_1, b_1)$ is w_1 which is the normalized area between interpolated point P and its diagonally opposite pixel $I(a_2, b_2)$.

In the process of up-sampling, an interpolating kernel is chosen which consists of the nearest 2×2 neighborhood of known pixel values surrounding the unknown pixels. It can be implemented using the triangle kernel which can be mathematically formulated as

$$f(x) = \begin{cases} 1 - |x| & |x| < 1 \\ 0 & \text{elsewhere} \end{cases}. \tag{10.2}$$

Bilinear interpolation is averaging in nature and at the boundaries of each grid square the interpolated function values change discontinuously. This causes a small decrease in resolution and blurring.

10.3.2.2 Bicubic multiresolution

Bicubic multiresolution goes one phase beyond the bilinear technique by computing a higher order derivative to further increase the accuracy of the interpolating function. To compute the bicubic interpolation within the grid square, we have to compute the gradient in both the x- and y-direction and the cross derivative at each of the four corners of the square. The output pixel in the destination image is the weighted average of the 16 (4 × 4) closest neighboring pixels' values to point P in the input image.

Given a point (x, y) in the destination image and point (c, d) in the input image, the intensity value $I(x, y)$ at the interpolated point P at (x, y) in the image can be estimated as

$$I(x, y) = \sum_{m=c-1}^{c+2} \sum_{n=d-1}^{d+2} I(m, n) \cdot w(m - c - dx) \cdot (dy - n + d), \tag{10.3}$$

where $c = \lfloor x \rfloor$ and $d = \lfloor y \rfloor$. Here, $dx = x - c$ and $dy = y - d$, respectively. The weighted function $w(x)$ is mathematically defined as

$$w(x) = \frac{1}{6}\left[g(x + 2)^3 - 4g(x + 1)^3 + 6g(x)^3 - 4g(x - 1)^3\right], \tag{10.4}$$

where $g(x) = \begin{cases} x & x > 0 \\ 0 & x \leqslant 0 \end{cases}$.

In the process of up-sampling, a convolution kernel is chosen which is composed of a piecewise cubic polynomial. The output pixel is the weighted sum of the nearest 4 × 4 neighborhoods of known pixel values surrounding the unknown pixels. The convolution kernel can be mathematically formulated as

$$f(x) = \begin{cases} (a + 2)|x|^3 - (a + 3)|x|^2 + 1; & x \leqslant 1 \\ a\,|x|^3 - 5a\,|x|^2 + 8a|x| - 4a; & 1 < x < 2 \\ 0; & \text{otherwise} \end{cases}. \tag{10.5}$$

Here, $a = -0.5$. Using a polynomial surface, the bicubic multiresolution approximates the local intensity values; hence better performance is achieved at the cost of time.

10.3.2.3 Discrete wavelet multiresolution

Let us first understand the basic concept of the discrete wavelet transform (DWT) [37]. The DWT can be implemented with well-defined filter coefficients. During forward DWT, the input discrete signal $a(m)$ at sampling point m is filtered by a low-pass filter by $\tilde{e}(j)$ and a high-pass filter by $\tilde{f}(j)$ at window translation points j. Sub-sampling is then performed by dropping alternate output samples to produce the low-pass $x_L(m)$ and the high-pass $x_H(m)$ degree of membership. For windowed Fourier transform, a discrete signal $a(m)$ is multiplied with these low-pass and high-pass filters. The procedure is then repeated for translated versions of the windows. For given m-taps analysis, the output low-pass and high-pass signals $x_L(m)$ and $x_H(m)$ can be computed as

$$x_L(m) = \sum_{j=0}^{l_L-1} \{\tilde{e}(j) \cdot a(2m - j)\} \tag{10.6}$$

$$x_H(m) = \sum_{j=0}^{l_H-1} \{\tilde{f}(j) \cdot a(2m - j)\}, \tag{10.7}$$

where l_L and l_H are the lengths of the low-pass and high-pass filter, respectively. Inverse transform can be performed by up-sampling and filtering, respectively. The resultant signal can be obtained by adding both x_L and x_H.

A 2D image generally has smooth and sharp variations. Low-frequency components contain smooth details and high-frequency components contain sharp details. By using DWT, we can separate these two details by applying the 1D DWT row-wise and later the same 1D DWT column-wise. When the 1D DWT is applied row-wise, L and H subbands in each row are produced. Later, when the same 1D DWT is applied column-wise, four subbands LL1, LH1, HL1 and HH1 are obtained. Here, the LL1 subband represents an approximate version of the original at half the resolution and is used as a down-sampled image for further analysis. The LH1, HL1 and HH1 subbands contain the vertical, horizontal and diagonal edge information.

In the process of up-sampling, a virtual DWT image is first generated. The new LL subband of the virtual DWT image is the exact LL1 subband from the down-sampled image. The new HH subband is set to all zeros while the new HL and LH subbands are generated by inserting zeros in the alternate rows and columns. The desired up-sampled image of the same size as the original image can be obtained by applying the inverse DWT on this virtual DWT image [27].

Wavelet is a lossless multiresolution technique and provides a smooth approximation. It has the capability to determine both frequency and location information (temporal resolution). The discrete wavelet technique preserves most of the sharp edge features and avoids color artifacts. Moreover, it is simple to implement, both in terms of hardware and software [27].

10.3.2.4 Lanczos multiresolution

An ideal filter is one which has a gain of 0 dB between the frequencies of 0 and 1 (known as the passband) and $-\infty$ beyond 1 (known as the stopband). Roll off in the passband causes blurriness, and leakage in the stopband causes aliasing effects. Among many popular filters, the sinc function is the ideal low-pass filter [38], which never goes to zero, but approaches it slowly. The sinc function can be defined mathematically as

$$\text{sinc}(x) = \frac{\sin(\pi x)}{\pi x}. \tag{10.8}$$

A practical filter can be implemented by multiplying a 'window', such as Hamming or Hann, with a sinc function of finite size. A Lanczos filter can be obtained by multiplying the sinc function with the Lanczos window. A two-lobed and three-lobed Lanczos-windowed sinc function can be shown as

$$\text{Lanczos2}(x) = \begin{cases} \dfrac{\text{sinc}(x) \cdot \text{sinc}(x/2)}{\pi x \cdot \pi x/2} & |x| < 2 \\ 0 & |x| \geqslant 2 \end{cases} \tag{10.9}$$

$$\text{Lanczos3}(x) = \begin{cases} \dfrac{\text{sinc}(x) \cdot \text{sinc}(x/3)}{\pi x \cdot \pi x/3} & |x| < 3 \\ 0 & |x| \geqslant 3 \end{cases}. \tag{10.10}$$

Lanczos multiresolution has the property of preserving details and avoiding aliasing effects and hence is particularly useful in graphics applications [31]. As the number of neighboring pixels varies with the order of the kernel, the size of the Lanczos window also changes the order of the convolution kernel. If the order is selected as two, 16 pixels are considered while if the order is three, 36 neighboring pixels are used for interpolation. We have tested both the orders (two and three) for the Lanczos technique in our study. By changing the order of the filter, we observed a negligible change. We have chosen the third order as it is the default algorithm and further, our earlier group [39] used the Lanczos third order filter and demonstrated optimal results.

In the process of up-sampling an interpolating kernel is chosen which can be mathematically formulated as

$$f(x) = \begin{cases} \text{sinc}(x) \cdot \text{sinc}\left(\dfrac{x}{3}\right) & |x| < 3 \\ 0 & \text{elsewhere} \end{cases}. \tag{10.11}$$

The Lanczos multiresolution technique is slower compared to previously described techniques since it uses 36 surrounding pixels to interpolate the center pixel. Although it is slower, Lanczos produces the best results because it includes more of the passband and excludes more of the stopband by using a series of overlapping sinc waves.

10.3.2.5 Gaussian pyramid multiresolution

For down-sampling any image, image reduction can be performed by using sub-sampling. Since sub-sampling alone will cause aliasing effects, the first step in the Gaussian pyramid involves low-pass filtering of the original image to obtain a smooth image. Image smoothing gets rid of the fast changes by removing the fast frequency components. Suppose the image is represented by the array f_0 with columns (C) and rows (R) number of pixels. Each pixel signifies the light intensity, between 0 and $q-1$. For the Gaussian pyramid, this image becomes the reference or zero level. Level 1 contains a reduced or low-pass filtered version of f_0 say image f_1. Each value is computed as a weighted average of values within a 5×5 window.

The low-pass filtering is followed by down-sampling of the image pixel. The averaging process is implemented by the function REDUCE, shown as

$$f_q = \text{REDUCE}(f_{q-1}). \tag{10.12}$$

For level $0 < l \leqslant M$ and nodes m, n, where, $0 \leqslant m \leqslant C_1$, $0 \leqslant n \leqslant R_1$,

$$f_l(m, n) = \sum_{i=-2}^{2} \sum_{j=-2}^{2} w(i, j) \cdot f_{l-1}(2m + i, 2n + j), \tag{10.13}$$

where M refers to the number of levels, while C_l and R_l are the dimensions of the first level. Note that $w(i, j)$ is known as the weighting function which corresponds to the weight w at the pixel location (i, j) that is used to generate each pyramid array from its predecessor.

In the process of up-sampling (expanding) the reduced image, new node values are interpolated between the values. The up-sampling of the image pixel is implemented by the function EXPAND, shown as

$$f_{q-1} = \text{EXPAND}(f_q). \tag{10.14}$$

For level $0 < l \leqslant M$ and nodes m, n, where $0 \leqslant m \leqslant C_{l-n}$, $0 \leqslant n \leqslant R_{l-n}$,

$$f_l(m, n) = 4 \sum_{i=-2}^{2} \sum_{j=-2}^{2} w(i, j) \cdot f_{l+1}\left(\left(\frac{i-m}{2}\right), \left(\frac{j-n}{2}\right)\right), \tag{10.15}$$

where M refers to the number of levels, while C_{l-n} and R_{l-n} are the dimension of the first reduced level.

10.3.2.6 Bias correction

Due to the low quality of the input images, there are certain regions in the image which are of poor quality and degrade the high-frequency component. After down-sampling, these tiny regions that are close to each other are segmented and interpreted to be calcium. Thus, these tiny regions are categorized into false calcium. This causes a slight overestimation in calcium volume. Thus, bias correction is necessary after the down-sampling of the IVUS grayscale ring images generated from different multiresolution techniques (as shown in figure 10.1).

In this study, we adopted an erosion process for correcting the bias due to the multiresolution paradigm. Grayscale erosion generally darkens the image as bright sections enclosed by dark sections shrink in size, and dark sections enclosed by bright sections grow in size. This preserves the original shape of the bright calcium, yielding correct calcium detection and volume computation.

Let $X(u, v)$ be the input (original) image and $Y(u, v)$ be a structuring element. The grayscale erosion of $X(u, v)$ is given as

$$XY(u, v) = \min \{X(u + u', v + v') - Y(u', v') | (u', v') \in DB\}, \tag{10.16}$$

where DB is the domain of the structuring element Y and $X(u, v)$ is assumed to be $+\infty$ outside the domain of the image.

If every pixel in the structuring element matches the image region, the center pixel is replaced with the number 1 [40]. We have also analyzed our results by adopting grayscale dilation for bias correction. In our study, incorrect results are obtained in dilation as it generally brightens the image leading to the incorrect estimation of calcium.

In this study, we have used a disk to perform the bias correction using MATLAB 2013a software. The disk shape was chosen because the calcium present in IVUS images is curved in nature. We have analyzed our results by increasing and decreasing the size of the disk. Incorrect results are obtained in both cases as decreased disk size yields negligible change and increased disk size causes blurring of the biased image. After several trials, we empirically chose a structuring element of a disk with size 3 × 3 to perform the bias correction:

$$Y(u, v) = \begin{bmatrix} 0 & 1 & 0 \\ 1 & 1 & 1 \\ 0 & 1 & 0 \end{bmatrix}. \tag{10.17}$$

10.3.3 Four segmentation methods

There are several common approaches to perform medical image segmentation (the threshold approach, region growing approach, classifiers, clustering, MRF, artificial neural networks, deformable models and the atlas-guided approach) [41]. Both segmentation and clustering techniques use a pixel-based strategy for classification. Although they look quite similar, they are not quite the same. Segmentation based on pixel classification groups the pixels using similarity criteria, whereas segmentation based on clustering is the process of finding similarities prior to grouping [41]. Practically it is difficult to implement all the segmentation methods. We hypothesize that widely used segmentation methods such as K-means, FCM and HMRF can characterize and isolate calcium lesions in individual IVUS frames. Further, by adopting the conventional Simpson's rule, calcium volume can be computed for each patient, provided the region of the coronary wall is *a priori* known, the so-called vessel wall region.

10.3.3.1 Multi-thresholding segmentation

Due to a significant difference in acoustic impedance, the echo produced between calcium and the lumen is very strong. Thus, calcification regions appear very bright in the IVUS images and segmentation can be achieved by grouping all the pixels with neighboring intensities into one class based on some intensity value called the threshold.

Otsu [42] in 1979 proposed a nonparametric and unsupervised method for automatic threshold selection in any image. The concept was based on the maximization of the separability of the resultant class.

Let the pixels of any given image be represented by M gray levels $[1, 2, \ldots, M]$. The number of pixels at gray level i is denoted by n_i and the total number of pixels by N. After normalization, the gray level histogram may be regarded as a probability distribution:

$$p_i = \frac{n_i}{N}, \qquad p_i \geqslant 0, \ \sum_{i=1}^{M} p_i = 1. \tag{10.18}$$

Now suppose that the pixels are separated into two classes, C_0 and C_1 (object and background or vice versa), by a threshold at gray level k; C_0 denotes pixels with gray levels $[1, \ldots, k]$ and C_1 denotes pixels with gray levels $[k + 1, \ldots, M]$. Then, the probabilities of class occurrence and the class mean gray levels, respectively, are given by

$$w_0 = P_r(C_0) = \sum_{i=1}^{k} p_i = w(k) \tag{10.19}$$

$$w_1 = P_r(C_1) = \sum_{i=k+1}^{M} p_i = 1 - w(k) \tag{10.20}$$

and

$$\mu_0 = \sum_{i=1}^{k} iP_r(i|C_0) = \sum_{i=1}^{k} ip_i/w_0 = \mu(k)/w(k) \tag{10.21}$$

$$\mu_1 = \sum_{i=k+1}^{M} iP_r(i|C_1) = \sum_{i=k+1}^{M} ip_i/w_1 = \frac{\mu_T - \mu(k)}{1 - w(k)}, \tag{10.22}$$

where

$$w(k) = \sum_{i=1}^{k} p_i$$

$$\mu(k) = \sum_{i=1}^{k} ip_i$$

and

$$\mu_T = \mu(M) = \sum_{i=1}^{M} ip_i.$$

The optimum threshold is defined as the value that maximizes the between-class variance

$$\sigma_B^2(k) = \frac{[\mu_T w(k) - \mu(k)]^2}{w(k)[1 - w(k)]}. \tag{10.23}$$

Thus, the optimal threshold k_* is given by

$$\sigma_B^2(k_*) = \max_{1 \leqslant k < M} \sigma_B^2. \tag{10.24}$$

As the gray levels of calcified regions and wall in a series of IVUS images are not constant, a single threshold may not be sufficient to obtain the desired segmentation of the calcified region. To solve this problem, Santo *et al* [16] in 2008 proposed a multi-thresholding method, which uses successive application of the Otsu method and hence is used in our work. The various steps of Santo's method are as follows:

Step 1. First, we compute a histogram of the original IVUS image.

Step 2. Then, we compute the optimum threshold that maximizes the between-class variance.

Step 3. Now, we re-compute the histogram for $i \geqslant k_*$.

Step 4. Finally, we repeat step 2 and 3 until segmentation sufficiently close to the region of calcification is obtained.

After several tests, we have observed that three iterations of the above algorithm were sufficient to obtain the desired region of calcification.

10.3.3.2 The K-means algorithm

K-means [43] is a well-known unsupervised clustering method that can group the pixels into the desired number of classes based on their inherent distance from each other and hence is used in our work. The K-means algorithm can classify n points into m clusters so that the within-cluster sum of squares is minimized. The points are clustered around centroids $\mu_i \ \forall \ i = 1 \ldots m$ which are obtained by minimizing the objective function,

$$V_{\text{obj}} = \sum_{i=1}^{m} \sum_{x_j \in S_i} (x_j - \mu_i)^2, \tag{10.25}$$

where there are m clusters S_i, $i = 1, 2, \ldots, m$ and μ_i is the centroid or mean point of all the points $x_j \in S_i$. Various steps of the algorithm are as follows:

Step 1. First, we compute a histogram of the original IVUS image.

Step 2. Then, we initialize the centroids with m random intensities.

Step 3. Now, we cluster the point based on the distance of their intensities from the centroid intensities:

$$c^{(i)} = \arg\min \ \|x^{(i)} - \mu_j\|^2 . \tag{10.26}$$

Step 4. Now, we compute the new centroid for each of the clusters,

$$\mu_i = \frac{\sum_{i=1}^{m} I\{c^{(i)} = j\} x^{(i)}}{\sum_{i=1}^{m} I\{c^{(i)} = j\}}, \tag{10.27}$$

where m is a parameter of the algorithm (the number of clusters to be found), I iterates over all the intensities, j iterates over all the centroids and μ_i are the centroid intensities.

Step 5. Finally, we should repeat the preceding steps until the cluster labels of the image do not change anymore.

10.3.3.3 The fuzzy c-means algorithm

The FCM algorithm [18] is an extension of the K-means algorithm. In FCM, instead of finding the absolute membership of a data point to one of the clusters, we assign a degree of membership to each data point corresponding to each cluster center based on the distance between the cluster center and the data point. Now, based on the required accuracy, appropriate tolerance can be fixed. The FCM algorithm provides a generalization to the K-means algorithm and hence is used in our work. The various steps of the algorithm are as follows:

Step 1. Given the dataset X, we first choose the number of clusters C, the weighting exponent m and the termination tolerance ε. The weighting exponent w, where $1 < w < \infty$, determines how much the clusters can overlap with each other. For C clusters, we assume C initial points as far away from each other as possible.

Step 2. Now, we compute the cluster centers:

$$c_j = \frac{\sum_{i=1}^{D} \mu_{ij}^{w} x_i}{\sum_{i=1}^{D} \mu_{ij}^{w}}, \tag{10.28}$$

where D is the number of data points, and μ_{ij} is the degree of membership for the ith data point x_i in cluster j. Note that at the start of the algorithm, the degree of membership for the ith data point x_i in cluster j is initialized with a random value which lies between 0 and 1, such that $\sum_{j}^{C} \mu_{ij} = 1$.

Step 3. Now, with each iteration, the objective function is minimized:

$$V_{\text{obj}} = \sum_{i=1}^{D} \sum_{j=1}^{C} \mu_{ij} \, \|x_i - c_j\|^2. \tag{10.29}$$

The norm $\|x_i - c_j\|$ measures the similarity of the data point x_i to the cluster vector c_j of cluster j.

Step 4. Finally, we will repeat the following steps until the difference between the degree of membership at iterations, say k and $k + 1$, is less than the termination tolerance ε. For this study, we have fixed ε as 0.001.

10.3.3.4 The hidden Markov random field algorithm

The HMRF algorithm [44] is a statistical model which takes into account the spatial information in a probabilistic way such that segmenting any pixel is conditioned on the neighboring surrounding pixels. For segmenting the pixels into different labels, we require the model parameters. In a supervised framework, these model parameters are known *a priori*. But in an unsupervised framework, to obtain these model parameters an iterative scheme called the expectation–maximization (EM) algorithm was suggested, hence is used in our work. The various steps of the HMRF algorithm are as follows:

Step 1. First we start with initial parameters set $\emptyset^{(0)}$ generated by initial segmentation using the K-means algorithm.

Step 2. Now, we calculate likelihood distribution $P^{(t)}(y_i|x_i, \theta_{xi})$.

Step 3. Now, the current parameter set $\text{Ø}^{(t)}$ is used to estimate labels by MAP estimations:

$$\begin{aligned} x^* &= \arg\min_{x \in x} \{P(y|x, \text{Ø}^{(t)})P(x)\} \\ &= \arg\min_{x \in x} \{U(y|x, \text{Ø}^{(t)}) + U(x)\} \end{aligned} \tag{10.30}$$

where the prior probability $P(x) = \frac{1}{Z}\exp(-U(x))$.

$U(x)$ is the prior energy function and we need to solve x^* that minimizes the total posterior energy. For this, an iterative algorithm was used as shown below.

Step 3.1. From the initial segmentation using the K-means algorithm we have an initial estimate $x^{(0)}$.

Step 3.2. Provided $x^{(k)}$, for all $1 \leqslant i \leqslant N$, we find

$$x_i^{(k+1)} = \arg\min \left\{ U(y_i|l) + \sum_{j \in N_i} V_c\left(l, x_j^{(k)}\right) \right\}, \tag{10.31}$$

where $V_c(x)$ is the clique potential.

In the image domain, we assume that one pixel has at most four neighbors, then the clique potential defined on the neighboring pixel is

$$V_c(x_i, x_j) = \frac{1}{2}(1 - I_{xi,xj}), \tag{10.32}$$

where

$$I_{xi,xj} = \begin{cases} 0 & \text{if } x_i \neq x_j \\ 1 & \text{if } x_i = x_j \end{cases}.$$

Step 3.3. We repeat the above step until $U((y|x), \text{Ø}) + U(x)$ converges or a maximum k is achieved.

Step 4. Now, the posterior distribution is calculated for all $l \in L$ and all pixels y_i:

$$P^{(t)}(l|y_i) = \frac{G(y_i; \theta_l)P\left(l|x_{N_i}^{(t)}\right)}{P^{(t)}(y_i)}, \tag{10.33}$$

where $x_{N_i}^{(t)}$ is the neighborhood configuration of $x_i^{(t)}$ and $P^{(t)}(y_i) = \sum_{l \in L} G(y_i; \theta_l)P\left(l|x_{N_i}^{(t)}\right)$ where we have $P\left(l|x_{N_i}^{(t)}\right) = \frac{1}{Z}\exp\left(-\sum_{j \in N_i} V_c\left(l, x_j^{(t)}\right)\right)$.

Step 5. $P^{(t)}(l|y_i)$ is used to update the parameters

$$\mu_l^{(t+1)} = \frac{\sum_i P^{(t)}(l|y_i)y_i}{\sum_i P^{(t)}(l|y_i)} \tag{10.34}$$

$$\left(\sigma_l^{(t+1)}\right)^2 = \frac{\sum_i P^{(t)}(l|y_i)\left(y_i - \mu_l^{(t+1)}\right)^2}{\sum_i P^{(t)}(l|y_i)}. \tag{10.35}$$

In our experimentation, we have considered a maximum of ten iterations for both the EM and mapping algorithm [44]. We utilized a sigma value of 0.02 for the EM algorithm due to the fact that a high sigma value can result in excessive smoothing, producing incorrect results [44].

In our experimentation for the proposed three segmentation methods (K-means, FCM and HMRF), we observed that the desired region of calcification was observed when the number of classes was chosen as four. We have also tried the above algorithm for three and five classes. In both cases, unsatisfactory results were obtained. When the number of classes was fixed as three, overestimation of the region of calcification was observed. However, when the number of classes was fixed as five, underestimation of the region of calcification was observed.

10.3.3.5 Calcium volume computation
In calculus, the volume can be determined from sequential areas a known distance apart. Scott *et al* [45], in 2000, proposed a similar method to compute the surface area from a sequential circumference a known distance apart. In our study, we have adopted the same strategy to compute the calcium volume. Later, calcium volume (V) is computed using a conventional integration method formulated as shown below:

$$V = \sum_{f=1}^{F} A_f, \tag{10.36}$$

where A_f represents the total calcium area of the binary regions representing the multi-focal calcium lesions in the frame f of the IVUS video, consisting of a total of F frames. It is important to note that, even though the videos have F number of frames, it is not necessary that all the patients have the same number of frames or that all the frames have calcium.

The proposed method uses the same concept of calcium volume computation as used by Araki *et al* [12], but has better volume accuracy, as now the calcium lesions area is measured in the entire IVUS video, whereas the previous study detected only the largest calcium in the entire video.

10.4 Results

10.4.1 Calcium detection

We tested all five multiresolution techniques described above on a dataset of $N = 19$ coronary patients with an average of about $F = 2040$ frames per patient representing one coronary artery. Figure 10.6 shows typical examples of calcium region detection with and without multiresolution. The first row shows the calcium detection

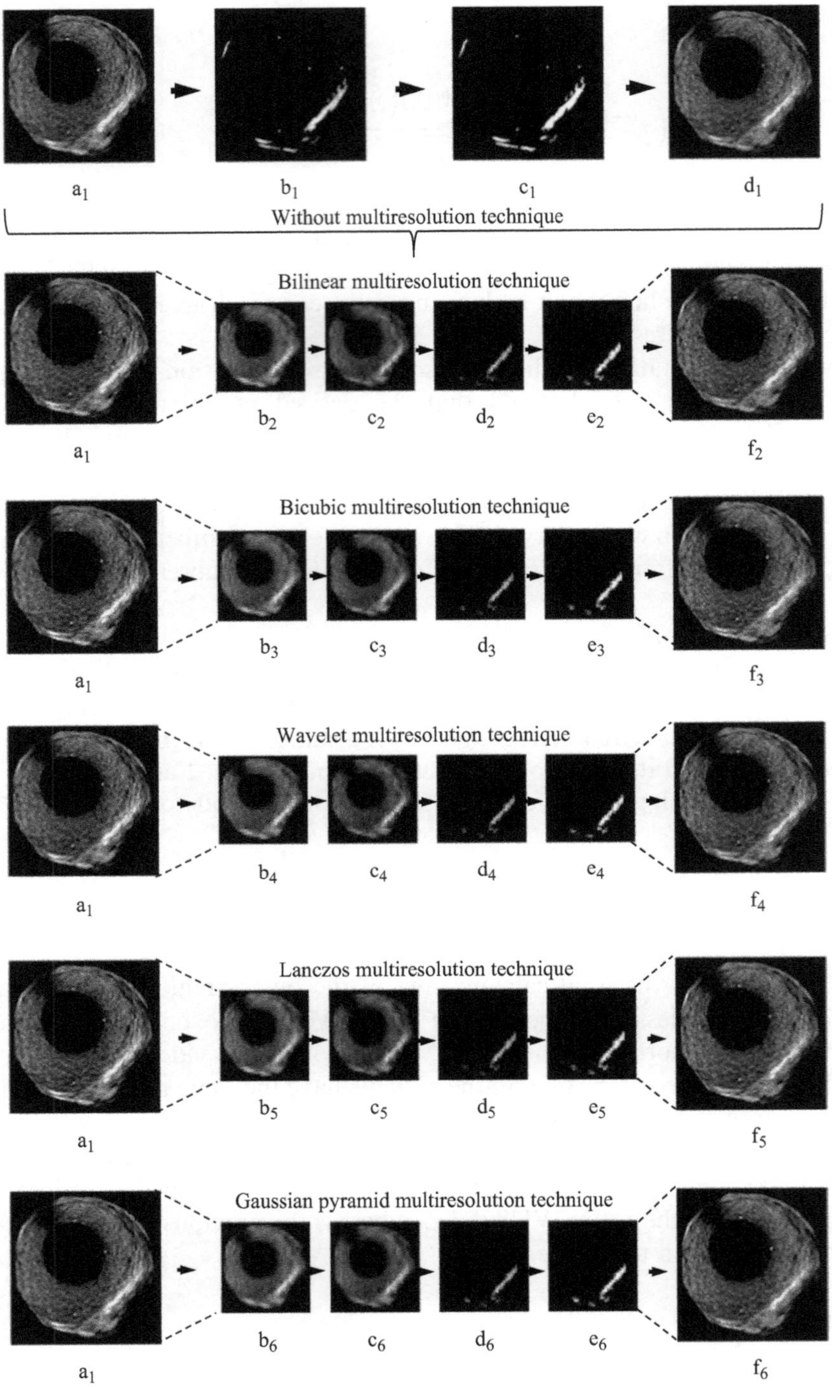

Figure 10.6. Sample result of the detection of the multi-focal calcified plaques taking FCM as a segmentation method and using different multiresolution techniques with bias correction. a_1 is the grayscale ring image, b_1–d_1 are segmented, binary and calcium detected images without using any multiresolution techniques. b_2–b_6, c_2–c_6, d_2–d_6, e_2–e_6 and f_2–f_6 are down-sampled, bias-corrected, segmented, binary and calcium detected images of multiresolution techniques 1–5, respectively.

approach without using any multiresolution technique: a_1 is the grayscale ring image; b_1–d_1 are segmented, binary and calcium detected images, respectively. The five multiresolution techniques are shown in the second and sixth rows. The second, third, fourth, fifth and sixth rows show the results of five different multiresolution techniques. Each row has a sequence of images in the following order: down-sampled image, bias-corrected image, segmented image, binary image and calcium detected image. We can observe that the calcium region detected without multiresolution (row 1, last column) and with all the five multiresolution techniques (row 2–row 6, last column) are very similar in nature.

10.4.2 Volume measurement

In the current study, we have evaluated the calcium volume measurement at the patient level. Let $A_{s_j}^{0}(n, f)$ be the area corresponding to frame f in patient n and $TA_{s_j}^{0}(n)$ be the total area over F frames corresponding to the segmentation method without multiresolution for patient n, then the calcium volume for the segmentation method s_j without multiresolution, $V_{s_j}^{0}(n)$, can be mathematically given as

$$V_{s_j}^{0}(n) = TA_{s_j}^{0}(n) = \sum_{f=1}^{F} A_{s_j}^{0}(n, f). \tag{10.37}$$

Let $A_{s_j}^{m_i}(n, f)$ be the area corresponding to frame f in patient n and $TA_{s_j}^{m_i}(n)$ be the total area over F frames corresponding to the segmentation method with multiresolution m_i for patient n, then the calcium volume for segmentation method s_j with multiresolution, $V_{s_j}^{m_i}(n)$, can be mathematically given as

$$V_{s_j}^{m_i}(n) = TA_{s_j}^{m_i}(n) = \sum_{f=1}^{F} A_{s_j}^{m_i}(n, f). \tag{10.38}$$

Table 10.1 shows the volume computed using 20 combinational systems using five different segmentation methods with and without using multiresolution techniques. For all the segmentation methods, the number of classes is fixed to four. The results of table 10.1 are used later for computing the precision-of-merit (PoM) in the performance evaluation section (section 10.5)

10.4.3 Percentage mean time improvement

For computing the percentage mean time improvement of each segmentation method with and without multiresolution techniques, we calculate the mean time difference between segmentation with and without the multiresolution method. Thus, the ratio is computed between the difference in mean time between the multiresolution method and non-multiresolution method to non-multiresolution method. All the algorithms were implemented on the same hardware and computer settings to avoid any bias in time computations.

Table 10.1. Volume computation using a combination of four segmentation and five multiresolution (with MR) techniques against without multiresolution (w/o MR).

	With MR (mm^3)					
Segmentation methods	Bilinear	Bicubic	Wavelet	Lanczos	Gaussian pyramid	w/o MR (mm^3)
Threshold	45.65	47.04	45.69	45.34	42.49	43.94
K-means	35.76	38.12	35.01	34.38	34.00	32.46
FCM	47.06	48.40	47.35	46.75	43.62	46.30
HMRF	60.57	64.09	62.30	62.21	60.88	56.20
Mean	47.26	49.41	47.59	47.17	45.25	44.73
SD	10.20	10.80	11.23	11.45	11.27	9.75

10.4.3.1 General definition of mean time

If $t_{s_j}^{m_i}(n, f)$ represents the time consumed for multiresolution technique m_i in segmentation paradigm s_j taken for the IVUS frame f in patient n and let $t_{s_j}^{m_i}$ be the mean time over F frames and N patients, then $t_{s_j}^{m_i}$ can be mathematically given as

$$t_{s_j}^{m_i} = \frac{\sum_{n=1}^{N}\sum_{f=1}^{F} t_{s_j}^{m_i}(n, f)}{N \cdot F}. \tag{10.39}$$

Representing i as the total number of multiresolution techniques and j as the total number of segmentation methods, our implementation thus consists of a total of 20 combinations with possible indices i and index j, respectively, as $i = \{1, 2, 3, 4, 5\}$ and $j = \{1, 2, 3, 4\}$. Mean times for all possible 20 combinations are shown in appendix B.

10.4.3.2 Mean time over non-multiresolution techniques over S *segmentation methods*

Let $\bar{t}_{s_j}^{0}$ be the mean time using the segmentation method s_j without multiresolution. Using this formulation, one can express the mean for the first, second, third and fourth segmentation methods without using multiresolution techniques, expressed using symbols $\bar{t}_{s_1}^{0}$, $\bar{t}_{s_2}^{0}$, $\bar{t}_{s_3}^{0}$ and $\bar{t}_{s_4}^{0}$, respectively. If $\bar{t}_{\forall S}^{0}$ is the mean of all the S-segmentation methods without using multiresolution techniques, then $\bar{t}_{\forall S}^{0}$ is mathematically given as

$$\bar{t}_{\forall S}^{0} = \frac{\sum_{j=1}^{S} \bar{t}_{s_j}^{0}}{S}, \tag{10.40}$$

where S represents the total number of four segmentation methods.

10.4.3.3 Mean time for M *multiresolution techniques over* S *segmentation methods*

Let $\bar{t}^{m_i}_{\forall S}$ be the mean time of multiresolution technique m_i averaged over all S-segmentation methods, then $\bar{t}^{m_i}_{\forall S}$ can be mathematically given as

$$\bar{t}^{m_i}_{\forall S} = \frac{\sum_{j=1}^{S} t^{m_i}_{s_j}}{S}. \tag{10.41}$$

Over five multiresolution techniques, this can be symbolized as $\bar{t}^{m_1}_{\forall S}$, $\bar{t}^{m_2}_{\forall S}$, $\bar{t}^{m_3}_{\forall S}$, $\bar{t}^{m_4}_{\forall S}$ and $\bar{t}^{m_5}_{\forall S}$, in table 10.2 expressed as mean MR (s) and mathematically shown in appendix A, equation (A.1).

10.4.3.4 Percentage improvement in mean time for M *multiresolution techniques*

The percentage improvement in mean time (PTI) for multiresolution technique m_i can be computed using equations (10.13) and (10.14), expressed as the last row in table 10.2 (PTI by MR (%)) and mathematically expressed as follows:

$$\mathrm{PTI}^{m_i}_{\forall S} = \left[\frac{\bar{t}^{m_i}_{\forall S} - \bar{t}^{0}_{\forall S}}{\bar{t}^{0}_{\forall S}}\right] \cdot 100. \tag{10.42}$$

Similarly, we can compute the percentage mean time improvement for the first, second, third, fourth and fifth multiresolution techniques averaged over all segmentation methods as $\mathrm{PTI}^{m_1}_{\forall S}$, $\mathrm{PTI}^{m_2}_{\forall S}$, $\mathrm{PTI}^{m_3}_{\forall S}$, $\mathrm{PTI}^{m_4}_{\forall S}$ and $\mathrm{PTI}^{m_5}_{\forall S}$,, respectively, as shown in appendix A, equation (A.2).

10.4.3.5 Percentage mean time improvement for S *segmentation methods over* M *multiresolution techniques*

We can use the same above concept to compute the percentage mean time improvement by each segmentation method without using any multiresolution

Table 10.2. Percentage mean improvement in time using 20 different kinds of systems using five different segmentation methods, with and without using multiresolution techniques, using bias correction.

	Multiresolution technique (w/ MR)							
Segmentation method	Bilinear	Bicubic	Wavelet	Lanczos	Gaussian pyramid	Mean seg. (s)	w/o MR	PTI by seg. (%)
Threshold (s)	$\bar{t}^{m_1}_{S1}$	$\bar{t}^{m_2}_{S1}$	$\bar{t}^{m_3}_{S1}$	$\bar{t}^{m_4}_{S1}$	$\bar{t}^{m_5}_{S1}$	$\bar{t}^{S1}_{\forall M}$	$\bar{t}^{0}_{S1}$	$\mathrm{PTI}^{S1}_{\forall M}$
K-means (s)	$\bar{t}^{m_1}_{S2}$	$\bar{t}^{m_2}_{S2}$	$\bar{t}^{m_3}_{S2}$	$\bar{t}^{m_4}_{S2}$	$\bar{t}^{m_5}_{S2}$	$\bar{t}^{S2}_{\forall M}$	$\bar{t}^{0}_{S2}$	$\mathrm{PTI}^{S2}_{\forall M}$
FCM (s)	$\bar{t}^{m_1}_{S3}$	$\bar{t}^{m_2}_{S3}$	$\bar{t}^{m_3}_{S3}$	$\bar{t}^{m_4}_{S3}$	$\bar{t}^{m_5}_{S3}$	$\bar{t}^{S3}_{\forall M}$	$\bar{t}^{0}_{S3}$	$\mathrm{PTI}^{S3}_{\forall M}$
HMRF (s)	$\bar{t}^{m_1}_{S4}$	$\bar{t}^{m_2}_{S4}$	$\bar{t}^{m_3}_{S4}$	$\bar{t}^{m_4}_{S4}$	$\bar{t}^{m_5}_{S4}$	$\bar{t}^{S4}_{\forall M}$	$\bar{t}^{0}_{S4}$	$\mathrm{PTI}^{S4}_{\forall M}$
Mean MR (s)	$\bar{t}^{m_1}_{\forall S}$	$\bar{t}^{m_2}_{\forall S}$	$\bar{t}^{m_3}_{\forall S}$	$\bar{t}^{m_4}_{\forall S}$	$\bar{t}^{m_5}_{\forall S}$	—	$\bar{t}^{0}_{\forall S}$	—
PTI by MR (%)	$\mathrm{PTI}^{m_1}_{\forall S}$	$\mathrm{PTI}^{m_2}_{\forall S}$	$\mathrm{PTI}^{m_3}_{\forall S}$	$\mathrm{PTI}^{m_4}_{\forall S}$	$\mathrm{PTI}^{m_5}_{\forall S}$	—	—	—

techniques against the mean time consumed by each segmentation method using the multiresolution techniques. If $t_{s_j}^{m_i}$ is the mean time of segmentation methods averaged over all multiresolution techniques, then

$$\bar{t}_{\forall M}^{s_j} = \frac{\sum_{i=1}^{M} t_{s_j}^{m_i}}{M}, \tag{10.43}$$

where M represents the total number of five multiresolution techniques.

Similarly, we can compute the mean time for the first, second, third and fourth segmentation method averaged over all multiresolution techniques as $\bar{t}_{\forall M}^{s_1}$, $\bar{t}_{\forall M}^{s_2}$, $\bar{t}_{\forall M}^{s_3}$ and $\bar{t}_{\forall M}^{s_4}$, respectively, shown in appendix A, equation (A.3).

Percentage mean time improvement for segmentation method s_j with respect to all M multiresolution techniques can be computed as

$$\mathrm{PTI}_{\forall M}^{s_j} = \left[\frac{\bar{t}_{\forall M}^{s_j} - \bar{t}_{s_j}^{0}}{\bar{t}_{s_j}^{0}}\right] \cdot 100. \tag{10.44}$$

Similarly, we can compute the percentage mean time improvement for the first, second, third and fourth segmentation methods, expressed as $\mathrm{PTI}_{\forall M}^{s_1}$, $\mathrm{PTI}_{\forall M}^{s_2}$, $\mathrm{PTI}_{\forall M}^{s_3}$ and $\mathrm{PTI}_{\forall M}^{s_4}$, respectively, shown in appendix A, equation (A.4).

A summary of all equations is provided in table 10.2.

Table 10.3 shows the time consumption using 20 different kinds of systems using five different segmentation methods with and without using multiresolution techniques. All multiresolution techniques result in a mean improvement in time [20]. It was observed that wavelet-based multiresolution experiences the highest mean improvement in time (in seconds) of 74.07% compared to other multiresolution techniques. From the experimental analysis, we observed that FCM takes the largest computational time to converge due to its large number of iterations. On the basis of the results shown in table 10.3, the FCM segmentation method with multiresolution

Table 10.3. Percentage of mean time improvement using 20 different kinds of systems using five different segmentation methods, with and without using multiresolution techniques, using bias correction.

	Multiresolution technique (w/ MR)							
Segmentation method	Bilinear	Bicubic	Wavelet	Lanczos	Gaussian pyramid	Mean seg. (s)	w/o MR	PTI by seg. (%)
Threshold (s)	0.0077	0.0076	0.0085	0.0073	0.0074	0.0077	0.0167	53.89%
K-means (s)	0.0490	0.0443	0.0530	0.0427	0.0432	0.0464	0.1104	57.97%
FCM (s)	4.0952	4.1987	4.1738	4.5423	4.1781	4.2376	18.5456	77.15%
HMRF (s)	2.1617	2.1231	1.9822	2.4405	2.0245	2.1464	5.3056	59.54%
Mean MR (s)	1.5784	1.5934	1.5544	1.7582	1.5633	—	5.9946	—
PTI by MR (%)	73.67	73.42	74.07	70.67	73.92	—	—	—

and bias correction experiences the highest mean improvement in time (in seconds) of 77.15% compared to the computation time without multiresolution.

10.5 Performance evaluation

10.5.1 Multiresolution error metrics against non-multiresolution technique

There are two fundamental paradigms where one can evaluate the performance of the multiresolution strategy when benchmarked against the non-multiresolution framework. The first paradigm is based on the way the interventional cardiologist mimics the volume computation. We can characterize this paradigm at the patient level. Since the cardiologist is mainly interested in getting all the numbers in the volumetric framework, it is necessary to compute the error metric at the patient level and simply compare the volumes for the multiresolution and non-multiresolution frameworks. Such techniques are very popular in CT/MR volumetric scans where the interventional cardiologist is interested in the calcium score on a volumetric basis.

Even though this method is giving a generalized approach to the error metric at a patient level, one can argue it aids in understanding the accuracy of calcium detection and its performance at the individual IVUS frame level between the multiresolution and non-multiresolution frameworks. This may sound to be more of a local approach, unlike the patient level as a global approach. Both these methods are valid, although their their advantages and disadvantages can be debated, we thus adopted both methods to evaluate our error metrics, labeled as the volume level-based error metric and frame-level-based error metric. We term this error metric as the precision-of-merit (PoM). Section 10.5.1.1 shows the PoM at the patient level using the volume paradigm and section 10.5.1.2 presents the PoM using the frame paradigm.

10.5.1.1 Precision-of-merit at the volume level (VL)

As shown in equations (10.10) and (10.11), if $V_{s_j}^{0}(n)$ is the volume for the segmentation method s_j without multiresolution, and $V_{s_j}^{m_i}(n)$ is the volume for the segmentation method s_j with multiresolution m_i, respectively, then the volume difference ratio $\mathrm{VDR}_{s_j}^{m_i}(n)\Big|_{\mathrm{VL}}$ with respect to the non-multiresolution method can be mathematically given as

$$\mathrm{VDR}_{s_j}^{mi}(n)\Big|_{\mathrm{VL}} = \left(\frac{\left| V_{S_j}^{m_i}(n) - V_{S_j}^{0}(n) \right|}{V_{S_j}^{0}(n)} \right). \tag{10.45}$$

Using the above equation, one can compute the mean volume difference ratio $\overline{\mathrm{VDR}_{s_j}^{mi}}\Big|_{\mathrm{VL}}$ for multiresolution techniques m_i and segmentation methods s_j taken over all 19 patients, as shown in appendix A, equation (A.5). Thus, the precision-of-merit $\mathrm{PoM}_{s_j}^{m_i}\Big|_{\mathrm{VL}}$ for multiresolution techniques m_i and segmentation methods s_j is mathematically represented as

Table 10.4. $\text{PoM}_{s_j}^{mi}\Big|_{\text{VL}}$ using 20 kinds of combinatorial systems.

Segmentation method	Bilinear	Bicubic	Wavelet	Lanczos	Gaussian pyramid
Threshold (%)	95.75 ± 2.31	93.19 ± 3.48	95.59 ± 2.52	96.35 ± 2.11	95.46 ± 2.80
K-means (%)	92.30 ± 9.28	85.94 ± 14.07	94.00 ± 6.79	95.45 ± 6.56	90.00 ± 8.76
FCM (%)	98.40 ± 1.32	95.84 ± 2.89	97.85 ± 1.26	98.96 ± 0.75	93.54 ± 3.72
HMRF (%)	86.98 ± 7.39	85.10 ± 4.78	88.79 ± 3.97	83.97 ± 6.85	91.27 ± 4.64
Mean ± SD (%)	93.36 ± 5.07	90.02 ± 6.31	94.06 ± 3.64	93.68 ± 4.07	92.57 ± 4.98

$$\text{PoM}_{s_j}^{mi}\Big|_{\text{VL}} = 100 - \left(\overline{\text{VDR}}_{s_j}^{mi} * 100\right). \tag{10.46}$$

The PoM at volume level computed for all 20 combinations of the four segmentation methods and five multiresolution techniques are given in table 10.4.

10.5.1.2 Precision-of-merit at the frame level (FL)

In this method, the so-called local approach, we take the difference in detected calcium area between automated segmentation with the multiresolution approach and segmentation using a non-multiresolution approach at the individual frame level. This means taking the difference of the detected calcium area for each frame of the video between the segmentation outputs with and without multiresolution. If $A_{s_j}^{0}(n, f)$ represents the area computed without multiresolution, for the segmentation method s_j corresponding to frame f in patient n, and $A_{s_j}^{m_i}(n, f)$ represents the area computed using the multiresolution m_i, for the segmentation method s_j corresponding to frame f in patient n, then the absolute area difference $\text{AD}_{s_j}^{m_i}(n, f)$ between the segmentation s_j outputs with and without multiresolution m_i, corresponding to frame f in patient n can be generalized as

$$\text{AD}_{s_j}^{m_i}(n, f) = \left| A_{s_j}^{m_i}(n, f) - A_{s_j}^{0}(n, f) \right|. \tag{10.47}$$

Let $\text{VD}_{s_j}^{m_i}(n)$ be the volume difference between the segmentation outputs with and without multiresolution m_i, for the segmentation method s_j for patient n. Correspondingly, let $V_{s_j}^{0}(n)$ be the volume for the segmentation method without multiresolution, then for each patient at the frame level, the volume difference ratio $\text{VDR}_{s_j}^{m_i}(n)\Big|_{\text{FL}}$ can be represented as

$$\text{VDR}_{s_j}^{mi}(n)\Big|_{\text{FL}} = \left(\frac{\text{VD}_{s_j}^{m_i}(n)}{V_{S_j}^{0}(n)}\right), \tag{10.48}$$

where

$$\text{VD}_{s_j}^{m_i}(n) = \sum_{f=1}^{F} \text{AD}_{s_j}^{m_i}(n, f). \tag{10.49}$$

The mean volume difference ratio $\overline{\mathrm{VDR}}_{s_j}^{mi}\Big|_{\mathrm{FL}}$, computed as the mean of all 19 patients for multiresolution techniques m_i and segmentation methods s_j, is shown in appendix A, equation (A.6). The precision-of-merit $\mathrm{PoM}_{s_j}^{m_i}\Big|_{\mathrm{FL}}$ for multiresolution techniques m_i and segmentation methods s_j is mathematically represented as

$$\mathrm{PoM}_{s_j}^{mi}\Big|_{\mathrm{FL}} = 100 - \left(\overline{\mathrm{VDR}}_{s_j}^{mi} * 100\right). \tag{10.50}$$

Note that $\mathrm{PoM}_{s_j}^{m_i}\Big|_{\mathrm{FL}}$ is computed for all 20 sets of combinations consisting of the four segmentation methods and five multiresolution techniques and is shown in table 10.5.

10.5.1.3 Difference between volume level and frame level mean PoMs

The percentage difference between mean PoM at the volume level $\overline{\mathrm{PoM}}_{s_j}^{mi}\Big|_{\mathrm{VL}}$ and mean PoM at the frame level $\overline{\mathrm{PoM}}_{s_j}^{m_i}\Big|_{\mathrm{FL}}$ is mathematically represented as follows, and is shown in table 10.6:

$$\%\mathrm{PoM}_{\mathrm{diff}} = \left(\frac{\overline{\mathrm{PoM}}_{s_j}^{mi}\Big|_{\mathrm{VL}} - \overline{\mathrm{PoM}}_{s_j}^{m_i}\Big|_{\mathrm{FL}}}{\overline{\mathrm{PoM}}_{s_j}^{m_i}\Big|_{\mathrm{FL}}}\right) * 100. \tag{10.51}$$

The mean percentage difference for the PoM ($\%\mathrm{PoM}_{\mathrm{diff}}$)is 15.93%. Thus, the mean PoM at the frame level is 15.93% lower compared to the mean PoM at the volume level. This is attributed to the fact that very small bright regions are detected as calcium due to the presence of noise. Note that both metrics can be adopted for evaluation of multiresolution methods, as both methods are above a threshold of acceptable performance, keeping in consideration the effect of motion of the heart, breathing condition and processing of QuickTime video rather than DICOM-based high-resolution image formats. Further research is required to study the effects of these factors. A full section (section 10.6) has been devoted discussion of this area.

Table 10.5. $\mathrm{PoM}_{s_j}^{mi}\Big|_{\mathrm{FL}}$ using 20 kinds of combinatorial systems.

Segmentation method	Bilinear	Bicubic	Wavelet	Lanczos	Gaussian pyramid
Threshold (%)	81.16 ± 20.33	78.81 ± 20.04	81.17 ± 20.26	81.66 ± 20.27	79.28 ± 19.16
K-means (%)	73.22 ± 21.71	70.56 ± 21.59	73.88 ± 20.02	74.96 ± 19.90	71.77 ± 19.12
FCM (%)	82.72 ± 21.48	80.35 ± 21.07	82.79 ± 21.52	83.18 ± 21.49	79.32 ± 19.37
HMRF (%)	86.95 ± 3.61	83.79 ± 4.15	87.53 ± 3.37	79.57 ± 7.34	87.13 ± 5.53
Mean ± SD (%)	81.01 ± 16.78	78.38 ± 16.71	81.34 ± 16.29	79.84 ± 17.25	79.38 ± 15.79

Table 10.6. Difference between mean PoM at the volume level and mean PoM at the frame level using 20 different kinds of systems.

Mean precision-of-merit ($\overline{\text{PoM}}$)	Bilinear	Bicubic	Wavelet	Lanczos	Gaussian pyramid	Mean
Mean PoM at the volume level $\left(\overline{\text{PoM}}_{s_j}^{mi}\Big\|_{\text{VL}}\right)$	93.36	90.02	94.06	93.68	92.57	—
Mean PoM at the frame level $\left(\overline{\text{PoM}}_{s_j}^{mi}\Big\|_{\text{FL}}\right)$	81.01	78.38	81.34	79.84	79.38	—
Difference in mean PoM ($\%\text{PoM}_{\text{diff}}$)	15.24	14.85	15.63	17.33	16.62	15.93

10.5.1.4 A note on comparison between volume level and frame level PoMs

Tables 10.4 and 10.5 show the comparative results of PoM for mean calcium volume similarity computed using 20 different kinds of systems using the two methods: the volume and frame levels. It is observed that wavelet interpolation presents the highest mean PoMs of 94.06% ± 3.64% and 81.34% ± 16.29% compared to the other multiresolution techniques for both the volume level and frame level, respectively.

For improving PoM accuracy, we perform bias correction on 16 patients after down-sampling of the IVUS grayscale ring images generated from different multiresolution techniques. We show an exemplary concept of bias removal by taking one of the three segmentation methods from our set-up. We selected K-means as it was proven to be the best segmentation method (according to expert scoring) compared to the other three segmentation methods (Santos, FCM and HMRF) in our recent publication [15].

Our goal was to understand the effect of bias correction on all five multiresolution techniques when applied to a K-means segmentation method. With bias correction, we observe an increase in mean calcium volume similarity between the multiresolution and non-multiresolution methods of 2.43%, 4.13%, 0.99%, 0.84% and 0.99%, corresponding to the bilinear, bicubic, discrete wavelet, Lanczos and Gaussian pyramid techniques, respectively. Among all five multiresolution techniques, it is observed that bicubic interpolation showed the largest mean calcium volume similarity increase of 4.13%, compared to the other multiresolution techniques.

10.5.2 The mean Jaccard index (JI) and Dice similarity coefficient (DSC)

We used the JI and DSC, as these are the simplest and most direct ways of measuring the overlap accuracy of the system [46]. The JI and DS parameters quantify the degree of similarity between the calcium regions in two binary frames (say p and q) of the IVUS video. If A is the ground truth binary image and B is the segmented binary image, then the Jaccard index formula is generalized as

$$J_{pq} = \frac{|A \cap B|}{|A \cup B|}. \tag{10.52}$$

Here the Jaccard distance always lies between the values of 0 and 1, i.e. $0 \leqslant J_{pq} \leqslant 1$. Similarly, we can compute the Dice similarity coefficient in terms of A and B as follows:

$$D_{pq} = \frac{2|A \cap B|}{|A| + |B|}. \tag{10.53}$$

Here the Dice similarity coefficient always lies between the value of 0 and 1, i.e. $0 \leqslant D_{pq} \leqslant 1$.

Our objective is to benchmark the JI and DSC computed using five multiresolution approaches against non-multiresolution methods. Thus, we need to compute the mean JI and DSC spanned over the four segmentation methods corresponding to each multiresolution method. Note that since JI and DS are computed at the frame level, these metrics are taken independent of the patients, and thus is computed over the entire population of frames, i.e. 38 760 frames. Further, note that because there are four segmentation methods, we thus take the mean over all the segmentation methods. This is exactly shown in figure 10.7 corresponding to five multiresolution methods. As shown in the figure, we observe that our third multiresolution technique (i.e. the wavelet-based technique) shows the highest JI and DS corresponding to values of 0.70 and 0.82, respectively.

10.5.3 Manual scoring of detected calcium by a radiologist

Blinded manual scoring was performed to evaluate the performance of the detected calcium. The radiologist had experience of 20 years and was an expert in atherosclerosis disease management. As discussed before, in this study we have

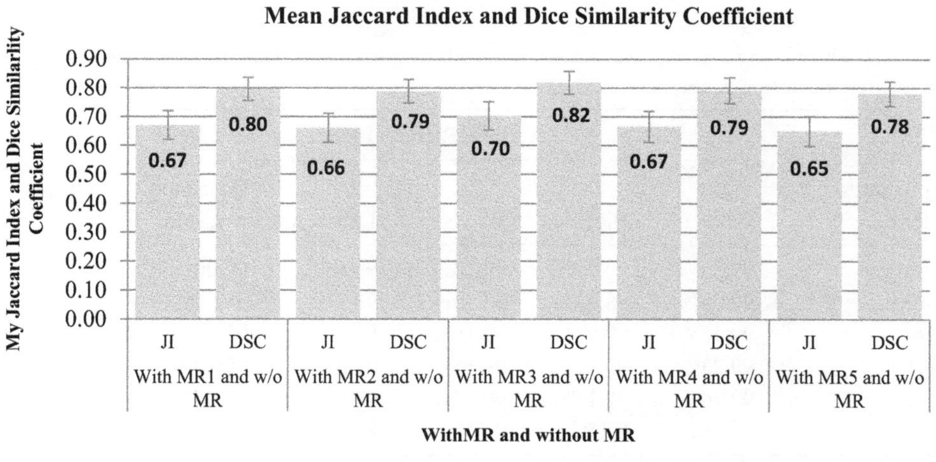

Figure 10.7. Bar plot of mean JI and DSC: MR1, MR2, MR3, MR4 and MR5 correspond to the bilinear, bicubic, wavelet, Lanczos and Gaussian pyramid multiresolution techniques, respectively.

analyzed the complete coronary IVUS videos taken from 19 patients. Because of the pullback speed of 0.5 mm s^{-1} during image acquisition, one video produced by the IVUS scanner consisted of approximately 2040 frames per video, amounting to about 38 760 frames. Since there are 20 combinatorial systems, this leads to 775 200 frames. It is practically too tedious for clinicians to give a clinical score on such a large number of frames. We thus took a representative sample of 100 frames per patient. For each frame that was processed by 20 combinational techniques, we computed the corresponding calcium detected overlay images computed from these 20 systems, and this resulted in 3800 frames for evaluation. These 100 frames per video were randomly selected from the mid-region of the coronary IVUS videos for the best representation. The scoring scale criteria were as follows: scores were given between 0 and 5, where 0 represented false calcium detection and 5 represented true calcium detection. Later, we averaged all the scores taken from all the patients to obtain the mean scores. The percentage accuracy of the system was computed as the ratio of the mean observed score (taken by our radiologist) to mean ideal score. This process was repeated for all 20 combinations consisting of five multiresolution and four segmentation methods.

Table 10.7 shows the mean scores on the proposed calcium detection methods using 20 combinations. The highest score was given to threshold segmentation combined with Lanczos multiresolution technique (99.65%), which was also close to FCM-based segmentation combined with wavelet-based multiresolution (99.18%). If we take the mean of the scores of all segmentation methods, we observe that the Lanczos multiresolution technique showed the highest score (99.31%). The percentage mean score by taking all four segmentation methods with multiresolution (99.20%, 99.13%, 99.14%, 99.31% and 99.04%) was close to that obtained without multiresolution (99.05%). The high scores for all the combinations justify our hypothesis that the inclusion of multiresolution techniques with segmentation methods resulted in low computational cost.

Table 10.7. Scoring on proposed calcium detection methods using a combination of four segmentation and five multiresolution techniques.

	With MR (mm^3)					
Segmentation method	Bilinear	Bicubic	Wavelet	Lanczos	Gaussian pyramid	Without MR (mm^3)
Threshold	99.35%	99.29%	99.28%	99.65%	99.00%	98.95%
K-means	99.05%	98.95%	98.95%	98.92%	98.91%	99.41%
FCM	99.16%	99.08%	99.18%	99.28%	99.04%	98.92%
HMRF	99.23%	99.19%	99.16%	99.38%	99.20%	98.92%
Mean	99.20%	99.13%	99.14%	99.31%	99.04%	99.05%
SD	0.12%	0.15%	0.14%	0.30%	0.12%	0.24%

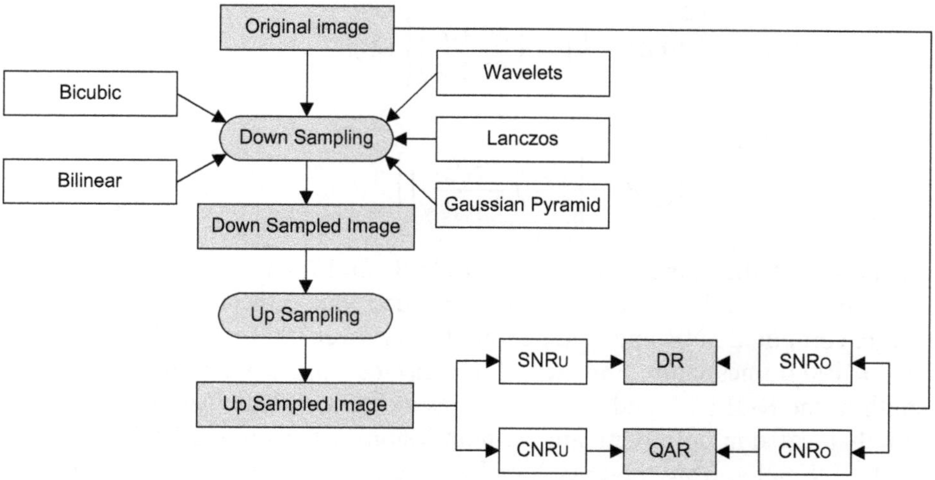

Figure 10.8. DR and QAR algorithm.

10.5.4 Degradation ratio and quality assessment ratio

Figure 10.8 is the second part of the main block diagram of the proposed system explaining the performance evaluation performed using signal-to-noise ratio (SNR) and contrast-to-noise ratio (CNR). We use the following formula for SNR computation [47]:

$$\mathrm{SNR}(j, k) = \frac{S_L(j, k) - S_{MB}(j, k)}{\sqrt{2} \cdot \sigma_B(j, k)}, \tag{10.54}$$

where $S_L(j, k)$ is the mean signal strength in the region-of-interest (ROI) with a lesion at the location (j, k), $S_{MB}(j, k)$ and $\sigma_B(j, k)$ are the mean signal strength and standard deviation of the background ROI without a lesion at the location the (j,k), respectively.

We use the following expression for CNR computation:

$$\mathrm{CNR} = \sqrt{\frac{(\mu(t) - \mu(n))^2}{\sigma(t) + \sigma(n)}}, \tag{10.55}$$

where $\mu(t)$ and $\mu(n)$ are the mean of the signal strength in the target ROI and the background noise, respectively, and $\sigma(t)$ and $\sigma(n)$ are the variance of the signal strength in the target ROI and the background noise, respectively.

In this article, we have analyzed the performance of all the five multiresolution techniques on two different parameters, i.e. degradation ratio (DR) and quality assessment ratio (QAR), computed from SNR and CNR.

SNR_U and CNR_U are the signal-to-noise ratio and contrast-to-noise ratio computed for the up-sampled image, and SNR_O and CNR_O are the signal-to-noise ratio and contrast-to-noise ratio computed for the original image, respectively. Mathematically, DR and QAR are given as

$$\mathrm{DR} = \left[1 - \left(\frac{\mathrm{SNR}_U}{\mathrm{SNR}_O}\right)\right] \cdot 100 \tag{10.56}$$

$$\mathrm{QAR} = \left[1 - \left(\frac{\mathrm{CNR}_U}{\mathrm{CNR}_O}\right)\right] \cdot 100. \tag{10.57}$$

To demonstrate the concept of SNR and CNR, and for reasons of simplicity, we choose only ten random frames from each video. Since there are 19 patients, we therefore computed SNR and CNR computations over 190 frames (19 patients × 10 frames = 190 frames) only. For each of the selected frames, we computed the signal strength in the ROI with and without lesions at ten locations. All ten locations are randomly located in both (with and without lesion) regions in the ROI, as shown in figure 10.9. Finally, from these ten locations, we compute the mean and variance of the signal strength in the ROI with and without lesions.

Using the SNR and CNR values, the last step is to compute the DR and QAR. These computations follow the same concept of frame level and then mean statistics

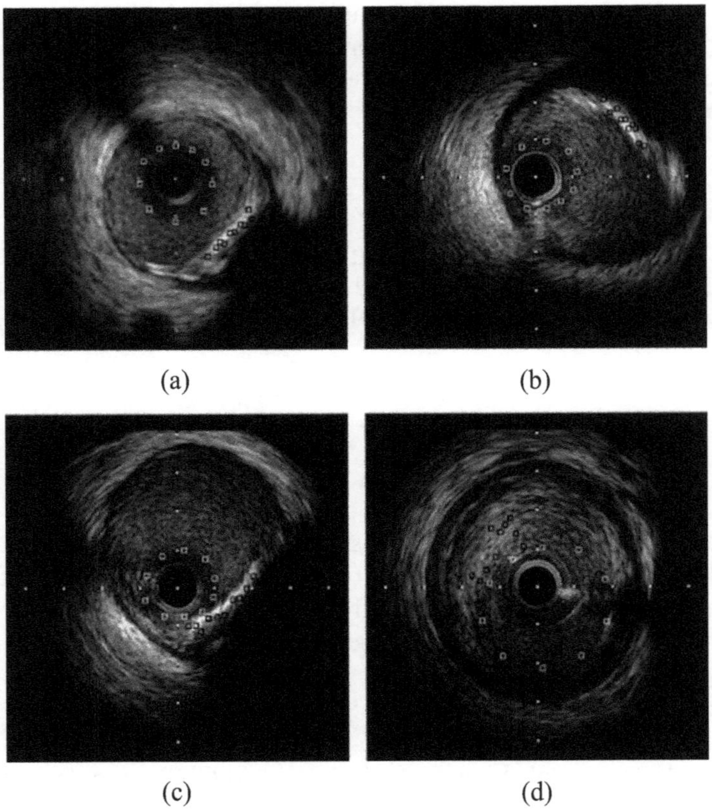

Figure 10.9. Signal and noise window locations on input (original) images with solid calcium (a)–(c) and soft calcium (d).

taken over all 19 patients. The process of DR and QAR is repeated for all four segmentation methods applied to all five multiresolution techniques. The mean statistics are shown in figures 10.10 and 10.11, respectively.

For analyzing the accuracy of the results, we compute the performance of merit for both DR and QAR. Mathematically, it is given as

$$\mathrm{PoM_{DR}}(\%) = 100 - \left[\left(\frac{\mathrm{SNR}_M - \mathrm{SNR}_O}{\mathrm{SNR}_O}\right) \cdot 100\right] \tag{10.58}$$

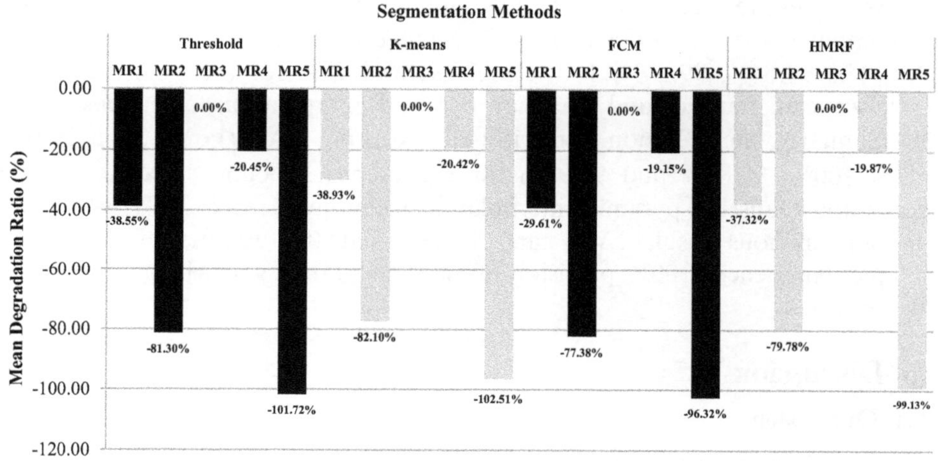

Figure 10.10. Bar plot of mean degradation ratio: MR1, MR2, MR3, MR4 and MR5 correspond to the bilinear, bicubic, wavelet, Lanczos and Gaussian pyramid multiresolution techniques, respectively.

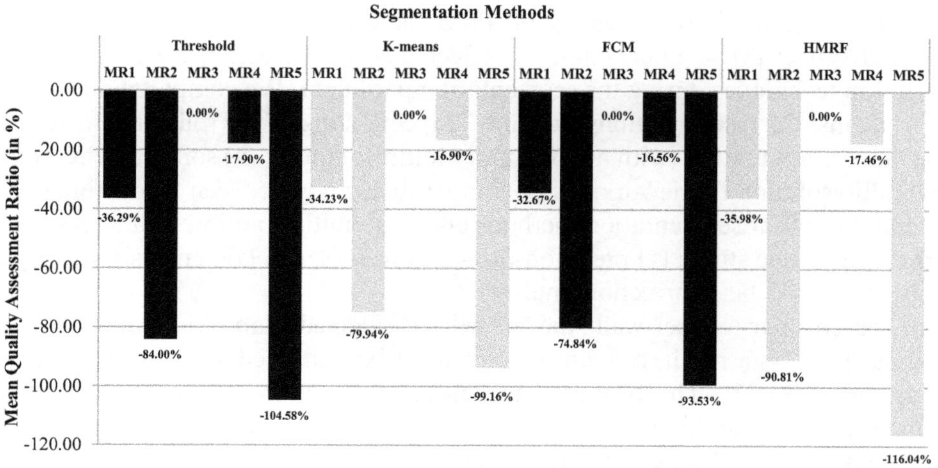

Figure 10.11. Bar plot of mean quality assessment ratio: MR1, MR2, MR3, MR4 and MR5 correspond to the bilinear, bicubic, wavelet, Lanczos and Gaussian pyramid multiresolution techniques, respectively.

$$\mathrm{PoM_{QAR}}(\%) = 100 - \left[\left(\frac{\mathrm{CNR}_M - \mathrm{CNR}_O}{\mathrm{CNR}_O} \right) \cdot 100 \right]. \tag{10.59}$$

On the basis of the experimental results obtained in figures 10.10 and 10.11, it is observed that wavelet interpolation experiences no DR and QAR. This happens because the wavelet decomposition technique is a lossless interpolation technique [27]. Among the other four lossy interpolation techniques, even if Lanczos interpolation is slow [48], it experiences the least DR and QAR among the other three multiresolution techniques (bicubic, bilinear and Gaussian pyramid). This is because it uses 36 surrounding pixels to interpolate a pixel to produce the best fit. Negative values are accounted for by low signal-to-noise ratio and contrast-to-noise ratio computed for the IVUS images. The precision-of-merit analyzed using DR ($\mathrm{PoM_{DR}}$) is 99.64%, 99.20%, 100%, 99.80%, and 99% for the bilinear, bicubic, discrete wavelet, Lanczos and Gaussian pyramid interpolation techniques, respectively. Similarly, the precision-of-merit analyzed using QAR ($\mathrm{PoM_{QAR}}$) is 99.65%, 99.18%, 100%, 99.83% and 99.97% for the bilinear, bicubic, discrete wavelet, Lanczos and Gaussian pyramid interpolation techniques, respectively, which supports our previous results. We can observe that the wavelet multiresolution technique experiences 100% precision-of-merit in both cases, which supports the above results.

10.6 Discussion

10.6.1 Our system

The main objectives of this study were: (1) to design and develop four segmentation methods (threshold, FCM, K-means and HMRF) embedded with five multiresolution techniques (bilinear, bicubic, wavelets, Lanczos and Gaussian pyramid) for calcium volume measurement in the coronary artery, and to further know which combination is the best among all 20 combinations; (2) develop a performance evaluation methodology consisting of: PoM, percentage mean time improvement, DR and QAR; (3) to understand the effect of bias correction on all five multiresolution techniques during the segmentation paradigm. We compared our current work against the recently published study [15] consisting of the following attributes: (i) volume computation (with and without multiresolution); (ii) soft segmentation in the multiresolution framework; (iii) a comparative study between 20 combinatorial systems from four segmentation methods and five multiresolution techniques; (iv) a comparative time study; (v) precision-of-merit analysis; (vi) DR analysis; (vii) QAR analysis; and (x) bias correction analysis.

In the coronary vessel wall, the growth of calcified plaque is multi-focal and irregular [13]. As a result, calcium volume must be estimated in an entire coronary artery video [9, 11]. The proposed technique in the current work computes the volume using a multiresolution framework, which offered the advantage of improving segmentation computation time. Even though the multiresolution techniques were analyzed prior [11, 20, 24, 33, 49–52], the comparison of four segmentation methods embedded with five kinds of multiresolution methods utilizing the entire videos of the coronary artery is innovative. We validated our assumption that multiresolution improved the computation time in all 20 combination systems.

10.6.2 Comparison of our down-sampling methods against other methods

Multiresolution methods are not restricted to what we used in this article. Methods such as short-time Fourier transform (STFT) [53], empirical mode decomposition (EMD) [54], and a combination of empirical mode decomposition and Hilbert transform (EMD + HT) [55] are also being used. For non-stationary processes, the amplitude and spectral content mostly vary with time. Analysis of such time-dependent characteristics requires time–frequency techniques such as STFT [53], EMD [54] and EMD + HT [55]. In STFT, the non-linearity is captured using a fixed window. Due to the above limitation, this might result in limited frequency resolutions [56]. In EMD, there is a decomposition of the spectral process into a finite set of intrinsic mode functions (IMFs) [54]. EMD makes no assumptions *a priori* about the composition of the signal. For successively tracing the IMFs, it uses spline interpolation between the maxima and minima. To alleviate the limitations in EMD, the ensemble EMD (EEMD) has been recently proposed [57]. Compared with the older time–frequency analysis schemes described above, combined EMD/EEMD with HT gives an adaptive approach towards its local behavior and presents a new perspective in interpreting nonlinear and non-stationary data. There are several common approaches to perform down-sampling such as pyramid methods, the differences of the low-pass (DOLP) transform, the tree matching algorithm and edge focusing [58]. Collectively, each method has its own advantages and disadvantage and can be used for different kinds of applications. It is, however, not possible to explore all the techniques in the current set-up and this is outside the scope of the study.

Our assumption was to perform an in-depth investigation to compare the results of various multiresolution techniques applied to entire IVUS videos. All these techniques were applied to the 19 complete patient IVUS videos, with about 2040 frames per patient (around 38 760 IVUS frames). This chapter utilized the data from Araki *et al* [9, 11, 15].

10.6.3 A note on gating and registration

The fundamental challenges during cardiac image acquisition are due to movement of the heart and lungs. Respiratory artifacts can be removed by the patient holding their breath during imaging, while the problem of heart motion can be solved using the gated strategy [59].

In the process of coronary IVUS video acquisition, the catheter is pulled out at a speed of 0.5 mm s^{-1}. As presented in the data acquisition section (section 10.2.2) of our current study, since the speed of the catheter is constant, our acquisition has two basic assumptions: (a) the gap between the frames is nearly zero and (b) the repetition of frames (vessel sections) is very low. The second assumption relates to the motion of the coronary artery due to the motion of the heart [24]. These two assumptions (zero frame gap and motion assumption) can affect the diagnostic accuracy during calcium volume computation. While relaxing these assumptions and assuming, say five percent of the IVUS frames are repeated, the detection accuracy can differ up to 5% from the expected value. Following the regulatory criteria for diagnostic applications, our model can be inaccurate within a tolerance range, but still acceptable for an accuracy–speed pilot study. While our multiresolution techniques and segmentation methods are within the bounds of accuracy, our recent model on IVUS segmentation [15] and registration [24] showed negligible improvement for plaque burden characterization.

For accurate IVUS segmentation, gating methods require a uniform cardiac motion along IVUS sequences [60]. Even though there can be an improvement in accuracy, the volume computation, gating and registration techniques offer many challenges [61, 62], especially if the above two assumptions are relaxed. This can happen when there is a large gap between the frames due to sudden catheter movements. Gated image acquisition also consumes more time compared to a conventional motorized transducer [63]. Thus, gating introduces limitations which restrict its use in patients with severe coronary stenosis. Further, it may be difficult in patients with arrhythmias and even impossible in the presence of arterial fibrillation. Finally, the computational burden on the gating and registration methods increases with an increase in IVUS video length. The above limitations pose a challenge in IVUS imaging and there is a trade-off between accuracy–speed and complexity. Our current clinical set-up did not have a gating method for IVUS acquisition, even though our two assumptions were taken into consideration during our approaches for volume computation. This kind of implementation is outside the scope of the current work and possibly can be implemented in the future.

10.6.4 Bias correction

Due to nonlinear operation during the multiresolution protocol, the pixels lose the image resolution component, thereby degrading the image quality. Even though the speed of processing is improved, the image quality degradation introduces a bias during this stage that needs to be corrected. This bias correction needs to be applied prior to the calcium detection process using soft classifiers.

In this chapter, we have used a disk of size 3×3 to perform the bias correction using MATLAB 2013a software. A disk shape was chosen because the calcium present in IVUS images is curved in nature. We have analyzed our results by increasing and decreasing the size of the disk. Incorrect results are obtained in both cases as decreased disk size yields negligible change and increased disk size causes blurring of the biased image. In this article, we adopted an erosion process for correcting the bias due to the multiresolution paradigm. Grayscale erosion generally darkens the image as bright sections enclosed by dark sections shrink in size, and dark sections enclosed by bright sections grow in size. This preserves the original shape of the bright calcium yielding correct calcium detection and volume computation. We have also analyzed our results by adopting grayscale dilation for bias correction. In our study, incorrect results are obtained in dilation as it generally brightens the image leading to incorrect estimation of calcium.

Out of 19 patients, three patients' coronary arteries contained soft calcium. IVUS images with soft calcium are gray in nature and the background is black. When bias correction is applied to such IVUS images, it increases the area containing the calcium, thereby increasing the volume and yielding incorrect results. An empirically derived greedy approach is applied with the exclusion of those three patients. We have studied the effect of bias correction on only one segmentation method, i.e. K-means. This method was proven to be the best segmentation method (according to expert scoring) when compared to the other three segmentation methods (Santos, FCM and HMRF) [15]. Relatively, we observed that the average time consumed by performing the bias correction was 0.006 s. Mean calcium volume similarity using bias correction was observed to increase for all multiresolution techniques. Increase in the mean calcium volume similarity justifies our hypothesis that bias correction is needed in the multiresolution techniques due to non-linearity in operation. In the future, we intend to add more robust measures to remove the bias. However, the current system shows an encouraging result.

10.6.5 A note on time complexity and precision-of-merit

Multiresolution is a crucial operation in improving the time of processing by shrinking the size of large images [20]. For all the four segmentation methods, multiresolution brings a mean improvement in time. From the experimental analysis, we observed that due to a large number of iterations, FCM took the largest computational time to converge. With multiresolution and bias correction, FCM also experienced the highest mean time improvement compared to computation time without multiresolution.

For all the IVUS videos tested, the mean precision-of-merit was found to be highest for the wavelet interpolation technique. In this chapter, for volume computation, interpolation techniques were used to down-sample the IVUS ring image and up-sample the binary calcium detected frames obtained using different segmentation methods. Due to low image quality, there are certain regions in the input image that degrade the high-frequency components. After down-sampling using multiresolution, these tiny regions that are close to each other are segmented and interpreted to be calcium. Thus, these tiny regions are categorized into false calcium. This causes a slight overestimation of volume. This is more prominent in the performance evaluation during the frame-level paradigm as discussed in subsection 10.5.1.2. Further, studies need to be conducted for validating this bias correction hypothesis.

For all the four different segmentation methods, the Jaccard index and Dice similarity coefficient were also computed to quantify the degree of similarity between calcium regions in two binary images obtained with and without multiresolution. Mean Jaccard index and Dice similarity coefficient were also the highest for wavelet multiresolution.

Note that our JI and DS computations in this article use the reference of the non-multiresolution-based outcome. This means JI and DSC values use the pipeline, as shown in figure 10.5. It is important to note that even though the block of multiresolution brings in considerable speed and simplicity, the poor resolution IVUS videos and amplification of noise due to non-linearity introduces false calcium detection, thereby affecting the JI and DS values. If we compare this study with the study conducted in [15], there is another key noticeable difference. The concept of JI and DS in [15] uses one of their own segmentation methods as a reference while computing JI and DS, and the concept of the multiresolution intruder is not prevalent. Thus, the study [15] shows higher JI and DS values at the expense of multiresolution and choice of reference. More work is required where one can compare the JI and DS metrics against their own segmentation references.

For performance analysis, we computed DR and QAR from SNR and CNR by locating ten windows of 7 × 7 size in both the signal and noise regions. Wavelet multiresolution technique experienced nil DR and QAR with 100% precision-of-merit in both the cases that supported the above results. We analyzed that the Gaussian pyramid multiresolution technique exhibited the worse performance with the lowest JI and DSC and the lowest DR and QAR.

10.6.6 Benchmarking

A review of previous multiresolution studies is summarized in table 10.8. Our five multiresolution techniques were benchmarked against the Zhang *et al* [33] and Lazrag and Naceur [52] methods. Zhang *et al* [33] proposed an automatic calcium segmentation method using a contourlet transform for the detection of multi-focal calcified plaques. They did not implement volume computation, as they used selective frames from a limited number of patients. They also did not perform five multiresolution technique comparisons within their study. Lazrag and Naceur [52] worked on a speckle reduction in IVUS images using different wavelet filters. The authors did not compare their results with other multiresolution techniques. Furthermore, the technique did not highlight the number of frames utilized for computing the effective calcium. Note that our calcium volume computation utilized two key assumptions: a gap in frames due to artery curvature and a negligible contribution of volume change due to motion effect, and we also argued that this relaxation compromises the volume accuracy which is under the regulatory error tolerance for this pilot study.

Calcium volume estimation is very common in several modalities such as CT, x-rays and ultrasound. None of the methods discussed above can be directly used for benchmarking against our methods. This is because of the following attributes: (i) the data type is IVUS video; (ii) the focus of the study is calcium segmentation; (iii) the type of soft classifier used; (iv) the multiresolution paradigm used; (v) the usage of the entire video; and (vi) the hardware set-up. Thus, to compare our work against any other study, one would require at least these six attributes to be resolved. There have been studies that have been conducted in the non-coronary area using multiresolution paradigms [24, 64–67] that cannot be compared; however, we

Table 10.8. Survey of calcium detection techniques and volume measurements.

C1 Authors	C2 #N	C3 †F/video	C4 Objective	C5 Multiresolution technique	C6 Performance evaluation	C7 Accuracy	C8 Benchmark	C9 Multiresolution comparisons	C10 Bias correction
Liang *et al* [49]	Not mentioned	Not mentioned	Arterial wall strain	Gaussian pyramid	Correlation coefficient	N/A	None	None	None
Srinivas *et al* [50]	Not mentioned	Not mentioned	Pre-rupture plaque	Wavelets	None	N/A	None	None	None
Zhang *et al* [33]	11	≈8	Calcium area	Contourlet transform	MD, RMD, MSD, RDD, ARD, TD, LD, RARD, RTD, and RLD	N/A	Santo *et al* [16]	None	None
Katouzian *et al* [9]	5	1158 frames	LI border	Wavelets	TP, FP, RMSE	N/A	None	None	None
Lazrag and Naceur [51]	Not mentioned	20 simulated, 30 real frames	Vascular boundary	Contourlet transform	None	N/A	Zhang *et al* [33]	None	None
Lazrag and Naceur [52]	Not mentioned	Not mentioned	Speckle reduction	Wavelets	PSNR*, NMSE*	N/A	None	None	None
Molinari *et al* [68]	Not mentioned	365 frames	Segmentation and IMT* measurement	Bicubic	LI and MA error and IMT bias	N/A	None	None	None
Puertas *et al* [20]	Not mentioned	Not mentioned	Stacked sequential learning	Gaussian pyramid	Accuracy, overlapping, recall and precision	N/A	None	None	None

Proposed	19	≈ 2040	Calcium volume	Bilinear, bicubic, wavelets, Lanczos and Gaussian pyramid	Percentage mean time improvement, precision-of-merit, mean volume similarity, degradation ratio and quality assessment ratio	$PTI^{*}_{\forall S}$ = 74.07%, $PTI^{*}_{\forall M}$ = 77.15%, $PoM_{VL}{}^{*}$ = 94.06%, $PoM_{FL}{}^{*}$ = 81.34%, $MVSIBC^{*}$ = 4.13%, DR^{*} = 0, QAR^{*} = 0, $PoM_{DR}{}^{*}$ = 100, and $PoM_{QAR}{}^{*}$ = 100	Zhang *et al* [33], Lazrag and Naceur [52]	Yes	Yes

[#] *N*: number of patients.

[†] F/Video: number of IVUS frames per video.

[*] IMT: intima–media thickness; PSNR: peak signal-to-noise ratio; NMSE: normalized mean square error; MD: mean distance error; RMD: relative mean distance error; MSD: mean signed distance error; RDD: relative difference degree; ARD: arc difference; TD: thickness difference; LD: length difference; RARD: relative arc difference; RTD: relative thickness difference; RLD: relative length difference; TP: true positive; FP: false positive; RMSE: root mean square error; NA: not available; $PTI^{*}_{\forall S}$: percentage mean time improvement for multiresolution technique; $PTI^{*}_{\forall M}$: percentage mean time improvement for segmentation method; PoM: precision-of-merit; VL: volume level; FL: frame level; MVSIBC: mean volume similarity increase with bias correction; DR: degradation ratio; QAR: quality assessment ratio.

already compared our proposed work with our previous work [15]. We do want to emphasize that we evaluated the times for two scientific engines for our current hardware set-up as isolated pieces and the results are as follows: the average computation times per frame for the four segmentation methods (threshold, FCM, K-means and HMRF) were 0.0071 s, 4.2083 s, 0.0373 s and 2.5796 s, respectively, while the computation times per frame for the five multiresolution techniques (bilinear, bicubic, wavelet, Lanczos and Gaussian pyramid) were 0.0057 s, 0.0021 s, 0.0046 s, 0.0026 s and 0.0020 s, respectively.

10.6.7 Strengths, weaknesses and extensions

The main strength of this study is the ability to decrease the computational time of extracting calcium in the entire coronary artery while ensuring a high mean precision-of-merit. The second strength of this study is the characterization and comparison of five different multiresolution frameworks, i.e. bilinear, bicubic, wavelets, Lanczos and Gaussian pyramid in four segmentation methods.

There are some limitations which can be removed: (a) an empirically based and greedy bias correction approach is used in the study. We anticipate an improvement in mean volume similarity by choosing a better bias correction strategy and making the technique automated. (b) The number of patients can be increased over time. (c) More validation can be performed to improve the reliability of the system.

10.7 Conclusion

This study presents five multiresolution methods (bilinear, bicubic, discrete wavelet, Lanczos and Gaussian pyramid) embedded into four segmentation paradigms (threshold, FCM, K-means and HMRF) leading to a combination of 20 systems. All the systems showed an improvement in computation time for volume computation when compared to the non-multiresolution method. Among the 20 combinations, FCM embedded with wavelet was the fastest. The system had comprehensive performance evaluation and data analysis based on degradation ratio and quality assessment ratio. The results of the design and development demonstrated encouraging results and require a larger population study for clinical validation.

Acknowledgments

Reprinted with permission from Banchhor S K, Araki T, Londhe N D, Ikeda N, Radeva P, El-Baz A, Saba L, Nicolaides A, Shafique S, Laird J R and Suri J S 2016 Five multiresolution-based calcium volume measurement techniques from coronary IVUS videos: a comparative approach *Comput. Methods Programs Biomed.* **134** 237–58. Copyright 2016 Elsevier.

Funding

This research did not receive any specific grant from funding agencies in the public, commercial or not-for-profit sectors.

Conflicts of interest

The authors declare no conflict of interest.

Appendix A Tables

Mean time for multiresolution techniques over *S* segmentation methods

$$\begin{aligned}&\bar{t}^{m_1}_{\forall S}=\frac{\sum_{j=1}^{S}t^{m_1}_{s_j}}{S},\ \bar{t}^{m_2}_{\forall S}=\frac{\sum_{j=1}^{S}t^{m_2}_{s_j}}{S},\\&\bar{t}^{m_3}_{\forall S}=\frac{\sum_{j=1}^{S}t^{m_3}_{s_j}}{S},\ \bar{t}^{m_4}_{\forall S}=\frac{\sum_{j=1}^{S}t^{m_4}_{s_j}}{S},\ \bar{t}^{m_5}_{\forall S}=\frac{\sum_{j=1}^{S}t^{m_5}_{s_j}}{S}\end{aligned}\tag{A.1}$$

Percentage time improvement for *M* multiresolution techniques over *S* segmentation methods

$$\begin{aligned}&\mathrm{PTI}^{m_1}_{\forall S}=\left[\frac{\bar{t}^{m_1}_{\forall S}-\bar{t}^{0}_{\forall S}}{\bar{t}^{0}_{\forall S}}\right]\cdot 100,\ \mathrm{PTI}^{m_2}_{\forall S}=\left[\frac{\bar{t}^{m_2}_{\forall S}-\bar{t}^{0}_{\forall S}}{\bar{t}^{0}_{\forall S}}\right]\cdot 100,\\&\mathrm{PTI}^{m_3}_{\forall S}=\left[\frac{\bar{t}^{m_3}_{\forall S}-\bar{t}^{0}_{\forall S}}{\bar{t}^{0}_{\forall S}}\right]\cdot 100\ \mathrm{PTI}^{m_4}_{\forall S}=\left[\frac{\bar{t}^{m_4}_{\forall S}-\bar{t}^{0}_{\forall S}}{\bar{t}^{0}_{\forall S}}\right]\cdot 100,\\&\mathrm{PTI}^{m_5}_{\forall S}=\left[\frac{\bar{t}^{m_5}_{\forall S}-\bar{t}^{0}_{\forall S}}{\bar{t}^{0}_{\forall S}}\right]\cdot 100\end{aligned}\tag{A.2}$$

Percentage mean time improvement for *S* segmentation methods over *M* multiresolution techniques

$$\begin{aligned}&\bar{t}^{s_1}_{\forall M}=\frac{\sum_{i=1}^{M}t^{m_i}_{s_1}}{M},\ \bar{t}^{s_2}_{\forall M}=\frac{\sum_{i=1}^{M}t^{m_i}_{s_2}}{M},\\&\bar{t}^{s_3}_{\forall M}=\frac{\sum_{i=1}^{M}t^{m_i}_{s_3}}{M},\ \bar{t}^{s_4}_{\forall M}=\frac{\sum_{i=1}^{M}t^{m_i}_{s_4}}{M}\end{aligned}\tag{A.3}$$

$$\begin{aligned}&\mathrm{PTI}^{s_1}_{\forall M}=\left[\frac{\bar{t}^{s_1}_{\forall M}-\bar{t}^{0}_{s_1}}{\bar{t}^{0}_{s_1}}\right]\cdot 100,\ \mathrm{PTI}^{s_2}_{\forall M}=\left[\frac{\bar{t}^{s_2}_{\forall M}-\bar{t}^{0}_{s_2}}{\bar{t}^{0}_{s_2}}\right]\cdot 100\\&\mathrm{PTI}^{s_3}_{\forall M}=\left[\frac{\bar{t}^{s_3}_{\forall M}-\bar{t}^{0}_{s_3}}{\bar{t}^{0}_{s_3}}\right]\cdot 100,\ \mathrm{PTI}^{s_4}_{\forall M}=\left[\frac{\bar{t}^{s_4}_{\forall M}-\bar{t}^{0}_{s_4}}{\bar{t}^{0}_{s_4}}\right]\cdot 100\end{aligned}\tag{A.4}$$

Mean volume difference ratio based on volume level

$$\overline{\mathrm{VDR}}_{s_j}^{mi}\Big|_{\mathrm{VL}} = \frac{\sum_{n=1}^{N} \mathrm{VDR}_{s_j}^{mi}(n)|_{\mathrm{VL}}}{N} \tag{A.5}$$

Mean volume difference ratio based on frame level

$$\overline{\mathrm{VDR}}_{s_j}^{mi}\Big|_{\mathrm{FL}} = \frac{\sum_{n=1}^{N} \mathrm{VDR}_{s_j}^{mi}(n)|_{\mathrm{FL}}}{N} \tag{A.6}$$

Appendix B Mean times of 20 combinations

Using these 20 combinations, we can estimate the following mean times

$$t_{s_1}^{m_1}, t_{s_1}^{m_2}, t_{s_1}^{m_3}, t_{s_1}^{m_4}, t_{s_1}^{m_5}, t_{s_2}^{m_1}, t_{s_2}^{m_2}, t_{s_2}^{m_3}, t_{s_2}^{m_5},$$
$$t_{s_3}^{m_1}, t_{s_3}^{m_2}, t_{s_3}^{m_3}, t_{s_3}^{m_4}, t_{s_3}^{m_5}, t_{s_4}^{m_1}, t_{s_4}^{m_2}, t_{s_4}^{m_3}, t_{s_4}^{m_4} \text{ and } t_{s_4}^{m_5}.$$

References

[1] Mozaffarian D *et al* 2016 Executive summary: heart disease and stroke statistics—2016 update: a report from the American Heart Association *Circulation* **133** 447–54

[2] Libby P, Ridker P M and Hansson G K 2011 Progress and challenges in translating the biology of atherosclerosis *Nature* **473** 317–25

[3] Ross R 1995 Cell biology of atherosclerosis *Annu. Rev. Physiol.* **57** 791–804

[4] Rose G 1991 ABC of vascular diseases. Epidemiology of atherosclerosis *BMJ* **303** 1537–9

[5] Banchhor S K *et al* 2016 Five multiresolution-based calcium volume measurement techniques from coronary IVUS videos: a comparative approach *Comput. Methods Programs Biomed.* **134** 237–58

[6] Ramani K, Judd R M, Holly T A, Parrish T B, Rigolin V H, Parker M A, Callahan C, Fitzgerald S W, Bonow R O and Klocke F J 1998 Contrast magnetic resonance imaging in the assessment of myocardial viability in patients with stable coronary artery disease and left ventricular dysfunction *Circulation* **98** 2687–94

[7] Coutts S B, Modi J, Shiel K, Patel S K, Demchuk A M, Goyal M and Hill M D 2012 CT/CT angiography and MRI findings predict recurrent stroke after transient ischemic attack and minor stroke results of the prospective CATCH study *Stroke* **43** 1013–7

[8] Schoenhagen P and Nissen S 2002 Understanding coronary artery disease: tomographic imaging with intravascular ultrasound *Heart* **88** 91–6

[9] Katouzian A, Angelini E, Sturm B, Konofagou E, Carlier S G and Laine A F 2012 Applications of multiscale overcomplete wavelet-based representations in intravascular ultrasound (IVUS) images *Ultrasound Imaging* 313–36

[10] Giannoglou G D *et al* 2007 A novel active contour model for fully automated segmentation of intravascular ultrasound images: *in vivo* validation in human coronary arteries *Comput. Biol. Med.* **37** 1292–302

[11] Araki T, Ikeda N, Molinari F, Dey N, Acharjee S, Saba L and Suri J S 2014 Link between automated coronary calcium volumes from intravascular ultrasound to automated carotid IMT from B-mode ultrasound in coronary artery disease population *Int. Angiol.* **33** 392–403

[12] Araki T, Ikeda N, Dey N, Acharjee S, Molinari F, Saba L, Godia E C, Nicolaides A and Suri J S 2015 Shape based approach for coronary calcium lesion volume measurement on intravascular ultrasound imaging and its association with carotid intima–media thickness *J. Ultrasound Med.* **34** 469–82

[13] Suri J S, Kathuria C and Molinari F 2010 *Atherosclerosis Disease Management* (Berlin: Springer)

[14] Saba L, Sanches J M, Pedro L M and Suri J S (ed) 2014 *Multi-Modality Atherosclerosis Imaging and Diagnosis* (New York: Springer)

[15] Araki T *et al* 2016 Reliable and accurate calcium volume measurement in coronary artery using intravascular ultrasound videos *J. Med. Syst.* **40** 1–20

[16] Santos F E, Saijo Y, Tanaka A and Yoshizawa M 2008 Detection and quantification of calcifications in intravascular ultrasound images by automatic thresholding *Ultrasound Med. Biol.* **34** 160–5

[17] Jain A K and Dubes R C 1988 *Algorithms for Clustering Data* (Englewood Cliffs, NJ: Prentice-Hall)

[18] Zadeh L A 1965 Fuzzy sets *Inf. Control* **8** 338–53

[19] Li S Z 1994 Markov random field models in computer vision *Proc. of the Third European Conf. on Computer Vision* pp 361–70

[20] Puertas E, Escalera S and Pujol O 2013 Generalized multi-scale stacked sequential learning for multi-class classification *Pattern Anal. Appl.* **18** 247–61

[21] Mahajan S H and Harpale V K 2015 Adaptive and non-adaptive image interpolation techniques *IEEE Int. Conf. on Computing Communication Control and Automation (ICCUBEA)* pp 772–5

[22] Marqués F and Gasull A 1992 *Multiresolution Image Segmentation Based On Compound Random Fields: Application to Image Coding. THD Dpto. TSC-UPC* (Barcelona: Universitat Politècnica de Catalunya) pp 86–114

[23] Suri J S and Laxminarayan S 2003 *Angiography and Plaque Imaging: Advanced Segmentation Techniques* (Boca Raton, FL: CRC Press)

[24] Araki T *et al* 2015 A comparative approach of four different image registration techniques for quantitative assessment of coronary artery calcium lesions using intravascular ultrasound *Comput. Methods Programs Biomed.* **118** 158–72

[25] Sanz-Requena R, Moratal D, García-Sánchez D R, Bodí V, Rieta J J and Sanchis J M 2007 Automatic segmentation and 3D reconstruction of intravascular ultrasound images for a fast preliminary evaluation of vessel pathologies *Comput. Med. Imaging Graph.* **31** 71–80

[26] Harb S M E, Isa N A M and Salamah S 2015 New adaptive interpolation scheme for image upscaling *Multimedia Tools Appl.* **75** 7293–325

[27] Acharya T and Tsai P S 2007 Computational foundations of image interpolation algorithms *ACM Ubiquity* **8** 1–17

[28] Press W H, Flannery B P, Teukolsky S A, Vetterling W T and Flannery B P 1989 *Numerical Recipes: The Art of Scientific Computing (FORTRAN)* (New York: Cambridge University Press)

[29] De Boor C 1978 A practical guide to splines *Mathematics of Computation* (New York: Springer)

[30] Mallat S G 1989 A theory for multiresolution signal decomposition: the wavelet representation *IEEE Trans. Pattern Anal. Mach. Intell.* **11** 674–93

[31] Turkowski K 1990 Filters for common resampling tasks *Graphics Gems* (New York: Academic) pp 147–65

[32] Adelson E and Burt P 1981 Image data compression with the Laplacian pyramid *Proc. Pattern Recognition Information Processing Conf.* pp 218–23

[33] Zhang Q, Wang Y, Wang W, Ma J, Qian J and Ge J 2010 Automatic segmentation of calcifications in intravascular ultrasound images using snakes and the contourlet transform *Ultrasound Med. Biol.* **36** 111–29

[34] Araki T, Nakamura M, Utsunomiya M and Sugi K 2013 Visualization of coronary plaque in arterial remodelling using a new 40-MHz intravascular ultrasound imaging system *Catheter. Cardiovasc. Interv.* **81** 471–80

[35] Kim S W *et al* 2014 DICOM-based intravascular ultrasound signal intensity analysis. Echoplaque medical imaging bench study *Coron. Artery Dis.* **25** 236–41

[36] Banchhor S K, Londhe N D, Araki T, Saba L, Radeva P, Laird J R and Suri J S 2017 Well-balanced system for coronary calcium detection and volume measurement in a low resolution intravascular ultrasound videos *Comput. Biol. Med.* **2017** 168–81

[37] Daubechies I 1992 *Ten Lectures on Wavelets* vol 61 (Philadelphia: Society for Industrial and Applied Mathematics) pp 198–202

[38] Parker J A, Kenyon R V and Troxel D E 1983 Comparison of interpolating methods for image resampling *IEEE Trans. Med. Imaging* **2** 31–9

[39] Ye Z, Suri J S, Sun Y and Janer R 2005 Four image interpolation techniques for ultrasound breast phantom data acquired using Fischer's full field digital mammography and ultrasound system (FFDMUS): a comparative approach *IEEE Int. Conf. on Image Processing (ICIP)* **vol 2** pp I261–4

[40] Elamaran V, Upadhyay H N, Narasimhan K and Priestley J J 2015 A case study of impulse noise reduction using morphological image processing with structuring elements *Asian J. Sci. Res.* **8** 291–303

[41] Pham D L, Xu C and Prince J L 1998 Current methods in medical image segmentation *Annu. Rev. Biomed. Eng.* **2** 315–37

[42] Otsu N 1979 A threshold selection method from gray-level histograms *IEEE Trans. Syst. Man Cybern.* **9** 62–6

[43] Hartigan J A and Wong M A 1979 Algorithm AS 136: a K-means clustering algorithm *J. R. Stat. Soc. Ser. C Appl. Stat.* **28** 100–8

[44] Wang Q 2012 *HMRF-EM-image: implementation of the hidden Markov random field model and its expectation-maximization algorithm* (arXiv: 1207.3510)

[45] Scott D S *et al* 2000 Pathologic validation of a new method to quantify coronary calcific deposits *in vivo* using intravascular ultrasound *Am. J. Cardiol.* **85** 37–40

[46] Sampat M P, Wang Z, Gupta S, Bovik A C and Markey M K 2009 Complex wavelet structural similarity: a new image similarity index *IEEE Trans. Image Process.* **18** 2385–401

[47] Suri J, Guo Y, Coad C, Danielson T, Elbakri I and Janer R 2005 Image quality assessment via segmentation of breast lesion in x-ray and ultrasound phantom images from Fischer's full field digital mammography and ultrasound (FFDMUS) system *Technol. Cancer Res. Treat.* **4** 83–92

[48] Stensland H K *et al* 2014 Bagadus. an integrated real-time system for soccer analytics *ACM Trans. Multimedia Comput. Commun. Appl.* **10** 14:1–21

[49] Liang Y, Oakeson K D, Zhu H and Friedman M H 2006 Estimation of arterial wall strain based on IVUS image registration *Conf. Proc.: Annu. Int. Conf.of the IEEE Engineering in Medicine and Biology Society* pp 3218–21
[50] Srinivas S, Chandrana C, Roy S and Fleischman A 2009 Multiresolution analysis of intravascular ultrasound harmonic signals to image pre-rupture plaques *IEEE Int. Ultrasonics Symp. (IUS)* pp 2363–6
[51] Lazrag H and Naceur M S 2012 Combination of the level-set methods with the contourlet transform for the segmentation of the IVUS images *Int. J. Biomed. Imaging* **2012** 439597
[52] Lazrag H and Naceur M S 2012 Wavelet filters analysis for speckle reduction in intravascular ultrasound images *6th Int. Conf. on IEEE in Sciences of Electronics, Technologies of Information and Telecommunications (SETIT)* pp 375–9
[53] Gabor D 1946 Theory of communication *J. Inst. Electron. Eng. Radio Commun. Eng.* **93** 429–57
[54] Li H, Kwong S, Yang L, Huang D and Xiao D 2011 Hilbert–Huang transform for analysis of heart rate variability in cardiac health *IEEE/ACM Trans. Comput. Biol. Bioinform.* **8** 1557–67
[55] Huang N E, Shen Z, Long S R, Wu M C, Shih H H, Zheng Q, Yen N C, Tung C C and Liu H H 1971 The empirical mode decomposition and the Hilbert spectrum for nonlinear and non-stationary time series analysis *Proc. R. Soc. Lond.* A **454** 903–95
[56] Cohen L 1995 *Time Frequency Analysis: Theory and Applications* (Englewood Cliffs, NJ: Prentice Hall)
[57] Wu Z and Huang N E 2009 Ensemble empirical mode decomposition: a noise-assisted data analysis method *Adv. Adap. Data Anal.* **1** 1–41
[58] Lifshitz L M and Pizer S M 1990 A multiresolution hierarchical approach to image segmentation based on intensity extrema *IEEE Trans. Pattern Anal. Mach. Intell.* **12** 529–40
[59] Maso Talou G D, Larrabide I, Blanco P J, Guedes Bezerra C, Lemos P A and Feijóo R A 2015 Improving cardiac phase extraction in IVUS studies by integration of gating methods *IEEE Trans. Biomed. Eng.* **62** 2867–77
[60] Hernàndez-Sabaté A, Gil D, Garcia-Barnés J and Martí E 2011 Image-based cardiac phase retrieval in intravascular ultrasound sequences *IEEE Trans. Ultrason. Ferroelectr. Freq. Control* **58** 60–72
[61] Gatta C, Pujol O, Leor O R, Ferre J M and Radeva P 2009 Fast rigid registration of vascular structures in IVUS sequences *IEEE Trans. Inf. Technol. Biomed.* **13** 1006–11
[62] Zhang L, Wahle A, Chen Z, Zhang L, Downe R W, Kovarnik T and Sonka M 2015 Simultaneous registration of location and orientation in intravascular ultrasound pullbacks pairs via 3D graph-based optimization *IEEE Trans. Med. Imaging* **34** 2550–61
[63] Von Birgelen C, de Vrey E A, Mintz G S, Nicosia A, Bruining N, Li W, Slager C J, Roelandt J R, Serruys P W and de Feyter P J 1997 ECG-gated three-dimensional intravascular ultrasound feasibility and reproducibility of the automated analysis of coronary lumen and atherosclerotic plaque dimensions in humans *Circulation* **96** 2944–52
[64] Meiburger K M, Molinari F, Zeng G, Saba L and Suri J S 2011 Carotid automated ultrasound double line extraction system (CADLES) via edge-flow *Annu. Int. Conf. of the IEEE in Engineering in Medicine and Biology Society (EMBC)* ((Piscataway, NJ: IEEE) 575–8
[65] Molinari F, Liboni W, Pantziaris M and Suri J S 2011 CALSFOAM-completed automated local statistics based first order absolute moment' for carotid wall recognition, segmentation

and IMT measurement: validation and benchmarking on a 300 patient database *Int. Angiol.* **30** 227–41

[66] Molinari F, Acharya R, Zeng G and Suri J S 2011 CARES: Completely Automated Robust Edge Snapper for carotid ultrasound IMT measurement on a multi-institutional database of 300 images: a two stage system combining an intensity-based feature approach with first order absolute moments *SPIE Medical Imaging* (Bellingham: International Society for Optics and Photonics) 796210

[67] Araki T *et al* 2016 PCA-based polling strategy in machine learning framework for coronary artery disease risk assessment in intravascular ultrasound: a link between carotid and coronary grayscale plaque morphology *Comput. Methods Programs Biomed.* **128** 137–58

[68] Molinari F, Pattichis C S, Zeng G, Saba L, Acharya U R, Sanfilippo R, Nicolaides A and Suri J S 2012 Completely Automated Multiresolution Edge Snapper—a new technique for an accurate carotid ultrasound IMT measurement: clinical validation and benchmarking on a multi-institutional database *IEEE Trans. Image Process.* **21** 1211–22

Chapter 11

A cloud-based smart lumen diameter measurement tool for stroke risk assessment during multicenter clinical trials

Luca Saba, Sumit K Banchhor, Narendra D Londhe, Tadashi Araki, John R Laird, Ajay Gupta, Andrew Nicolaides and Jasjit S Suri

This pilot study presents a completely automated, novel, smart, cloud-based, point-of-care system for (a) carotid lumen diameter (LD); (b) stenosis severity index (SSI); and (c) total lumen area (TLA) measurement using B-mode ultrasound. The proposed system was (i) validated against manual readings taken by a neurologist and (ii) benchmarked against a commercially available system. Ultrasounds scans of the left/right common carotid artery of one hundred patients (73 M/27 F, mean age: 68 ± 11 years, institutional review board (IRB) approved, written informed consent), 200 ultrasound scans in total, were acquired using a 7.5 MHz linear transducer. The measured mean LDs for the left and right carotids were (in millimeters) 6.49 ± 1.77 and 6.66 ± 1.70 for the proposed system and 6.29 ± 1.79 and 6.45 ± 1.63 for the manual system, respectively, and the coefficient of correlation between the cloud-based automated system against the manual system was 0.98 ($p < 0.0001$), respectively. The corresponding TLA error, precision-of-merit and figure-of-merit when measured against the manual system were 4.56% ± 3.54%, 96.18% ± 3.21% and 96.85%, respectively. The area under curve for the receiving operating characteristics for the cloud-based system was 1.0. Four statistical tests, the two-tailed z-test, Mann–Whitney test, Kolmogorov–Smirnov (KS) and one-way ANOVA, were performed to demonstrate consistency and reliability. The proposed system is a reliable, accurate, fast, completely automated, anytime-anywhere solution for multicenter clinical trials and routine vascular screening.

doi:10.1088/2053-2563/ab0820ch11

11.1 Introduction

Cardiovascular-related diseases are the number one killer of people worldwide, accounting for 17.3 million deaths per year, a number that is expected to grow to more than 23.6 million by 2030 [1]. Cardiovascular disease is caused by the progression of atherosclerosis disease [2] which leads to the blockage of arteries (stenosis) [3], limiting the flow of oxygen-rich blood to the coronary artery and brain (figure 11.1), which in turn leads to myocardial infarction and stroke [4–6]. Unlike carotid intima–media thickness (cIMT), used for preventive healthcare screening, rapid and accurate stenosis detection and computation of the LD of the carotid arteries is vital for planning surgical procedures such as carotid artery stenting (CAS) and endarterectomy (CEA) [7, 8]. Since the plaque distribution is independent of the wall side (proximal versus distal) and multi-focal [9], robust LD/stenosis methods need to be developed for better characterization for risk assessment [10]. The main objective of this study is to propose a completely automated, smart, point-of-care system for LD/stenosis measurement using B-mode ultrasound (BMUS).

Several medical imaging modalities have been adopted for imaging arterial stenosis, such as ultrasound (US), computed tomography (CT) and magnetic resonance imaging (MRI) [5]. US offers several advantages, such as its ability to image lesions non-invasively, the absence of radiation, its user-friendliness and real-time operation [11, 12]. Doppler ultrasound is currently used to compute the percentage of stenosis severity. This is based on pulse wave velocity, which in turn uses the assumption that arteries have constant stiffness [13, 14]. This assumption may not always be valid, and therefore may not be accurate for LD/stenosis measurement. Further, it does not use image-based features, which are a true representation for image reconstruction. Another way of computing LD/stenosis is manual tracing, which is tedious and shows large intra- and inter-observer

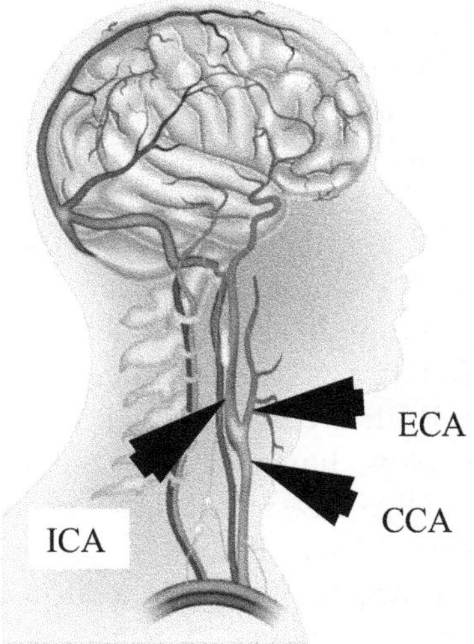

Figure 11.1. Illustration of plaque formation in the carotid artery. (Courtesy of AtheroPoint™, Roseville, CA, USA.)

variability [11, 15]. Innovations such as compound and harmonic imaging in recent years have made it possible to automatically compute LD/stenosis [16].

Recently, automated image-based LD detection techniques have started to emerge. These techniques can be broadly classified into three categories: region-based [10, 17–20], boundary-based [21], and a fusion of both region- and boundary-based techniques [16].

In 2013, Santo *et al* [17] proposed a region-based automated lumen segmentation algorithm which was based on the pixel classification strategy. The strategy used the row-wise intensity distribution model from 2D histograms to identify the lumen region from the carotid artery images. To attenuate the speckle noise, an anisotropic diffusion filter was applied. Finally, the Chan–Vese geometrical model was used to detect the lumen borders which was followed by contour smoothing by cubic spline interpolation. Using 256 gray level longitudinal B-mode ultrasound images, the study showed a mean overlap area of 96.78% between the manually and automatically contoured results.

In the same year, Ikeda *et al* [18] investigated LD detection for computing the location of (a) the carotid bulb and (b) common carotid artery (CCA) bifurcation. The study assumed that bulb edge points for the near and far wall could be an excellent reference source for the computation of cIMT and LD. First, the region-of-interest (ROI) was extracted by removing the non-tissue information. The lumen–intima borders were then estimated by employing the K-means classification approach. Finally, transition points between the bulb and the CCA were located based on the curvature characteristics. By analyzing 155 ultrasound bulb images, the study achieved 100% accuracy in the detection of the bulb. The mean lumen–intima error was observed to be 0.0133 mm and the precision against manual tracing was observed to be 98.92%. The detected bulb edge points were further verified and validated using a combination of five different local image processing methods: (i) lumen–intima shapes, (ii) bulb slopes, (iii) bulb curvature, (iv) mean lumen thickness and its variations, and (v) geometric shape fitting. The proposed technique also showed a high processing speed of 9 s per image. Dey *et al* [19] also used the same strategy for computing the LD all along the carotid artery and studied the effect of watermarking algorithms and their recovery of the watermarks for stroke applications.

In 2016, Saba *et al* [20] tried to estimate the relationship between carotid LD and inter-adventitial diameter (IAD) with a plaque score (PS). A database of 404 CCA B-mode sonographic images from 202 patients' left and right carotid arteries was retrospectively analyzed. Initially, a higher order Gaussian derivative filter was applied. For extracting the ROI, a column-based approach was used to trace the enhanced adventitial edges. Now, lumen borders are computed by using the pixel classification paradigm. Finally, LD was computed automatically by using the polyline distance measurement (PDM). Similarly, IAD was computed by measuring the distance between the near and the far wall media–adventitia interfaces. PS was also computed by summing the maximal thickness of plaques from the internal carotid artery (ICA), bulb and CCA. Carotid IAD was found to be more strongly correlated to PS than carotid LD. The study did not mention stenosis measurement.

In the same year, Araki *et al* [10] presented AtheroEdge™ 2.0 for automated carotid LD measurement along with stenosis severity index (SSI) computation. By assuming that the far wall adventitia is the brightest and blood flow has a constant density, the study automatically detects the LD boundaries. The study combines spectral analysis with tissue classification for LD measurement. By analyzing 404 images obtained from 202 patients, the study achieved an accuracy of 98.05% and 99.03% w.r.t. to two different manual observers. The SSI also showed a high 97% accuracy. As the region-based strategy is independent of the gradient information, the proposed techniques are also independent of the cases in which the carotid artery boundary contains gaps.

A boundary-based automated approach for the detection of the lumen axis of the CCA in B-mode ultrasound images was proposed by Rocha *et al* [21] in 2014. At first, the image is smoothed using a Gaussian filter to attenuate the noise. Then a dynamic programming scheme extracts the dominant paths of local minima of the intensity. By assuming that jugular veins are always present above the CCA, the study was tested on 199 images and showed an accuracy of 99.5%. The study only considered the far wall as near wall visibility in 2D B-modes images is usually much worse. Hence, the study is limited to the detection of the lumen axis and cannot measure the carotid LD, which requires the segmentation of both the near wall and far wall borders.

Golemati *et al* [22] used Hough's transform (HT) to segment ultrasound images of longitudinal and transverse sections for approximating the carotid lumen from sequences of B-mode ultrasound images. The proposed study was based on the assumption that straight lines and circles can be used for the segmentation of longitudinal and transverse carotid images, respectively. By analyzing ten normal carotid arteries, the study observed a high accuracy of 0.96% for both longitudinal and transverse sections. Even though the accuracy is high, the study poses two challenges. First, the algorithm was not tested for the segmentation of arteries with atherosclerotic plaques. Second, since the shape of the arteries is not always straight and the orientation can depend on the acquisition procedure [23], the assumption may not be always valid. Motivated by the above work [22], Araki *et al* [16] in 2016 proposed two automated techniques for the detection of the carotid lumen and LD measurement. Both algorithms were designed as a two-stage process. The first stage involved the extraction of the ROI using the scale-space-based approach with spectral analysis. This stage is common for both algorithms. The second stage involves the extraction of the lumen interface. Algorithm 1 employed a regional-based strategy using a K-means classifier and algorithm 2 used a boundary-based strategy by fusing HT with the level set approach for benchmarking. The algorithm was tested on two different databases: the Japan database (202 patients, 404 images) and Hong Kong database (50 patients, 300 images). The study observed an accuracy of 97.4% and 98.0% with respect to two manual tracings for the Japan database and 99.7% and 97.9% with respect to two manual tracings for the Hong Kong database.

The systems discussed above are prone to challenges since they lack user-control, which is highly desirable in low-resolution or poor image acquisition protocols. It is crucial to have mitigation tools controlled by the physician or sonographer, thus the concept of semi-automated system design is missing, in particular when there are spikes or bumps in the wall interfaces [16]. Second, previous methods did not focus on image-based SSI measurement and lacked accurate blood flow regions such as total lumen area (TLA). Third, the above systems are not suitable for telemedicine-based models [24] and as a result are confined to the operating and emergency rooms or preventive healthcare facilities.

The proposed study removes all of the above weaknesses by adopting a web-based model designed previously for plaque burden measurement [6]. In the same spirit, we design and develop a combined image and web-based LD detection and LD measurement system (so-called AtheroCloud), followed by SSI and TLA computation. The image-based LD system is more challenging in the sense that the near wall interfaces are crucial for estimation of LD, SSI and TLA, which introduces the novelty of this study. The current study is focused on: (a) design and development; (b) validation against manual readings (as traced by a neurologist); and (c) benchmarking against desktop-based systems. Note that the overall system provides a complete experimental protocol and statistical analysis as developed in the spirit of the cloud-based cIMT system in [6]. Further, based on the NASCET criteria, the current AtheroCloud system computes the SSI in both the automated and semi-automated combined image and web-based frameworks. Finally, our study presents comprehensive performance evaluation against the gold standard and benchmarking against a commercially available desktop system.

11.2 Materials and methods

Two hundred and four (204) patients underwent B-mode carotid ultrasound scans from July 2009 to December 2010. A total of 407 ultrasound scans (one patient had one image missing) from both the left and right CCA artery were obtained from Toho University, Japan (ethics approved by the IRB). For this pilot study, due to manual tracing and cost considerations, we randomly selected 100 patients (200 CCA ultrasound scans). No special criteria were adopted in choosing these 100 patients (73 M/27 F, mean age: 68 ± 11 years). Of these, 50 patients had a proximal lesion location, 29 a middle location and 21 a distal location. These 100 patients (42 smokers) had a mean HbA1c, low-density lipoprotein (LDL), high-density lipoprotein (HDL) and total cholesterol of 6.40 ± 1.2 (mg dl^{-1}), 103.96 ± 31.34 (mg dl^{-1}), 51.17 ± 14.04 (mg dl^{-1}) and 179.60 ± 38.61 (mg dl^{-1}), respectively.

All the patients, who provided written informed consent, were scanned using an ultrasound scanner (Aplio XV) equipped with a 7.5 MHz linear array transducer from Toshiba, Inc., Tokyo, Japan. The same sonographer (with 15 years' experience) scanned all the patients. The average resolution was 0.0529 mm/pixel.

11.2.1 Manual lumen diameter reading

For performance evaluation of the AtheroCloud software system, a gold standard was adopted by manually tracing the lumen–intima (LI) interface of the near and far walls of the carotid artery. These manual borders were traced by an experienced neurologist (L S) using the commercial software package ImgTracer™ (courtesy of

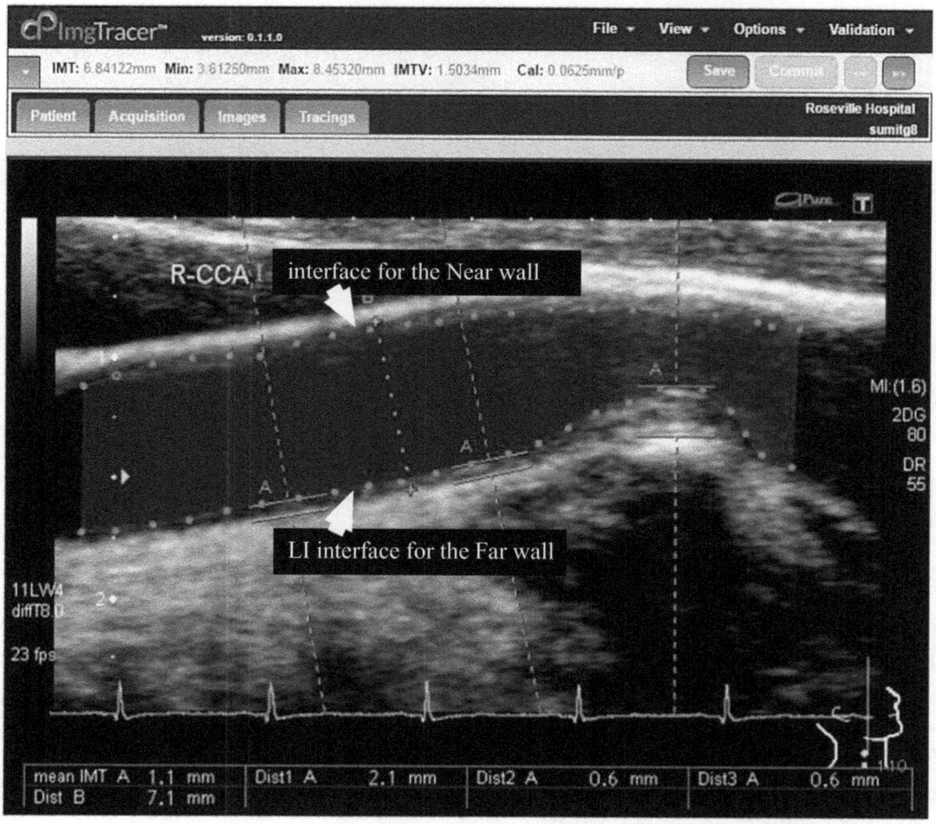

Figure 11.2. Manual tracings (red) of the carotid LD region showing the LI interface for the near wall and far wall using ImgTracer software. (Courtesy of AtheroPoint™, Roseville, CA, USA.)

AtheroPoint™, Roseville, CA, USA), as shown in figure 11.2. The top red line indicates the LI interface for the near wall and the bottom red line indicates the LI interface for the far wall. The ImgTracer™ software has been successfully used for several anatomical applications all around the world. Recently Suri and his team [6] used this software (ImgTracer™) to generate the ground truth by manually tracing the LI interfaces of carotid arteries.

11.2.2 Workflow architecture of the AtheroCloud ultrasound system

This pilot study is the first of its kind in which LD, TLA and SSI are automatically measured in carotid ultrasound scans in cloud-based settings. The workflow is shown in figure 11.3. It is a three-layered architecture. (i) The graphical user interface layer, where the doctor can interact with the AtheroCloud software, accessing ultrasound scans stored in the cloud-based server using a laptop or a desktop. (ii) The business logic layer, where the scientific engine performs measurements automatically. (iii) The persistence or database layer, which is used for storage of digital results. The database server tier performs the data storage functions and is implemented using industry-leading relational database management systems (RDBMSs). The application

Figure 11.3. The workflow of AtheroCloud and its components. The tower represents the server in the cloud. The arrows represent the bi-directional flow of information. (Courtesy of AtheroPoint™, Roseville, CA, USA.)

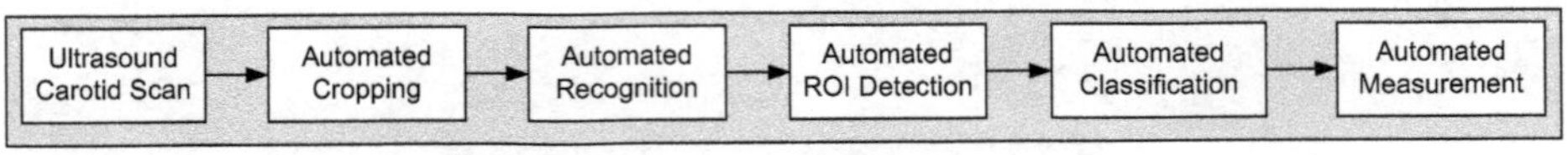

Figure 11.4. Block diagram of the overall engineering components of the design.

components access the RDBMSs and support a wide range of reporting options. An RDBMS is a program that lets you create, update and administer a relational database. Most commercial RDBMSs use the Structured Query Language to access the database. As regards patient privacy, the AtheroCloud system uses the Amazon Web Services Business Model, which has inbuilt Health Insurance Portability and Accountability Act compliance and all RDBMSs are fully patient privacy protected. Further, all reports are password protected.

11.2.3 Engineering design of the AtheroCloud ultrasound system

The block diagram for the overall engineering components of the design is shown in figure 11.4. It consists of (a) the automated cropping phase (block #2) [25], (b) the automated carotid artery recognition phase (block #3) [26], (c) the automated near and far wall LI interface and ROI detection phase (block #4) [9], (d) the automated classification phase (block #5) [16, 20, 27], and (e) the automated LI extraction of near and far wall LD [28, 29], SSI [30] and TLA measurement phase (block #6) [31–33].

Automated cropping is performed on the ultrasound carotid scans to avoid interference of the patient text information [25]. Automated recognition of the near and far walls is based on the hypothesis that adventitia region intensities are the highest in carotid ultrasound scans [26]. In the automated ROI detection step, the near and far wall media–adventitia (MA) interfaces all along the carotid artery are estimated using an iterative signal processing paradigm [9]. In the classification step, the automated classifiers detect the lumen region and lumen borders [16, 20, 27]. Finally, LD borders are fed into the PDM system for LD computation [28, 29] (see appendix A). SSI is computed using the NASCET criteria [30]. Further, TLA is computed by simply summing all the pixels in the lumen region. TLA is a secondary indicator of arterial blockage along the long axis of the carotid artery [31–33].

Our internet speed was 180 Mbps. The proposed software AtheroCloud can effectively compute LD/stenosis/TLA in variably sized ultrasound images, which can vary from database-to-database. In our database, on an average, small images (684 pixels × 504 pixels) took less than three seconds and large images (1054 pixels × 772 pixels) took less than five seconds for both uploading and measurement protocols.

11.2.4 Two application modes of AtheroCloud: routine mode and pharma trial mode

The proposed AtheroCloud software can run in two modes: (i) the 'routine' mode and (ii) the 'pharma trial' mode. In the routine mode, the measurements can be computed on one ultrasound scan at a time, whereas in the pharma trial mode a batch (up to a maximum of 10 000 images) of ultrasound scans can be automatically processed in sequence. An example of the routine mode is shown in figure 11.5. After loading the image using the AtheroCloud system in the routine mode, with a click of the button 'Auto Trace', the LI-near/far wall borders are computed and the LD region representing total blood flow area is filled in with red (figure 11.5). As our carotid scans database did not have any bulbs, no reference point was chosen, and the LD/stenosis was determined all along the carotid artery. In the AtheroCloud software, the user can also upload the pair of near/far LI wall interfaces previously computed using AtheroCloud and corresponding manual tracings traced by an expert, and such superimposition can be used for comparison. We can validate our

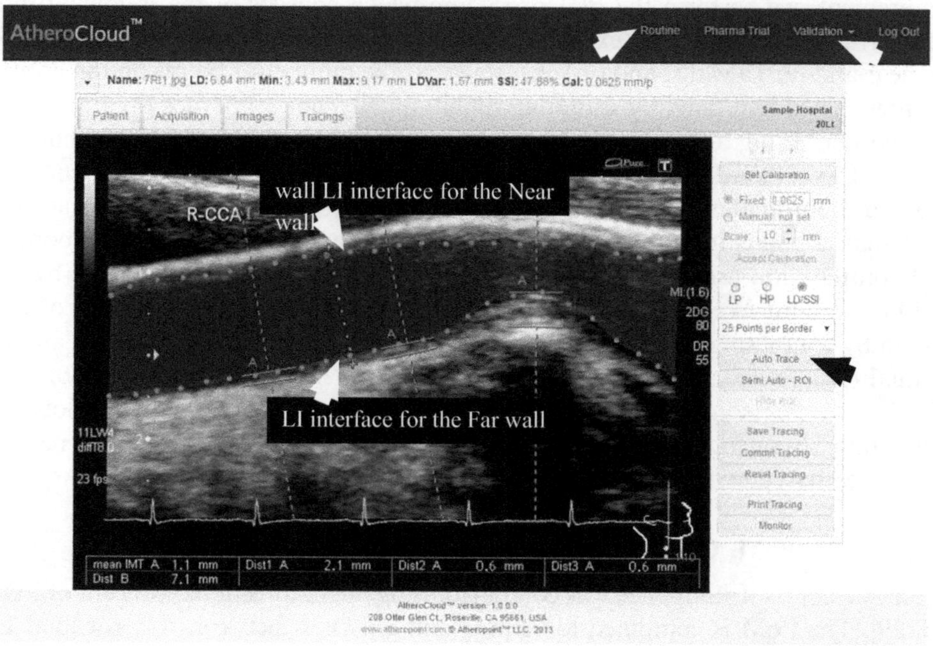

Figure 11.5. Routine trial mode automated tracings (red) of the carotid LD region showing the LI interface for the near wall and far wall using AtheroCloud software. (Courtesy of AtheroPoint™, Roseville, CA, USA.)

AtheroCloud results using the 'Validation' button in the top-right corner of the AtheroCloud front panel (figure 11.5). An example of pharma trial mode is as shown in appendix B, figure B1.

11.3 Results

11.3.1 Measurements and visualization

Table 11.1 shows the mean and standard deviation of the computed carotid LD over 200 carotid scans using: (i) the automated AtheroCloud software (both routine and pharma trial mode) and (ii) manual (gold standard) readings. Our observations show that the AtheroCloud mean LD/stenosis readings for both routine and pharma trial mode have a percentage difference of less than 4%. This proves that the proposed cloud-based system can perform nearly equivalently to manual tracings by a neurologist and thus can be adapted for clinical trials. Figure 11.6 shows a comparison of the near/far LI wall interfaces computed by the proposed AtheroCloud software (solid line) with manual expert tracings (dotted line). The near/far LI wall interfaces show very slight deviations between AtheroCloud and manual tracings.

11.3.2 Performance evaluation of the AtheroCloud ultrasound system

For validating AtheroCloud results against the manual (gold standard) results, we need to first match the near/far LI wall borders computed by AtheroCloud with the near/far LI wall borders traced by the expert neurologist. To achieve the above requirement, we perform the BCI procedure which consists of the following three steps: (i) B-spline fitting, (ii) extraction of borders which have common support for the near and far LI walls, and (iii) ensuring that the number of points on the borders are preselected and fixed by using interpolation. B-spline smoothes the near/far LI wall borders using a B-spline technique. Common support confirms that both the near/far LI wall borders of AtheroCloud and the manual borders have the same start and ends coordinates. Finally, we perform interpolation to achieve 100 equidistant interpolated points in both AtheroCloud and the manual near/far LI wall borders. BCI computation used in this study is similar to a recent study carried out by Suri and his team [6]. We evaluated the efficiency and performance of the AtheroCloud system by computing the following analysis between AtheroCloud readings and the manual (gold standard) readings: (1) precision-of-merit (PoM); (2) figure-of-merit (FoM); (3) Pearson correlation coefficient (CC); (4) Bland–Altman plots; (5) cumulative frequency distribution of LD errors; (6) TLA error; (7) receiver operating characteristic (ROC); and (8) statistical tests.

11.3.3 PoM, FoM, CC and Bland–Altman plots

In the current pilot study, LD was computed using the bi-directional concept of PDM [28, 29]. The PoM is computed to measure the closeness between AtheroCloud LD and manual LD readings. In the current study, the PoM for AtheroCloud is computed

Table 11.1. Mean LD ± SD using AtheroCloud (routine mode versus pharma trial mode) and manual (gold standard) reading.

	AtheroCloud vs Manual		
Neck side	Routine mode (mm)	Manual (mm)	Percentage difference
Left carotid LD	6.49 ± 1.77	6.29 ± 1.79	3.17% ± 1.11%
Right carotid LD	6.66 ± 1.70	6.45 ± 1.63	3.25% ± 4.29%
	Pharma mode (mm)	**Manual (mm)**	—
Left carotid LD	6.48 ± 1.77	6.29 ± 1.79	3.02% ± 1.11%
Right carotid LD	6.66 ± 1.70	6.45 ± 1.63	3.25% ± 4.29%

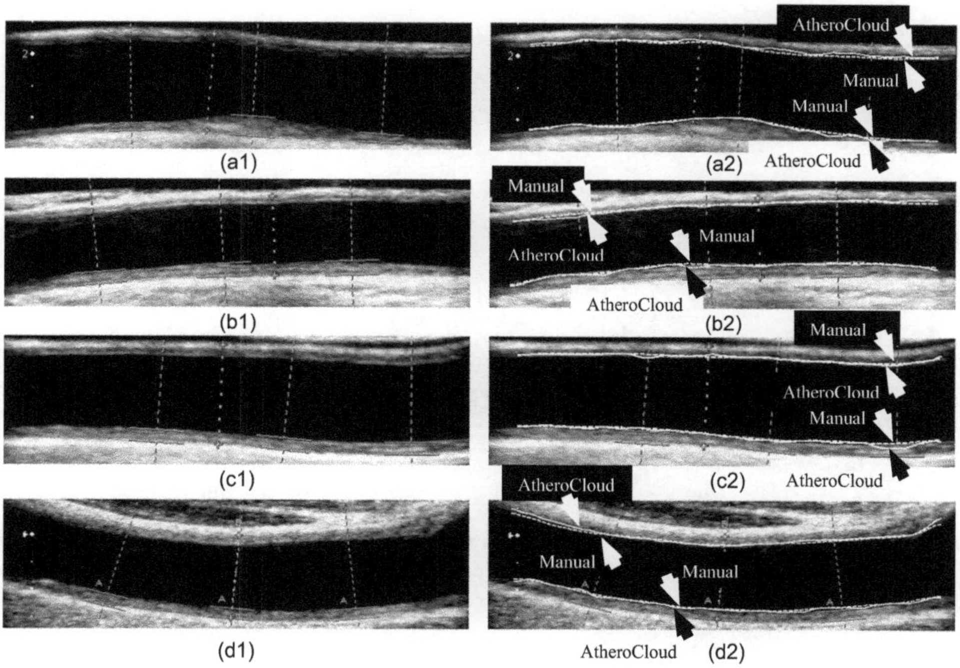

Figure 11.6. (a1), (b1), (c1) and (d1) Original carotid artery images. (a2), (b2), (c2) and (d2) Near/far wall overlays using AtheroCloud software (solid line) and manual tracings (dotted line).

as shown in appendix A. We have also computed the CC to check the relationship between the two quantities. Table 11.2 shows the PoM, FoM and CC between AtheroCloud and manual LD readings. The regression curves and Bland–Altman plots between AtheroCloud and manual readings are shown in figures 11.7(a)–(d), respectively. Bland–Altman plots display the average bias or the average of the differences between two readings [6].

Table 11.2. Performance parameters of the AtheroCloud system against the manual system (gold standard).

Parameter	PoM (%)	FoM (%)	CC
AtheroCloud against manual	96.18 ± 3.21	96.85	0.99

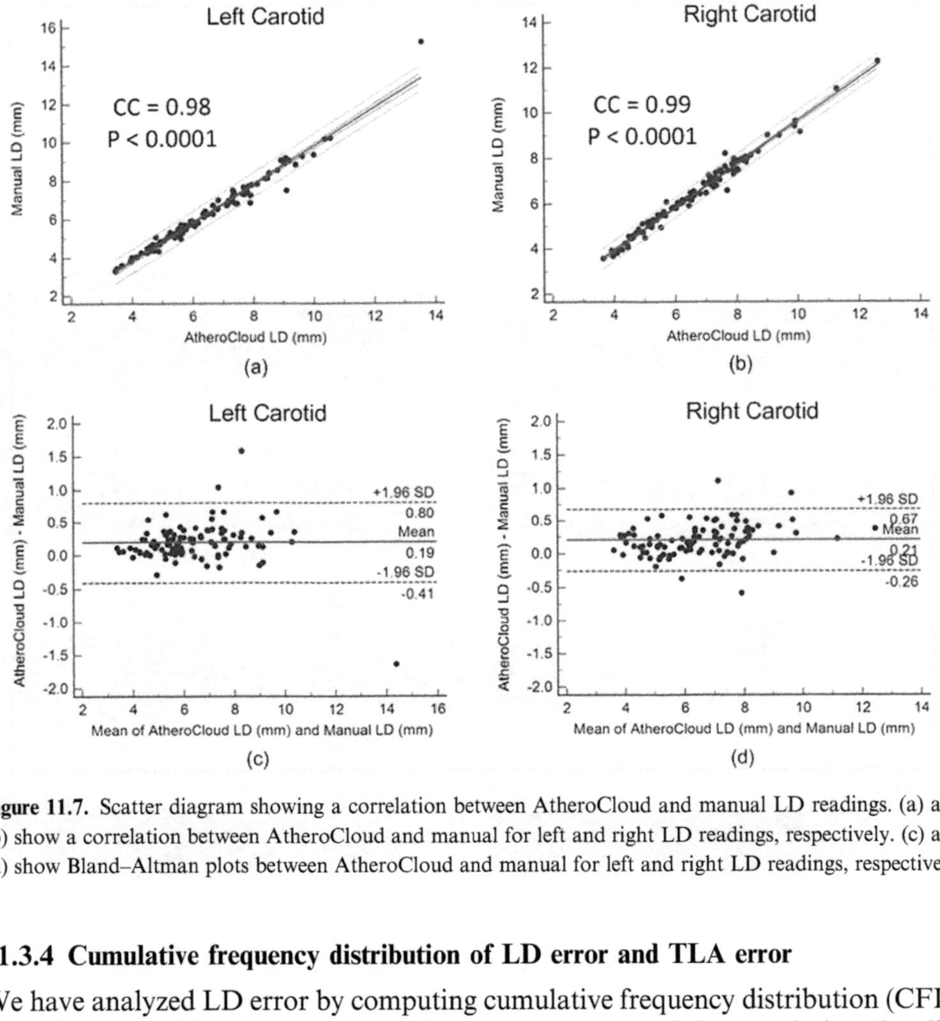

Figure 11.7. Scatter diagram showing a correlation between AtheroCloud and manual LD readings. (a) and (b) show a correlation between AtheroCloud and manual for left and right LD readings, respectively. (c) and (d) show Bland–Altman plots between AtheroCloud and manual for left and right LD readings, respectively.

11.3.4 Cumulative frequency distribution of LD error and TLA error

We have analyzed LD error by computing cumulative frequency distribution (CFD) plots. Using a CFD plot, we can define the percentage of the population that lies above (or below) a specific LD threshold value. In general, a CFD plot indicates how well the system performs. Keeping the LD error threshold as < 5%, we observed that 72.64% of the population met the error threshold criteria. The TLA error between AtheroCloud and manual LD readings was observed to be 4.56% ± 3.54%, lower than the 5% error threshold cut-off as defined by the regulatory bodies, where the medical device shows a degradation of less than 10% in performance [34].

11.3.5 Receiver operating characteristic

Sensitivity and specificity are generally used to quantify the diagnostic accuracy of a test. They are measures to identify how well the patients are classified as diseased or non-diseased, and can be computed using four parameters: true positive (TP), true negative (TN), false positive (FP) and false negative (FN). TP and TN are defined as the number of times LD is correctly identified with respect to manually computed LD. Similarly, FP and FN are defined as the number of times LD is incorrectly identified with respect to manually computed LD. Sensitivity and specificity can be mathematically defined as

$$\text{Sensitivity} = \frac{\text{TP}}{(\text{TP} + \text{FN})} \tag{11.1}$$

$$\text{Specificity} = \frac{\text{TN}}{(\text{TN} + \text{FP})}. \tag{11.2}$$

The ROC is a graphical representation between the sensitivity (TP rate) and 100-specificity (FP rate). The choice of a risk threshold affects both the sensitivity and specificity. The area under the ROC curve (AUC) is a predictor of the diagnostic accuracy. The results of the ROCs for different LD risk thresholds are shown in table 11.3. Since our database has an average mild stenosis (10%) and this corresponds to an average LD of 6 mm, we therefore took a threshold of 6 mm to demonstrate the ROC analysis. This is shown in figure 11.8 which compares AtheroCloud against AtheroEdge™ (a commercially available system) and figure 11.9 shows the stratification of the population into low- and high-risk bins.

11.3.6 Statistical tests

A statistical test measures the reliability and stability of the system. A statistical test will accept or reject the null hypothesis. The null hypothesis is generally an assumption that the two quantities are associated with each other. In this pilot study, all statistical analyses were performed using MedCalc 16.0 software (Osteen, Belgium). All statistical tests were performed with 0.05 as the level of significance. We performed a two-tailed z-test as we have more than 30 readings. The Mann–Whitney test, KS test and one-way ANOVA test were used to identify the significance difference between the variables. Table 11.4 shows the results of the two-tailed z-test and Mann–Whitney test between AtheroCloud and manual readings for the left and right LD readings. The negative z-score shows that the raw result is less than the mean. The normality of each continuous variable group was confirmed by the KS test. Finally, the association between AtheroCloud and manual readings was further assessed using the one-way ANOVA test.

Table 11.3. Performance of the AtheroCloud system using 200 US scans: ROC analysis (first five columns) for the AtheroCloud system for different LD thresholds; benchmarking of AC against AE for LD measurements (6th and 7th column); and benchmarking of AC against AE for SSI measurements (8th and 9th columns).

C1	C2	C3	C4	C5	C6	C7	C8	C9
LD threshold (mm)	AUC	*p*-value	Sensitivity (%)	Specificity (%)	Mean LD ± SD (mm)		Mean SSI ± SD (%)	
					AC	AE	AC	AE
6.0	0.997	< 0.0001	98.89	97.27	5.07 ± 0.70	5.05 ± 0.70	9.42 ± 6.93	9.42 ± 6.93
6.2	0.998	< 0.0001	96.04	100.00	5.20 ± 0.76	5.17 ± 0.75	9.54 ± 6.84	9.54 ± 6.84
6.4	0.998	< 0.0001	98.15	97.83	5.28 ± 0.80	5.25 ± 0.80	9.78 ± 6.81	9.78 ± 6.81
6.6	0.996	< 0.0001	98.29	98.80	5.40 ± 0.88	5.37 ± 0.88	9.52 ± 6.67	9.52 ± 6.67
6.8	0.995	< 0.0001	96.67	100.00	5.44 ± 0.91	5.41 ± 0.90	9.58 ± 6.71	9.58 ± 6.71
7.0	0.988	< 0.0001	95.45	95.59	5.61 ± 1.02	5.57 ± 1.01	9.48 ± 6.44	9.48 ± 6.44
7.2	0.992	< 0.0001	93.48	93.39	5.68 ± 1.05	5.65 ± 1.04	9.34 ± 6.35	9.34 ± 6.34
7.4	0.991	< 0.0001	93.88	98.11	5.79 ± 1.11	5.76 ± 1.11	9.21 ± 6.20	9.21 ± 6.20
7.6	0.986	< 0.0001	90.57	100.00	5.95 ± 1.21	5.91 ± 1.20	9.07 ± 6.02	9.07 ± 6.02
7.8	0.992	< 0.0001	99.39	94.29	6.02 ± 1.24	5.98 ± 1.23	9.04 ± 5.94	9.04 ± 5.94
8.0	0.990	< 0.0001	97.69	96.30	6.12 ± 1.29	6.08 ± 1.28	8.97 ± 5.83	8.97 ± 5.83

LD: lumen diameter; AUC: area under the curve; AC: AtheroCloud; AE: AtheroEdge; SSI: stenosis severity index; SD: standard deviation.

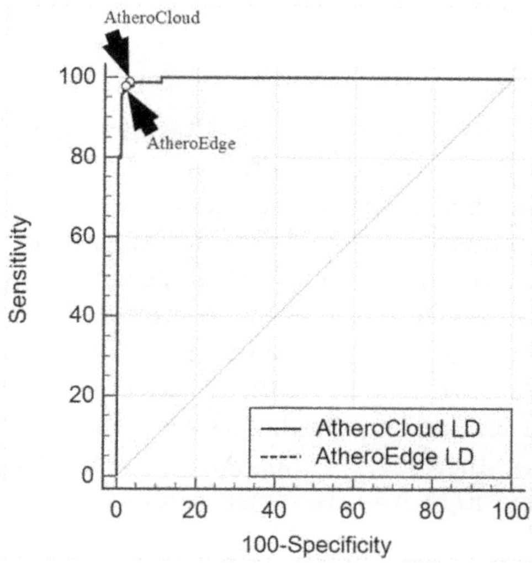

Figure 11.8. ROC curve analysis for AtheroCloud versus AtheroEdge™. The AUC is 0.997 for both the AtheroCloud and AtheroEdge™ systems.

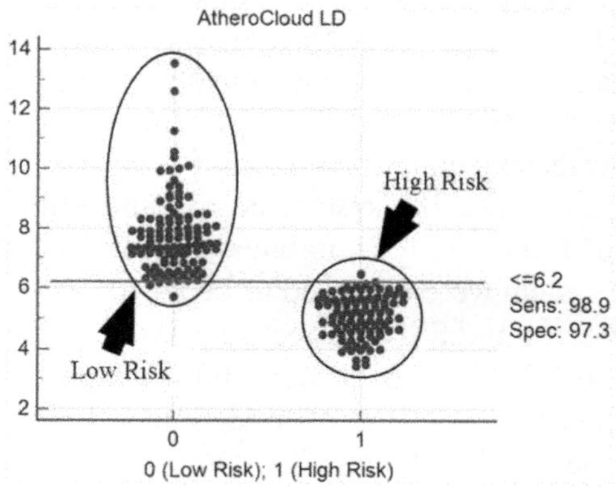

Figure 11.9. Interactive dot diagram for AtheroCloud LD and the respective threshold, sensitivity and specificity. Stratification of images into high-risk and low-risk bins.

Table 11.4. Two statistical tests (z-test and Mann–Whitney test) between AtheroCloud and manual LD readings.

AtheroCloud against manual	Two-tailed z-test		Mann–Whitney
	z	p-value	p-value
Left carotid LD	−0.7945	< 0.4269	= 0.3602
Right carotid LD	−0.8916	< 0.3726	= 0.3400

11.4 Discussion

11.4.1 Our system

The major objectives of this pilot study were to (i) design and develop a screening tool using automated/semi-automated cloud-based software for LD/stenosis/TLA measurement in routine and pharma trial mode in ultrasound carotid scans; (ii) validate the proposed AtheroCloud results against the manual (gold standard) LD/stenosis/TLA readings; (iii) evaluate the performance of the AtheroCloud system by computing precision parameters, such as PoM, FoM, CC, Bland–Altman plots, CFD plots and ROC curves; (iv) measure the reliability and stability of the system through statistical tests, such as the two-tailed *z*-test, Mann–Whitney test, KS test and one-way ANOVA test; and (v) benchmark the proposed AtheroCloud software against the standard, AtheroEdge™ 2.0 (AtheroPoint, Roseville, CA, USA), a desktop-based LD/stenosis/TLA measurement system.

11.4.2 Benchmarking AtheroCloud against AtheroEdge™ 2.0

Since this study is the first of its kind and no previous cloud-based system for LD/stenosis/TLA measurement system was available, we have therefore chosen AtheroEdge™ 2.0, a desktop-based system, for comparison against our proposed AtheroCloud system. In the AtheroEdge™ 2.0 system, the same pool of databases was run in the pharmaceutical trial mode run (as shown in figure 11.10) and is compared against the AtheroCloud LD results (as shown in table 11.5). The obtained LD errors were less than 5% (typically error less than 10% is adopted for regulatory purposes), hence the system is safe and effective to use in clinical settings. The PoM, FoM and CC for combined (left and right) carotid arteries for the AtheroEdge™ 2.0 system were observed to be 99.40% ± 0.29%, 99.42% and 1.0, respectively. The TLA error and SSI error between AtheroCloud and AtheroEdge™ 2.0 were observed to be 0.32% ± 1.38% and 0.00% ± 0.02%, respectively, showing the high accuracy of the proposed AtheroCloud system. Figure 11.11 shows that for different values of the SSI (in %), AtheroCloud and AtheroEdge™ 2.0 show similar results for TLA (in %). Figure 11.12 shows the cumulative frequency distribution of the SSI relating the number of patients in our database for different values of the SSI. The two-tailed *z*-test, Mann–Whitney test, KS test and one-way ANOVA test were also performed between the AtheroCloud and AtheroEdge™ 2.0 LD results. The statistical results demonstrate the high consistency and reliability of the proposed AtheroCloud software system.

11.4.3 Strengths, weaknesses and extensions

This is the first system of its kind that computes LD/SSI/TLA using carotid B-mode US in an automatic web-based framework. It enables multi-tenancy and can run on private clouds such as IBM or Amazon. However, internet speed is still geography-dependent, but over time cloud-technology will become more economical. This is a pilot research study, so no cost study was performed. Factors such as the effects of lighting conditions and manual tracer experience can be considered an extension to this project. In the future, the proposed software could be explored on a different

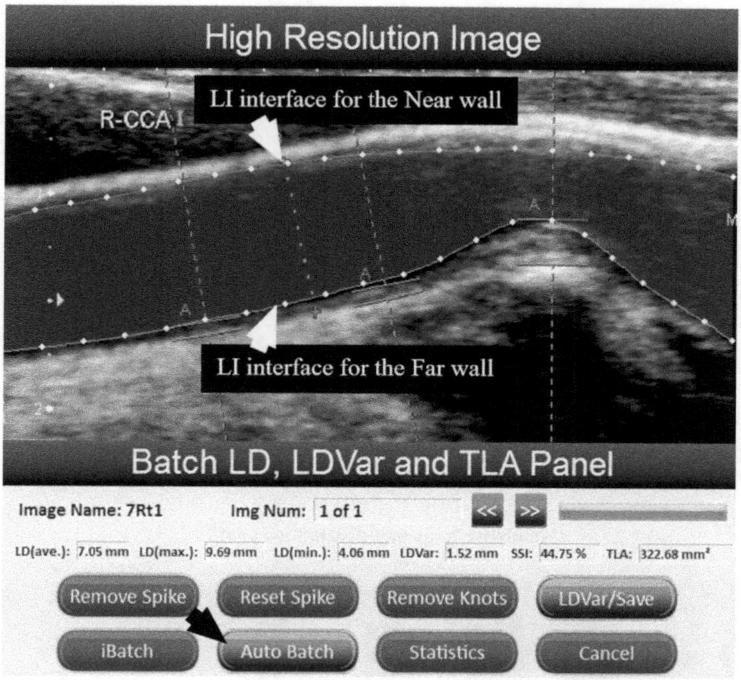

Figure 11.10. Pharmaceutical trial mode automated tracings of the carotid LD region showing the lumen–intima (LI) interface for the near wall (red) and far wall (green) using AtheroEdge™ software. (Courtesy of AtheroPoint™, Roseville, CA, USA.)

Table 11.5. Benchmarking (percentage difference) of AtheroCloud (routine and pharmaceutical mode) against AtheroEdge 2.0 (desktop-based LD measurement system).

	AtheroCloud vs AtheroEdge 2.0		
Neck side	Routine mode (mm)	AtheroEdge 2.0 (mm)	Percentage difference
Left carotid LD	6.49 ± 1.77	6.45 ± 1.76	0.62% ± 0.56%
Right carotid LD	6.66 ± 1.70	6.62 ± 1.69	0.06% ± 0.59%
	Pharma mode (mm)	**AtheroEdge 2.0 (mm)**	
Left carotid LD	6.48 ± 1.77	6.45 ± 1.76	0.62% ± 0.56%
Right carotid LD	6.66 ± 1.70	6.62 ± 1.69	0.06% ± 0.59%

database and intra- and inter-operator reproducibility and variability could also be tested, but the current results are truly encouraging.

11.5 Conclusion

We have designed and developed AtheroCloud—a completely automated, reliable, fast, accurate, reproducible, anytime-anywhere solution for multicenter clinical

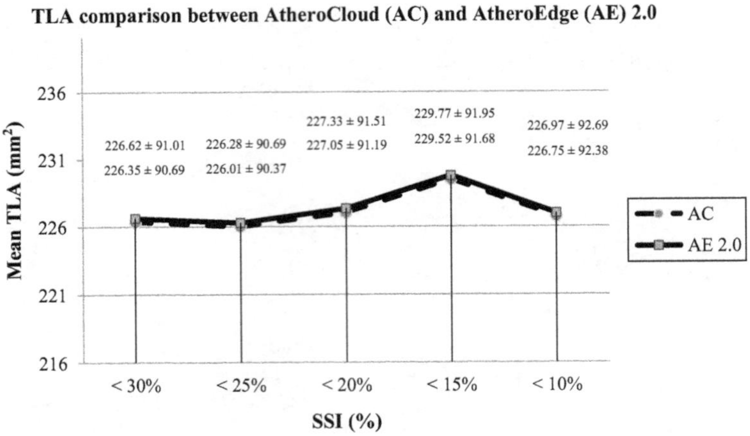

Figure 11.11. Performance of AtheroCloud and AtheroEdge™ TLA with a change in the SSI.

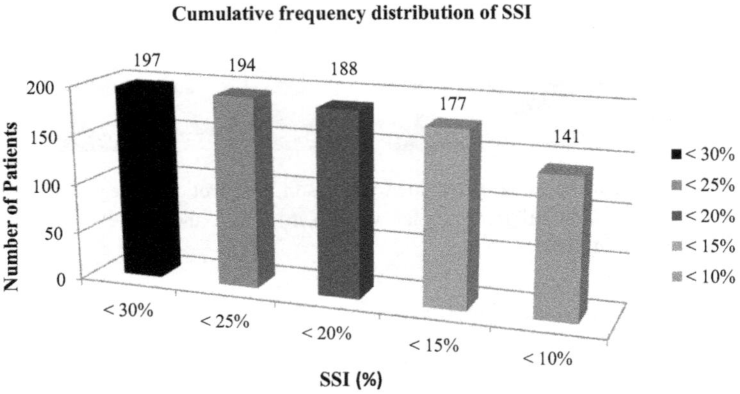

Figure 11.12. The relationship between the numbers of patients and the SSI.

trials and routine vascular screening. It can automatically compute the lumen diameter, stenosis and total lumen area of carotid ultrasound scans in cloud-based settings. AtheroCloud was compared against the manual LD/stenosis readings and the results were very encouraging. With the applicability of two different modes (routine and pharma trial), AtheroCloud was benchmarked against the standard, AtheroEdge™ 2.0, a desktop-based LD/stenosis measurement system. Comprehensive statistical tests demonstrate the consistency, reliability and accuracy of the AtheroCloud system. Although more multicenter studies are required, the current results indicate that the system can be adapted to clinical settings for clinical routines or multicenter pharmaceutical trials.

Acknowledgments

Reprinted with permission from Saba L, Banchhor S K, Londhe N D, Araki T, Laird J R, Gupta A, Nicolaides A and Suri J S 2017 Web-based accurate measurements of carotid lumen diameter and stenosis severity: an ultrasound-based clinical tool for stroke risk assessment during multicenter clinical trials *Comput. Biol. Med.* **91** 306–317. Copyright 2017 Elsevier.

Funding

This research did not receive any specific grant from funding agencies in the public, commercial, or not-for-profit sectors.

Conflicts of interest

The authors declare no conflict of interest.

Appendix A Precision-of-merit and figure-of-merit for AtheroCloud LD measurements

Precision-of-merit (PoM). If $\mathrm{FarWall}_{\mathrm{AtheroCloud}}$ and $\mathrm{NearWall}_{\mathrm{AtheroCloud}}$ are the LI interfaces computed using the AtheroCloud automated method, we can compute the AtheroCloud LD using the PDM, and it is given as

$$\mathrm{LD}_{\mathrm{AtheroCloud}} = \mathrm{PDM}(\mathrm{FarWall}_{\mathrm{AtheroCloud}}, \mathrm{NearWall}_{\mathrm{AtheroCloud}}). \tag{A.1}$$

Similarly, we can compute the LD measurements using manual tracings, given as

$$\mathrm{LD}_{\mathrm{Manual}} = \mathrm{PDM}(\mathrm{FarWall}_{\mathrm{Manual}}, \mathrm{NearWall}_{\mathrm{Manual}}). \tag{A.2}$$

Let N be the number of images in the overall database, then the system's performance at the frame level can be computed using the PoM in percentage as

$$\mathrm{PoM}_{\mathrm{AtheroCloud}}(\%) = \frac{1}{N}\sum_{i=1}^{N} 100 - \left[\left(\frac{\left|\mathrm{LD}_{\mathrm{AtheroCloud}(i)} - \mathrm{LD}_{\mathrm{Manual}(i)}\right|}{\mathrm{LD}_{\mathrm{Manual}(i)}}\right) * 100\right]. \tag{A.3}$$

Figure-of-merit (FoM). Let $\mathrm{LD}_{\mathrm{AtheroCloud}(i)}$ be the LD value automatically computed by the proposed AtheroCloud system on the ith image of the database of N images. Now the overall mean AtheroCloud LD can be computed as

$$\overline{\mathrm{LD}}_{\mathrm{AtheroCloud}} = \frac{1}{N}\sum_{i=1}^{N} \mathrm{LD}_{\mathrm{AtheroCloud}(i)}. \tag{A.4}$$

Correspondingly, if $\mathrm{LD}_{\mathrm{Manual}(i)}$ is the LD value computed from the radiologist's traced manual measurements on the ith image of the database of N images, then the overall mean manual LD can be computed as

$$\overline{\mathrm{LD}}_{\mathrm{Manual}} = \frac{1}{N}\sum_{i=1}^{N} \mathrm{LD}_{\mathrm{Manual}(i)}. \tag{A.5}$$

The system's performance at volume level can be computed using the FoM in percentage as

$$\mathrm{FoM}_{\mathrm{AtheroCloud}}(\%) = 100 - \left[\left(\frac{|\overline{\mathrm{LD}}_{\mathrm{AtheroCloud}} - \overline{\mathrm{LD}}_{\mathrm{Manual}}|}{\overline{\mathrm{LD}}_{\mathrm{Manual}}}\right) * 100\right]. \tag{A.6}$$

Appendix B Figures

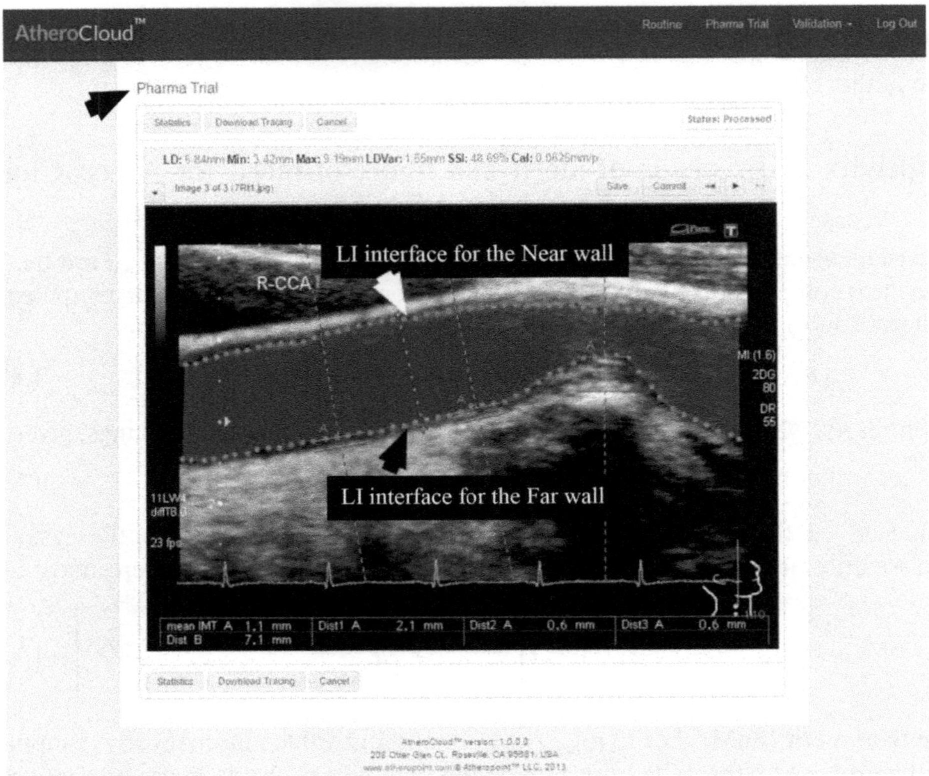

Figure B1. Pharma trial mode automated tracings (red) of the carotid LD region showing the LI interface for the near wall and far wall using AtheroCloud. (Courtesy of AtheroPoint™, Roseville, CA, USA.)

References

[1] Mozaffarian D *et al* 2016 Executive summary: heart disease and stroke statistics—2016 update: a report from the American Heart Association *Circulation* **133** 447

[2] Ross R 1995 Cell biology of atherosclerosis *Annu. Rev. Physiol.* **57** 791–804

[3] Libby P, Ridker P M and Hansson G K 2011 Progress and challenges in translating the biology of atherosclerosis *Nature* **473** 317–25

[4] Strandness D E and Eikelboom B C 1998 Carotid artery stenosis—where do we go from here? *Eur. J. Ultrasound* **7** S17–26

[5] Silver F L *et al* 2011 Safety of stenting and endarterectomy by symptomatic status in the Carotid Revascularization Endarterectomy Versus Stenting Trial (CREST) *Stroke* **42** 675–80

[6] Saba L *et al* 2016 Accurate cloud-based smart IMT measurement, its validation and stroke risk stratification in carotid ultrasound: a web-based point-of-care tool for multicenter clinical trial *Comput. Biol. Med.* **75** 217–34
[7] Davies K N and Humphrey P R 1993 Complications of cerebral angiography in patients with symptomatic carotid territory ischemia screened by carotid ultrasound *J. Neurol. Neurosurg. Psychiatry* **56** 967–72
[8] Featherstone R L, Brown M M and Coward L J 2004 International carotid stenting study: protocol for a randomised clinical trial comparing carotid stenting with endarterectomy in symptomatic carotid artery stenosis *Cerebrovasc. Dis.* **18** 69–74
[9] Naim C *et al* 2014 Vulnerable atherosclerotic carotid plaque evaluation by ultrasound, computed tomography angiography, and magnetic resonance imaging: an overview *Can. Assoc. Radiol. J.* **65** 275–86
[10] Araki T *et al* 2016 Ultrasound-based automated carotid lumen diameter/stenosis measurement and its validation system *J. Vasc. Ultrasound* **40** 120–34
[11] Saba L *et al* 2014 *Multi-Modality Atherosclerosis Imaging and Diagnosis* (New York: Springer)
[12] Sanches J M, Laine A F and Suri J S 2012 *Ultrasound Imaging* (New York: Springer)
[13] Greenwald S E 2002 Pulse pressure and arterial elasticity *QJM* **95** 107–12
[14] Patel A K, Suri H S and Singh J *et al* 2016 A review on atherosclerotic biology, wall stiffness, physics of elasticity, and its ultrasound-based measurement *Curr Atheroscler Rep.* **18** 83
[15] Saba L *et al* 2012 Comparison between manual and automated analysis for the quantification of carotid wall by using sonography. A validation study with CT *Eur. J. Radiol.* **81** 911–8
[16] Araki T *et al* 2016 Two automated techniques for carotid lumen diameter measurement: regional versus boundary approaches *J. Med. Syst.* **40** 1–19
[17] Santos A M *et al* 2013 Automatic segmentation of the lumen of the carotid artery in ultrasound B-mode images *Med. Imaging* **8670** 867031
[18] Ikeda N *et al* 2014 Automated and accurate carotid bulb detection, its verification and validation in low quality frozen frames and motion video *Int. Angiol.* **33** 573–89
[19] Dey N *et al* 2016 Effect of watermarking on diagnostic preservation of atherosclerotic ultrasound video in stroke telemedicine *J. Med. Syst.* **40** 1–4
[20] Saba L *et al* 2016 Carotid inter-adventitial diameter is more strongly related to plaque score than lumen diameter: an automated tool for stroke analysis *J. Clin. Ultrasound* **44** 210–20
[21] Rocha R, Silva J and Campilho A 2014 Automatic detection of the carotid lumen axis in B-mode ultrasound images *Comput. Methods Programs Biomed.* **115** 110–8
[22] Golemati S *et al* 2007 Using the Hough transform to segment ultrasound images of longitudinal and transverse sections of the carotid artery *Ultrasound Med. Biol.* **33** 1918–32
[23] Molinari F, Zeng G and Suri J S 2010 An integrated approach to computer-based automated tracing and its validation for 200 common carotid arterial wall ultrasound images: a new technique *J. Ultrasound Med.* **29** 399–418
[24] Sicari R *et al* 2011 The use of pocket-size imaging devices: a position statement of the European Association of Echocardiography *Eur. J. Echocardiogr.* **12** 85–7
[25] Molinari F *et al* 2009 Automatic computer-based tracings (ACT) in longitudinal 2-D ultrasound images using different scanners *J. Mech. Med. Biol.* **9** 481–505
[26] Molinari F *et al* 2012 Carotid artery recognition system: a comparison of three automated paradigms for ultrasound images *Med. Phys.* **39** 378–91

[27] Molinari F *et al* 2012 Hypothesis validation of far-wall brightness in carotid-artery ultrasound for feature-based IMT measurement using a combination of level-set segmentation and registration *IEEE Trans. Instrum. Meas.* **61** 1054–63
[28] Saba L *et al* 2012 What is the correct distance measurement metric when measuring carotid ultrasound intima–media thickness automatically? *Int. Angiol.* **31** 483–9
[29] Molinari F, Zeng G and Suri J S 2010 Intima–media thickness: setting a standard for a completely automated method of ultrasound measurement *IEEE Trans. Ultrason. Ferroelectr. Freq. Control* **57** 1112–24
[30] Fox A J 1993 How to measure carotid stenosis *Radiology* **186** 316–8
[31] Zarins C K *et al* 1983 Carotid bifurcation atherosclerosis. Quantitative correlation of plaque localization with flow velocity profiles and wall shear stress *Circ. Res.* **53** 502–14
[32] Mintz G S *et al* 1997 Determinants and correlates of target lesion calcium in coronary artery disease: a clinical, angiographic and intravascular ultrasound study *J. Am. Coll. Cardiol.* **29** 268–74
[33] Sangiorgi G *et al* 1998 Arterial calcification and not lumen stenosis is highly correlated with atherosclerotic plaque burden in humans: a histologic study of 723 coronary artery segments using nondecalcifying methodology *J. Am. Coll. Cardiol.* **31** 126–33
[34] Saba L *et al* 2016 Inter-observer variability analysis of automatic lung delineation in normal and diseased patients *J. Med. Syst.* **40** 142

Section V

Micro-electro-mechanical-system (MEMS)

Chapter 12

A MEMS-based manufacturing technique of vascular bed

Arindam Bit and Jasjit Suri

A micro-electro-mechanical-system (MEMS) is a kind of miniaturized system commonly used in the domain of sensor technology and drug delivery devices in the healthcare industry. Although there is a significant amount of potential in the manufacturing routes of MEMS synthesis, its use tends to be limited to the semiconductor device industries. In this chapter, a very careful amalgamation has been carried out to form a process between the synthesis routes of MEMS and its application to synthesize multi-layered vascular bed with micro-scale level refinement. This technique can be used as a potential method for re-defining the construction of multi-layered tissues of many organs.

12.1 Introduction

MEMSs are largely confined to the semiconductor industries. However, from an application perspective, they also provide solutions to a wider spectrum of healthcare industry challenges. A spectrum view of MEMS applications in the healthcare industry is shown in figure 12.1 [1–4]. A disposable blood pressure transducer (DPT) uses a MEMS-based transducer. It has a huge market size of 20 million units per year in the United States alone. Intra-uterine pressure sensors are also made of piezoelectric-transducer-based MEMS devices. They are used to monitor intra-uterine pressure during delivery. Angioplasty pressure sensors are also made of MEMSs. They are installed at the tip of the intra-balloon wall to measure the pressure inside the angioplastic balloon inside the blood vessel. This has direct implications in the management of cardiovascular disorders (CVDs), which affect 37% of the world's population. Another form of MEMS application is infusion pump pressure sensors. These pumps are used for regulating intravenous fluid flow in order to produce effective mixing of drugs in one flow channel [5–7]. MEMSs are

doi:10.1088/2053-2563/ab0820ch12

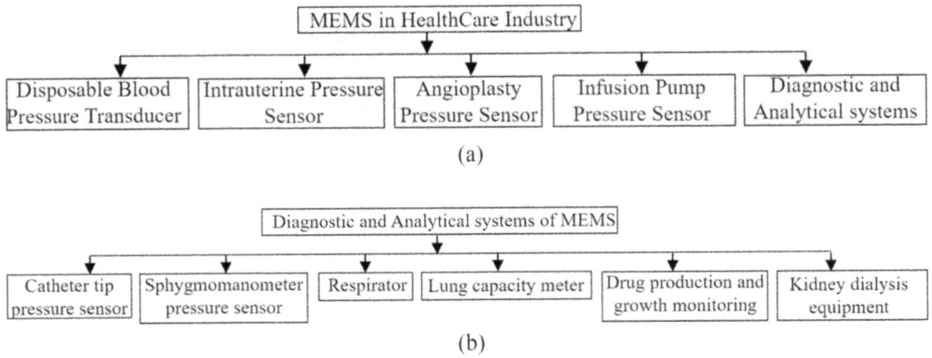

Figure 12.1. (a) Flowchart of MEMS applications in the healthcare industry and (b) diagnostic and analytical systems of MEMS.

also being used to fabricate many other healthcare support systems, including catheter tip pressure sensors, sphygmomanometers, respirators, lung capacity meters, drug production under the preview of bacteria culture and dialysis filter monitoring sensors. A flowchart is given in figures 12.1(a) and (b) with structured details of MEMS applications in healthcare engineering [8–11].

This chapter primarily includes a comprehensive illustration of the microstructural details of the vascular bed, followed by modeling aspects of blood vessel microstructure. The modeling aspects are further illustrated with the design principles of acoustic wave mechanosensors, pressure mechanosensors, microvalves and micropumps. This is followed by a description of the scaling law, and its physical implications in the design of the microstructural vascular bed. The scaling law includes an illustration of geometric scaling and fluid dynamics scaling within the blood vessels. Then, various microfabrication techniques are overviewed for the development of blood vessels, including soft lithography, self-assembly and sputtering techniques. Various design parameters are also illustrated thereafter. The last section of this chapter includes a systemic description of the finite element modeling of pressure sensors for balloon angioplasty intervention in biological and synthetic blood vessels.

12.2 Microstructural anatomy of blood vessels

12.2.1 Arteries and veins

Arteries are blood vessels which carry blood away from the heart to different parts of the body. Except for the pulmonary and umbilical arteries, all arteries carry oxygenated blood, while the pulmonary and umbilical arteries carry deoxygenated blood. The pulmonary artery carries blood to the lungs to remove excess carbon dioxide and for oxygenation.

The pulse that is felt under the skin is due to the expansion and contraction of an artery when the heart flushes blood into the arteries via the aorta. Arterial pressure varies with the maximum pressure of the heart during contraction and the minimum pressure during relaxation when the heart expands and fills with blood, called

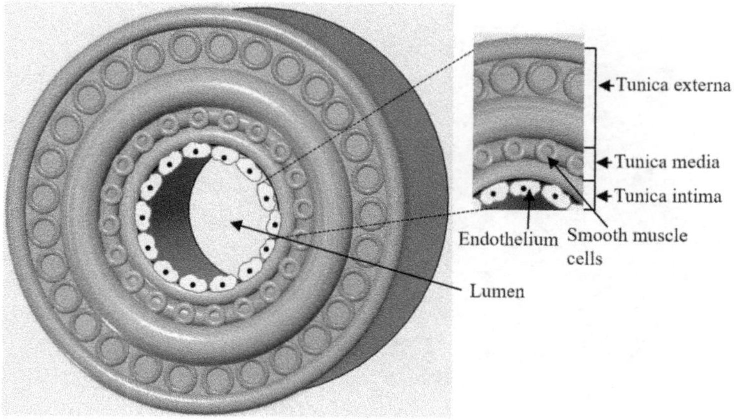

Figure 12.2. Schematic diagram of arterial structure.

systolic and diastolic pressure, respectively. The variation of pressure that occurs within the artery produces the pulse that represents the activity of the heart. The pressure in the arterial system will be highest in the aorta and lowest in the venous system when the blood returns to the heart after the supply of oxygen to the tissues in systemic circulation [12–14].

The largest artery is the aorta, arising from the left ventricle of the heart. The aorta further gives rise to small arteries call arterioles, which further connect to minute vessels called capillaries that carry oxygen to tissues and then take away carbon dioxide through veins. The aorta arches slightly upwards before going downwards near the backbone. From this arch, the arteries arise which supply blood to the head, neck and arms. As it travels down, the aorta gives rise to the blood vessels supplying blood to the internal organs in the thorax region. At the abdomen region, the aorta divides into two branches supplying blood to each leg.

An artery has three layers, i.e. the tunica intima, tunica media and tunica adventitia. The tunica intima is lined by a network of connective tissue and elastic fibers. The middle layer, the tunica media, consists of smooth muscles and elastic fibers arranged in a spiral manner [15–18]. The top layer, the tunica adventitia, is made up of mainly collagen fibers, which provide support to the artery. The thickness of all these layers varies with the size of the artery. A schematic of aortic structure is given in figure 12.2.

Veins are the blood vessels that carry deoxygenated blood and return it to the heart. The deoxygenated blood that flows through the veins is collected by the capillaries. Veins are thinner than arteries with thin walls and a larger lumen. Veins consist of valves that help in unidirectional flow of the blood. However, valves are absent in veins less than 2 mm in diameter, the venae cavae and in some hepatic, renal, ovarian and pulmonary veins. Compared to the arteries, the elastic and muscular tissue content in the walls of the veins is less, which in turn results in low venous pressure. Veins are made up of three layers, similar to the arteries, the tunica intima, tunica media and tunica adventitia. The muscle and elastic content of the layers is low. The tunica intima is a single layer made up of endothelial cells and

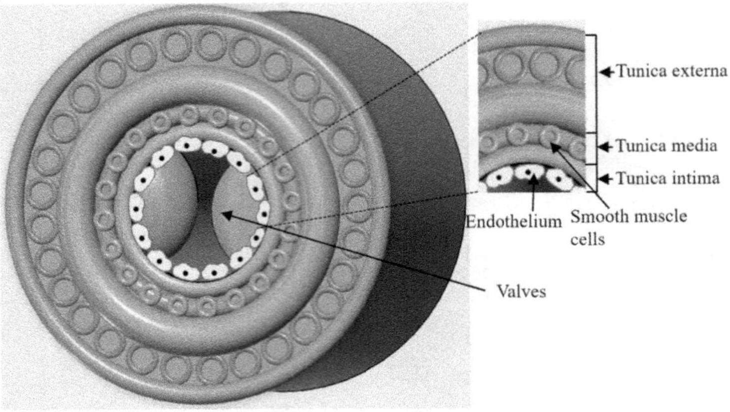

Figure 12.3. Schematic diagram of vein structure.

connective tissues. This layer will have valves in some conditions which help in preventing backward flow of the blood, particularly in the arms and legs. In the tunica media, the amount of collagen fibers is greater than the amount of elastic and muscle fibers. The tunica adventitia is the thickest and most developed of all the three layers and is made up of connective tissue. It has minute blood vessels called the vasa vasorum, which supply blood to the walls of veins [19–22]. Depending on their location and function, veins are divided into two types, i.e. pulmonary and systemic veins. A schematic representation of vein structure is given in figure 12.3.

Pulmonary veins carry deoxygenated blood from the heart to the lungs. Once the blood is oxygenated in the lungs, the veins carry the oxygenated blood back to the heart. There are four pulmonary veins that carry oxygenated blood to the heart. All the remaining veins will only carry deoxygenated blood. Systemic veins carry deoxygenated blood back from the body to the heart and enter the pulmonary veins. Most of the veins are systemic veins. Systemic veins are further classified into deep veins, superficial veins and connecting veins. Deep veins are usually found in muscles or through bones. The tunica intima layer in these veins will have valves preventing the backward flow of blood. The muscles present nearby also compress the vein. Superficial veins are located in the fatty layers of the skin and have a one-way valve in the tunica intima layer to help achieve slower movement of blood than in the deep veins. Blood flowing through superficial veins travels into deep veins through small blood vessels called connecting veins [23–25]. The valves present in these veins allow blood to flow in one direction from the superficial veins to the deep veins.

12.2.2 Capillaries

Capillaries can be defined as minute blood vessels in the circulatory system connecting the arterioles to the venules. They are the most abundant and smallest blood vessels in the body. They are very small in order to penetrate into the body tissues and allow the exchange of oxygen, nutrients and waste products between the

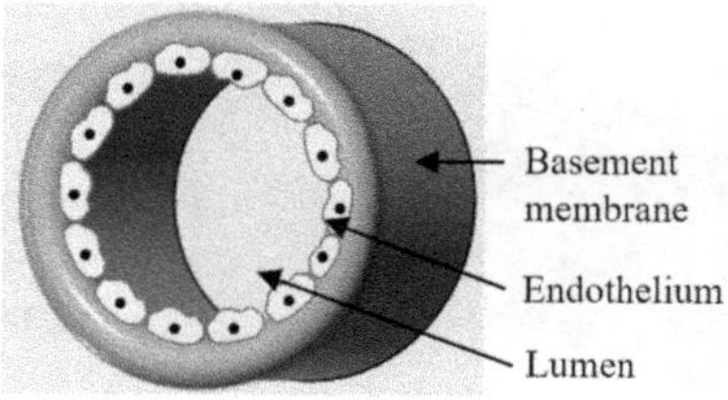

Figure 12.4. Schematic diagram of capillary structure.

blood and the tissues. The exchange of all these materials is achieved through passive diffusion and pinocytosis, which means cells ingesting fluids. During infections or any damage, white blood cells enter the targeted tissues through capillaries [26–28].

Mostly capillaries are 3–4 μm and some are up to 40 μm in diameter. The tunica intima layer of capillaries is made up of epithelial cells. The epithelial cells are partially covered by an incomplete layer of pericytes. These pericytes regulate blood pressure during contraction in capillaries, improving the efficiency of the exchange of blood between capillaries and the surrounding tissues [29–32]. Blood flow in the capillaries is regulated by smooth muscles called pre-capillary sphincters. The structural details of capillaries are shown in figure 12.4.

There are three types of capillaries, namely continuous, fenestrated and sinusoidal. (i) Continuous capillaries have a continuous endothelial lining. Continuous capillaries are normally found in the nervous system and in fat and muscle tissues. In the nervous system, they form the blood–brain barrier controlling the flow of cells and molecules in the blood and interstitial fluid around the brain. (ii) Fenestrated capillaries are mostly found in the kidneys, endocrine glands and small intestine, where there is exchange of large molecules. They play a vital role in the kidneys, particularly in the glomerulus where blood is filtrated in the process of urine formation. The capillaries have small openings called fenestrates which open into their endothelium, with a 80–100 nm diameter. (iii) Sinusoidal capillaries are often called sinusoids or discontinuous capillaries, and have multiple fenestrations (openings) with endothelial lining [33–36]. They are around 30–40 nm in diameter. They have no diaphragm, allowing serum and blood cells to pass through them. They are mainly found in the liver between hepatocytes and endothelial cells, spleen lymph nodes, the bone marrow and in endocrine glands.

12.3 Modeling of blood vessels as a microsystem

12.3.1 Acoustic wave mechanosensors

The primary application of an acoustic wave sensor is to measure the chemical constituents of a gaseous mixture. The sensor is composed of a piezoelectric crystal which converts electrical inputs into mechanical waves and vice versa. A schematic diagram of an acoustic wave sensor is shown in figure 12.5.

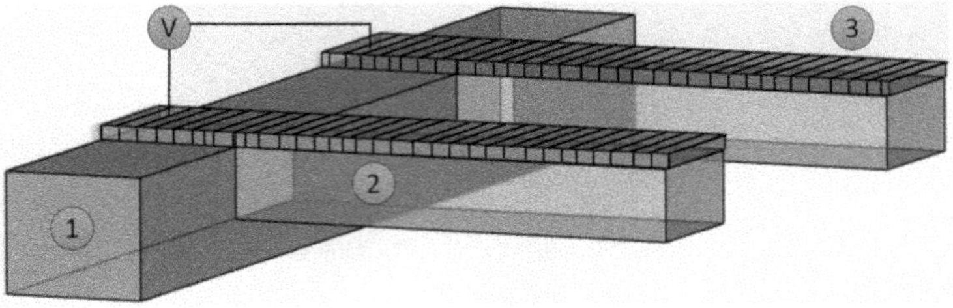

Figure 12.5. A cantilever beam type acoustic wave mechanosensor: (1) substrate layer, (2) piezoelectric crystal layer and (3) Si/GaAs cantilever beam. The piezoelectric crystal layer and Si/GaAs cantilever beam are excited by an excitation voltage through an external connection.

The chemical constituents present in the gaseous mixture will elevate the gaseous pressure based on the enthalpy of each gaseous constituent. Based on the enthalpy, each gaseous constituent can differentially create mechanical waves within the mixture. Acoustic wave devices are also being used to actuate fluid flow in microfluidic systems [37–39]. These microfluidics-based acoustic wave devices are used for cell-separation technology, DNA–RNA separation technology, antigen–antibody binding and reaction modulation, and cellular differentiation.

12.3.2 Pressure mechanosensors

A pressure sensor works on the principle of the deformation of a thin diaphragm under the influence of applied pressure. Stress developed within the deformed diaphragm is evaluated for calibrating the pressure. There are two types of pressure sensors: absolute and gage pressure sensors. Absolute pressure sensors have an evacuated cavity on one side of the diaphragm, whereas the vacuum is absent in the case of gage pressure sensors. There are two ways to apply pressure to the diaphragm: back-side pressurization and front-side pressurization. In the former, there is no interference with signal transduction, as there is no piezo-resistor, and the measurand is directly interacting with the diaphragm. In the case of front-side pressurization, the interference of the pressurizing medium with the signal transducer is high due to the presence of a piezo-resistive element.

The sensing element is made of a micro-layered silicon die. A microfabrication technique is used to create a cavity on one side of the die. The cavity is covered with a thin diaphragm which deforms through external stress applied by the measurand fluid. The diaphragm is also made of silicon and is a few microns thin. The silicon die is supported by a constraint base which is either made of metal or ceramics. A deformed diaphragm induces an electrical signal by a transduction technique. Four piezo-resistors (r1, r2, r3, r4) are implanted beneath the surface of the silicon die. They convert the induced stress of the silicon diaphragm into a dynamic change in electrical resistance, which is further converted into output voltage through a Wheatstone bridge circuit (as shown in figure 12.6).

These piezo-resistors are miniaturized strain gages. According to the figure, the resistances r_1 and r_3 are elongated on application of induced stress, which causes an

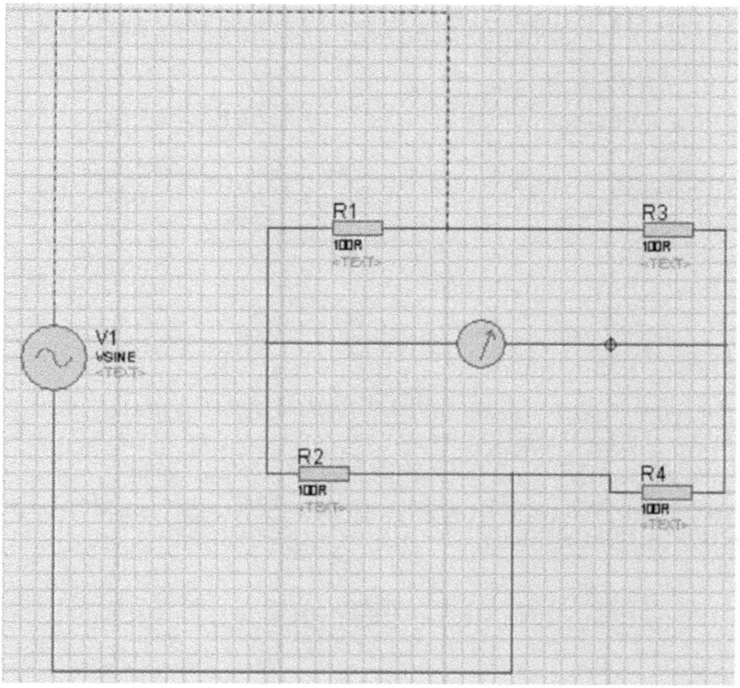

Figure 12.6. A Wheatstone bridge circuit of piezo-resistors is shown, for measuring the output voltage with respect to change in applied external pressure. Resistances r_1, r_2 and r_4 are fixed resistances, whereas resistance r_3 corresponds to the piezo-resistor.

increase in the resistance of these piezo-resistors. However, r_2 and r_4 experience opposite stress.

These changes in the resistance value of the piezo-resistors under the influence of external mechanical stress can be measured by the dynamic deflection operation of the Wheatstone bridge:

$$V_0 = V_{in}\left(\frac{r_1}{r_1 + r_4} - \frac{r_3}{r_2 + r_3}\right), \tag{12.1}$$

where V_0 and V_{in} are the measured voltage and supplied voltage to the Wheatstone bridge, respectively.

Another mode of micro-pressure sensing is by utilizing capacitance change. In this case, both the top and bottom surface of the thin diaphragm can be covered with a thin metallic plate. The diaphragm will act as a dielectric medium, and any deformation of the diaphragm under the influence of measured fluid influx will alter the gap between the two plates, and thus a change of capacitance will be recorded. This method is superior to the piezo-resistive method, because in this case thermal stresses will not alter the capacitance, which was true in the previous situation [40–44]. If the thickness of the diaphragm is d, then the capacitance between two plates will be expressed as

Table 12.1. Relative permittivity of biocompatible dielectric materials.

Materials	Relative permittivity
Cellulose paper	3.2
Porcelain	6.4
Silicon	12
Pyrex	4.7

$$C = \epsilon_r \epsilon_0 \frac{A}{d}, \tag{12.2}$$

where ε_r is the relative permittivity of the dielectric medium and ε_0 is the permittivity of free space ($\varepsilon_0 = 8.85$ F m^{-1}).

The relative permittivity of several dielectric substance is given in table 12.1. When arranged in a Wheatstone bridge (figure 12.7), the variable capacitance of these capacitive microsystems can be measured by the output voltage V_0 as

$$V_0 = \frac{\Delta C}{2(2C + \Delta C)} V_{\text{in}}, \tag{12.3}$$

where ΔC is the change in the capacitance, C is the capacitance of the reference capacitors of the bridge and V_m is the supplied voltage of the bridge.

12.3.3 Microvalves and micropumps

Microvalves are high-precision devices used for regulating the blood flow in an artery. These valves work on the principle of micro-actations. Jerman [45] reported a primary design for a micro-valve (as shown in figure 12.8). The two electrical resistor rings at the top of the diaphragm, when heated, cause downward movement of the diaphragm, closing the valves. The heating rings are made of aluminum (5 μm thick). These valves have a capacity of 300 cm^3 min^{-1} at a fluid pressure of up to 100 psi [46–48].

A micropump can be developed by implementing the electrostatic actuation of a diaphragm, as illustrated in figure 12.9.

The silicon diaphragm is deformable under the influence of an external electric field. Therefore, when a voltage is applied between the silicon diaphragm and the metallic electrode, the diaphragm deforms and reduces the gap between the diaphragm and the electrode. This results in the increase of the volume of the pumping chamber. The increased volume leads to a decrease in pressure inside the chamber, and thus the inlet check valve is opened, and the influx of fluid takes place. Now, when the voltage between the electrode and diaphragm increases and reaches the cut-off voltage, the diaphragm immediately deforms in the negative polarity, and comes back to its original position. This results in a sudden decrease in

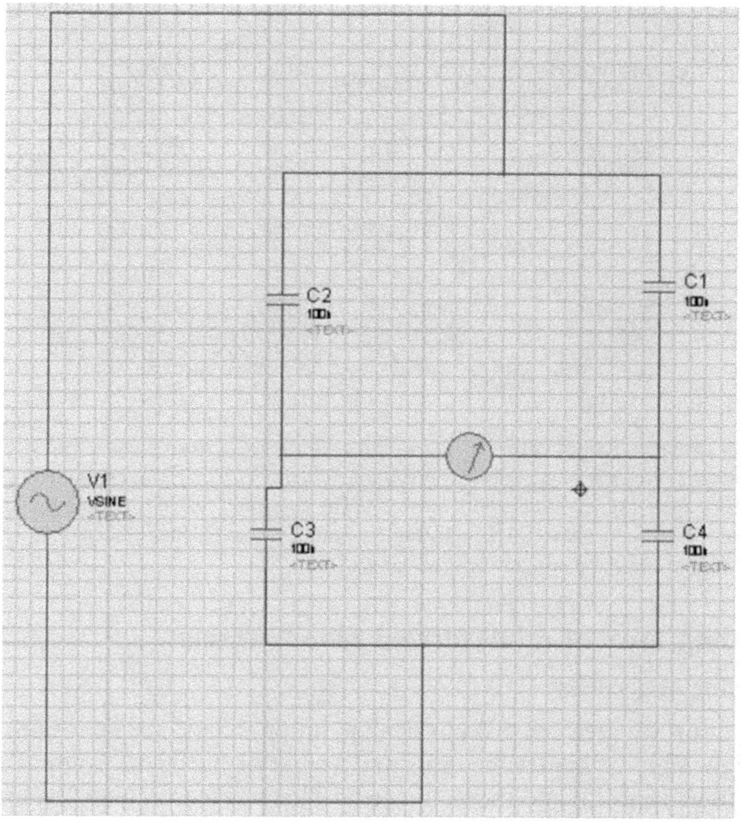

Figure 12.7. A Wheatstone bridge circuit of a piezo-capacitor is shown to measure the output voltage with respect to change in applied external pressure. Capacitances C_1, C_2 and C_4 are fixed capacitors, whereas, capacitor C_3 corresponds to the piezo-capacitor.

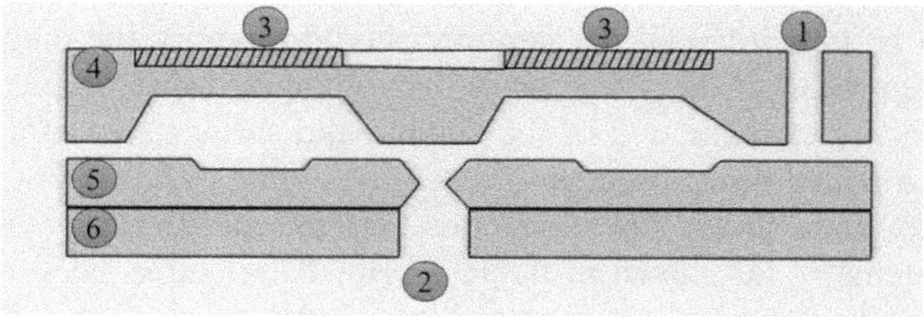

Figure 12.8. A schematic representation of the working principles of microvalves: (1) the inlet, (2) the outlet, (3) the resistive heating ring and (4) the upper diaphragm, made of silicon. The upper diaphragm is connected with the base (5), which is made of a Si/GaAs composite. The base layer (6) is coated over the substrate.

the volume of the pumping chamber, thereby an abrupt increase in the internal pressure of the chamber filled with fluid [49–52]. This elevated pressure resists the opening of the inlet check valve and opens the outlet check valve, causing the high-pressurized fluid efflux out of the pumping chamber.

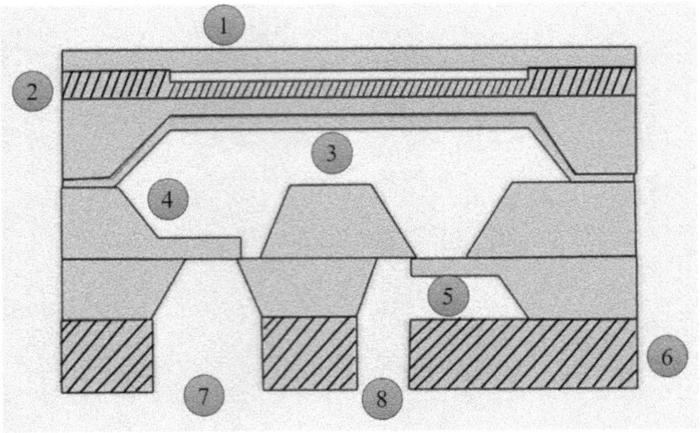

Figure 12.9. A schematic representation of the working principles of micropumps. The electrode (1) and silicon diaphragm (2) are externally excited by excitation voltage (V). The pumping chamber (3) is regulated by the inlet check valve (4) and outlet check valve (5). The inlet check valve is opened by the flow of fluid through the inlet (7), embedded through the substrate base of the pump (6), and opening of the outlet valve will drive the fluid flow out of the pump through the outlet (8).

12.4 Scaling laws of miniaturized blood vessels

12.4.1 Scaling in geometry

Volume and surface area are two physical quantities which are important for micro-device design. Volume relates the mass and weight of the device components, and is correlated to mechanical and thermal inertia. Thermal inertia gives a measure of the specific heat of the material, and is of prime importance, in particular when designing thermal actuators. Surface area, on the other hand, is correlated with interfacing physical parameters, such as normal stress, hydrostatic pressure and buoyant force in the case of fluid interaction. It can also be correlated with heat absorption or dissipation in the case of convective heat transfer. When a physical quantity is to be miniaturized, the designer should quantify the magnitude of the tentative consequences of the reduction of the volume and surface area of the device [53–57]. Proportionate reduction of the volume and surface area of a device is not achievable in a scaled-down process.

In order to understand the above fact, let us consider the following solid rectangular geometry (as shown in figure 12.10). It has three sides $a1 > a2 > a3$. It has a volume $V = a1\ a2\ a3$ and surface area $S = 2 \times (a1a2 + a2a3 + a3a1)$. Let us assume that L represents the linear dimension of the box, then volume $V \propto L^3$ and surface area $S \propto L^2$:

$$S/V = L^{-1}. \tag{12.4}$$

The scaling formula can be concluded from the above equations which state that a reduction of size of 10 times ($L = 0.1$) will produce a $10^3 = 1000$ times reduction in volume, and a $10^2 = 100$ times reduction in surface area. A reduction in volume by 1000 times also means a 1000 times reduction of weight inertia, whereas in the same

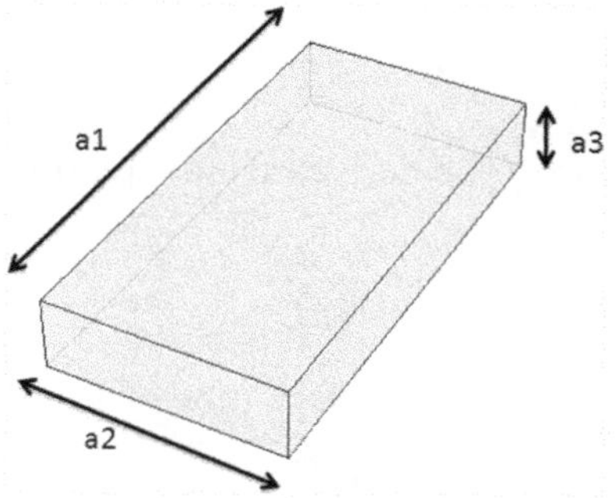

Figure 12.10. A solid rectangle with length ($a1$), width ($a2$) and height ($a3$).

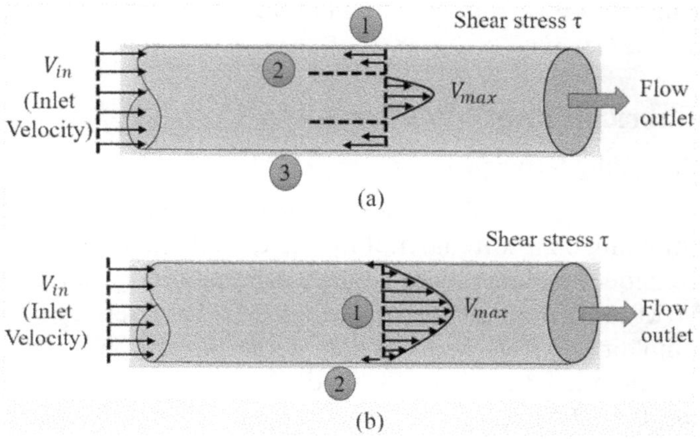

Figure 12.11. (a) The flow of blood through larger vessels with centreline velocity profile (1) is diminished at the boundary wall due to the no-slip boundary (2). (b) The flow profile of blood in micro-capillaries with centreline velocity profile (1) is diminished at the boundary wall due to lubricating layer formation (2) and the no-slip boundary (3).

conditions, a reduction of surface area by 100 times reduces the interfacing properties by 100 times. Therefore, there is a resultant increase of the interfacing properties of miniaturized solid matter by ten times under such circumstances [58, 59].

12.4.2 Scaling in fluid dynamics

In most biological matter, fluid flows in a closed system, and thus their flow is highly influenced by shear forces. At the same time, at sub-micron and nano-scale capillaries, the absence of convective acceleration and momentum parameters simplifies the Navier–Stokes equation. However, capillary flow does not scale down in sub-micron level. Details of this statement are illustrated below.

Figures 12.11(a) and (b) illustrate the fluid flow through cylindrical conduits of length L and radius r.

The pressure drop ΔP over the length L of the cylindrical conduit can be obtained by the Hagen–Poiseuille equation

$$\Delta P = \frac{8\mu LQ}{\pi r^4}. \tag{12.5}$$

The volumetric flow rate Q can be derived from equation (12.5) as

$$Q = \frac{\pi r^4 \Delta P}{8\mu L}. \tag{12.6}$$

The pressure drop for a section of capillary tube of length L can be evaluated by

$$\Delta P = \frac{8\mu V_{\text{avg}} L}{r^2}. \tag{12.7}$$

From equation (12.5), the scaling laws for fluid flow in capillary tubes are derived as $Q \propto r^4$ for volumetric flow, from equation (12.6), the scaling law for pressure drop per unit length is $\Delta P/L \propto r^{-2}$.

12.5 Microfabrication of blood vessels

12.5.1 Soft lithography techniques

Soft lithography is a non-photolithography technique which uses tools such as self-assembly and molding on a soft material for nano- and microfabrication. Some soft lithography techniques include micro-contact printing (μCP), solvent assist micromolding (SAMIM), micromolding in capillaries (MIMIC), replica molding (REM), microtransfer molding (μTM) and capillary force lithography (CFL). An elastometric stamp with micropatterning is used as the casting case. This stamp is fabricated using the REM technique. Alternatively, chemically patterned surfaces can be fabricated by the μCP technique. Fabrication of 3D structures can be performed using the μTM technique [60–63]. Figure 12.12 shows a schematic representation of replica-molding-based photolithography.

This method was developed by Whiteside's group in the mid-1990s. In this fabrication process, the polymer to be molded is known as the precursor and is used in a liquid state. The precursor is poured over the master cast plate which contains the relief pattern. Sylgard 184 based PDMS is used as pre-polymer. Pouring of the co-polymer totally covers the master. Then, the pre-polymer replicates the contours of the original stamp. The molten pre-polymer is subsequently transformed into solid elastomer by annealing due to polymer cross-linking in the presence of a cross-linker. The solidified elastomer is then peeled off the master cast in the presence of silane vapor or solution, a process known as silanized. Silane is applied in order to reduce the surface energy of the elastomer, ensuring a distortion-free peel-off process. The quality of sharp edges, micro-channels, micropatterns, etc, created over the mold is highly influenced by the van der Waals interactions and wetting properties of the master material [64–67]. Therefore, the viscosity of the pre-polymer solution is very important for governing the properties of the master cast. However, cross-linked PDMS undergoes about 2% shrinkage during curing, resulting in distortion of the dimensions of the mold. Also, on exposure to non-polar organic solvent, this elastomer swells. Finally, the elastic and the thermal mis-match of the cross-linker

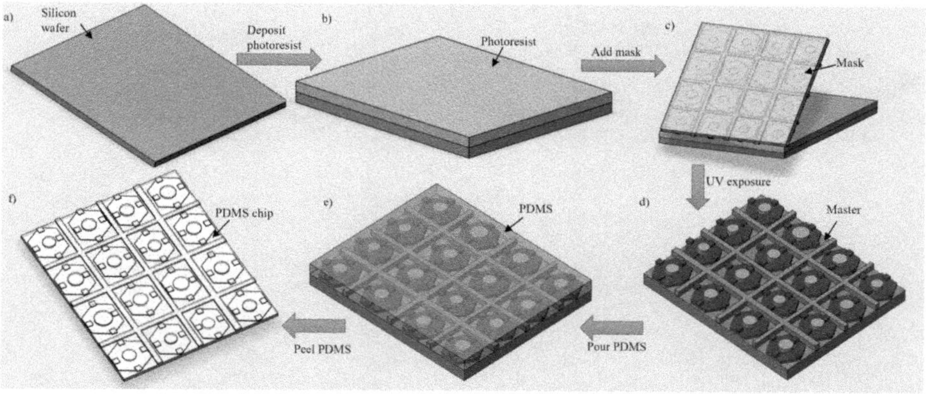

Figure 12.12. A schematic diagram of soft lithography techniques for the fabrication of blood vessel with micro-details. Silicon wafer (a) coated with photoresist (b), followed by masking with UV treatment (c) and (d), then the mask is filled with PDMS (e) and subjected to heat treatment, followed by peeling off the PDMS layer, forming a PDMS chip (f).

from that of elastomeric pre-polymer makes the micropattern highly unstable. Also, the softness of the elastomer prevents achieving the desired aspect ratio.

12.5.2 Self-assembly techniques

Self-assembled monolayers (SAMs) have primarily been used for self-assembly of non-biological self-assembling systems. A substrate is immersed in a solution containing ligands which are reactive towards the surface of the substrate. The same binding of the ligand–substrate complex can also be achieved by exposing the substrate to ligand vapors. The surface properties as well as the thickness of the substrate can be regulated by altering the alkyl chain. Selectively binding the ligands on the surface of the substrate is a challenging task. SAMs possess low densities of defects in the final structure, and are anneable to achieve interfacial properties such as electrochemical and biochemical properties. This process uses Au and Ag as the most stable substrates. Alkanethiolates on the Au surface increase the likelihood of ligand binding on the substrate [68–71]. During this process, the sulfur atoms of alkanethiolates form a commensurate over-layer on the Au.

In this process of micropatterning using self-assembly phenomena, the PDMS mold is wetted with a biological ink, and is brought into contact with the Au surface for 10–20 s. The hexadecanethiol transfers from the mold of the gold during active contact, and thus forms a coat of hexadecanethiolate, thus generating patterns of SAMs on the surface of the gold [72–74]. A schematic of self-assembly technology is shown in figure 12.13.

12.5.3 Sputtering techniques

Sputtering is a low-temperature and high-vacuum supported process. In this technique, metallic films are deposited over the surface of the substrate using metallic vapor. The fabricated films are 100 Å thick. These metallic layers form the

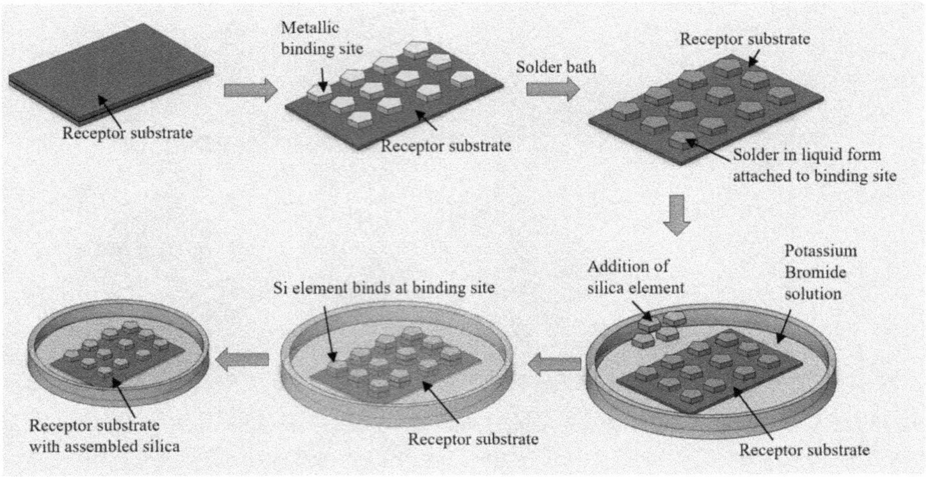

Figure 12.13. Self-assembly techniques for the fabrication of blood vessels on a petri plate. It is formed by using a receptor substrate patterned with metallic binding sites. Then, it is coated with liquid solder, followed by silica element treatment over the mounted sites. When the Si elements are bonded at the desired sites, the metallic binding sites are dissolved with UV treatment, leaving the Si bonded assembly over the receptor surface.

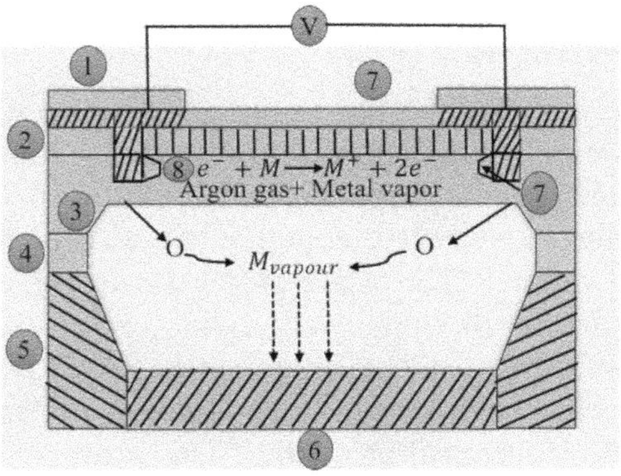

Figure 12.14. A schematic representation of the working principles of a sputtering unit. Metal electrodes (1) are mounted over a piezo-resistor cantilever (2), which covers the pressurized medium chamber (3), which in turn is connected with the strain base (5) through a die bond adhesive (4). The constraint base is mounted over substrate (6). A sandwiched insulating layer (7) is formed at the bottom of the metallic electrodes embedded through the pressurized chamber, and it forms the base of the target plate (8) at the medium chamber.

conducting interface between the source and sensors. The sputtering process is executed in the presence of plasma at a high-vacuum condition of 5×10^{-7} Torr. Positively charged ionic gases of the metallic substrate bombard the surface of the target in an inert gas (argon) chamber. The high velocity of bombardment leads to the transfer of momentum to the receiver surface, and results in evaporation of metallic ions from the surface of the receiver. This evaporated vapor is then deposited on the surface of the substrate following condensation. Thus, this process is also known as the physical deposition process [75, 76]. A schematic of sputtering principles is shown in figure 12.14.

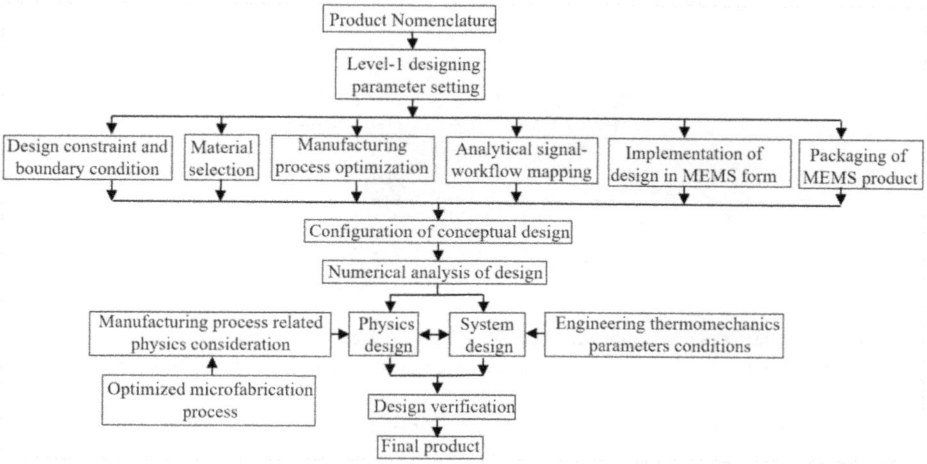

Figure 12.15. Flow diagram of the design considerations for generic MEMS devices.

12.6 Microvessel design

12.6.1 Design consideration

The design consideration outline for any arbitrary microsystem is shown in figure 12.15. After defining the specification of the product, a few points need to be reviewed. They include the constraints of the design, material selection, manufacturing process optimization, mapping of signals, mapping of transduction, the physical boundary conditions of electromechanical systems and product packaging. Once the above parameters are optimized, the product can be considered for manufacturing. Designing constraints for MEMS systems includes various aspects, such as the fabrication facility, application, physical size and weight constraints, environmental conditions, customer demand, time to market and, finally, budget. Material selection can be either of two types: a passive substrate material for support only (polymers, plastics, ceramics) or an active substrate material (Si, GaAs, quartz) for sensing or micro-actuating. Manufacturing processes for MEMS fabrication include bulk micromachining, surface micromachining and higher aspect ratio processes. Signal transduction is necessary for both micro-sensors and actuators. Signal transduction techniques available in MEMS system fabrication include piezoresistors, piezoelectric material, piezo-capacitors, resonant vibrators, electroresistant heating and shape memory alloys [77–79].

12.6.2 Mechanical design of a balloon angioplasty pressure sensor using finite element methods

The operating principle of a capacitive pressure sensor is to measure the change in capacitance between two electrodes when a change in pressure displaces one of the electrodes, located on a thin diaphragm. The diaphragm separates a reference compartment kept at vacuum pressure and a pressurized compartment, as shown in figure 12.16.

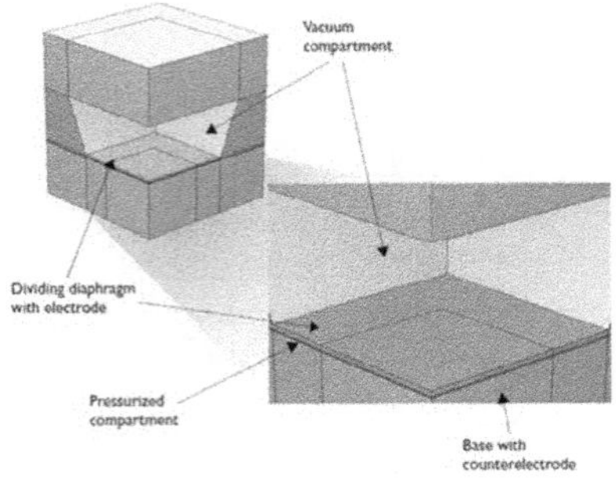

Figure 12.16. One quarter of a pressure sensor designed with a designated vacuum compartment, a segregating diaphragm with mounted electrodes, a pressurized compartment and a base with a counter electrode.

At the bottom of the pressurized compartment is a fixed base (with one electrode), while the diaphragm (with a counter electrode) is located at its top. As the pressure changes, the diaphragm that separates the two compartments is displaced, and the change in separation between the two electrodes results in a corresponding change in the capacitance [80].

Although the deformation of the sensor is primarily caused by the applied pressure, any initial stresses in the material also affect the deformation. Therefore, the manufacturing process and the selected materials directly influence sensor operation. For example, in some structures the membrane and cavities are engraved onto silicon and sealed with layers of glass. Because the materials are bonded together at a high temperature, cooling them down to the sensor's normal operating temperature produces undesirable stresses in the material that affect device performance.

The sensor in this case measures the static pressures of a magnitude from zero to atmospheric pressure. The model first computes the initial stresses from the manufacturing process, then it accounts for the structure's mechanical deformation resulting from an applied pressure [81, 82]. It finally calculates the sensor's capacitance for the deformed shape; the 2D model calculates the capacitance from a computed electric field, whereas the 3D model simply integrates infinitesimal capacitance contributions over the electrode boundary (figure 12.17).

12.6.2.1 Complex parallel plate capacitive sensor structure

A parallel plate high-sensitivity MEMS capacitive absolute pressure sensor is proposed here. In this model, a polysilicon ultrathin diaphragm is used to sense the pressure. It has a fixed top plate and movable bottom plate. The movable bottom plate is attached to the center of the diaphragm with a small die separator. The fabrication process of the sensor structure is more complex [83].

12.6.2.2 Complex comb drive finger parallel plate capacitive sensor structure

Seo and Shandas proposed a capacitance sensor with comb drive finger plates, focused on solving non-linear capacitive sensitivity in membrane-type capacitive pressure sensors. They developed a comb drive capacitor to solve the problem.

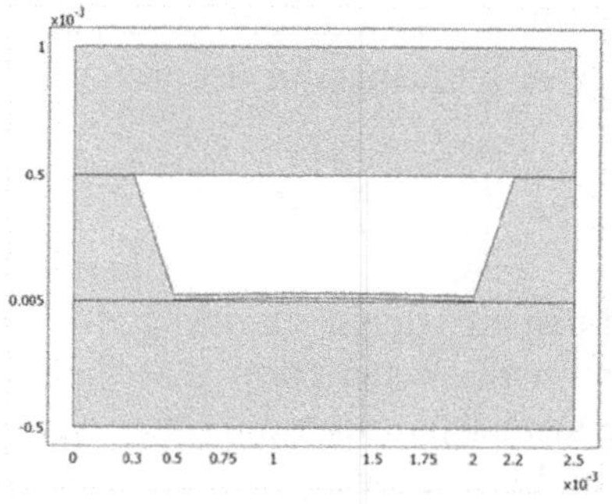

Figure 12.17. 2D lateral view of the designed pressure sensor with a vacuum chamber 0.5 mm in height.

12.6.2.3 Model geometry

The pressure sensor consists of a silicon structure that includes a micrometer-thick diaphragm situated between two glass layers. In addition, two 1 mm^2 rectangular plates at the pressurized compartment's top (applied potential) and bottom (grounded) form the electrodes.

12.6.2.4 Stress and deformation

During manufacturing, the sensor is bonded together in a vacuum and at a high temperature before it is cooled down. Therefore, during this process no external forces act on the sensor's boundaries, but internal stresses appear because the two materials have different coefficients of thermal expansion. This process also produces a vacuum in the upper cavity that serves as the reference pressure.

During regular operation the sensor is fixed to a solid surface, and ambient pressure pushes on all outer boundaries. The temperature also changes, which produces extra stresses due to thermal expansion.

For a linear elastic material, the stress–strain relationship—taking into account initial stress, σ_0, initial strain, ε_0, and thermal strain, ε_{th}, is

$$\sigma = D_{\varepsilon_{el}} = D(\varepsilon - \varepsilon_{th} - \varepsilon_0) + \sigma_0, \tag{12.8}$$

where D is the elasticity tensor, and the six-dimensional vectors σ and ε give the normal and shear values of the stresses and strains.

12.6.2.5 Capacitance

To compute the sensor's capacitance, the 2D model solves for the electric field in the deformed geometry (or frame), which is defined by the moving mesh (ALE) application mode. Using a port boundary condition, the capacitance is obtained from the energy of the electric field from the equation

$$C = \frac{2}{U^2} \int We d\Omega, \tag{12.9}$$

where U is the potential difference between the plates ($U = 1$ V for the port boundary condition) and We is the electric energy density. The area Ωd corresponds to the narrow air gap in the sensor.

The 3D model uses another method for the capacitance calculation, namely integrating over the surface of the capacitor according to

$$C = \varepsilon \int \frac{1}{h} dA, \tag{12.10}$$

where h denotes the local distance across the capacitor and ε is the permittivity of air. This calculation rests on the assumption that the lower glass block does not deform much, so that the local separation only depends on the initial distance and the diaphragm deformation. In the model this expression is further multiplied by 4 to obtain the full capacitance of the model.

To model in COMSOL one has to define the geometry first, which can either be designed or imported. Since thrust is being dealt with here, it means our structure can be anything depending on the requirements of the environment and proper material selection. The cavity should a vacuum, the base should be made of steel and the subjected area should be made of silicon. The following parameters need to be defined prior to simulation:

1. *Pressure, die bonding temperature and reference temperature*: These are used to decide the amount of pressure and thermal stress which have to be taken into account while modeling. We fix the maximum value and steps.
2. *Operators*: These are used to define which operators we are using on which area, dimension or boundary of the geometry. We have to apply only the average and integration operators in our project.
3. In addition, we have to fix other operators such as the difference, explicit, symmetry and other operators in order to have an accurate simulation.
4. *Mesh analysis*: This is used to create a mesh in different segments of the sensor to examine the effect of thrust minutely. This is the reason why we choose the mesh size of the silicon part to be so small.
5. *Materials*: In the study section we have to choose the stationary sub-sets and then materials have to be selected for different parts of the geometry.

While modeling the sensor, one has to assign values in different studies, and then graphs of displacement and capacitance with applied stress will be analyzed. These studies can be made in two cases, one in the presence of thermal stress and another without thermal stress. Stress at different levels in different sections of geometry will be evaluated, where without the thermal effect stress will be negligible for the whole body and significant at the points where we are giving thrust. If thermal stress is taken into account, then there will be stress on the whole body, while the cavity experiences the greatest stress in the presence of external thrust. A contour plot of stress levels is shown in figure 12.18. Colors are used to show the different levels of stress.

12.6.2.6 Simulation without thermal stress

In another case, thermal stress was not considered, and only thrust is applied on the cavity so that deformation can take place on a small localized area and graphs can

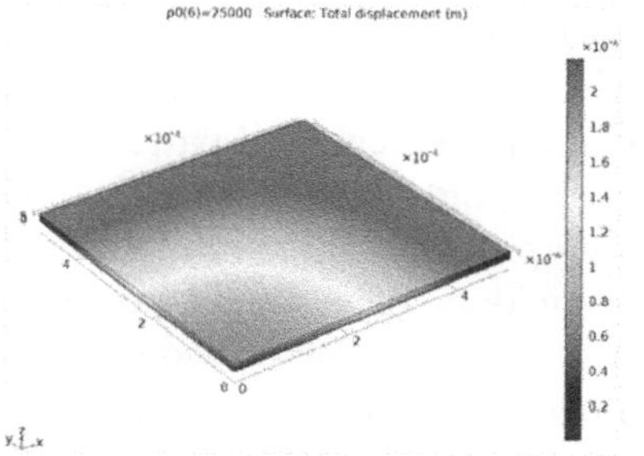

Figure 12.18. Stress distribution over the Si wafer in the presence of thermal stress. The same is plotted with a color scale, red indicating zone of maximal stress, whereas blue represents the least magnitude of distributive stress over the region of the piezo-resistive plate.

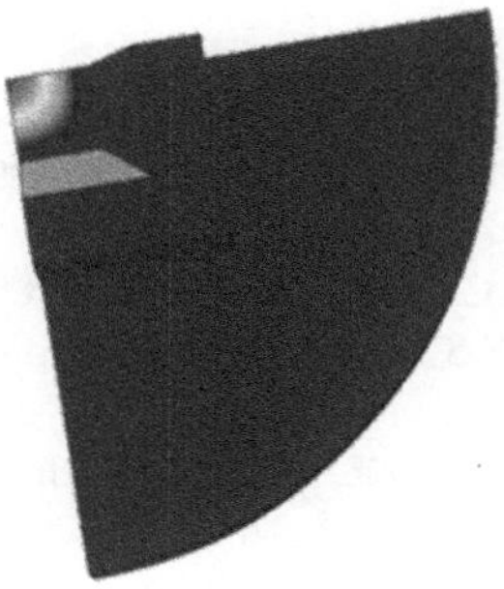

Figure 12.19. Device responses to thrust at 200 Pa in the absence of thermal stress.

be obtained. Pressure levels are varied accordingly and stress distribution can be computed. Four pressure levels are chosen and simulations are performed on these. The pressures are 200 Pa, 2000 Pa, 10 000 Pa and 20 000 Pa. Minute variations in capacitances and displacements with changes in pressure were analyzed. At different pressure levels there is slight change, which is in the range of nanometers, observed in the displacement of the cavity. In the case of capacitance this value is in picofarads or femtofarads. These results show us that we can design specific capacitive pressure sensors for use according to the environment.

12.6.2.7 Cavity pressure at 200 Pa

With 200 Pa, which is a low pressure, being applied on the cavity only, the change in stress levels at different points varies significantly. The stress is maximum at the center of the cavity and decreases radially outward. Figure 12.19 shows the effect of pressure on displacement, where one step or 50 Pa pressure is needed to displace it by 3 nm.

12.7 Conclusion

The vascular bed is an independent living entity in the field of tissue engineering. It forms a complex and defined vascular network. The complexity of this network arises from the presence of cyclic stress, neuro-vascular coupling, dynamicity in pressure variance and structural integrity of the host tissue. Thus, merely fabricating the tubular structure of a blood vessel may not be sufficient to address the above-mentioned factors. Instead, highly precise and patterned vascular network synthesis is required. Since blood vessels, independent of their dimensions, are highly transductive in nature, converting pressure-waves into neural signals and regulating pH and hormonal signal transduction, they should no longer be considered as mere cylindrical facilitators, rather, they can be considered as bioMEMS devices which are responsible for the flow of matter (fluid) as well as signals (chemical and electrical). In connection to the above context, this chapter summarizes the MEMS-based manufacturing techniques of vascular bed. A numerical solution is also provided for designing a baro-receptor on the inner lumen of the blood vessel in the form of a MEMS.

References

[1] Abbott N J, Rönnbäck L and Hansson E 2006 Astrocyte–endothelial interactions at the blood–brain barrier *Nat. Rev. Neurosci.* **7** 41–53

[2] Achyuta A K H, Conway A J, Crouse R B, Bannister E C, Lee R N, Katnik C P, Behensky A A, Cuevas J and Sundaram S S 2013 A modular approach to create a neurovascular unit-on-a-chip *Lab Chip* **13** 542–53

[3] Ahearne M, Yang Y and Liu K 2008 Mechanical characterisation of hydrogels for tissue engineering applications *Tissue Eng.* **4** 1–16

[4] Ahearne M, Yang Y, El Haj A J, Then K Y and Liu K-K 2005 Characterizing the viscoelastic properties of thin hydrogel-based constructs for tissue engineering applications, *J. R. Soc. Interface* **2** 455–63

[5] Anseth K S, Bowman C N and Brannon-Peppas L 1996 Mechanical properties of hydrogels and their experimental determination *Biomaterials* **17** 1647–57

[6] Ahmadi B B, Namdari M and Mobarakeh H 2014 Erectile dysfunction as a predictor of early stage of coronary artery disease *J. Tehran Univ. Heart Cent.* **9** 70

[7] Bertassoni L E *et al* 2014 Hydrogel bioprinted microchannel networks for vascularization of tissue engineering constructs *Lab Chip* **14** 2202–11

[8] Booth R and Kim H 2012 Characterization of a microfluidic *in vitro* model of the blood-brain barrier (μBBB) *Lab Chip* **12** 1784

[9] Bozzuto A M 2015 Liposomes as nanomedical devices *Int. J. Nanomed.* **10** 975–99

[10] Chen Q, Suki B and An K-N 2004 Dynamic mechanical properties of agarose gels modeled by a fractional derivative model *J. Biomech. Eng.* **126** 666–71

[11] Cho H, Seo J H, Wong K H K, Terasaki Y, Park J, Bong K, Arai K, Lo E H and Irimia D 2015 Three dimensional blood-brain barrier model for *in vitro* studies of neurovascular pathology *Sci. Rep.* **5** 15222

[12] Chrobak K M, Potter D R and Tien J 2006 Formation of perfused, functional microvascular tubes *in vitro Microvasc. Res.* **71** 185–96

[13] Corot P, Robert J M, Idée M and Port M 2006 Recent advances in iron oxide nanocrystal technology for medical imaging *Adv. Drug Deliv. Rev.* **58** 1471–504

[14] Daraee H, Eatemadi A, Abbasi E, Fekri Aval S, Kouhi N and Akbarzadeh A 2016 Application of gold nanoparticles in biomedical and drug delivery *Artif. Cells Nanomed. Biotechnol.* **44** 410–22

[15] Eatemadi A, Daraee H, Karimkhanloo H, Kouhi M, Zarghami N, Akbarzadeh A, Abasi M, Hanifehpour Y and Joo S W 2014 Carbon nanotubes: properties, synthesis, purification, and medical applications *Nanoscale Res. Lett.* **9** 1–13

[16] Eatemadi A, Daraee H, Zarghami N, Melat Yar H and Akbarzadeh A 2016 Nanofiber: synthesis and biomedical applications *Artif. Cells Nanomed. Biotechnol.* **44** 111–21

[17] Eatemadi A, Darabi M, Afraidooni L, Zarghami N, Daraee H, Eskandari L, Mellatyar H and Akbarzadeh A 2016 Comparison, synthesis and evaluation of anticancer drug-loaded polymeric nanoparticles on breast cancer cell lines *Artif. Cells Nanomed. Biotechnol.* **44** 1008–17

[18] Errachid A, Ivorra A, Aguiló J, Villa R, Zine N and Bausells J 2001 New technology for multi-sensor silicon needles for biomedical applications *Sens. Actuators* **B 78** 279–84

[19] Goudarzi G, Tahmasbi F, Anbari K and Ghafarzadeh M 2016 Distribution of genes encoding resistance to macrolides among staphylococci isolated from the nasal cavity of hospital employees in Khorramabad, Iran *Iran. Red Crescent Med. J.* **18** e25701

[20] Goudarzi G, Ghafarzadeh M, Shakib P and Anbari K 2015 Culture and real-time PCR based maternal screening and antibiotic susceptibility for group B *Streptococcus*: an Iranian experience *Glob. J. Health Sci.* **7** 233

[21] Griep L M *et al* 2013 BBB on CHIP: microfluidic platform to mechanically and biochemically modulate blood–brain barrier function *Biomed. Microdevices* **15** 145–50

[22] Gu B K, Choi D J, Park S J, Kim M S, Kang C M and Kim C-H 2016 3-dimensional bioprinting for tissue engineering applications *Biomater. Res.* **20** 12

[23] Daraee H, Etemadi A, Kouhi M, Alimirzalu S and Akbarzadeh A 2016 Application of liposomes in medicine and drug delivery *Artif. Cells Nanomed. Biotechnol.* **44** 381–91

[24] Mahmoudvand H, Fallahi S, Mahmoudvand H, Shakibaie M, Harandi M F and Dezaki E S 2015 Efficacy of *Myrtus communis L.* to inactivate the hydatid cyst protoscoleces *J. Investig. Surg.* **29** 137–43

[25] Mellatyar H, Akbarzadeh A, Rahmati M, Ghalhar M G, Etemadi A, NejatiKoshki K, Zarghami N and Barkhordari A 2014 Comparison of inhibitory effect of 17-DMAG nanoparticles and free 17-DMAG in HSP90 gene expression in lung cancer *Asian Pac. J. Cancer Prev.* **15** 8693–8

[26] Danenberg H D, Fishbein I, Gao J, Mönkkönen J, Reich R, Gati I, Moerman E and Golomb G 2002 Macrophage depletion by Clodronate-containing liposomes reduces neointimal formation after balloon injury in rats and rabbits *Circulation* **106** 599–605

[27] Aiyelabegan H T, Zaidi S S Z, Fanuel S, Eatemadi A, Ebadi M T K and Sadroddiny E 2016 Albumin-based biomaterial for lungs tissue engineering applications *Int. J. Polym. Mater. Polym. Biomater.* **44** 45–50

[28] Heidari H and Taylor H 2018 A multi-layer, self-aligning hydrogel micro-molding process offering a fabrication route to perfusable 3D *in-vitro* microvasculature *bioRxiv*

[29] Herland A, Van Der Meer A D, FitzGerald E A, Park T E, Sleeboom J J F and Ingber D E 2016 Distinct contributions of astrocytes and pericytes to neuroinflammation identified in a 3D human blood–brain barrier on a chip *PLoS One* **11** 1–21

[30] Kim J H, Heller D A, Jin H, Barone P W, Song C, Zhang J, Trudel L J, Wogan G N, Tannenbaum S R and Strano M S 2009 The rational design of nitric oxide selectivity in single-walled carbon nanotube near-infrared fluorescence sensors for biological detection *Nat. Chem.* **1** 473–81

[31] Dubach J M, Das S, Rosenzweig A and Clark H A 2009 Visualizing sodium dynamics in isolated cardiomyocytes using fluorescent nanosensors *Proc. Natl Acad. Sci. USA* **106** 16145–50

[32] Janani F, Kohan S, Taleghani F and Ghafarzadeh M 2015 Challenges to implementing physiologic birth program (PBP): a qualitative study of midwives opinion in Iran *Acta Med. Mediterr.* **31** 1373–80

[33] Shin K H, Moon C R, Lee T H, Lim C H and Kim Y J 2005 Flexible wireless pressure sensor module *Sens. Actuators* A **123** 30–5

[34] Kageyama T, Kakegawa T, Osaki T, Enomoto J, Ito T, Nittami T and Fukuda J 2014 Rapid engineering of endothelial cell-lined vascular-like structures in *in situ* crosslinkable hydrogels *Biofabrication* **6** 25006

[35] Kim J A, Kim H N, Im S-K, Chung S, Kang J Y and Choi N 2015 Collagen-based brain microvasculature model *in vitro* using three-dimensional printed template *Biomicrofluidics* **9** 1–15

[36] Kim S P, Jae H L, Yong Y J and Jon S 2007 Antibiofouling polymer-coated gold nanoparticles as a contrast agent for *in vivo* x-ray computed tomography imaging *J. Am. Chem. Soc.* **129** 7661–5

[37] Landers R, Hübner U, Schmelzeisen R and Mülhaupt R 2002 Rapid prototyping of scaffolds derived from thermoreversible hydrogels and tailored for applications in tissue engineering *Biomaterials* **23** 4437–47

[38] Levy A F, Zayats M, Guerrero-Cazares H, Quiñones-Hinojosa A and Searson P C 2014 Influence of basement membrane proteins and endothelial cell-derived factors on the morphology of human fetal-derived astrocytes in 2D *PLoS One* **9** 1–7

[39] Liu J, Zheng H, Poh S P, Machens H-G and Schilling F A 2015 Hydrogels for engineering of perfusable vascular networks *Int. J. Mol. Sci.* **16** 15997–6016

[40] Ahmadinejad M, Aliepour A, Anbari K, Kaviani M, Ganjizadeh H, Nadri S, Foroutani N, Meysami M and Almasi V 2015 Fine-needle aspiration, touch imprint, and crush preparation cytology for diagnosing thyroid malignancies in thyroid nodules *Indian J. Surg.* **77** 480–3

[41] Ghafarzadeh M, Eatemadi A and Fakhravar Z 2016 Human amniotic fluid derived mesenchymal stem cells cause an anti-cancer effect on breast cancer cell line *in vitro Cell Mol. Biol.* **2016** 102–6

[42] Ghafarzadeh M, Moeininasab S and Namdari M 2015 Effect of early amniotomy on dystocia risk and cesarean delivery in nulliparous women: a randomized clinical trial *Arch. Gynecol. Obstet.* **292** 321–5

[43] Nahrendorf M, Zhang H, Hembrador S, Panizzi P, Sosnovik D E, Aikawa E, Libby P, Swirski F K and Weissleder R 2008 Nanoparticle PET-CT imaging of macrophages in inflammatory atherosclerosis *Circulation* **117** 379–87

[44] Namdari M, Ghafarzadeh M and Nikoo M A 2011 Efficacy of intramuscular methyl prednisolone in preventing restenosis after coronary artery stenting with bare-metal stainless steel stent: a double-blind, randomised, controlled clinical trial *Cardiovasc. J. Afr.* **22** 67

[45] Jerman J H 1991 Semiconductor microactuator *US Patent* 5,069,419

[46] Rouhanizadeh M, Tang T, Li C, Hwang J, Zhou C and Hsiai T K 2006 Differentiation of oxidized low density lipoproteins by nanosensors *Sens. Actuators B Chem.* **114** 788–98

[47] Kaneda M M, Caruthers S, Lanza G M and Wickline S A 2009 Perfluorocarbon nano-emulsions for quantitative molecular imaging and targeted therapeutics *Ann. Biomed. Eng.* **37** 1922–33

[48] Mohamadi M N, Rasoulian B, Raoufi A and Nazari A 2016 Pre-exposure to normobaric hyperoxia has no effect on myocardial injury biomarkers after percutaneous transluminal coronary angioplasty *Iranian Heart J.* **17** 18–26
[49] Mohammadian A E 2016 Drug loading and delivery using nanofibers scaffolds *Artif. Cells Nanomed. Biotechnol.* **45** 881–8
[50] Murphy S V and Atala A 2014 3D bioprinting of tissues and organs *Nat. Biotechnol.* **32** 773–85
[51] Devaraj N K, Keliher E J, Thurber G M, Nahrendorf M and Weissleder R 2009 18F labeled nanoparticles for *in vivo* PET-CT imaging *Bioconjug. Chem.* **20** 397–401
[52] Normand V, Lootens D L, Amici E, Plucknett K P and Aymard P 2000 New insight into agarose gel mechanical properties *Biomacromolecules* **1** 730–8
[53] Novosel E C, Kleinhans C and Kluger P J 2011 Vascularization is the key challenge in tissue engineering *Adv. Drug. Deliv. Rev.* **63** 300–11
[54] Oberheim N A *et al* 2010 Uniquely hominid features of adult human astrocytes *Nancy* **29** 4707–8
[55] Namdari P, Daraee H and Eatemadi A 2016 Recent advances in silicon nanowire biosensors: synthesis methods, properties, and applications *Nanoscale Res. Lett.* **11** 406
[56] Shakib P, Ghafourian S, Goudarzi G, Ghafarzadeh M and Noruzian H 2013 Prevalence of *Listeria* monocytogenes in pregnant women in Khoram Abad, Iran *J. Pure Appl. Microbiol.* **7** 475–7
[57] Singh P, Devi A R, Sharma H R and Sharma H M 2015 *Int. J. Curr. Res. Acad. Rev.* **3** 102–14
[58] Pan M P *et al* 2009 Molecular photoacoustic tomography with colloidal nanobeacons *Angew. Chem. Int. Ed.* **48** 4170–3
[59] Placone A F, McGuiggan P M, Bergles D E, Guerrero-Cazares H, Quiñones-Hinojosa A and Searson P C 2015 Human astrocytes develop physiological morphology and remain quiescent in a novel 3D matrix *Biomaterials* **42** 134–43
[60] Raub C B, Putnam A J, Tromberg B J and George S C 2010 Predicting bulk mechanical properties of cellularized collagen gels using multiphoton microscopy *Acta Biomater.* **6** 4657–65
[61] Banai S, Chorny M, Gertz S D, Fishbein I, Gao J, Perez L, Lazarovichi G, Gazit A, Levitzki A and Golomb G 2005 Locally delivered nanoencapsulated Tyrphostin (AGL-2043) reduces neointima formation in balloon-injured rat carotid and stented porcine coronary arteries *Biomaterials* **26** 451–61
[62] Beiranvand S, Eatemadi A and Karimi A 2016 New updates pertaining to drug delivery of local anesthetics in particular Bupivacaine using lipid nanoparticles *Nanoscale Res. Lett.* **11** 1–10
[63] Jaferian S, Negahdari B and Eatemadi A 2016 Colon cancer targeting using conjugates biomaterial 5-Flurouracil *Biomed. Pharmacother.* **84** 780–8
[64] Mukundan S, Ghaghada K B, Badea C T, Kao C Y, Hedlund L W, Provenzale J M, Johnson G A, Chen E, Bellamkonda R V and Annapragada A 2006 A liposomal nanoscale contrast agent for preclinical CT in mice *Am. J. Roentgenol.* **186** 300–7
[65] Nadri S, Mahmoudvand H and Eatemadi A 2016 Magnetic nanogel polymer of Bupivacaine for ankle block in rats *J. Microencapsul.* **33** 656–62
[66] Nadri S, Mahmoudvand H, Ahmadinejad M, Farrokhfar M and Obeidavi Z 2016 Effect of oral Gabapentinon post-operative pain, morphine consumption, nausea and vomiting in patients with urgent appendectomy: a randomized double-blind placebo-controlled trial *Int. J. Adv. Biol. Biomed. Res.* **4** 202–7

[67] Vahabi S and Kazemi A H 2011 Effects of Clonidine as a premedication on plasma renin activity, serum and urine electrolytes and body fluids in general anaesthesia *Middle East J. Anaesthesiol.* **21** 71–6

[68] Vahabi S and Akhlaghi J 2011 T606 comparison of local anesthetic effects of Tramadol and Lidocaine used subcutaneously in minor surgeries with local anesthesia *Eur. J. Pain* **5** 82

[69] Vahabi S, Heidari M, Ahmadinejad M, Akhlaghi J and Birjandi M 2011 Comparison of local anesthetic effects of Tramadol and Lidocaine used subcutaneously in minor surgeries with local anesthesia *Middle East J. Anesthesiol.* **21** 9–14

[70] Vogt S, Troitzsch D, Spath S and Moosdorf R 2004 Efficacy of ion-selective probes in early epicardial *in vivo* detection of myocardial ischemia *Physiol. Meas.* **25** N21–6

[71] Zaitsev S, Cartier R, Vyborov O, Sukhorukov G, Paulke B R, Haberland A, Parfyonova Y, Tkachuk V and Böttger M 2004 Polyelectrolyte nanoparticles mediate vascular gene delivery *Pharm. Res.* **21** 1656–61

[72] Seidi A E, Mansoori B, Jahanban-Esfahlan R and Farajzadeh D 2014 Nanomagnet-based detoxifying machine: an alternative/complementary approach in HIV therapy *J. AIDS Clin. Res.* **5**

[73] Seto Y, Inaba R, Okuyama T, Sassa F, Suzuki H and Fukuda J 2010 Engineering of capillary-like structures in tissue constructs by electrochemical detachment of cells *Biomaterials* **31** 2209–15

[74] Stephan B G, San H, Schubnel M, Nabel G J and Nabel E G 1997 Direct gene transfer in the rat kidney *in vivo Arch. Mal. Coeur Vaiss.* **90** 1127–30

[75] Ji T, Rai P, Jung S and Varadan V K 2008 *In vitro* evaluation of flexible pH and potassium ion-sensitive organic field effect transistor sensors *Appl. Phys. Lett.* **10** 33–41

[76] Tolabi T, Vanaki Z, Memarian R and Namdari M 2012 Quality of nursing documentations in CCU by hospital information system (HIS) *J. Crit. Care Nurs.* **5** 53–62

[77] Takei T, Yamaguchi S, Sakai S, Ijima H and Kawakami K 2007 Novel technique for fabricating double layered tubular constructs consisting of two vascular cell types in collagen gels used as templates for three-dimensional tissues *J. Biosci. Bioeng.* **104** 435–8

[78] Volterra A and Meldolesi J 2009 The role of astroglia in neuroprotection *Dialogues Clin. Neurosci.* **11** 281–96

[79] Chen W, Vucic E, Leupold E, Mulder W J M, Cormode D P, Briley-Saebo K C, Barazza A, Fisher E A, Dathe M and Fayad Z A 2008 Incorporation of an apoE-derived lipopeptide in high-density lipoprotein MRI contrast agents for enhanced imaging of macrophages in atherosclerosis *Contrast Media Mol. Imaging* **3** 233–42

[80] Michalet X, Pinaud F F, Bentolila L A, Tsay J M, Doose S, Li J J, Sundaresan G, Wu A M, Gambhir S S and Weiss S 2005 Quantum dots for live cells, *in vivo* imaging, and diagnostics *Science* **307** 538–44

[81] Yin X, Fu Y, Yutani C, Ikeda Y, Enjyoji K and Kato H 2009 HVJ-AVE liposome-mediated tissue factor pathway inhibitor (TFPI) gene transfer with recombinant TFPI (rTFPI) irrigation attenuates restenosis in atherosclerotic arteries *Int. J. Cardiol.* **135** 245–8

[82] Cui Y, Wei Q, Park H and Lieber C M 2001 Nanowire nanosensors for highly sensitive and selective detection of biological and chemical species *Science* **293** 1289–92

[83] Yeon J H, Na D, Choi K, Ryu S W, Choi C and Park J K 2012 Reliable permeability assay system in a microfluidic device mimicking cerebral vasculatures *Biomed. Microdevices* **14** 1141–8

Lightning Source UK Ltd.
Milton Keynes UK
UKHW030625011119
352580UK00014B/327/P

9 780750 320009